Geometry of Nonholonomically Constrained Systems

ADVANCED SERIES IN NONLINEAR DYNAMICS*

Editor-in-Chief: R. S. MacKay *(Univ. Warwick)*

Published

Vol. 9 Transport, Chaos and Plasma Physics 2
S. Benkadda, F. Doveil & Y. Elskens

Vol. 10 Renormalization and Geometry in One-Dimensional and Complex Dynamics
Y.-P. Jiang

Vol. 11 Rayleigh–Bénard Convection
A. V. Getling

Vol. 12 Localization and Solitary Waves in Solid Mechanics
A. R. Champneys, G. W. Hunt & J. M. T. Thompson

Vol. 13 Time Reversibility, Computer Simulation, and Chaos
W. G. Hoover

Vol. 14 Topics in Nonlinear Time Series Analysis – With Implications for EEG Analysis
A. Galka

Vol. 15 Methods in Equivariant Bifurcations and Dynamical Systems
P. Chossat & R. Lauterbach

Vol. 16 Positive Transfer Operators and Decay of Correlations
V. Baladi

Vol. 17 Smooth Dynamical Systems
M. C. Irwin

Vol. 18 Symplectic Twist Maps
C. Gole

Vol. 19 Integrability and Nonintegrability of Dynamical Systems
A. Goriely

Vol. 20 The Mathematical Theory of Permanent Progressive Water-Waves
H. Okamoto & M. Shoji

Vol. 21 Spatio-Temporal Chaos & Vacuum Fluctuations of Quantized Fields
C. Beck

Vol. 22 Energy Localisation and Transfer
eds. T. Dauxois, A. Litvak-Hinenzon, R. MacKay & A. Spanoudaki

Vol. 23 Geometrical Theory of Dynamical Systems and Fluid Flows
T. Kambe

Vol. 24 Microscopic Chaos, Fractals and Transport in Nonequilibrium Statistical Mechanics
R. Klages

Vol. 25 Smooth Particle Applied Mechanics – The State of the Art
W. G. Hoover

Vol. 26 Geometry of Nonholonomically Constrained Systems
by R. H. Cushman, J. Śniatycki & H. Duistermaat

*For the complete list of titles in this series, please visit
http://www.worldscibooks.com/series/asnd_series.shtml

ADVANCED SERIES IN NONLINEAR DYNAMICS
VOLUME 26

Geometry of Nonholonomically Constrained Systems

Richard Cushman
University of Calgary, Canada

Hans Duistermaat
University of Utrecht, The Netherlands

Jędrzej Śniatycki
University of Calgary, Canada

World Scientific
NEW JERSEY • LONDON • SINGAPORE • BEIJING • SHANGHAI • HONG KONG • TAIPEI • CHENNAI

Published by

World Scientific Publishing Co. Pte. Ltd.

5 Toh Tuck Link, Singapore 596224

USA office: 27 Warren Street, Suite 401-402, Hackensack, NJ 07601

UK office: 57 Shelton Street, Covent Garden, London WC2H 9HE

Library of Congress Cataloging-in-Publication Data

Cushman, Richard H., 1942–
Geometry of nonholonomically constrained systems / by Richard H. Cushman, Jędrzej Śniatycki & Hans Duistermaat.
p. cm. -- (Advanced series in nonlinear dynamics ; v. 26)
Includes bibliographical references and index.
ISBN-13: 978-981-4289-48-1 (hardcover : alk. paper)
ISBN-10: 981-4289-48-5 (hardcover : alk. paper)
1. Nonholonomic dynamical systems. 2. Geometry, Differential. I. Śniatycki, Jędrzej. II. Duistermaat, Hans. III. Title.
QA614.833.C87 2009
516.3'6--dc22

2009024384

British Library Cataloguing-in-Publication Data

A catalogue record for this book is available from the British Library.

Printed by FuIsland Offset Printing (S) Pte Ltd. Singapore

Dedicated to our wives

Acknowledgments

This book started many years ago about 1994 from a conversation I (RC) had with Hans (JJD) at the University of Utrecht about a book on nonholonomically constrained systems – a topic which we were currently (and still are) interested in. As I usually spent the summers at the University of Calgary, I mentioned my interest to Jedrzej (JS) and he said he would like to participate. Many years passed as our mathematical ideas clarified. The general theory contains a discussion of all known approaches to derive the equations of motion for a nonholonomically constrained system. The description as a distributional Hamiltonian system had to be invented. Theory of symmetry reduction for such systems under went a major evolution, when it was realized that even in the regular case generalized distributions were needed. To handle the singular case, the concept of a differential space proved very useful. The notion of a vector field on a differential space had to be invented. The realization that all the geometry of the reduced system is contained in the almost Poisson structure of the algebra of smooth functions on the reduced space is quite recent. All of the above new mathematics required many rewrites until the text became stable.

As this book took a long time to write many thanks need to be given. First, the Department of Mathematics at the University of Utrecht, where I worked until I retired. Second, the Department of Mathematics and Statistics at the University of Calgary under the headship of Prof. Ted Bisztriczky, where I am currently working as an adjunct professor. Without the support of the departments and of my coauthor JS this project would never have been completed. JS work was partially supported by a Killam and an NSERC grant. JJD used some of his time as an Academy Professor to work on this book. His manuscripts on most of the chapters were of great importance in preparing the final text. Finally, I would like to thank Dr. David Kemppainen of Mount Royal College in Calgary for creating the figures in chapters 5 and 6 and figures 7.10 and 7.11 in chapter 7. All the remaining figures were produced by JJD.

In spite of our best efforts, there may be (hopefully few and inessential) mistakes. If you find any, please contact me at rcushman(at)ucalgary.ca .

Richard Cushman
University of Calgary
July 12, 2009

Foreword

This book is aimed mainly at researchers and graduate students in the field of geometric mechanics, especially the theory of nonholonomic constraints. A wider audience may consist of pure mathematicians who are interested in applications of modern differential geometry to mechanics.

The first part of the book is devoted to the general theory of systems with linear nonholonomic constraints. In chapter 1 we review different approaches to the description of dynamics of nonholonomically constrained systems appearing in the literature. Our only addition to this literature is the systematic use of the Levi Civita connection of the kinetic energy metric. We discuss restrictions on motions of nonholonomically constrained systems provided by conserved quantities and by accessible sets of the constraint distribution.

In chapter 2 we treat the basic properties of the action of a Lie group on a smooth manifold, concentrating especially on the case when the action is proper. We also discuss the differential geometry of the space of orbits of a proper action using the concept of a differential space.

In chapter 3 we use the results presented in chapter 2 to discuss reduction of a proper action of the symmetry group of a nonholonomically constrained system. We begin with a discuss symmetry and reduction for a general dynamical system. Next, we define the notion of symmetry of a nonholonomically constrained system and then show how to reduce it, when the action of the symmetry group is proper. We specialize our results to the case when the action of the symmetry group is free and proper. We then compare the above results to different approaches to reduction of systems with nonholonomic constraints.

In chapter 4 we discuss reconstruction of solutions of the original equations of motion from solutions of the reduced equations. We also study relative equilibria and relative periodic orbits of nonholonomically constrained systems.

The second part of the book contains a comprehensive analysis of concrete systems with nonholonomic constraints. In chapter 5 we discuss the classical nonholonomically constrained system known as Carathéodory's sleigh. In order to illustrate the theory given in the preceding chapters, we derive its equations of motion in five different ways, construct the reduced system in three different ways, and carry out reconstruction explicitly.

In chapter 6 we treat the example of a smooth strongly convex rigid body rolling without slipping on a horizontal plane under the influence of a constant vertical gravitational force. We use traditional notation, which sometimes clashes with that of the preceding chapters.

In chapter 7 we give a comprehensive analysis of the motion of a rolling disk, which is a body of revolution whose edge rolls on a horizontal plane under the influence of a constant vertical gravitational field. The rim of the disk is a planar circle with its center at the center of mass. We assume that during the motion the lowest point of the rim remains in contact with the horizontal plane, which prevents the disk from taking off into space. The rolling disk has a symmetry group $E(2) \times S^1$, where $E(2)$ is the Euclidean motion group of the plane, and S^1 is the group of internal symmetries of the disk. We reduce these symmetries, solve the reduced equations of motion and then reconstruct the reduced motion. We give a complete qualitative analysis of the motion of the disk. We obtain a global gyroscopic stabilization principle, namely, relative equilibria are stable (= elliptic) if their energy is larger than a fixed number. For exceptional values of the parameters, the disk falls flat in finite time. We give an asymptotic analysis of the situation when the disk nearly falls flat and then rises up again. A surprising result of this analysis is the existence of a universal constant change in the angle of the point of contact.

Contents

Geometry of Nonholonomically Constrained Systems

Chapter 1

Nonholonomically constrained motions

1.1 Newton's equations

An English translation[1] of Newton's formulation of his *second law* reads: "A change in motion is proportional to the motive force impressed and takes place along a straight line in which that force is impressed." The current formulation reads: mass times acceleration is equal to the force. It uses the term *acceleration* for change of motion.

We will use a suitably generalized version of Newton's second law as our starting point in our presentation of the dynamics of nonholonomically constrained systems on manifolds. The main point of our generalization of Newton's equations is giving an appropriate definition of the notion of acceleration.

When configuration space is $\mathbb{R}^3$, the motion of a particle is described by giving its position vector $\mathbf{q}$ as a function of time, namely, $t \mapsto \mathbf{q}(t)$. In this case, its velocity $\mathbf{v}(t) = \frac{\mathrm{d}\mathbf{q}(t)}{\mathrm{d}t}$ and acceleration $\mathbf{a}(t) = \frac{\mathrm{d}\mathbf{v}(t)}{\mathrm{d}t}$ are well defined vectors in $\mathbb{R}^3$. If one interprets force $\mathbf{F}$ as a vector in $\mathbb{R}^3$, then the usual formulation of the second law reads

$$m\,\mathbf{a}(t) = \mathbf{F}, \tag{1}$$

where m is the mass of the particle. The force $\mathbf{F}$ acting on the particle may depend on $\mathbf{q}(t)$, $\mathbf{v}(t)$, and t. The kinetic energy of a particle with mass m and velocity $\mathbf{v}$ is $k = \frac{1}{2}\,m\,\langle \mathbf{v}, \mathbf{v}\rangle$, where $\langle\ ,\ \rangle$ is the standard Euclidean metric on $\mathbb{R}^3$. Thus k is one half of length of $\mathbf{v}$ with respect to the *kinetic energy metric* $\mathrm{k} = m\,\langle\ ,\ \rangle$.

For a system of n interacting particles in $\mathbb{R}^3$, the configuration space is $(\mathbb{R}^3)^n = \mathop{\times}\limits_{\alpha=1}^{n} \mathbb{R}^3_\alpha$. The position $\mathbf{q}$ of the system, its velocity $\mathbf{v}$ and acceleration

[1] See page 417 of [83].

$\mathbf{a}$ are vectors in $(\mathbb{R}^3)^n$. In this case the second law reads

$$m_\alpha\, \mathbf{a}_\alpha(t) = \mathbf{F}_\alpha, \quad \text{for } \alpha = 1, \ldots, n. \tag{2}$$

Here the subscript α labels the individual particles having mass m_α, acceleration $\mathbf{a}_\alpha$, and subjected to a force $\mathbf{F}_\alpha$. In this case the kinetic energy of the system is one half the square of the length of the velocity with respect to the kinetic energy metric

$$\mathrm{k} = \sum_{\alpha=1}^{n} m_\alpha \langle\, ,\, \rangle_\alpha, \tag{3}$$

where $\langle\, ,\, \rangle_\alpha$ is the standard Euclidean metric on $\mathbb{R}^3{}_\alpha$. If we interpret the force $\mathbf{F}$ imposed on the system as a covector on $(\mathbb{R}^3)^n$, then the left hand side of (2) can be interpreted as the covector given by evaluating the kinetic energy metric k, given by (3), on the vector $\mathbf{a}(t)$. Since the coefficients of k are *constant*, the derivative of the velocity $\mathbf{v}(t)$ coincides with the covariant derivative of $\mathbf{v}(t)$ with respect to the *Levi-Civita connection* ∇ associated to the kinetic energy metric k. This interpretation enables us to extend Newton's equations to manifolds.

Consider now a mechanical system with configuration space Q, which is a smooth manifold. The kinetic energy of the system defines a Riemannian metric k on Q such that for each $v \in T_qQ$, the kinetic energy of the motion of the system with velocity v is $k(v) = \frac{1}{2}\,\mathrm{k}(v,v)$. We shall refer to k as the *kinetic energy metric on the manifold* Q of the system. The metric k gives rise to the vector bundle isomorphisms $\mathrm{k}^\flat : TQ \to T^*Q$ and $\mathrm{k}^\sharp : T^*Q \to TQ$ where

$$\langle \mathrm{k}^\flat(v) \mid u \rangle = \mathrm{k}(v,u) \quad \text{for every } u, v \in T_qQ, \tag{4}$$

and $\mathrm{k}^\sharp = (\mathrm{k}^\flat)^{-1}$.

Let $t \mapsto q(t)$ be a smooth curve on Q which describes a motion of our system. Let $t \mapsto v(t) = \dot{q}(t)$ be its tangent lift to TQ. In other words, $v(t)$ is the *velocity* of the system at time t. Define the *acceleration* of the system as the covariant derivative of its velocity with respect to the Levi-Civita connection ∇ associated to the kinetic energy metric k. In other words, the acceleration $a(t)$ of the system at time t is

$$a(t) = \nabla_v v(t) = \ddot{q}(t). \tag{5}$$

A connection is needed to interpret acceleration as a tangent vector. Throughout the rest of this chapter we will use the Levi-Civita connection associated to the kinetic energy metric.

Let $\pi_Q : T^*Q \to Q$ and $\tau_Q : TQ \to Q$ be the cotangent and tangent bundle projection maps, respectively and observe that $\tau_Q = \pi_Q \circ \mathrm{k}^\flat$. A *force* acting on a mechanical system, which is in the configuration $q \in Q$, is a *covector* in T_q^*Q. Consider a map $\varphi : \mathbb{R} \times TQ \to T^*Q$ such that $\pi_Q \circ \varphi = \tau_Q \circ \mathrm{pr}_2$, where $\mathrm{pr}_2 : \mathbb{R} \times TQ \to TQ : (t, v_q) \mapsto v_q$. Then φ describes the dependence of the force on time, position and velocity of the system. In other words, if $t \in \mathbb{R}$ and $v_q \in T_qQ$, then $\varphi(t, v_q)$ is the force acting on the system at time t, position q, and velocity v_q. *Newton's equations* for the motion $t \mapsto q(t)$ of the system are

$$\mathrm{k}^\flat(\ddot{q}(t)) = \varphi(t, \dot{q}(t)). \tag{6}$$

In what follows we shall show that our formulation of Newton's equations motion for a mechanical system subject to a linear nonholonomic constraint is equivalent to all other standard formulations of its equations of motion.

1.2 Constraints

Constraints in dynamics are restrictions on positions and velocities of the system. Phenomenological constraints are introduced instead of unknown forces to describe observed motions. For example, a rigid body is a system of material points with fixed distances between each pair of points. Another example is the no slip condition in the motion of a rolling body. This constraint requires that the relative velocity of the point of contact of the rolling body vanishes. In the first example, the constraint depends only on the position of the material point. Such constraints are called *holonomic*. In the second example, the no slip condition is a linear relation on the velocity of the motion. Such conditions are called *linear nonholonomic constraints.* Nonlinear nonholonomic constraints appear in control theory, but they will not be considered here, see [74]. Neither shall we consider inhomogeneous linear nonholonomic constraints, which appear in the problem of a sphere rolling on a turntable when regarded in rotating coordinates [89].

There are two ways of dealing with holonomic constraints. First, we can modify the configuration space by imposing the holonomic constraints. Alternatively, we can present the holonomic constraints as linear nonholonomic constraints by extending the tangent bundle of the constraint manifold to a distribution on configuration space. Recall that a *distribution* on a smooth manifold M is a smooth vector subbundle of the tangent bundle TM of M. The original (holonomic) constraint manifold is then an *integral manifold* of this distribution.

Let D be a distribution on Q describing the linear nonholonomic constraints under consideration. A motion $t \mapsto q(t)$ of the system is *dynamically allowed* if its velocity $\dot{q}(t)$ at time t lies in $D_{q(t)}$. If D is involutive,[2] then we are dealing with a family of holonomically constrained systems. Here each integral manifold M of D describes a holonomic constraint.

Our nonholonomically constrained mechanical system is acted on by an *external force* φ_{ext} and a *reaction force* of the constraints $\varphi_{\mathrm{constr}}$. We have hypothesized that the constraints do not depend on time. In particular, the reaction force of the constraint depends only on the velocity and has no explicit dependence on time. We assume that the external force is *conservative*, that is, there is a *potential function* $V \in C^\infty(Q)$ such that $\varphi_{\mathrm{ext}} = -\,\mathrm{d}V$. Then *Newton's equation for the nonholonomically constrained system* is

$$\mathrm{k}^\flat(\ddot{q}(t)) = -\pi_Q^*\,\mathrm{d}V(q(t)) + \varphi_{\mathrm{constr}}(\dot{q}(t)). \tag{7}$$

In order to be able to solve equation (7) we have to make an assumption on the nature of the reaction force of the constraints. Let $D^0 \subseteq T^*Q$ be the set of all covectors which annihilate the corresponding vector in the constraint distribution D. In other words, for each $q \in Q$,

$$D_q^0 = D^0 \cap T_q^*Q = \{p \in T_q^*Q \,\big|\, \langle p \mid v\rangle = 0 \quad \text{for every } v \in D_q\}. \tag{8}$$

Interpreting covectors in T^*Q as forces, we see that D^0 is the set of all forces which do no work on virtual displacements in D. We now make the following

Hypothesis 1.2.1. *The work of the reaction force of the constraints on virtual displacements in D vanishes. In other words,*

$$\varphi_{\mathrm{constr}}(v) \in D^0_{\tau_Q(v)} \quad \textit{for every } v \in D.$$

Constraints satisfying this hypothesis are called *perfect*. A motion given by a curve $t \mapsto q(t)$ in Q with $\dot{q}(t) \in D_{q(t)}$, which satisfies Newton's equation (7) with constraint force $\varphi_{\mathrm{constr}}$, is called *dynamically admissible*. We summarize the above discussion as

[2]Frobenius' theorem states that if D is involutive, then D is integrable.

Theorem 1.2.2. (d'Alembert principle) *A smooth curve $q : (t_0, t_1) \subseteq \mathbb{R} \to Q : t \mapsto q(t)$ is a dynamically admissible motion of a mechanical system with kinetic energy $k(v) = \frac{1}{2}\,\mathrm{k}(v, v)$ and potential energy V subject to the linear nonholonomic constraint $D \subseteq TQ$ if and only if*

$$\mathrm{k}^\flat(\ddot{q}(t)) + \mathrm{d}V(q(t)) \in D^0_{q(t)} \text{ and } \dot{q}(t) \in D_{q(t)}, \tag{9}$$

for all $t \in (t_0, t_1)$.

1.3 Lagrange-d'Alembert equations

The formulation of the dynamics of constrained motion given in (9) can be expressed in terms of the *Lagrangian* $\ell = k - \tau_Q^* V : TQ \to \mathbb{R}$ for the unconstrained system. Here $\tau_Q^* V = V \circ \tau_Q$ is the pull back of V by the tangent bundle projection map τ_Q. Let $c : \mathbb{R} \to Q : t \mapsto c(t)$ be a smooth curve on Q and let $\dot{c} : \mathbb{R} \to TQ : t \mapsto Tc(t)$ be its tangent lift to TQ. The Lagrange derivative of ℓ at $\dot{c}(t)$ is a 1-form on $T_{c(t)}Q$, which is defined as follows. Let $\{q^i\}$ be local coordinates on Q and let $\{q^i, v^i\}$ be the corresponding local coordinates on TQ. Then the curves c and $\dot{c}$ are given by $t \mapsto q(t) = (q^i(t))$ and $t \mapsto (q(t), \dot{q}(t)) = (q^i(t), v^i(t))$, respectively. The Lagrangian ℓ along $\dot{c}$ is a function of $(q(t), \dot{q}(t))$. Its *Lagrange derivative* $\delta\ell$ at $(q(t), \dot{q}(t))$ is the 1-form on $T_{q(t)}Q$

$$\delta\ell(q(t), \dot{q}(t)) = \left\{ \frac{\mathrm{d}}{\mathrm{d}t}\frac{\partial \ell}{\partial \dot{q}^i} - \frac{\partial \ell}{\partial q^i} \right\} \mathrm{d}q^i, \tag{10}$$

where summation is performed over repeated indices. The Lagrange derivative is a natural operator which takes the same form in all coordinate systems, see [65], [60].[3]

Lemma 1.3.3. *For every smooth curve $c : \mathbb{R} \to Q : t \mapsto q(t)$ on Q, the Lagrange derivative of the kinetic energy $k(v) = \frac{1}{2}\,\mathrm{k}(v, v)$ along $\dot{c}$ is*

$$\delta k(q(t), \dot{q}(t)) = \mathrm{k}(q(t))^\flat(\nabla_{\dot{q}}\dot{q}(t)), \tag{11}$$

for all t.

---

[3]What we mean by this is the following. Let Q and $\widetilde{Q}$ be configuration spaces which are diffeomorphic by the diffeomorphism φ. The map φ induces a diffeomorphism $T^*\varphi$ of the cotangent bundle $T^*\widetilde{Q}$ onto the cotangent bundle T^*Q, which is defined as follows. For every $\widetilde{q} \in \widetilde{Q}$ and every $\widetilde{\alpha}_{\widetilde{q}} \in T^*_{\widetilde{q}}\widetilde{Q}$, we have $T^*\varphi(\widetilde{\alpha}_{\widetilde{q}})$ is the cotangent vector $(T_{\widetilde{q}}\varphi^{-1})^T(\widetilde{\alpha}_{\widetilde{q}})$ to Q at q. Here the superscript T denotes transpose. If $\ell : TQ \to \mathbb{R}$ and $\widetilde{\ell} : T\widetilde{Q} \to \mathbb{R}$ are Lagrangians, where $\widetilde{\ell} = \ell \circ T\varphi$, then $\delta\widetilde{\ell} = \delta\ell \circ T^*\varphi$. As Lagrange [65] observed, the preceding formula follows from the interpretation of the Lagrange derivative $\delta\ell$ as the variational derivative of $\int \ell\, \mathrm{d}t$.

Proof. We compute

$$\begin{aligned}\delta k(q(t),\dot q(t)) &= \left\{ \frac{\mathrm{d}}{\mathrm{d}t}\frac{\partial k}{\partial \dot q^i} - \frac{\partial k}{\partial q^i} \right\} \mathrm{d}q^i = \left\{ \frac{\mathrm{d}}{\mathrm{d}t}(k_{ij}\dot q^j) - \tfrac{1}{2}\frac{\partial k_{j\ell}}{\partial q^i}\dot q^j \dot q^\ell \right\} \mathrm{d}q^i \\ &= \left\{ k_{ij}\frac{\mathrm{d}}{\mathrm{d}t}\dot q^j + \frac{\partial k_{ij}}{\partial q^\ell}\dot q^\ell \dot q^j - \tfrac{1}{2}\frac{\partial k_{j\ell}}{\partial q^i}\dot q^j \dot q^\ell \right\} \mathrm{d}q^i \\ &= \left\{ k_{ij}\frac{\mathrm{d}}{\mathrm{d}t}\dot q^j + \tfrac{1}{2}\left(\frac{\partial k_{jn}}{\partial q^\ell} + \frac{\partial k_{j\ell}}{\partial q^n} - \frac{\partial k_{n\ell}}{\partial q^j} \right) \dot q^n \dot q^\ell \right\} \mathrm{d}q^i \\ &= k_{ij}\left\{ \frac{\mathrm{d}}{\mathrm{d}t}\dot q^j + \Gamma^j_{n\ell}\dot q^n \dot q^\ell \right\} \mathrm{d}q^i = k_{ij}\ (\nabla_{\dot q}\dot q)^j .\end{aligned}$$

Here

$$k_{ij}\Gamma^j_{n\ell} = \tfrac{1}{2}\left(\frac{\partial k_{jn}}{\partial q^\ell} + \frac{\partial k_{j\ell}}{\partial q^n} - \frac{\partial k_{n\ell}}{\partial q^j} \right) \tag{12}$$

are the *Christoffel symbols* of the Levi-Civita connection ∇ of the kinetic energy metric k on Q. □

From (9) it follows that if $\ell = k - \tau^* V$, then

$$\begin{aligned}\delta \ell(q(t),\dot q(t)) &= \delta(k - \tau_Q^* V)(q(t),\dot q(t)) = \delta k(q(t),\dot q(t)) + \mathrm{d}V(q(t)) \\ &= \mathrm{k}^\flat(\ddot q(t)) + \mathrm{d}V(q(t)).\end{aligned}$$

Hence, we are led to the Lagrangian form of the equations of motion for a nonholonomically constrained system.

Theorem 1.3.4. (Lagrange-d'Alembert principle) *A smooth curve* $c : (t_0, t_1) \to Q : t \mapsto q(t)$ *is a dynamically admissible motion of a nonholonomically constrained mechanical system with constraint distribution* D *and Lagrangian* $\ell = k - \tau_Q^* V$ *if and only if for every* $t \in (t_0, t_1)$

$$\delta\ell(q(t),\dot q(t)) \in D^0_{q(t)} \ \text{ and } \ \dot q(t) \in D_{q(t)}. \tag{13}$$

Corollary 1.3.5. *The total energy function* $h = k + \tau_Q^* V$ *is constant on all dynamically admissible motions.*

Proof. Since $\langle \frac{\partial \ell}{\partial \dot q} \mid \dot q\rangle = \mathrm{k}(q)(\dot q,\dot q) = k(q)(\dot q)$, we may write

$$h(q,\dot q) = \langle \frac{\partial \ell}{\partial \dot q} \mid \dot q\rangle - \ell(q,\dot q). \tag{14}$$

Let $t \mapsto q(t)$ be an admissible motion. Then differentiating (14) gives

$$\begin{aligned}\frac{\mathrm{d}h(q(t),\dot q(t))}{\mathrm{d}t} &= \langle \frac{\mathrm{d}}{\mathrm{d}t}\left(\frac{\partial \ell}{\partial \dot q}\right) \mid \dot q\rangle + \langle \frac{\partial \ell}{\partial \dot q} \mid \ddot q\rangle - \langle \frac{\partial \ell}{\partial q} \mid \dot q\rangle - \langle \frac{\partial \ell}{\partial \dot q} \mid \ddot q\rangle \\ &= \langle \delta\ell(q(t),\dot q(t)) \mid \dot q(t)\rangle = 0,\end{aligned}$$

using (13). □

1.4 Lagrange derivative in a trivialization

We now show how to compute the Lagrange derivative in a trivialization of TQ.

Suppose that Q is an n-dimensional smooth manifold. We say that the tangent bundle $\tau_Q : TQ \to Q$ of Q is *trivial* if there is a smooth mapping $\kappa : TQ \to Q \times \mathbb{R}^n$ such that $\kappa | T_qQ$ is a linear isomorphism with $\mathbb{R}^n$. The map κ is called a *trivialization*. Recall that the tangent bundle of Q is locally trivial, that is, about every point of Q there is an open neighborhood U such that the bundle $\tau | \tau_Q^{-1}(U) : \tau_Q^{-1}(U) \to U$ is trivial. Let $\lambda : Q \times \mathbb{R}^n \to TQ$ be the inverse of the trivialization κ. Using the standard basis $\{e_i\}_{i=1}^n$ of $\mathbb{R}^n$, we see that

$$X_i : Q \to TQ : q \mapsto \lambda(q, e_i)$$

is a smooth vector field on Q such that $\{X_i(q)\}_{i=1}^n$ is a basis of T_qQ for every $q \in Q$. In other words, $\{X_i\}_{i=1}^n$ is a *moving frame* on Q. Conversely, suppose that $\{X_i\}_{i=1}^n$ is a moving frame on Q. For each $c \in \mathbb{R}^n$ define a vector field c^Q on Q by

$$c^Q(q) = \sum_{i=1}^n c_i \, X_i(q). \tag{15}$$

Then $X_i = e_i^Q$ and $c \mapsto c^Q$ is a linear mapping.[4] Therefore the map

$$\lambda : Q \times \mathbb{R}^n \to TQ : (q, c) \mapsto c^Q(q) = c_q^Q \tag{16}$$

is the inverse of a trivialization of the tangent bundle. In this way we see that moving frames on Q and trivializations of TQ are equivalent objects.

Let $\ell : TQ \to \mathbb{R}$ be a smooth function, called a *Lagrangian*, and let $\lambda : Q \times \mathbb{R}^n \to TQ$ be the inverse of a trivialization of TQ. Define the *Lagrangian in the trivialization* λ^{-1} to be $\mathcal{L} = \ell \circ \lambda : Q \times \mathbb{R}^n \to \mathbb{R}$. The following proposition gives a formula for the Lagrange derivative of $\mathcal{L}$.[5]

Proposition 1.4.6. *Let γ be a smooth curve on Q. Define a smooth curve $c : \mathbb{R} \to \mathbb{R}^n : t \mapsto c(t)$ by requiring that $\lambda(\gamma(t), c(t)) = \dot{\gamma}(t)$, which lies in $T_{\gamma(t)}Q$. For any $a \in \mathbb{R}^n$ we have*

[4] In §30 of chapter II Whittaker [118] calls the c_i quasi-coordinates.

[5] Formula (17) is due to Poincaré [92], see next page.

$$\begin{aligned}
&\langle \delta\mathcal{L}(\gamma(t), c(t)) \mid a \rangle \\
&\quad = \langle \frac{\mathrm{d}}{\mathrm{d}t}\frac{\partial \mathcal{L}}{\partial c}(\gamma(t), c)\Big|_{c=c(t)} \mid a \rangle - \langle \frac{\partial \mathcal{L}}{\partial q}(q, c(t))\Big|_{q=\gamma(t)} \mid a^Q_{\gamma(t)} \rangle \\
&\qquad - \langle \frac{\partial \mathcal{L}}{\partial c}(\gamma(t), c)\Big|_{c=c(t)} \mid \lambda^{-1}_{\gamma(t)}\big([c(t)^Q, a^Q]_{\gamma(t)}\big) \rangle. \qquad (17)
\end{aligned}$$

Here $[c(t)^Q, a^Q]$ *denotes the Lie bracket of the vector fields* $c(t)^Q$ *and* a^Q *on* Q.

Proof. We compute

$$\langle \frac{\partial \mathcal{L}}{\partial c}(q, c) \mid a \rangle = \langle \frac{\partial \ell}{\partial c}(q, c^Q_q) \mid a^Q_q \rangle = \langle \frac{\partial \ell}{\partial \dot{q}}(q, \dot{q})\Big|_{\dot{q}=c^Q_q} \mid a^Q_q \rangle. \qquad (18)$$

We now use local coordinates on Q. Substituting $q = \gamma(t)$, $c = c(t)$, $\dot{q} = \dot{\gamma}(t) = c^Q_{\gamma(t)}$ into (18), and then differentiating the result with respect to t gives

$$\begin{aligned}
&\langle \frac{\mathrm{d}}{\mathrm{d}t}\frac{\partial \mathcal{L}}{\partial c}(\gamma(t), c)\Big|_{c=c(t)} \mid a \rangle \\
&\quad = \langle \frac{\mathrm{d}}{\mathrm{d}t}\frac{\partial \ell}{\partial \dot{q}}(\gamma(t), \dot{q})\Big|_{\dot{q}=\dot{\gamma}(t)} \mid a^Q_{\gamma(t)} \rangle + \langle \frac{\partial \ell}{\partial \dot{q}}(\gamma(t), \dot{q})\Big|_{\dot{q}=\dot{\gamma}(t)} \mid \mathrm{D}a^Q_{\gamma(t)}\, c(t)^Q_{\gamma(t)} \rangle.
\end{aligned}$$

But

$$\frac{\partial \mathcal{L}}{\partial q}(q, c) = \frac{\partial \ell}{\partial q}(q, c^Q_q) = \frac{\partial \ell}{\partial q}(q, \dot{q})\Big|_{\dot{q}=c^Q_q} + \frac{\partial \ell}{\partial \dot{q}}(q, \dot{q})\Big|_{\dot{q}=c^Q_q} \circ \mathrm{D}c^Q_q.$$

Consequently,

$$\begin{aligned}
&\langle \frac{\mathrm{d}}{\mathrm{d}t}\frac{\partial \mathcal{L}}{\partial c}(\gamma(t), c)\Big|_{c=c(t)} \mid a \rangle - \langle \frac{\partial \mathcal{L}}{\partial q}(q, c(t)) \mid a^Q_{\gamma(t)} \rangle = \langle \delta\ell(\gamma(t), \dot{\gamma}(t)) \mid a^Q_{\gamma(t)} \rangle \\
&\qquad + \langle \frac{\partial \ell}{\partial \dot{q}}(\gamma(t), \dot{q})\Big|_{\dot{q}=\dot{\gamma}(t)} \mid \Big(\mathrm{D}a^Q_{\gamma(t)}\, c(t)^Q_{\gamma(t)} - \mathrm{D}c(t)^Q_{\gamma(t)} a^Q_{\gamma(t)}\Big) \rangle.
\end{aligned}$$

This implies (17) because $\delta\mathcal{L}(\gamma(t), c(t)) = \delta\ell(\gamma(t), \dot{\gamma}(t))$ and the Lie bracket of two vector fields X and Y in local coordinates is given by $[X, Y](q) = \mathrm{D}Y(q)X(q) - \mathrm{D}X(q)Y(q)$. □

We now look at the special case of proposition 1.4.6 when Q is a Lie group G with Lie algebra $\mathfrak{g}$. The *standard left trivialization* of TG is

$$\kappa : TG \to G \times \mathfrak{g} : (g, \dot{g}) \mapsto (g, \xi = T_e L_{g^{-1}} \dot{g}). \tag{19}$$

Here $L_g : G \to G : h \mapsto gh$ is multiplication on the left by g. Another way to define ξ is to take the derivative of the curve $t \mapsto g(0)^{-1} g(t)$ at $t = 0$, where $t \mapsto g(t)$ is a smooth curve in G with $g(0) = g$ and $\dot{g}(0) = \dot{g}$. The inverse of κ is given by

$$\lambda : G \times \mathfrak{g} \to TG : (g, \xi) \mapsto T_e L_g \xi. \tag{20}$$

Corollary 1.4.7. *Using the standard left trivialization λ^{-1} of TG, the Lagrangian derivative of $\mathcal{L} = \ell \circ \lambda$ is*

$$\begin{aligned} \delta\mathcal{L}(\gamma(t), c(t)) = {} & \frac{\mathrm{d}}{\mathrm{d}t} \frac{\partial \mathcal{L}}{\partial c}(\gamma(t), c)\bigg|_{c=c(t)} - \frac{\partial \mathcal{L}}{\partial c}(\gamma(t), c)\bigg|_{c=c(t)} \circ \mathrm{ad}_{c(t)} \\ & - \frac{\partial \mathcal{L}}{\partial q}(q, c(t))\bigg|_{q=\gamma(t)} \circ \lambda_{\gamma(t)}. \end{aligned} \tag{21}$$

Here $t \mapsto \gamma(t)$ is a smooth curve on G with the curve $c : \mathbb{R} \to \mathfrak{g} : t \mapsto c(t)$ defined by requiring that $\dot{\gamma}(t) = \lambda(\gamma(t), c(t))$. Moreover, $\mathrm{ad}_\xi : \mathfrak{g} \to \mathfrak{g} : \eta \mapsto [\xi, \eta]$ *for every $\xi \in \mathfrak{g}$.*

1.5 Hamilton-d'Alembert equations

In this section we derive the *Hamilton-d'Alembert* formulation of dynamics of a nonholonomically constrained system. Because the constraints are given by a distribution $D \subseteq TQ$, it is more convenient to work in *velocity space* TQ than in *momentum space* T^*Q.

The *canonical 1-form* θ_Q on T^*Q is defined by

$$\langle \theta_Q(p) \mid u_p \rangle = \langle p \mid T\pi_Q(u_p) \rangle \quad \text{for all } u_p \in T_p(T^*Q). \tag{22}$$

The *canonical symplectic form* on T^*Q is $\omega_Q = -\,\mathrm{d}\theta_Q$. We use the notation $\theta = (\mathrm{k}^\flat)^* \theta_Q$ and $\omega = (\mathrm{k}^\flat)^* \omega_Q$. Since $\mathrm{k}^\flat$ is a diffeomorphism, it follows that ω is a symplectic form on TQ. Moreover, $\omega = -\,\mathrm{d}\theta$.

Lemma 1.5.8. *For every $w \in T_u(TQ)$ with $u \in TQ$, we have $\langle \theta(u) \mid w \rangle =$* $\mathrm{k}(u, T\tau_Q(w))$.

Proof. Because $\pi_Q \circ \mathrm{k}^\flat = \tau_Q$ we get

$$\begin{aligned}\langle \theta(u) \mid w \rangle &= \langle (\mathrm{k}^\flat)^* \theta_Q(u) \mid w \rangle = \langle \theta_Q(\mathrm{k}^\flat(u)) \mid T\mathrm{k}^\flat(w) \rangle \\ &= \langle \mathrm{k}^\flat(u) \mid T\pi_Q(Tk^\flat(w)) \rangle = \mathrm{k}(u, T\tau_Q(w)).\end{aligned}$$

□

Since the fiber of $\tau_Q : TQ \to Q$ is a vector space, for each $q \in Q$ and each $u \in T_qQ$, we have a vector space isomorphism $\iota_u : T_qQ \to \ker T_u\tau_Q$ such that for every $v \in T_qQ$ and every $f \in C^\infty(TQ)$, the *fiber derivative* of f in the direction $\iota_u(v)$ is

$$\iota_u(v)f = \left.\frac{\mathrm{d}}{\mathrm{d}t}\right|_{t=0} f(u + tv). \tag{23}$$

Lemma 1.5.9. *For every $u, v \in T_qQ$, $w \in T_u(TQ)$ we have*

$$\omega(w, \iota_u(v)) = \mathrm{k}(T\tau_Q(w), v).$$

Proof. For each $q \in Q$, $u \in T_qQ$ and $w \in T_u(TQ)$,

$$\langle \theta(u) \mid w \rangle = \langle k^\flat(u) \mid T\tau_Q(w) \rangle. \tag{24}$$

Every vector $w \in T_u(TQ)$ can be extended to a vector field on TQ which projects to a vector field on Q under τ_Q. Therefore,

$$\begin{aligned}\omega(w, \iota_u(v)) &= \langle \iota_u(v) \lrcorner \, \mathrm{d}\theta(u) \mid w \rangle = \iota_u(v) \langle \theta(u) \mid w \rangle \\ &= \left.\frac{\mathrm{d}}{\mathrm{d}t}\right|_{t=0} \mathrm{k}(u + tv, T\tau_Q(w)) = k(v, T\tau_Q(w)).\end{aligned}$$

□

On the right hand sign of the first equality $\lrcorner$ denotes the left interior product (contraction). In other words, for every pair of vector fields X and X' on TQ, $\langle X \lrcorner \, \omega \mid X' \rangle = \omega(X, X')$.

The 2-form ω on TQ is *symplectic*. In other words, ω is nondegenerate and closed. For every function $h \in C^\infty(TQ)$, the *Hamiltonian vector field* corresponding to h is the vector field X_h on (TQ, ω) such that

$$X_h \lrcorner \, \omega = \mathrm{d}h. \tag{25}$$

Lemma 1.5.10. *The projection from (TQ, ω) to Q under τ_Q of an integral curve of the Hamiltonian vector field X_k of the kinetic energy $k(v) = \frac{1}{2}\mathrm{k}(v, v)$ is a geodesic of the Levi-Civita connection ∇ associated to* k.

Proof. In terms of local coordinates $\{q^i\}$ on Q and induced local coordinates $\{q^i, v^j = \dot{q}_j\}$ on TQ, we have

$$\omega = -\,\mathrm{d}(k_{ij}v^j) \wedge \mathrm{d}q^j = -k_{ij}\ \mathrm{d}v^i \wedge \mathrm{d}q^j - \frac{\partial k_{ij}}{\partial q^\ell} v^i\ \mathrm{d}q^\ell \wedge \mathrm{d}q^j$$

and

$$\mathrm{d}k = k_{ij}\, v^i\ \mathrm{d}v^j + \tfrac{1}{2}\, \frac{\partial k_{ij}}{\partial q^\ell}\ v^i v^j\ \mathrm{d}q^\ell.$$

If $X_k = \dot{q}^i \frac{\partial}{\partial q^i} + \dot{v}^i \frac{\partial}{\partial v^i}$, then

$$\begin{aligned} X_k \mathbin{\lrcorner} \omega &= -k_{ij}\dot{v}^i\ \mathrm{d}q^j + k_{ij}\dot{q}^j\ \mathrm{d}v^i - \frac{\partial k_{ij}}{\partial q^\ell} v^i \dot{q}^\ell\ \mathrm{d}q^j + \frac{\partial k_{ij}}{\partial q^\ell} v^i \dot{q}^j\ \mathrm{d}q^\ell \\ &= k_{ij}\dot{q}^j\ \mathrm{d}v^i - \left\{ k_{ij}\dot{v}^i + \frac{\partial k_{ij}}{\partial q^\ell} v^i \dot{q}^\ell - \frac{\partial k_{i\ell}}{\partial q^j} v^i \dot{q}^\ell \right\} \mathrm{d}q^j. \end{aligned}$$

Hence, $X_k \mathbin{\lrcorner} \omega = \mathrm{d}k$ implies that $\dot{q}^j = v^j$ and

$$k_{ij}\dot{v}^i + \frac{\partial k_{ij}}{\partial q^\ell} v^i \dot{q}^\ell - \frac{\partial k_{i\ell}}{\partial q^j} v^i \dot{q}^\ell = -\tfrac{1}{2}\, \frac{\partial k_{i\ell}}{\partial q^j} v^i v^\ell.$$

Therefore,

$$\begin{aligned} \dot{v}^j &= k^{im} \left\{ -\tfrac{1}{2}\, \frac{\partial k_{n\ell}}{\partial q^m} v^n v^\ell + \frac{\partial k_{n\ell}}{\partial q^m} v^n v^\ell - \frac{\partial k_{nm}}{\partial q^\ell} v^n v^\ell \right\} \\ &= k^{im} \left\{ \tfrac{1}{2}\, \frac{\partial k_{n\ell}}{\partial q^m} v^n v^\ell - \frac{\partial k_{nm}}{\partial q^\ell} v^n v^\ell \right\} = -\Gamma^i_{n\ell} v^n v^\ell, \end{aligned}$$

where $k^{im}k_{mj} = \delta_{ij}$ and Γ^i_{nk} is the Christoffel symbol (12). Consequently,

$$X_k(v) = v^i \frac{\partial}{\partial q^i} - \Gamma^i_{n\ell} v^n v^\ell \frac{\partial}{\partial v^i}. \tag{26}$$

If $t \mapsto v(t)$ is an integral curve of X_k and $t \to q(t) = \tau_Q(v(t))$ is its projection to Q, then

$$\frac{\mathrm{d}}{\mathrm{d}t}\dot{q}^i(t) + \Gamma^i_{n\ell}(q(t))\dot{q}^n(t)\dot{q}^\ell(t) = 0.$$

Thus $t \to q(t)$ is a geodesic of the Levi-Civita connection ∇ associated to the kinetic energy metric k. □

Lemma 1.5.11. *Let $(t_0, t_1) \to Q : t \mapsto q(t)$ be a smooth integral curve of X_k and let $t \mapsto u(t) = \dot{q}(t)$ be its tangent lift to TQ, that is, $\tau_Q(u(t)) = q(t)$. Then*

$$\dot{u}(t) - X_k(\dot{q}(t)) = \iota_{\dot{q}(t)}(\nabla_{\dot{q}}\dot{q}(t)) \tag{27}$$

for all $t \in (t_0, t_1)$.

Proof. Since $u(t) = \dot{q}(t)$ it follows that $\dot{u}(t) = v^i(t)\frac{\partial}{\partial q^i} + \dot{v}^i(t)\frac{\partial}{\partial v^i}$. Equation (26) implies that

$$\begin{aligned}\dot{u}(t) - X_k(\dot{q}(t)) &= \left(\dot{v}^i + \Gamma^i_{nk} v^n v^k\right)\frac{\partial}{\partial v^i} \\ &= \left(\frac{\mathrm{d}}{\mathrm{d}t}\dot{q}^i(t) + \Gamma^i_{nk}(q(t))\dot{q}^n(t)\dot{q}^k(t)\right)\frac{\partial}{\partial v^i} \\ &= \nabla_{\dot{q}}\dot{q}(t)\frac{\partial}{\partial v^i} = \iota_{\dot{q}(t)}(\nabla_{\dot{q}}\dot{q}(t)).\end{aligned}$$

□

The set

$$F = \{w \in T(TQ) \mid T\tau_Q(w) \in D\} \tag{28}$$

is a distribution on Q. To see this let $u \in D$. Since the map $T_u\tau_Q : T_u(TQ) \to T_{\tau_Q(u)}Q$ is surjective, $(T_u\tau_Q)^{-1}D_{\tau_Q(u)}$ is a linear subspace of $T_u(TQ)$, which has constant dimension and depends smoothly on u. Therefore $\coprod_{u\in D}(T_u\tau_Q)^{-1}D_{\tau_Q(u)}$ is a smooth vector subbundle of $T_D(TQ)$, the bundle of tangent vectors to TQ with base point in D.

Denote by F^0 the *annihilator* of F, that is, for every $u \in TQ$,

$$F^0_u = \{p \in T^*_u(TQ) \mid \langle p \mid w\rangle = 0 \quad \text{for all } w \in F_u\}. \tag{29}$$

Theorem 1.5.12. (Hamilton-d'Alembert principle) *A curve* $(t_0, t_1) \to Q : t \mapsto q(t)$ *with lift* $t \mapsto u(t) = \dot{q}(t) \in D$ *is a dynamically admissible motion corresponding to the Hamiltonian* $h = k + \tau_Q^* V$ *subject to the nonholonomic constraint* D *if and only if*

$$\dot{u}(t) \lrcorner\, \omega - \mathrm{d}h(\dot{q}(t)) \in F^0_{\dot{q}(t)} \tag{30}$$

for every $t \in (t_0, t_1)$.

Proof. For each $w \in F_{\dot{q}(t)}$, from lemma 1.5.11 we obtain

$$\begin{aligned}\langle \dot{u}(t) \lrcorner\, \omega \mid w\rangle &= \omega(\dot{u}(t), w) = \omega(X_k(\dot{q}(t)), w) + \iota_{\dot{q}(t)}(\nabla_{\dot{q}}\dot{q}), w) \\ &= \omega(X_k(\dot{q}(t)), w) + \omega(\iota_{\dot{q}(t)}(\nabla_{\dot{q}}\dot{q}), w) \\ &= \langle \mathrm{d}k \mid w\rangle - \mathrm{k}(\nabla_{\dot{q}}\dot{q}, T\tau_Q(w)) \\ &= \langle \mathrm{d}h \mid w\rangle - \langle \mathrm{d}V \mid T\tau_Q(w)\rangle - \mathrm{k}(\nabla_{\dot{q}}\dot{q}, T\tau_Q(w)) \\ &= \langle \mathrm{d}h \mid w\rangle - \langle \mathrm{k}^\flat(\nabla_{\dot{q}}\dot{q}) + \mathrm{d}V \mid T\tau_Q(w)\rangle.\end{aligned}$$

Hence, $\dot{u}(t) \lrcorner\, \omega - \mathrm{d}h(\dot{q}(t)) \in F^0_{\dot{q}(t)}$ if and only if $\mathrm{k}^\flat(\nabla_{\dot{q}}\dot{q}(t)) + \mathrm{d}V(q(t)) \in D^0_{q(t)}$. By hypothesis $\dot{q}(t) \in D$. The theorem follows from d'Alembert's principle 1.2.2. □

Corollary 1.5.13. *The Hamiltonian function h is constant on every admissible motion.*

Proof. Because $\dot{u}(t) \in F_{\dot{q}(t)}$ we get

$$0 = \dot{u}(t) \lrcorner (\dot{u}(t) \lrcorner \omega - \mathrm{d}h) = \omega(\dot{u}, \dot{u}) - \mathrm{d}h(\dot{u}(t)) = -\mathrm{d}h(\dot{u}(t)).$$

□

1.6 Distributional Hamiltonian formulation

In this section we give a distributional Hamiltonian formulation of the equations of motion of a nonholonomically constrained system.

1.6.1 *The symplectic distribution* (H, ϖ)

Since D is a distribution on Q, it is a submanifold of TQ. We denote by $T^\omega D$ the *symplectic annihilator* of TD. In other words, for every $u \in D$,

$$T_u^\omega D = \{w \in T_u(TQ) \mid \omega(w, v) = 0 \quad \text{for all } v \in T_u D\}. \tag{31}$$

$T^\omega D$ is a distribution on D being a vector subbundle of $T_D(TQ)$, which is the bundle of tangent vectors to TQ whose base point lies in D. Similarly, we denote the symplectic annihilator of the distribution F (28) by F^ω, that is, for every $u \in D$,

$$F_u^\omega = \{w \in T_u(TQ) \mid \omega(w, v) = 0 \quad \text{for all } v \in F_u\}. \tag{32}$$

Theorem 1.6.1.14. *The vector bundle $T_D(TQ)$ has two direct sum decompositions*

$$T_D(TQ) = F^\omega \oplus TD = F \oplus T^\omega D. \tag{33}$$

Proof. First we show that $F^\omega \cap TD = \{0\}$. Because $\ker T\tau_Q$ is a *Lagrangian distribution* on (TQ, ω), that is, $\omega | \ker T\tau_Q = 0$ and $\dim \ker T\tau_Q = \frac{1}{2} \dim T(TQ)$, and $\ker T\tau_Q \subseteq F$, it follows that $F^\omega \subseteq \ker T\tau_Q$. Hence,

$$F^\omega \cap TD = F^\omega \cap (TD \cap \ker T\tau_Q).$$

Let $u \in D$ and $w \in T_u D \cap F_u^\omega$. Then $w \in TD \cap \ker T\tau_Q$ and is of the form $w = \iota_u(v)$ for some $v \in D_{\tau_Q(u)}$. Moreover, $w \in F_u^\omega$ which implies that $\omega(\iota_u(v), w') = \omega(w, w') = 0$ for every $w' \in F_u$. Using lemma 1.5.9 we get $\mathrm{k}(v, T\tau_Q(w')) = 0$ for all $w' \in F_u$. Since $T\tau_Q(F) = D$, we can choose w' so that $T\tau_Q(w') = v$. Thus $\mathrm{k}(v, v) = 0$, which implies that $v = 0$ and $w = \iota_u(v) = 0$. Therefore, $F^\omega \cap TD = \{0\}$.

If d is the *rank* of the distribution D and $n = \dim Q$, we have $\dim F_u = n + d$, $\dim F_u^\omega = 2n - \dim F_u = n - d$, and $\dim T_u D = n + d$. Hence, $\dim F_u^\omega + \dim T_u D = 2n = \dim T_p(TQ)$. Since $F_u^\omega \cap T_u D = \{0\}$, it follows that $T_u(TQ) = F_u^\omega \oplus T_u D$. Taking symplectic annihilators we get $T_u(TQ) = F_u \oplus T_u^\omega D$. □

Let H be the intersection of F (28) and the tangent bundle of D, that is,

$$H = F \cap TD. \tag{34}$$

H is a distribution on D.

Proposition 1.6.1.15. *For each $u \in D$,*

$$H_u \cap \ker T\tau_Q = \{\iota_u(v) \mid v \in D_{\tau_Q(u)}\}.$$

Proof. For every $u, v \in D_q$, $u + sv \in D_q$ for all $s \in \mathbb{R}$. Hence, $\iota_u(v) \in TD$. Since $\iota_u(v) \in \ker T\tau_Q \subseteq F$, it follows that $\iota_u(v) \in H_u$. Consequently,

$$\{\iota_u(v) \mid v \in D_{\tau_Q(u)}\} \subseteq H_u \cap \ker T\tau_Q.$$

Conversely, suppose that $w \in H_u \cap \ker T\tau_Q$. Then $T\tau_Q(w) = 0$ which implies that there exists $v \in T_{\tau_Q(u)}Q$ such that $w = \iota_u(v)$. If $v \notin D_{\tau_Q(u)}$, then there exists a 1-form η on Q with values in D^0 such that $\langle \eta \mid v \rangle \neq 0$. The hypothesis that η has values in D^0 implies that the function $f : TQ \to \mathbb{R} : v' \mapsto \langle \eta \mid v' \rangle$ vanishes on D. But

$$\langle \mathrm{d}f \mid w \rangle = \left.\frac{\mathrm{d}}{\mathrm{d}s}\right|_{s=0} \langle \eta \mid u + sv \rangle = \langle \eta \mid v \rangle \neq 0$$

which contradicts the assumption that $w = \iota_u(v) \in H \subseteq TD$. □

For each $u \in D$ write

$$\varpi_u = \omega_u \Big|_{H_u \times H_u}. \tag{35}$$

Corollary 1.6.1.16. *(H, ϖ) is a* symplectic distribution *on D, that is, for every $u \in D$ the 2-form ϖ_u is nondegenerate.*

Proof. Let $\{v_1, ..., v_d\}$ be a k-orthonormal basis of D_q. For each $u \in D_q$, let $w_i \in T_u D$ be a lift of v_i, that is $T\tau_Q(w_i) = v_i$. Then $\{w_1, ..., w_d, \iota_u(v_1), ..., \iota_u(v_d)\}$ is a basis of H_u.

Suppose that $w = \sum_i (a^i w_i + b^i \iota_u(v_i))$ is in the kernel of the restriction of ω to H_u, that is, $\omega(w, w') = 0$ for all $w' \in H_u$. Taking $w' = \iota_u(v_i)$, from lemma 1.5.9 we obtain

$$0 = \omega(w, \iota_u(v_i)) = \mathrm{k}(v_i, T\tau_Q(\sum_j (a^j w_j + b^j \iota_u(v_j)))) = \mathrm{k}(v_i, \sum_j a^j v_j) = a_i.$$

Hence, $w = \sum_i b^i \iota_u(v_i)$. Similarly, taking $w' = w_i$, we get

$$0 = \omega(w, w_i) = \omega(\sum_j b^j \iota_u(v_j), w_i) = \sum_j b^j \omega(\iota_u(v_j), w_i)$$
$$= -\sum_j b^j \mathrm{k}(v_j, T\tau_Q(w_i)) = -\sum_j b^j \mathrm{k}(v_j, v_i) = -b_i.$$

Therefore $w = 0$, that is, the restriction of ω to H_u is nondegenerate. In other words, (H, ϖ) is a symplectic distribution on D. □

In order to get a clearer idea of the distributions D, F, and H defined above, we express them in local coordinates. Let $q = \{q^i\}$ be local coordinates on Q. Then $(q, v = \dot{q}) = \{q^i, v^i = \dot{q}^i\}$ and $(q, v, \dot{q}, \dot{v}) = \{q^i, v^i, \dot{q}^i, \dot{v}^i\}$ are induced local coordinates on TQ and $T(TQ)$, respectively. The tangent bundle projection $\tau_Q : TQ \to Q$ is $\tau_Q(q, v) = q$ and its tangent $T\tau_Q : T(TQ) \to TQ$ is $T\tau_Q(q, v, \dot{q}, \dot{v}) = (q, \dot{q})$.

We have $(q, v) \in D_q$ if and only if locally on Q there are ℓ linearly independent 1-forms $\{\alpha_j\}_{j=1}^{\ell}$ such that $\langle \alpha_j(q) | v \rangle = 0$ for every $1 \le j \le \ell$. From the definition (28) of the distribution F we see that

$$F_{(q,v)} = \{(\dot{q}, \dot{v}) \in T_{(q,v)}(TQ) \big| (q, \dot{q}) \in D_q \subseteq T_q Q\}.$$

Consequently, the distribution H (34) is the set of $(\dot{q}, \dot{v}) \in T_{(q,v)}(TQ)$ such that

$$\begin{aligned} 0 &= \langle \alpha_j(q) | \dot{q} \rangle = \langle (\alpha_j(q), 0) | (\dot{q}, \dot{v}) \rangle \\ 0 &= \langle \alpha_j(q) | \dot{v} \rangle + \langle D\alpha_j(q)\dot{q} | v \rangle = \langle ((D\alpha_j(q))^t v, \alpha_j(q)) | (\dot{q}, \dot{v}) \rangle \end{aligned} \tag{36}$$

for $j = 1, \ldots, \ell$, where $D\alpha_j(q)$ denotes the derivative of α_j at q. Clearly, $H_{(q,v)}$ is a linear subspace of $T_{(q,v)}(TQ)$ of codimension 2ℓ and $\coprod_{(q,v)\in D} H_{(q,v)}$ is a smooth vector subbundle of $T_D(TQ)$. In local coordinates the symplectic form ω on TQ is

$$\omega_{(q,v)}((\dot{q}, \dot{v}), (\dot{q}', \dot{v}')) = \langle \mathrm{k}^\flat(\dot{v}') | \dot{q} \rangle - \langle \mathrm{k}^\flat(\dot{v}) | \dot{q}' \rangle = \langle (-\mathrm{k}^\flat(\dot{v}), \mathrm{k}^\flat(\dot{q})) | (\dot{q}', \dot{v}') \rangle,$$

where k is a Riemannian metric on Q.

We now give an alternative proof of corollary 1.6.1.16.

Proof of corollary 1.6.1.16. Suppose that $(\dot{q}, \dot{v}) \in H_{(q,v)}$ and that for every $(\dot{q}', \dot{v}') \in H_{(q,v)}$ we have $0 = \omega(q, v)((\dot{q}, \dot{v}), (\dot{q}', \dot{v}'))$. From (36) and the local expression for the symplectic form ω, it follows that for $1 \le j \le \ell$ there are $(\lambda_j, \mu_j) \in \mathbb{R}^2$ such that

$$(-\mathrm{k}^\flat(\dot{v}), \mathrm{k}^\flat(\dot{q})) = \sum_{j=1}^{\ell} \left[\lambda_j(\alpha_j(q), 0) + \mu_j((D\alpha_j(q))^t v, \alpha_j(q)) \right],$$

that is,

$$\mathrm{k}^\flat(\dot{q}) = \sum_{j=1}^{\ell} \mu_j\, \alpha_j(q) \text{ and } -\mathrm{k}^\flat(\dot{v}) = \sum_{j=1}^{\ell} \big[\lambda_j\, \alpha_j(q) + \mu_j\, (D\alpha_j(q))^t\big]. \quad (37)$$

Since $(\dot{q}, \dot{v}) \in H_{(q,v)}$, the first equation in (36) reads

$$0 = \langle \alpha_i(q) | \mathrm{k}^\sharp(\sum_{j=1}^{\ell} \mu_j\, \alpha_j(q))\rangle = \sum_{j=1}^{\ell} \mu_j\, \langle \alpha_i(q) | \mathrm{k}^\sharp(\alpha_j(q))\rangle \quad (38)$$

for every $1 \le i \le \ell$, using the first equation in (37). But k is a nondegenerate symmetric bilinear form. Thus its $\ell \times \ell$ matrix $(\langle \alpha_i(q)|\mathrm{k}^\sharp(\alpha_j(q))\rangle)$ is invertible. Hence (38) implies that $\mu_j = 0$ for every $1 \le j \le \ell$. Since $(\dot{q}, \dot{v}) \in H_{(q,v)}$, using the second equation in (37) and the fact that $\mu_j = 0$ for every $1 \le j \le \ell$, the second equation in (36) reads

$$0 = -\sum_{j=1}^{\ell} \lambda_i\, \langle \alpha_i(q) | \mathrm{k}^\sharp(\alpha_j(q))\rangle, \quad 1 \le i \le \ell.$$

Since k is nondegenerate, we obtain $\lambda_j = 0$ for every $1 \le j \le \ell$. Thus $(\dot{q}, \dot{v}) = (0, 0)$, using (37). □

1.6.2 *H and ϖ in a trivialization*

In this subsection we determine the distribution H and the symplectic form ϖ in a trivialization.

First we define a trivialization of the constraint distribution D on Q. Suppose that $\{X_i\}_{i=1}^d$ is a moving frame on the distribution D. Then

$$\lambda_D : Q \times \mathbb{R}^d \to D \subseteq TQ : (q, c = (c_1, \ldots c_d)) \mapsto \big(q, \sum_{i=1}^{d} c_i\, X_i(q)\big) \quad (39)$$

is the inverse of a trivialization of the distribution D, thought of as a vector subbundle of TQ whose fibers have dimension d.

Second we determine what the distribution H on D is in the trivialization λ_D^{-1}. From (34) we see that the vector subbundle H of TD is defined by $H_u = T_uD \cap (T_u\tau_Q)^{-1}(D_q)$ for every $u \in D$. Here $q = \tau_Q(u)$, where $\tau_Q : TQ \to Q$ is the tangent bundle projection map. Let

$$\pi_1 = \tau_Q \circ \lambda_D : Q \times \mathbb{R}^d \to Q : (q, c) \mapsto q. \quad (40)$$

Then

$$T_{(q,c)}\pi_1 = T_u\tau_Q \circ T_{(q,c)}\lambda_D : T_qQ \times T_c\mathbb{R}^d \to T_qQ : (v_q, \xi) \mapsto v_q,$$

where $u = \lambda_D(q,c)$. From the fact that λ_D is a diffeomorphism, we obtain

$$\begin{aligned}(T_{(q,c)}\lambda_D)^{-1}H_u &= (T_{(q,c)}\lambda_D)^{-1}\big(T_uD \cap (T_u\tau_Q)^{-1}(D_q)\big)\\ &= T_{(q,c)}(Q\times\mathbb{R}^d) \cap (T_{(q,c)}\pi_1)^{-1}(D_q)\\ &= D_q \times T_c\mathbb{R}^d = D_q\times\mathbb{R}^d. \qquad (41)\end{aligned}$$

Thus the distribution H in the trivialization λ_D^{-1} is the distribution $H_{Q\times\mathbb{R}^d}$ defined by

$$(H_{Q\times\mathbb{R}^d})_{(q,c)} = D_q \times \mathbb{R}^d. \qquad (42)$$

Finally we determine what the symplectic form ϖ on D restricted to the distribution H is in the trivialization λ_D^{-1}. Let k be a Riemannian metric on TQ. Then on TQ we have the symplectic form $\omega = (\mathrm{k}^\flat)^*\omega_Q$, where ω_Q is the canonical 2-form $\omega_Q = -\,\mathrm{d}\theta_Q$ on T^*Q and θ_Q is the canonical 1-form, see (22). Pulling ω back by λ_D gives the 2-form $\mu^*\omega_Q$ on $Q\times\mathbb{R}^d$ where

$$\mu = \mathrm{k}^\flat \circ \lambda_D : Q\times\mathbb{R}^d \to T^*Q : (q, c=(c_1,\ldots,c_d)) \mapsto (q, \sum_{i=1}^d c_i\,\alpha_i(q)) \quad (43)$$

and $\alpha_i(q) = \mathrm{k}^\flat(q)(X_i(q))$. In other words, the 1-forms $\{\alpha_i\}_{i=1}^d$ define a moving coframe on the vector subbundle $D^* = \mathrm{k}^\flat(D)$ of T^*Q.

To determine the 2-form $\omega_{Q\times\mathbb{R}^d} = \mu^*\omega_Q$ on $Q\times\mathbb{R}^d$ we compute

$$\begin{aligned}\mu^*\theta_Q(q,c) &= \theta_Q(q,c)\,T_{(q,c)}\mu\\ &= \sum_{i=1}^d c_i\,\alpha_i(q)\circ T_{\mu(q,c)}\pi_Q \circ T_{(q,c)}\mu, \quad \text{using (43)}\\ &= \sum_{i=1}^d c_i\,\alpha_i(q)\circ T_{(q,c)}\pi_1 = \sum_{i=1}^d c_i\,\pi_1^*\alpha_i(q),\end{aligned}$$

where we view c_i as a function on $Q\times\mathbb{R}^d$. Therefore we get

$$\omega_{Q\times\mathbb{R}^d} = \mu^*(-\,\mathrm{d}\theta_Q) = -\,\mathrm{d}(\mu^*\theta_Q) = -\sum_{i=1}^d \big(\mathrm{d}c_i\wedge\pi_1^*\alpha_i + c_i\,\pi_1^*(\mathrm{d}\alpha_i)\big). \quad (44)$$

To give an explicit expression for $\omega_{Q\times\mathbb{R}^d}$ restricted to $H_{Q\times\mathbb{R}^d}$, we argue as follows. For $a = (a_1,\ldots,a_d) \in \mathbb{R}^d$ let $a^Q(q) = \sum_{i=1}^d a_i\,X_i(q)$, where $X_i(q) = \lambda_D(q,e_i)$. Then a^Q is a vector field on Q with values in D. For every $a\in\mathbb{R}^d$ define two *vector fields* $a^Q_{\rightarrow}$ and $a^Q_{\uparrow}$ on Q with values in $H_{Q\times\mathbb{R}^d} \subseteq D\times\mathbb{R}^d$ by

$$a^Q_{\rightarrow}(q) = (a^Q(q), 0) \text{ and } a^Q_{\uparrow}(q) = (0, a). \qquad (45)$$

From the definition of $a^Q_{\rightarrow}$, $a^Q_{\uparrow}$, and the mapping π_1 it follows that

$$T_{(q,c)}\pi_1\, a^Q_{\rightarrow}(q) = a^Q(q) \text{ and } T_{(q,c)}\pi_1\, a^Q_{\uparrow}(q) = 0. \tag{46}$$

Moreover,

$$\langle \mathrm{d}c_i(q,c) \mid a^Q_{\rightarrow}(q)\rangle = 0 \text{ and } \langle \mathrm{d}c_i(q,c) \mid a^Q_{\uparrow}(q)\rangle = a_i. \tag{47}$$

Also

$$\begin{aligned}\langle \pi_1^*\alpha_i(q) \mid a^Q_{\rightarrow}(q)\rangle &= \langle \alpha_i(q) \mid T_{(q,c)}\pi_1\, a^Q_{\rightarrow}(q)\rangle = \langle \alpha_i(q) \mid a^Q(q)\rangle, \quad \text{using (46)}\\ &= \mathrm{k}(q)(X_i(q), a^Q(q)), \quad \text{by definition of } \alpha_i.\end{aligned}$$

We are now in position to calculate $\varpi_{Q\times\mathbb{R}^d}$, which is the restriction of $\omega_{Q\times\mathbb{R}^d}$ to $H_{Q\times\mathbb{R}^d}\times H_{Q\times\mathbb{R}^d}$. Using (44) we find that for every $a, b, \widetilde{a}, \widetilde{b} \in \mathbb{R}^d$ we have

$$\begin{aligned}&-\varpi_{Q\times\mathbb{R}^d}(q,c)\big(a^Q_{\rightarrow}(q) + b^Q_{\uparrow}(q), \widetilde{a}^Q_{\rightarrow}(q) + \widetilde{b}^Q_{\uparrow}(q)\big)\\
&\quad = \sum_{i=1}^d (\mathrm{d}c_i \wedge \pi_1^*\alpha_i)(q,c)\big(a^Q_{\rightarrow}(q) + b^Q_{\uparrow}(q), \widetilde{a}^Q_{\rightarrow}(q) + \widetilde{b}^Q_{\uparrow}(q)\big)\\
&\qquad + \sum_{i=1}^d c_i\, \pi_1^*(\mathrm{d}\alpha_i)(q,c)\big(a^Q_{\rightarrow}(q) + b^Q_{\uparrow}(q), \widetilde{a}^Q_{\rightarrow}(q) + \widetilde{b}^Q_{\uparrow}(q)\big)\\
&\quad = -\sum_{i=1}^d \big(\langle \mathrm{d}c_i(q,c) \mid \widetilde{b}^Q_{\uparrow}(q)\rangle\, \langle \pi_1^*\alpha_i(q) \mid a^Q_{\rightarrow}(q)\rangle\\
&\qquad\qquad + \langle \mathrm{d}c_i(q,c) \mid b^Q_{\uparrow}(q)\rangle\, \langle \pi_1^*\alpha_i(q) \mid \widetilde{a}^Q_{\rightarrow}(q)\rangle\big)\\
&\qquad\qquad\qquad + \sum_{i=1}^d c_i\, \pi_1^*(\mathrm{d}\alpha_i)(q)(a^Q_{\rightarrow}(q), \widetilde{a}^Q_{\uparrow}(q)), \quad \text{using (47)}\\
&\quad = -\sum_{i=1}^d \widetilde{b}_i\, \mathrm{k}(q)(X_i(q), a^Q(q)) + \sum_{i=1}^d b_i\, \mathrm{k}(q)(X_i(q), \widetilde{a}^Q(q))\\
&\qquad\qquad + \sum_{i=1}^d c_i\, \mathrm{d}\alpha_i(q)(a^Q(q), \widetilde{a}^Q(q)).\end{aligned}$$

So the symplectic form ϖ on D restricted to the distribution H on D in the trivialization λ_D^{-1} is

$$\begin{aligned}&\varpi_{Q\times\mathbb{R}^d}(q,c)\big(a^Q_{\rightarrow}(q) + b^Q_{\uparrow}(q), \widetilde{a}^Q_{\rightarrow}(q) + \widetilde{b}^Q_{\uparrow}(q)\big)\\
&\qquad = \mathrm{k}(q)\big(a^Q(q), \widetilde{b}^Q(q)\big) - \mathrm{k}(q)\big(\widetilde{a}^Q(q), b^Q(q)\big)\\
&\qquad\qquad - \sum_{i=1}^d c_i\; \mathrm{d}\alpha_i(q)\big(a^Q(q), \widetilde{a}^Q(q)\big),\end{aligned} \tag{48}$$

where $\alpha_i(q) = \mathrm{k}^\flat(q)(X_i(q))$. Therefore $\mathrm{d}\alpha_i(q) = \mathrm{d}(\mathrm{k}^\flat(X_i))(q)$, which implies

$$\sum_{i=1}^{d} c_i \,\mathrm{d}\alpha_i(q) = \sum_{i=1}^{d} c_i \,\mathrm{d}(\mathrm{k}^\flat(X_i))(q)$$
$$= \mathrm{d}\mathrm{k}^\flat(\sum_{i=1}^{d} c_i\, X_i)(q) = \mathrm{d}(\mathrm{k}^\flat(c^Q))(q). \tag{49}$$

In other words, with respect to the basis $\{((X_i(q),0),(0,e_i))\}_{1\le i\le d}$ of the space $(H_{Q\times\mathbb{R}^d})_{(q,c)}$, the matrix of $\varpi_{Q\times\mathbb{R}^d}(q,c)$ is $\left(\begin{smallmatrix} B & A^t \\ -A & 0\end{smallmatrix}\right)$. Here $A = A(q)$ is the $d\times d$ positive definite symmetric matrix[6] $(A_{jk}) = \big(\mathrm{k}(q)(X_j(q),X_k(q))\big)$ and $B = B(q)$ is the $d\times d$ antisymmetric matrix $(B_{\ell,m})$, where

$$B_{\ell m} = \sum_{i=1}^{d} c_i \,\mathrm{d}\alpha_i(q)(X_\ell(q),X_m(q)) = L_{X_\ell}\big(\mathrm{k}(c^Q,X_m)\big)(q)$$
$$- L_{X_m}\big(\mathrm{k}(c^Q,X_\ell)\big)(q) - \mathrm{k}(q)\big(c^Q(q),[X_\ell,X_m](q)\big). \tag{50}$$

To obtain the last equality in (50), we have used the fact that for a 1-form β its exterior derivative evaluated on the vector fields X and Y is

$$\mathrm{d}\beta(X,Y) = L_X(Y \mathbin{\lrcorner} \beta) - L_Y(X \mathbin{\lrcorner} \beta) - \beta([X,Y]).$$

Using (49) we can rewrite (50) as

$$\mathrm{d}(\mathrm{k}^\flat(c^Q))(q)(X_\ell,X_m) = L_{X_\ell}(X_m \mathbin{\lrcorner} \mathrm{k}^\flat(c^Q))(q) - L_{X_m}(X_\ell \mathbin{\lrcorner} \mathrm{k}^\flat(c^q))(q)$$
$$- ([X_\ell,X_m] \mathbin{\lrcorner} \mathrm{d}(\mathrm{k}^\flat(c^Q)))(q),$$

which is bilinear in $X_\ell(q)$ and $X_m(q)$. Thus we obtain

$$\sum_{i=1}^{d} c_i \,\mathrm{d}\alpha_i\big(a^Q(q),\widetilde{a}^Q(q)\big) = \mathrm{d}(\mathrm{k}^\flat(c^Q))\big(a^Q(q),\widetilde{a}^Q(q)\big)$$
$$= L_{a^Q}(\widetilde{a}^Q \mathbin{\lrcorner} \mathrm{k}^\flat(c^Q))(q) - L_{\widetilde{a}^Q}(a^Q \mathbin{\lrcorner} \mathrm{k}^\flat(c^Q))(q)$$
$$- ([a^Q,\widetilde{a}^Q] \mathbin{\lrcorner} \mathrm{k}^\flat(c^Q))(q). \tag{51}$$

Consequently

$$\varpi_{Q\times\mathbb{R}^d}(q,c)\big(a^Q_{\rightarrow}(q) + b^Q_{\uparrow}(q), \widetilde{a}^Q_{\rightarrow}(q) + \widetilde{b}^Q_{\uparrow}(q)\big)$$
$$= \mathrm{k}(q)\big(a^Q(q),\widetilde{b}^Q(q)\big) - \mathrm{k}(q)\big(\widetilde{a}^Q(q),b^Q(q)\big)$$
$$- \mathrm{d}(\mathrm{k}^\flat(c^Q))\big(a^Q(q),\widetilde{a}^Q(q)\big) \tag{52}$$
$$= \mathrm{k}(q)\big(a^Q(q),\widetilde{b}^Q(q)\big) - \mathrm{k}(q)\big(\widetilde{a}^Q(q),b^Q(q)\big)$$
$$- L_{a^Q}(\widetilde{a}^Q \mathbin{\lrcorner} \mathrm{k}^\flat(c^Q))(q) + L_{\widetilde{a}^Q}(a^Q \mathbin{\lrcorner} \mathrm{k}^\flat(c^Q))(q)$$
$$+ ([a^Q,\widetilde{a}^Q] \mathbin{\lrcorner} \mathrm{k}^\flat(c^Q))(q). \tag{53}$$

[6]Note that $A = I$ if and only if $\{X_i(q)\}_{i=1}^d$ is a k-orthogonal basis of D_q.

Lemma 1.6.2.17. *For any $u \in D$ the form ϖ_u is a symplectic form on H_u.*

Proof. We have to show that $\varpi_{Q\times\mathbb{R}_d}(q,c)$ is nondegenerate for every $(q,c) \in Q \times \mathbb{R}^d$. This holds if and only if the matrix $\begin{pmatrix} B & A^t \\ -A & 0 \end{pmatrix}$ is invertible, that is, if and only if A is invertible. But A is invertible because it is positive definite. □

1.6.3 *Distributional Hamiltonian vector field*

We return to developing the general theory of distributional Hamiltonian systems.

Let H^ω be the symplectic annihilator of H in $(T(TQ), \omega)$, that is, for each $u \in D$,

$$H_u^\omega = \{w \in T_u(TQ) \mid \omega(w,v) = 0 \quad \text{for all } v \in H_u\}. \tag{54}$$

Since H is symplectic, it follows that H^ω is also symplectic. The bundle $T_D(TQ)$ has a direct sum decomposition

$$T_D(TQ) = H \oplus H^\omega. \tag{55}$$

Let ϖ be the restriction of the symplectic form ω on TQ to $H \times H$ and ϖ^ω the restriction of ω to $H^\omega \times H^\omega$. Both of these 2-forms are nondegenerate. Moreover,

$$\omega | T_D(TQ) = \varpi \oplus \varpi^\omega. \tag{56}$$

Similarly, for every $f \in C^\infty(TQ)$, we denote by $\partial_H f$ and $\partial_{H^\omega} f$ the restrictions of $\mathrm{d}f$ to H and H^ω, respectively. Thus we obtain the following decomposition

$$\mathrm{d}f | T_D(TQ) = \partial_H f \oplus \partial_{H^\omega} f. \tag{57}$$

Lemma 1.6.3.18. *$F \cap H^\omega = F^\omega$. Moreover, for every $u \in D$ each $w \in F_u^\omega$ is of the form $w = i_v(u)$, where v is k-orthogonal to D and $\langle \mathrm{d}f \mid w\rangle = 0$ for each $f \in C^\infty(TQ)$.*

Proof. Since $H = F \cap TD$, it follows that $H^\omega = F^\omega + T^\omega D$. From theorem 1.6.1.14, it follows that $F \cap T^\omega D = \{0\}$. Hence,

$$F \cap H^\omega = F \cap (F^\omega + T^\omega D) = F \cap F^\omega + F \cap T^\omega D = F \cap F^\omega = F^\omega \subseteq \ker T\pi_Q.$$

Therefore, each $w \in F_u \cap H_u^\omega$ is of the form $w = \iota_u(v)$ for some $v \in T_u Q$. Since $w \in H_u^\omega$, taking lemma 1.5.9 into account we get

$$0 = \omega(w', w) = \omega(w', \iota_u(v)) = k(v, T\tau_Q(w'))$$

for every $w' \in H_u$. However, $T\tau_Q(H_u) = D_{\tau_Q(u)}$. Therefore v is k-orthogonal to D. This implies that

$$\begin{aligned} \langle \mathrm{d}f \mid w \rangle = \langle \mathrm{d}f \mid \iota_u(v) \rangle &= \frac{\mathrm{d}}{\mathrm{d}s}\Big|_{s=0} \left(\tfrac{1}{2}\,\mathrm{k}(u+sv, u+sv) + V(\tau_Q(u)) \right) \qquad (58) \\ &= \mathrm{k}(u,v) = 0, \quad \text{because } u \in D. \end{aligned}$$

□

We now define *the distributional Hamiltonian vector field of* $h \in C^\infty(D)$ with respect to the *symplectic distribution* (H, ϖ) on D to be the unique *vector field* Y_h on D *with values in* H such that

$$Y_h \,\lrcorner\, \varpi = \partial_H h. \qquad (59)$$

Since the skew symmetric bilinear form ϖ on $H \subseteq TD$ is nondegenerate, it follows that Y_h is well defined for every $h \in C^\infty(D)$. Moreover, equation (33) and lemma 1.6.3.18 imply that for every extension $\widetilde{h} \in C^\infty(TQ)$ of $h \in C^\infty(D)$ the H-component of the restriction of the Hamiltonian vector field $X_{\widetilde{h}}$ on (TQ, ω) coincides with Y_h. If the symplectic distribution (H, ϖ) on D is understood, we shall refer to Y_h as the distributional Hamiltonian vector field of h.

Lemma 1.6.3.19. *Let $t \to u(t)$ be an integral curve of the distributional Hamiltonian vector field Y_h of h and $t \to q(t) = \tau_Q(u(t))$. Then $u(t) = \dot{q}(t)$ for all t.*

Proof. By proposition 1.6.1.15, $H_u \cap \ker T\tau_Q = \{\iota_u(v) \mid v \in D_{\tau_Q(u)}\}$. Hence lemma 1.5.9 ensures that $\omega(w, \iota_u(v)) = \mathrm{k}(v, T\tau_Q(w))$ for each $u \in D$ and each $v \in D_{\tau_Q(u)}$. Now

$$\begin{aligned} 0 = \langle Y_h \,\lrcorner\, \varpi - \partial_H h \mid \iota_u(v) \rangle &= \omega(Y_h(u), \iota_u(v)) - \langle \mathrm{d}h \mid \iota_u(v) \rangle \\ &= \mathrm{k}(v, T\tau_Q(Y_h(u))) - \mathrm{k}(v,u) = \mathrm{k}(v, T\tau_Q(Y_h(u)) - u). \end{aligned}$$

Thus $T\tau_Q(Y_h(u)) - u$ is k-orthogonal to every vector $v \in D_{\tau_Q(u)}$. But, $T\tau_Q(Y_h(u)) - u \in D_{\tau_Q(u)}$, which gives $T\tau_Q(Y_h(u)) - u = 0$. In other words, $T\tau_Q(Y_h(u)) = u$ for all $u \in D$. This implies that every integral curve $t \mapsto u(t)$ of Y_h is the lift of its projection $t \mapsto q(t)$ to Q by τ_Q, that is, $u(t) = \dot{q}(t)$ for all t. □

Theorem 1.6.3.20. *A curve $t \mapsto q(t)$ in Q is a dynamically admissible motion of a system on (TQ, ω) with Hamiltonian $h = k + \tau_Q^* V$ subject to the linear nonholonomic constraint D if and only if it is the projection to Q under τ_Q of an integral curve $t \mapsto u(t)$ of the distributional Hamiltonian vector field Y_h of h.*

Proof. Suppose that $t \mapsto q(t)$ in Q is a dynamically admissible motion of a system with Hamiltonian $h = k + \tau_Q^* V$ and nonholonomic constraint D. Let $u(t) = \dot{q}(t)$. Since $\dot{q}(t) \in D_{q(t)}$ it follows that $\dot{u}(t) \in TD$. Moreover, the motion $t \mapsto q(t) = \tau_Q(u(t))$ satisfies the constraint $\dot{q} \in D$. But $\dot{q}(t) = T\tau_Q(\dot{u}(t))$ which implies that $\dot{u}(t) \in F$. Hence $\dot{u}(t) \in H$.

Each $w \in F_{\dot{q}(t)}$ can be decomposed as $w = w_H + w_{H^\omega}$ with $w_H \in H$ and $w_{H^\omega} \in H^\omega$. Now

$$\begin{aligned}\langle \dot{u}(t) \lrcorner\, \omega - \mathrm{d}h \mid w \rangle &= \langle \dot{u}(t) \lrcorner\, \omega - \mathrm{d}h \mid w_H + w_{H^\omega} \rangle \\ &= \langle \dot{u}(t) \lrcorner\, \varpi - \partial_H h \mid w_H \rangle - \langle \partial_{H^\omega} h \mid w_{H^\omega} \rangle, \qquad (60)\end{aligned}$$

since $\dot{u}(t) \in H_{\dot{q}(t)}$. By the Hamilton-d'Alembert principle 1.5.12, $\dot{u}(t) \lrcorner\, \omega - \mathrm{d}h(\dot{q}(t)) \in F^0_{\dot{q}(t)}$. Because $H \subseteq F$, we obtain

$$\dot{u}(t) \lrcorner\, \varpi = \partial_H h, \qquad (61)$$

and

$$\langle \mathrm{d}h \mid u \rangle = 0 \quad \text{for all } u \in F \cap H^\omega, \qquad (62)$$

using (60). Equation (61) just says that $t \to u(t) = \dot{q}(t)$ is an integral curve of the distributional Hamiltonian vector field Y_h of h.

Conversely, if $t \to u(t)$ is an integral curve of the distributional Hamiltonian vector field Y_h of h, then equation (61) is satisfied. Moreover, by lemma 1.6.3.18, equation (62) is also satisfied. Hence, equation (60) implies that $\dot{u}(t) \lrcorner\, \omega - \mathrm{d}h \in F^0_{u(t)}$. In addition, if $t \to q(t) = \tau_Q(u(t))$ is the projection of $t \to u(t)$ to Q, then $\dot{q}(t) = u(t)$ for all t by lemma 1.6.3.19. Hence, the Hamilton-d'Alembert principle 1.5.12 ensures that the curve $t \to q(t)$ is a dynamically admissible motion. □

An equation for integral curves $t \to u(t)$ of the distributional Hamiltonian vector field Y_h of h can be written in the form

$$\dot{u}(t) \lrcorner\, \varpi = \partial_H h(u(t)). \qquad (63)$$

We shall refer to (63) as the *distributional form* of Hamilton's equations of motion. We shall refer to the quadruple (D, H, ϖ, h), where D is a manifold, (H, ϖ) is a symplectic distribution on D and $h \in C^\infty(D)$ as a *distributional Hamiltonian system.* The vector field Y_h is determined by the restriction of $\mathrm{d}h$ to the distribution H, which implies that it is determined by the restriction of h to the constraint distribution D. However, this last statement uses the symplectic form ϖ on H, which is *not* determined by the restriction of the Lagrangian ℓ, the kinetic energy, or the total energy to D.

We now give an alternative proof of theorem 1.6.3.20 using the local coordinates that we introduced in proving corollary 1.6.1.16.

Proof of theorem 1.6.3.20. Suppose that the curve $t \mapsto q(t)$ is an admissible motion of the nonholonomic system on (TQ, ω) with Hamiltonain h and constraint distribution D. Recall that in our local coordinates

1. (q,v) lies in D if and only if locally on Q there are linearly independent 1-forms $\{\alpha_j\}_{j=1}^{\ell}$ such that $\langle \alpha_j(q)|v\rangle = 0$ for $1 \le j \le \ell$.
2. $(\dot{q}, \dot{v})$ lies in $F_{(q,v)}$ if and only if $T\tau_Q(q,v,\dot{q},\dot{v}) = (q,\dot{q}) \in D_q$.
3. $(\dot{q}, \dot{v})$ lies in $H_{(q,v)}$ if and only if $\langle \alpha_j(q)|v\rangle = 0$ and $\langle \alpha_j(q)|\dot{v}\rangle + \langle D\alpha_j(q)\dot{q}|v\rangle = 0$ for $1 \le j \le \ell$.

Because $t \mapsto q(t)$ is an admissible motion, the curve $t \mapsto u(t) = (q(t), \dot{q}(t)) \in D$. Since

$$T\tau_Q(u(t), \dot{u}(t)) = T\tau_Q(q(t), \dot{q}(t), \dot{q}(t), \ddot{q}(t)) = (q(t), \dot{q}(t))$$

lies in D, it follows that $\dot{u}(t) \in F_{(q(t),\dot{q}(t))}$. Now $\dot{u}(t) = (\dot{q}(t), \ddot{q}(t)) \in H_{(q(t),\dot{q}(t))}$. To see this note that

$$0 = \langle \alpha_j(q(t))|\dot{q}(t)\rangle, \tag{64}$$

since $(q(t), \dot{q}(t)) \in D$. Differentiating (64) with respect to t gives

$$0 = \langle \alpha_j(q(t))|\ddot{q}(t)\rangle + \langle D\alpha_j(q(t))\dot{q}(t)|\dot{q}(t)\rangle. \tag{65}$$

By item 3 above, equations (64) and (65) imply that $\dot{u}(t) = (\dot{q}(t), \ddot{q}(t)) \in H_{(q(t),\dot{q}(t))}$. Since $t \mapsto q(t)$ is an admissible motion, the Hamilton-d'Alembert principle 1.5.12 holds, that is, for every $(\dot{q}', \dot{v}') \in F_{(q,\dot{q})}$ we have

$$((\dot{q}, \ddot{q}) \lrcorner\, \omega)(\dot{q}', \dot{v}') - \mathrm{d}h(q,\dot{q})(\dot{q}', \dot{v}') = 0. \tag{66}$$

But $H_{(q,\dot{q})} \subseteq F_{(q,\dot{q})}$. So (66) holds for all $(\dot{q}', \dot{v}') \in H_{(q,\dot{q})}$. Thus (66) reads

$$\dot{u} \lrcorner\, \varpi = \partial_H h. \tag{67}$$

If $(\dot{q}', \dot{v}') \in H^{\omega}_{(q,\dot{q})} \cap F_{(q,\dot{q})}$, then (66) reads $\mathrm{d}h(q,\dot{q})(\dot{q}', \dot{v}') = 0$ because $(\dot{q}, \ddot{q}) \in H_{(q,\dot{q})}$. In other words,

$$\langle \mathrm{d}h|u\rangle = 0 \quad \text{for all } u \in H^{\omega} \cap F. \tag{68}$$

This verifies the $\Longrightarrow$ implication of theorem 1.6.3.20.

We now prove the converse. Suppose that (67) and (68) hold. Note that

$$F = (F \cap H) \oplus (F \cap H^{\omega}) = H \oplus (F \cap H^{\omega}). \tag{69}$$

From (67) we see that (66) holds for every $(\dot{q}', \dot{v}') \in H_{(q,\dot{q})}$; while from (68) it follows that (66) holds for every $(\dot{q}', \dot{v}') \in H^{\omega}_{(q,\dot{q})} \cap F_{(q,\dot{q})}$. Thus from (69) we find that equation (66) holds for every $(\dot{q}', \dot{v}') \in F_{(q,\dot{q})}$. In other words, the Hamilton-d'Alembert principle holds for the curve $t \mapsto q(t)$.

Let $t \mapsto u(t) = (q(t), r(t))$. Then by (66) the curve $t \mapsto \dot{u}(t) = (\dot{q}(t), \dot{r}(t))$ satisfies

$$\omega(q,r)((\dot{q},\dot{r}),(\dot{q}',\dot{v}')) = \mathrm{d}h(q,r)(\dot{q}',\dot{v}') \tag{70}$$

for every $(\dot{q}', \dot{v}') \in F_{(q,r)}$. Using the local expression for the symplectic form ω and differentiating the Hamiltonian $h = k + \tau_Q^* V$, we see that (70) is equivalent to

$$\mathrm{k}(q)(\dot{q},\dot{v}') - \mathrm{k}(q)(\dot{r},\dot{q}') = \tfrac{1}{2}\left(D\mathrm{k}(q)\dot{q}'\right)(r,r) + \mathrm{k}(q)(r,\dot{v}') + DV(q)\dot{q} \tag{71}$$

for every $(\dot{q}', \dot{v}') \in F_{(q,r)}$. Set $\dot{q}' = 0$. Then (71) becomes

$$\mathrm{k}(q)(\dot{q},\dot{v}') = \mathrm{q}(r,\dot{v}') \quad \text{for every } \dot{v}'.$$

This implies that $r = \dot{q}$, since $\mathrm{k}(q)$ is nondegenerate. Hence $u(t) = (q(t), \dot{q}(t))$. But $(u(t), \dot{u}(t)) \in F_{(q(t),\dot{q}(t))}$. So $T\tau_Q(u(t), \dot{u}(t)) = (q(t), \dot{q}(t)) \in D$. In other words, the curve $t \mapsto (q(t), \dot{q}(t))$ lies in the constraint distribution D. Hence $t \mapsto q(t)$ is an admissible motion. □

1.7 Almost Poisson brackets

1.7.1 *Hamilton's equations*

In this section we define an almost Poisson bracket and give a formulation of Hamilton's equations of the distributional Hamiltonian system (D, H, ϖ, h) in terms of this bracket.

Define a bracket $\{\,,\,\}$ on $C^\infty(D)$ by

$$\{f,g\}(u) = \varpi(u)(Y_g(u), Y_f(u)) = \langle \mathrm{d}g(u) \mid Y_f(u)\rangle, \tag{72}$$

for every $f, g \in C^\infty(D)$ and every $u \in D$. Here Y_f and Y_g are the distributional Hamiltonian vector fields (59) corresponding to f and g, respectively. Because $\{f,g\} = L_{Y_f} g$, the bracket $\{\,,\,\}$ satisfies Leibniz' rule

$$\{f, g\cdot h\} = \{f,g\}\cdot h + g\cdot\{f,h\}, \tag{73}$$

for every $f, g, h \in C^\infty(D)$. Here $(f \cdot g)(u) = f(u)g(u)$ for every $u \in D$. Therefore the bracket $\{\,,\,\}$ defines an *almost Poisson structure* on $C^\infty(D)$, that is, an antisymmetric bilinear map $\{\,,\,\} : C^\infty(D) \times C^\infty(D) \to C^\infty(D)$.

Because the distribution D need not be integrable, the bracket $\{\, ,\, \}$ need *not* satisfy the Jacobi identity. Thus an almost Poisson bracket on $C^\infty(D)$ need not be a Poisson bracket on $C^\infty(D)$.

We can use the almost Poisson structure $\{\, ,\, \}$ on $C^\infty(D)$ to rewrite the distributional form of Hamilton's equations (63) as

$$\frac{\mathrm{d}f(u(t))}{\mathrm{d}t} = \{h, f\}(u(t)) \tag{74}$$

for every $f \in C^\infty(D)$.

Proposition 1.7.1.21. *A curve $u : \mathbb{R} \to D : t \mapsto u(t)$ satisfies equation (74) for every $f \in C^\infty(D)$ and all t if and only if it is an integral curve of the distributional Hamiltonian vector field Y_h, that is, $\frac{\mathrm{d}u(t)}{\mathrm{d}t} = Y_h(u(t))$.*

Proof. Suppose that $t \mapsto u(t)$ is an integral curve of Y_h and that $f \in C^\infty(D)$. Then

$$\frac{\mathrm{d}f(u(t))}{\mathrm{d}t} = (L_{Y_h} f)(u(t)) = \langle \mathrm{d}f(u(t)) \mid Y_h(u(t))\rangle = \{h, f\}(u(t)).$$

Conversely, suppose that (74) holds for every $f \in C^\infty(D)$. Recall that every tangent vector w_u to D at u can be identified with the derivation $f \mapsto \langle \mathrm{d}f(u) | w_u\rangle$. From the definition of the bracket $\{\, ,\, \}$, the derivation $f \mapsto \{h, f\}$ corresponds to L_{Y_h}. Hence the tangent vector to the curve $t \mapsto u(t)$ at $u(t)$ is $Y_h(u(t))$, that is $\frac{\mathrm{d}u(t)}{\mathrm{d}t} = Y_h(u(t))$. Therefore $t \mapsto u(t)$ is an integral curve of Y_h. □

Corollary 1.7.1.22. *The smooth function $h : D \to \mathbb{R}$ is constant on the integral curves of Y_h.*

Proof. We have

$$L_{Y_h} h = Y_h \,\lrcorner\, \mathrm{d}h = \{h, h\} = 0,$$

since $\{\, ,\, \}$ is skew symmetric. □

For each $u \in D$ we have the inclusion map $i_u : H_u \to T_uD$. Because i_u is injective, its transpose $i_u^t : T_u^*D \to H_u^*$ is surjective. Since the skew symmetric form $\varpi_u : H_u \times H_u \to \mathbb{R}$ is bilinear, there is a linear mapping $\varpi_u^\flat : H_u \to H_u^*$ defined by $\langle \varpi_u^\flat(v_u) \mid w_u\rangle = \varpi_u(v_u, w_u)$ for every $v_u, w_u \in H_u$. Because ϖ_u is nondegenerate, the map $\varpi_u^\flat$ is invertible. We denote its inverse by $\varpi_u^\sharp$. For each $u \in D$ define the map $\Pi_u : T_u^*D \times T_u^*D \to \mathbb{R}$ by

$$\Pi_u(\alpha_u, \beta_u) = \langle \beta_u \mid i_u \circ \varpi_u^\sharp \circ i_u^t(\alpha_u)\rangle,$$

for every $\alpha_u, \beta_u \in T_u^*D$. Then $\Pi_u : T_u^*D \times T_u^*D \to \mathbb{R}$ is a skew symmetric bilinear map. Moreover the map $u \mapsto \Pi_u$ is smooth and is called the *almost Poisson structure tensor field* Π associated to the almost Poisson bracket $(\{\,,\,\}, C^\infty(D))$. Because D is a finite dimensional smooth manifold, for every $u \in D$ we have $T_u^*D = \mathrm{span}\{\mathrm{d}f(u) \big| \text{ for all } f \in C^\infty(D)\}$. Therefore for every $f, g \in C^\infty(D)$ and every $u \in D$, we have

$$\begin{aligned}\Pi_u(\mathrm{d}f(u), \mathrm{d}g(u)) &= \langle \mathrm{d}g(u) \mid i_u(\varpi_u{}^\sharp(i_u^t(\mathrm{d}f(u))))\rangle \\ &= \langle \mathrm{d}g(u) \mid Y_f(u)\rangle = \{f, g\}(u). \end{aligned} \tag{75}$$

Since Π_u is bilinear, it induces a linear map $\Pi_u^\sharp : T_u^*D \to T_uD$ defined by $\langle \beta_u \mid \Pi_u^\sharp(\alpha_u)\rangle = \Pi_u(\alpha_u, \beta_u)$ for every $\alpha_u, \beta_u \in T_u^*D$. The map $u \mapsto \Pi_u^\sharp$ is smooth.

Lemma 1.7.1.23. *For every $u \in D$, we have*

$$H_u = \mathrm{im}\,\Pi_u^\sharp = \mathrm{span}\{Y_f(u) \big| \textit{ for every } f \in C^\infty(D)\}. \tag{76}$$

Proof. We show that the first equality in (76) holds. By definition $\Pi_u^\sharp(\mathrm{d}f(u)) = i_u(\varpi_u^\sharp(i_u^t(\mathrm{d}f(u)))) = Y_f(u)$ for every $f \in C^\infty(D)$. But $i_u^t : T_u^*D \to H_u^*$ is surjective and $\varpi_u^\sharp : H_u^* \to H_u$ is bijective. Therefore $\mathrm{im}\,\Pi_u^\sharp = H_u$.

Next we prove the second equality in (76). Suppose that $f \in C^\infty(D)$. Since $Y_f(u) \in H_u \subseteq T_uD$, it follows that $\mathrm{span}\{Y_f(u) \big|$ for every $f \in C^\infty(D)\} \subseteq H_u$. Conversely, suppose that $v_u \in H_u$. Then $\varpi_u^\flat(v_u) \in H_u^*$. Since $i_u^t : H_u^* \to T_u^*D$ is surjective and $T_u^*D = \mathrm{span}\{\mathrm{d}g(u) \big| \text{ for all } g \in C^\infty(D)\}$, there is a smooth function f on D such that $i_u^t(\mathrm{d}f(u)) = \varpi_u^\flat(v_u)$. In other words, $v_u = \varpi_u^\sharp(i_u^t(\mathrm{d}f(u))) = Y_f(u)$. Therefore $H_u \subseteq \mathrm{span}\{Y_f(u) \big| \text{ for every } f \in C^\infty(D)\}$. This proves the lemma. □

Now we show how to recover a symplectic generalized distribution starting from an almost Poisson structure tensor field. Suppose that M is a smooth manifold and that $\{\,,\,\}$ is an almost Poisson structure on $C^\infty(M)$, that is, $\{\,,\,\} : C^\infty(M) \times C^\infty(M) \to C^\infty(M)$, which is bilinear, skew symmetric, and satisfies Leibniz' rule

$$\{f, g \cdot h\} = \{f, g\} \cdot h + g \cdot \{f, h\}$$

for every $f, g, h \in C^\infty(M)$. For every $m \in M$ define the map

$$\Pi_m : T_m^*M \times T_m^*M \to \mathbb{R} : (\mathrm{d}f(m), \mathrm{d}g(m)) \mapsto \{f, g\}(m).$$

Recall that the cotangent space T_m^*M to M at m is equal to $\mathrm{span}\{\mathrm{d}f(m) \big| f \in C^\infty(M)\}$. Then Π_m is bilinear and skew symmetric.

Moreover, Π_m depends smoothly on m. We call the map $m \mapsto \Pi_m$ the *almost Poisson tensor field* Π on M associated to the almost Poisson structure $\{\,,\,\}$ on $C^\infty(M)$. Because Π_m is bilinear, there is an associated linear map $\Pi^\sharp_m : T^*_m M \to T_m M$ defined by $\langle \mathrm{d}g(m) \mid \Pi^\sharp_m(\mathrm{d}f(m))\rangle = \Pi_m(\mathrm{d}f(m), \mathrm{d}g(m))$, which depends smoothly on m. For $f \in C^\infty(M)$ let $P_f(m) = \Pi^\sharp_m(\mathrm{d}f(m))$. Then $m \mapsto P_f(m)$ is a smooth vector field on M called the *almost Poisson vector field* associated to f.

Lemma 1.7.1.24. *For each $m \in M$ let $H_m = \mathrm{im}\,\Pi^\sharp_m$. Then the map $m \mapsto H_m$ is a* generalized distribution *H on M, in the sense that it is locally spanned by smooth vectors on M.*

Proof. By definition $H_m = \mathrm{span}\{P_f(m) \in T_m M \,\big|\, f \in C^\infty(M)\}$. For each $m \in M$ choose $f_1, \ldots, f_{n_m} \in C^\infty(M)$ such that $\{P_{f_i}(m)\}_{i=1}^{n_m}$ spans H_m. Since the function $M \mapsto \mathbb{Z}_{>0} : m \mapsto n_m$ is lower semi-continuous, there is an open neighborhood U of m in M such that it attains a maximum N. Therefore for every $m \in U$ we have $H_m = \mathrm{span}\,\{P_{f_i}(m)\}_{i=1}^{N}$. Consequently, $m \to H_m$ is a generalized distribution on M. □

For each $m \in M$ by definition of H_m the linear map $\Pi^\sharp_m : T^*_m M \to H_m$ is surjective. Therefore the induced linear map $\widetilde{\Pi}^\sharp_m : T^*_m M/\ker \Pi^\sharp_m \to H_m$ is a bijective linear map.

Lemma 1.7.1.25. *The linear map*

$$j_m : T^*_m M/\ker \Pi^\sharp_m \to H^*_m : \alpha_m + \ker \Pi^\sharp_m \mapsto \alpha_m | H_m$$

is bijective.

Proof. To see that the map j_m is well defined it suffices to show that $\ker \Pi^\sharp_m | H_m = 0$. Let $\beta_m \in \ker \Pi^\sharp_m$ and let $v_m \in H_m$. Since $\Pi^\sharp_m : T^*_m M \to H_m$ is surjective, there is a $\gamma_m \in T^*_m M$ such that $\Pi^\sharp_m(\gamma_m) = v_m$. Therefore

$$\begin{aligned} \beta_m(v_m) &= \beta_m(\Pi^\sharp_m(\gamma_m)) = \Pi_m(\gamma_m, \beta_m) \\ &= -\Pi_m(\beta_m, \gamma_m), \text{ since } \Pi_m \text{ is skew symmetric} \\ &= -\gamma_m(\Pi^\sharp_m(\beta_m)) = 0, \text{ since } \beta_m \in \ker \Pi^\sharp_m. \end{aligned} \tag{77}$$

Therefore j_m is well defined. Clearly it is linear.

Since the inclusion map $i_m : H_m \to T_m M$ is injective, its transpose $i^t_m : T^*_m M \to H^*_m$ is surjective. Let $\beta_m \in H^*_m$. Then there is $\gamma_m \in T^*_m M$ such that $i^t_m(\gamma_m) = \beta_m$. For $v_m \in H_m$, we have

$$\beta_m(v_m) = (i^t_m(\gamma_m))(v_m) = \gamma_m(i_m(v_m)) = \gamma_m(v_m),$$

that is, $\gamma_m | H_m = \beta_m$. So $j_m(\gamma_m + \ker \Pi^\sharp_m) = \gamma_m | H_m = \beta_m$. Therefore j_m is surjective. Since $\widetilde{\Pi}^\sharp_m$ is a bijective linear map, it follows that $\dim T^*_m M / \ker \Pi^\sharp_m = \dim H_m = \dim H^*_m$. Therefore j_m is bijective. □

Let $\varpi^\flat_m = j_m \circ (\widetilde{\Pi}^\sharp_m)^{-1} : H_m \to H^*_m$. Then $\varpi^\flat_m$ is a bijective linear mapping associated to the skew symmetric bilinear map

$$\varpi_m : H_m \times H_m \to \mathbb{R} : (v_m, w_m) \mapsto \langle \varpi^\flat_m(v_m) \mid w_m \rangle.$$

Thus we have proved

Proposition 1.7.1.26. *The generalized distribution (H, ϖ) on M is symplectic.*

Corollary 1.7.1.27. *For every $f \in C^\infty(M)$ the Hamiltonian vector field Y_f with respect to the generalized symplectic distribution (H, ϖ) is equal to the almost Poisson vector field P_f.*

Proof. For every $m \in M$ we have

$$P_f(m) = \Pi^\sharp_m(\mathrm{d}f(m)) = \widetilde{\Pi}^\sharp_m(\mathrm{d}f(m) + \ker \Pi^\sharp_m) \in H_m,$$

by definition of H_m. So

$$\begin{aligned} \varpi^\flat_m(P_f(m)) &= \varpi^\flat_m(\widetilde{\Pi}^\sharp_m(\mathrm{d}f(m) + \ker \Pi^\sharp_m)) = j_m(\mathrm{d}f(m) + \ker \Pi^\sharp_m) \\ &= \mathrm{d}f(m)|H_m = \varpi^\flat_m(Y_f(m)). \end{aligned}$$

Therefore $P_f(m) = Y_f(m)$ because $\varpi^\flat_m$ is invertible. □

1.7.2 *Nonholonomic Dirac brackets*

Next we give a procedure for finding an almost Poisson bracket on D in terms of the Poisson bracket on (TQ, ω), defined by (72). Suppose that the nonholonomic constraint D is a distribution on Q given by $D_q = \bigcap_{i=1}^{\ell} \ker \varphi_i(q)$, where φ_i, $i = 1, \ldots, \ell$, are 1-forms on Q such that $\varphi_i(q)$ are linearly independent on $T^*_q Q$ for every $q \in Q$. Since D is a distribution, it is a submanifold of TQ, defined by the common zeroes of the *constraint functions*

$$c_i : TQ \to \mathbb{R} : u \mapsto \langle \varphi_i(\tau_Q(u)) | u \rangle, \qquad 1 \le i \le \ell. \tag{78}$$

Here $\tau_Q : TQ \to Q$ is the tangent bundle projection. Note that for every $v_u \in T_u(TQ)$

$$c_i(u) = \langle \tau^*_Q \varphi_i(u) \mid v_u \rangle, \tag{79}$$

because

$$\langle \tau_Q^* \varphi_i(u) \mid v_u \rangle = \langle \varphi_i(\tau_Q(u)) \mid T_u \tau_Q v_u \rangle \; = \; \langle \varphi_i(\tau_Q(u)) \mid u \rangle.$$

Now consider the *constraint map* $c : TQ \rightarrow \mathbb{R}^\ell : u \mapsto \big(c_1(u), \ldots \, c_\ell(u)\big)$. Then

Fact 1.7.2.28. *0 is a regular value of the map c.*

Proof. First we show that for every $u \in TQ$ the 1-forms $\{\mathrm{d}c_i(u)\}_{i=1}^{\ell}$ are linearly independent. Suppose that $0 = \sum_{i=1}^{\ell} \alpha_i \, \mathrm{d}c_i(u)$. Then $0 = \sum_{i=1}^{\ell} \alpha_i \, \mathrm{d}\tau_Q^* \varphi_i(u)$, using (79), which implies $0 = \tau_Q^* \big(\sum_{i=1}^{\ell} \alpha_i \, \mathrm{d}\varphi_i(q)\big)$, where $q = \tau_Q(u)$. Hence $0 = \sum_{i=1}^{\ell} \alpha_i \, \mathrm{d}\varphi_i(q)$. But then $\alpha_i = 0$ for all $1 \le i \le \ell$ because $\{\mathrm{d}\varphi_i(q)\}_{i=1}^{\ell}$ are linearly independent. Since $\{\mathrm{d}c_i(u)\}_{i=1}^{\ell}$ are linearly independent, the derivative of the map c is surjective. □

Next we show that

Fact 1.7.2.29. $c^{-1}(0) = D$.

Proof. To see this note that

$$\begin{aligned} c_i(u) = 0 \quad \text{for all } i &\Longleftrightarrow \langle \varphi_i(\tau_Q(u)) | u \rangle = 0 \quad \text{for all } i \\ &\Longleftrightarrow u \in \ker \varphi_i(q) \quad \text{for all } i, \text{ where } q = \tau(u) \\ &\Longleftrightarrow u \in D_q. \end{aligned}$$

□

We now prove

Fact 1.7.2.30. *For every $u \in D$ we have*

$$T_u^{\omega} D = \mathrm{span}\{ X_{c_1}(u), \ldots , X_{c_\ell}(u)\}. \tag{80}$$

Proof. For every $v_u \in T_u D$,

$$\omega_u(X_{c_i}(u), v_u) = \langle \mathrm{d}c_i(u) \mid v_u \rangle \; = \; 0.$$

So $X_{c_i}(u) \in T_u^{\omega} D$. Therefore $\{X_{c_i}(u)\}_{i=1}^{\ell} \subseteq T_u^{\omega} D$. Note that $\{X_{c_i}(u)\}_{i=1}^{\ell}$ are linearly independent because if

$$0 = \sum_{i=1}^{\ell} \alpha_i X_{c_i}(u) \; = \; \omega_u^{\sharp}\big(\sum_{i=1}^{\ell} \alpha_i \, \mathrm{d}c_i(u)\big),$$

then $0 = \sum_{i=1}^{\ell} \alpha_i \, \mathrm{d}c_i(u)$. But this implies $\alpha_i = 0$ for $1 \le i \le \ell$. From fact 1.7.2.28 it follows that $(T_u D)^{\circ} = \mathrm{span}\{\mathrm{d}c_i(u)\}_{i=1}^{\ell}$. Moreover, $\omega_u^{\sharp}$ is a bijective linear map from $(T_u D)^{\circ}$ onto $T_u^{\omega} D$. Therefore $\dim T_u^{\omega} D = \ell$, which completes the proof. □

By definition, F (28) is the distribution on TQ whose image under $T\tau_Q$ is the distribution D. In our situation we have

Fact 1.7.2.31. *For every* $u \in TQ$

$$F_u^\omega = \operatorname{span}\{\omega_u^\sharp(\tau_Q^*\varphi_i(u)) \,\big|\, i = 1, \ldots, \ell\}. \tag{81}$$

Proof. To see this we argue as follows. By definition, $\omega_u^\sharp(\tau_Q^*\varphi_i(u))$ is the unique vector in $T_u(TQ)$ such that for every $w \in T_u(TQ)$

$$\langle \varphi_i(q) | T\tau_Q w\rangle = \langle \tau_Q^*\varphi_i(u)|w\rangle \;=\; \omega(\omega_u^\sharp(\tau_Q^*\varphi_i(u)), w), \tag{82}$$

where $q = \tau_Q(u)$. Now

$$w \in F_u \Leftrightarrow T\tau_Q v \in D_q \Leftrightarrow \langle \tau_Q^*\varphi_i(u) \mid v\rangle = \langle \varphi_i(\tau_Q(u))|T\tau_Q v\rangle = 0 \quad \forall\, i.$$

From (82) it follows that $\omega_u^\sharp(\tau_Q^*\varphi_i(u)) \in F_u^\omega$.

Conversely, suppose that $w' \in F_u^\omega$. Let $q = \tau_Q(u)$. Since $w' \in T_D(TQ)$, lemma 1.6.3.18 implies that there is a $v' \in T_qQ$ k-orthogonal to D_q such that $w' = \iota_u(v')$. Consider the 1-form on Q defined by $\varphi(q) = k(q)^\flat(v')$. Then for every $u' \in D_q$, we have $\langle \varphi(q) \mid u'\rangle = k(v', u') = 0$ because v' is k-orthogonal to D_q. Thus $\varphi(q)$ is a linear combination of the 1-forms $\varphi_i(q)$. For every $w \in T_u(TQ)$ we have

$$\omega(u)(w', w) = k(q)(v', T\tau_Q w) \;=\; \langle \varphi_i(\tau_Q(u))|T\tau_Q w\rangle \;=\; \langle \tau_Q^*\varphi_i(u)|w\rangle,$$

which implies

$$w' \in \operatorname{span}\{\omega_u^\sharp(\tau_Q^*\varphi_i)(u) \,\big|\, i = 1, \ldots, \ell\}.$$

This proves (81). □

By definition of the symplectic distribution $H = F \cap TD$ (34) on D, using theorem 1.6.3.20 we have $F_u^\omega \cap T_u^\omega D = \{0\}$ for every $u \in D$. Consequently,

$$\begin{aligned} H_u^\omega &= T_u^\omega D \oplus F_u^\omega \\ &= \operatorname{span}\{X_{c_1}(u), \ldots, X_{c_\ell}(u), \omega_u^\sharp(\tau_Q^*\varphi_1(u)), \ldots, \omega_u^\sharp(\tau_Q^*\varphi_\ell)(u)\} \end{aligned} \tag{83}$$

is a ω_u-symplectic subspace of $(T_u(TQ), \omega_u)$ of dimension 2ℓ.

The Poisson bracket $[f_1, f_2]$ of $f_1, f_2 \in C^\infty(TQ)$, with respect to the symplectic form ω, can be expressed in terms of the differentials $\mathrm{d}f_1$ and $\mathrm{d}f_2$ as

$$[f_1, f_2] = \omega(\omega^\sharp(\mathrm{d}f_1), \omega^\sharp(\mathrm{d}f_2)).$$

In the above formula, we can replace differentials by arbitrary 1-forms and extend the notion of the Poisson bracket to one forms which need not be exact. In particular, we can write

$$\begin{aligned} [\tau_Q^*\varphi_i, \tau_Q^*\varphi_j] &= \omega(\omega^\sharp(\tau_Q^*\varphi_i), \omega^\sharp(\tau_Q^*\varphi_j)), \\ \{c_i, \tau_Q^*\varphi_j\} &= \omega(\omega^\sharp(\mathrm{d}c_i), \omega^\sharp(\tau_Q^*\varphi_j)). \end{aligned} \tag{84}$$

Since $T_D(TQ) = H \oplus H^\omega$, for every $f \in C^\infty(TQ)$, we can express the restriction of X_f to points in D in the form $X_{f|D} = Y_f + Z_f$, where Y_f has values in H and Z_f has values in H^ω. Similarly, for $g \in C^\infty(TQ)$, the restriction $\mathrm{d}g\ |_{T_D TQ}$ of $\mathrm{d}g$ to points in D can be written as $\mathrm{d}g\ |_{T_D(TQ)} = \partial_H g \oplus \partial_{H^\omega} g$, where $\partial_H g$ annihilates H^ω and $\partial_{H^\omega} g$ annihilates H. Hence, restriction of $[f, g]$ to D can be expressed as

$$\begin{aligned} [f,g]\Big|_D &= \langle dg \mid X_f\rangle\Big|_D \\ &= \langle \mathrm{d}g \mid Y_f + Z_f\rangle\Big|_D = \langle \partial_H g \oplus \partial_{H^\omega} g \mid Y_f\rangle + \langle \mathrm{d}g \mid Z_f\rangle \\ &= \langle \partial_H g \mid Y_f\rangle + \langle \mathrm{d}g \mid Z_f\rangle = \{f\Big|_D, g\Big|_D\} + \langle \mathrm{d}g \mid Z_f\rangle, \end{aligned}$$

where $\{f\Big|_D, g\Big|_D\}$ is the nonholonomic almost Poisson bracket of $f\Big|_D$ and $g\Big|_D$ in $C^\infty(D)$. From equation (83) that

$$\begin{aligned} Z_f(u) = {} & a_1(u) X_{c_1}(u) + \ldots + a_\ell(u) X_{c_\ell}(u) \\ & + a_{\ell+1}(u)\, \omega_u^\sharp(\tau_Q^*\varphi_1(u)) + \ldots + a_{2\ell}(u)\, \omega_u^\sharp(\tau_Q^*\varphi_\ell(u)) \end{aligned}$$

for every $u \in D$. We can express the coefficients $a_1(u), \ldots, a_\ell(u)$ and $b_1(u), \ldots, b_\ell(u)$ in terms of Poisson brackets of $c_1, ..., c_\ell$ and $\tau_Q^*\varphi_1, ..., \tau_Q^*\varphi_\ell$ with f as follows. Since $\langle \mathrm{d}c_j \mid Y_f\rangle = 0$ and $\langle \tau_Q^*\varphi_j \mid Y_f\rangle = 0$ for all $j = 1, \ldots, \ell$, we get

$$\begin{aligned} [f, c_j] &= \langle \mathrm{d}c_j \mid X_f\rangle = \langle \mathrm{d}c_j \mid Z_f\rangle \\ &= \sum_{i=1}^{\ell} a_i\, \langle \mathrm{d}c_j \mid X_{c_i}\rangle + \sum_{i=1}^{\ell} a_{\ell+i}\, \langle \mathrm{d}c_j \mid \omega^\sharp(\tau_Q^*\varphi_i)\rangle, \\ [f, \tau_Q^*\varphi_j] &= \langle \tau_Q^*\varphi_j \mid X_f\rangle = \langle \tau_Q^*\varphi_j \mid Z_f\rangle \\ &= \sum_{i=1}^{\ell} a_i\, \langle \tau_Q^*\varphi_j \mid X_{c_i}\rangle + \sum_{i=1}^{\ell} a_{\ell+i}\, \langle \tau_Q^*\varphi_j \mid \omega^\sharp(\tau_Q^*\varphi_i)\rangle. \end{aligned}$$

Using the definition of the extended Poisson brackets (84) we are led to a system of 2ℓ linear equations

$$\begin{aligned} \textstyle\sum_{i=1}^{\ell} a_i\, [c_i, c_j] + \sum_{i=1}^{\ell} a_{\ell+i}\, [\tau_Q^*\varphi_i, c_j] &= [f, c_j] \\ \textstyle\sum_{i=1}^{\ell} a_i\, [c_i, \tau_Q^*\varphi_j] + \sum_{i=1}^{\ell} a_{\ell+i} [\tau_Q^*\varphi_i, \tau_Q^*\varphi_j] &= [f, \tau_Q^*\varphi_j] \end{aligned} \tag{85}$$

for $a_1, ..., a_{2\ell}$. This system can be written in terms of matrices as $Aa = b$, where

$$A = \left(\begin{array}{c|c} ([c_j, c_i]) & ([c_j, \tau_Q^* \varphi_i]) \\ \hline ([\tau_Q^* \varphi_j, c_i]) & ([\tau_Q^* \varphi_j, \tau_Q^* \varphi_i]) \end{array} \right). \tag{86}$$

Claim 1.7.2.32. *The matrix A is invertible.*

Proof. We compute each block of A. First,

$$\omega_u(X_{c_j}(u), X_{c_i}(u)) = \langle \mathrm{d}c_j(u) \mid X_{c_i}(u) \rangle = [c_j, c_i]_u.$$

Next

$$\omega_u(\omega^\sharp(\tau_Q^* \varphi_i(u)), X_{c_j}(u)) = \langle \tau_Q^* \varphi_i(u) \mid X_{c_j}(u) \rangle = [c_j, \tau_Q^* \varphi_i]_u.$$

Finally

$$\omega_u(\omega^\sharp(\tau_Q^* \varphi_i(u)), \omega^\sharp(\tau_Q^* \varphi_i(u))) = \langle \tau_Q^* \varphi_i(u) \mid \omega^\sharp(\tau_Q^* \varphi_i(u)) \rangle = [\tau_Q^* \varphi_j, \tau_Q^* \varphi_i]_u.$$

Therefore the matrix $A = (A_{ij})$ is the matrix of the symplectic form ω with respect to the basis (83) of H^ω. Hence A is invertible because (H_u^ω, ω_u) is a symplectic subspace of $(T_u(TQ), \omega_u)$. □

Let $A^{-1} = (A^{ij}) = \begin{pmatrix} (b_{ij}) & (c_{ij}) \\ (-c_{ji}) & (d_{ij}) \end{pmatrix}$ be the inverse of the matrix A. Solving (85) gives

$$a_i = \sum_{j=1}^{\ell} \left(b_{ij} \, [f, c_j] + c_{ij} \, [f, \tau_Q^* \varphi_j] \right),$$

$$a_{\ell+i} = \sum_{j=1}^{\ell} \left(-c_{ji} \, [f, c_j] + d_{ij} [f, \tau_Q^* \varphi_j] \right),$$

when written out in components. Now

$$\begin{aligned} \langle \mathrm{d}g \mid Z_f \rangle &= \sum_{i=1}^{\ell} \left(a_i \langle \mathrm{d}g \mid X_{c_i} \rangle + a_{\ell+i} \langle \mathrm{d}g \mid \omega^\sharp(\tau_Q^* \varphi_i) \rangle \right) \\ &= \sum_{i=1}^{\ell} \left(a_i \, [g, c_i] + a_{\ell+i} \, [g, \tau_Q^* \varphi_i] \right) \\ &= \sum_{i=1}^{\ell} \sum_{j=1}^{\ell} \left(b_{ij} \, [f, c_j] + c_{ij} [f, \tau_Q^* \varphi_j] \right) [g, c_i] + \\ &\qquad \left(-c_{ji} \, [f, c_j] + d_{ij} \, [f, \tau_Q^* \varphi_j] \right) [g, \tau_Q^* \varphi_i] \\ &= \sum_{i=1}^{\ell} \sum_{j=1}^{\ell} [g, c_i] \, b_{ij} \, [f, c_j] + [g, c_i] \, c_{ij} \, [f, \tau_Q^* \varphi_j] \\ &\qquad - [g, \tau_Q^* \varphi_i] \, c_{ji} \, [f, c_j] + [g, \tau_Q^* \varphi_i] \, d_{ij} \, [f, \tau_Q^* \varphi_j]. \end{aligned}$$

This implies that

$$\{f\big|_D, g\big|_D\} = [f,g]\big|_D - \sum_{1\le i,j\le \ell} ([g,c_i],[g,\tau_Q^*\varphi_i])A^{ij}\begin{pmatrix}[f,c_j]\\ [f,\tau_Q^*\varphi_j]\end{pmatrix}. \tag{87}$$

In chapter 5 in order to find the equations of motion for Carathéodory's sleigh, we use the special case of (87) when $\ell = 1$. Written out explicitly, we get

$$\begin{aligned}&\{f\big|_D, g\big|_D\} = [f,g]\big|_D\\ &+\tfrac{1}{[c_1,\tau_Q^*\varphi_1]}([g,c_1],[g,\tau_Q^*\varphi_1])\begin{pmatrix}0 & [c_1,\tau_Q^*\varphi_1]\\ -[c_1,\tau_Q^*\varphi_1] & 0\end{pmatrix}\begin{pmatrix}[f,c_1]\\ [f,\tau_Q^*\varphi_1]\end{pmatrix}.\end{aligned} \tag{88}$$

1.8 Momenta and the momentum equation

In this section Q is a smooth manifold with a Riemannian metric k.

1.8.1 *Momentum functions*

Let Z be a smooth vector field on Q. The *momentum function* associated to Z is $\mu_Z : T^*Q \to \mathbb{R} : \alpha_q \mapsto \langle \alpha_q \mid Z(q)\rangle$. Let $\varphi_s : Q \to Q$ be the flow of Z. Then the flow of the *cotangent lift* Z_{T^*Q} of Z to T^*Q is

$$\widetilde{\varphi}_s : T^*Q \to T^*Q : \alpha_q \mapsto (T_q\varphi_{-s})^t\alpha_q.$$

Note that $T\pi_Q \circ Z_{T^*Q} = Z\circ \pi_Q$, where $\pi_Q : T^*Q \to Q$ is the cotangent bundle projection map.

Lemma 1.8.1.33. *Z_{T^*Q} is the Hamiltonian vector field X_{μ_Z} on (T^*Q, ω_Q) corresponding to the momentum function μ_Z. Here $\omega_Q = -\,\mathrm{d}\theta_Q$, where θ_Q is the canonical 1-form on T^*Q.*

Proof. First we show that θ_Q is invariant under the flow of Z_{T^*Q}. For $u \in T_qQ$, $\alpha \in T_q^*Q$, and $v_\alpha \in T_\alpha(T^*Q)$ we have

$$\begin{aligned}\langle \widetilde{\varphi}_s^*\theta_Q(\alpha) \mid v_\alpha\rangle &= \langle \theta_Q(\widetilde{\varphi}_s(\alpha)) \mid T_u\widetilde{\varphi}_s v_\alpha\rangle = \langle \widetilde{\varphi}_s(\alpha) \mid T_u(\tau_Q\circ\widetilde{\varphi}_s)v_\alpha\rangle\\ &\qquad\text{where } \tau_Q \text{ is the tangent bundle projection map}\\ &= \langle \widetilde{\varphi}_s(\alpha) \mid T\varphi_s(T_u\tau_Q v_\alpha)\rangle = \langle \alpha \mid u\rangle = \langle \theta_Q(\alpha) \mid v_\alpha\rangle.\end{aligned}$$

Therefore

$$0 = (L_{Z_{T^*Q}}\theta_Q)(\alpha) = (Z_{T^*Q} \mathbin{\lrcorner} \mathrm{d}\theta_Q)(\alpha) + \mathrm{d}\langle \theta_Q(\alpha) \mid Z_{T^*Q}(\alpha)\rangle,$$

that is,

$$\begin{aligned}(Z_{T^*Q} \lrcorner\, \omega_Q)(\alpha) &= \mathrm{d}\langle \theta_Q(\alpha) \mid Z_{T^*Q}(\alpha)\rangle \\ &= \langle \alpha \mid T\pi_Q\, Z_{T^*Q}(\alpha)\rangle = \langle \alpha \mid Z(\pi_Q(\alpha))\rangle = \mu_Z(\alpha).\end{aligned}$$

□

Using the vector bundle isomorphism $\mathrm{k}^\flat : TQ \to T^*Q$, we pull back the momentum function μ_Z to a smooth function P_Z on TQ. In other words, $P_Z(v_q) = \mathrm{k}(q)(v_q, Z(q))$ for every $q \in Q$ and every $v_q \in T_qQ$. P_Z is the *momentum function* on TQ associated to the vector field Z with respect to the Riemannian metric k. Let X_{P_Z} be the Hamiltonian vector field corresponding to P_Z on the symplectic manifold (TQ, ω), where $\omega = (\mathrm{k}^\flat)^*\omega_Q$. The *tangent lift* Z_{TQ} of the vector field Z is the vector field on TQ whose flow $\overline{\varphi}_s : TQ \to TQ$ is $\overline{\varphi}_s(v_q) = T_q\varphi_s\, v_q$ for every $q \in Q$. Note that $T\tau_Q \circ Z_{TQ} = Z \circ \tau_Q$.

Lemma 1.8.1.34. *We have*

$$T\tau_Q \circ X_{P_Z} = Z \circ \tau_Q. \tag{89}$$

Also $X_{P_Z} = Z_{TQ}$ *if and only if the flow of* Z *preserves the Riemannian metric* k.

Proof. Pulling back both sides of $X_{\mu_Z} \lrcorner\, \omega_Q = \mathrm{d}\mu_Z$ by $\mathrm{k}^\flat$ gives

$$\begin{aligned}(\mathrm{k}^\flat)^* X_{\mu_Z} \lrcorner\, \omega &= (\mathrm{k}^\flat)^* X_{\mu_Z} \lrcorner\, (\mathrm{k}^\flat)^*\omega_Q = \mathrm{k}^\flat(X_{\mu_Z} \lrcorner\, \omega_Q) \\ &= (\mathrm{k}^\flat)^*\, \mathrm{d}\mu_Z = \mathrm{d}((\mathrm{k}^\flat)^*\mu_Z) = \mathrm{d}P_Z.\end{aligned}$$

Therefore $X_{P_Z} = (\mathrm{k}^\flat)^* X_{\mu_Z}$. From $T\pi_Q \circ \mathrm{k}^\flat = T\tau_Q$ it follows that

$$\begin{aligned}T\tau_Q \circ X_{P_Z} &= T\pi_Q \circ \mathrm{k}^\flat \circ (\mathrm{k}^\flat)^* X_{\mu_Z} = T\pi_Q \circ X_{\mu_Z} \circ \mathrm{k}^\flat \\ &= Z \circ \pi_Q \circ \mathrm{k}^\flat = Z \circ \tau_Q.\end{aligned}$$

This proves (89).

Since $X_{P_Z} = (\mathrm{k}^\flat)^* X_{\mu_Z}$ and $X_{\mu_Z} = Z_{T^*Q}$, it follows that $X_{P_Z} = Z_{TQ}$ if and only if $Z_{TQ} = (\mathrm{k}^\flat)^* Z_{T^*Q}$. This equality holds if and only if $T\varphi_s = (\mathrm{k}^\flat)^{-1} \circ (T\varphi_{-s})^t \circ \mathrm{k}^\flat$, that is, for every $q \in Q$ and every $v_q \in T_qQ$ we have

$$\mathrm{k}^\flat(\varphi_s(q))(T_q\varphi_s v_q) = \mathrm{k}^\flat(q)(v_q) T_{\varphi_{-s}(q)}\varphi_s. \tag{90}$$

Evaluating both sides of (90) on $T_q\varphi_s w_q$, where $w_q \in T_qQ$, we get

$$\mathrm{k}(\varphi_s(q))(T_q\varphi_s v_q, T_q\varphi_s w_q) = \mathrm{k}(q)(v_q, w_q),$$

for every $q \in Q$ and every $v_q, w_q \in T_qQ$. In other words, the flow of Z is an isometry. □

1.8.2 *Momentum equations*

Let (D, H, ϖ, h) be a distributional Hamiltonian system with constraint distribution D on Q and symplectic distribution (H, ϖ) on D.

A smooth curve $\gamma : I \subseteq \mathbb{R} \to TQ$ satisfies the *second order equation condition* for $f \in C^\infty(Q)$ if and only if for every $t \in I$

$$\frac{\mathrm{d}\tau_Q^* f(\gamma(t))}{\mathrm{d}t} = \langle \mathrm{d}f(q(t)) \mid v(t) \rangle, \tag{91}$$

where $q(t) = \tau_Q(\gamma(t))$ and $v(t) = \gamma(t) \in T_{q(t)}Q$. Note that (91) is equivalent to

$$\langle \mathrm{d}f(\gamma(t)) \mid T_{\gamma(t)}\tau_Q(\dot\gamma(t)) \rangle = \langle \mathrm{d}f(\gamma(t)) \mid v(t) \rangle \tag{92}$$

for every $t \in I$.

Lemma 1.8.2.35. *For every $f \in C^\infty(Q)$ the following statements are equivalent.*

1. *Y_f defines a second order differential equation. In other words, if $I \subseteq \mathbb{R} \to D \subseteq TQ : t \mapsto (q(t), v(t))$ is an integral curve of Y_f, then $v(t) = \frac{\mathrm{d}q(t)}{\mathrm{d}t}$ for every $t \in I$.*

2. *There is a function $V \in C^\infty(Q)$ such that*

$$f(q, v_q) = \tfrac{1}{2}\,\mathrm{k}(q)(v_q, v_q) + V(q), \tag{93}$$

for every $(q, v_q) \in D$.

Proof. Using a local trivialization of D and the notation of §1.6.3, we may write $Y_f = a_\rightarrow + b_\uparrow$, see (45). Writing $u = (q, v) \in D_q$ we have $b_\uparrow(u) = \iota_u(a^Q(u))$. Therefore

$$\langle \mathrm{d}f(u) \mid \iota_u(z) \rangle = \mathrm{k}(q)(a^Q(u), z), \tag{94}$$

for every $z = b^Q(u) \in D_q$. Because $a^Q(u) = T_u\tau_Q Y_f(u)$, statement 1 is equivalent to $a^Q(u) = v$. Therefore (94) is equivalent to

$$\left.\frac{\mathrm{d}}{\mathrm{d}t}\right|_{t=0} f(u + tz) = \mathrm{k}(q)(v, z), \tag{95}$$

for every $z \in D_q$. But (95) is equivalent to (93). □

A smooth curve $\gamma : I \subseteq \mathbb{R} \to TQ$ satisfies the *momentum equation* for the smooth vector field Z on Q if and only if

$$\frac{\mathrm{d}P_Z(\gamma(t))}{\mathrm{d}t} = -L_{P_Z} h(\gamma(t)), \tag{96}$$

for every $t \in I$.

Theorem 1.8.2.36. *Every integral curve $I \subseteq \mathbb{R} \to D : t \mapsto u(t)$ of the distributional Hamiltonian vector field Y_h associated to the distributional Hamiltonian system (D, H, ϖ, h) satisfies the second order equation condition for every $f \in C^\infty(Q)$ and the momentum equation for every smooth vector field Z on Q, which takes values in the constraint distribution D. Conversely, let $\gamma : I \subseteq \mathbb{R} \to D \subseteq TQ : t \mapsto u(t) = (q(t), v(t))$ be a smooth curve. Let $\mathcal{F}$ be a collection of smooth functions f on Q such that* $\mathrm{span}\{\mathrm{d}f_{q(t)}|D_{q(t)}\} = D_{q(t)}$ *for every $t \in I$ and let $\mathcal{Z}$ be a collection of smooth vector fields on Q with values in D such that* $\mathrm{span}\{Z(q(t))\} = D_{q(t)}$ *for every $t \in I$. If γ satisfies the second order equation condition for every $f \in \mathcal{F}$ and the momentum equation for every $Z \in \mathcal{Z}$, then γ is an integral curve of Y_h.*

Proof. Suppose that $\gamma : I \subseteq \mathbb{R} \to D \subseteq TQ : t \mapsto u(t) = (q(t), v(t))$ is an integral curve of Y_h. From lemma 1.8.2.35 it follows that $T_{\gamma(t)}\tau_Q(\dot{\gamma}(t)) = v(t)$. Therefore (92) holds for all $t \in I$, that is, the second order equation condition holds for every $f \in C^\infty(Q)$. Let Z be a smooth vector field on Q with values in D. Since $T\tau_Q \circ X_{P_Z} = Z \circ \tau_Q$ and $Z(Q) \subseteq D$, it follows that X_{P_Z} has values in the distribution F (28). Therefore

$$\begin{aligned} L_{Y_h} P_Z &= Y_h \,\lrcorner\, \mathrm{d}P_Z = \omega(Y_h, X_{P_Z}) \\ &= -\omega(X_{P_Z}, \dot{\gamma}) = -X_{P_Z} \,\lrcorner\, \mathrm{d}h = -L_{P_Z} h. \end{aligned}$$

Consequently, the Hamiltonian h satisfies the momentum equation for Z.

To prove the converse we start by observing that because γ is a smooth curve in D, it follows that $v(t) \in D_{q(t)}$ and $T_{\gamma(t)}\dot{\gamma}(t) \in D_{q(t)}$ for every $t \in I$. Since (92) holds, we obtain $T_{\gamma(t)}\dot{\gamma}(t) = v(t)$. From $T_{\gamma(t)}\tau_Q(Y_h(\gamma(t))) = v(t)$, it follows that $T_{\gamma(t)}\tau_Q\lambda(t) = 0$ for every $t \in I$, where $\lambda(t) = \dot{\gamma}(t) - Y_h(\gamma(t))$. In addition, we have $\lambda(t) \in T_{\gamma(t)}D \cap T_{\gamma(t)}\tau_Q$, since $\dot{\gamma}(t)$ and $Y_h(\gamma(t))$ both lie in $T_{\gamma(t)}D$. Therefore for every $t \in I$ we see that $\lambda(t) \in D_{\gamma(t)}$, if we identify $\ker T_{\gamma(t)}\tau_Q$ with $T_{\gamma(t)}Q$. On the other hand, the assumption that the momentum equation holds for γ implies $Y_h \,\lrcorner\, \mathrm{d}P_Z = -L_{X_{P_Z}} h = \langle \mathrm{d}P_Z \mid \dot{\gamma} \rangle$. Consequently, $\langle \mathrm{d}P_Z \mid \lambda(t) \rangle = 0$ for every $Z \in \mathcal{Z}$. Because $\lambda(t)$ belongs to the tangent space of each fiber of the tangent bundle projection map τ_Q and the fiber derivative of P_Z is $\mathrm{k}^\flat(Z)$, see (95), it follows that $\mathrm{k}(\gamma(t))(Z(\gamma(t)), \lambda(t)) = 0$ for every $Z \in \mathcal{Z}$. Because $\mathrm{span}_{Z \in \mathcal{Z}}\{Z(\gamma(t))\} = D_{\gamma(t)}$, we obtain $\lambda(t) = 0$ for every $t \in I$. In other words, $\dot{\gamma}(t) = Y_h(\gamma(t))$ for every $t \in I$, that is, $t \mapsto \gamma(t)$ is an integral curve of Y_h. □

1.8.3 *Homogeneous functions*

Because we have assumed that the nonholonomic constraints are linear, the constraint manifold D is a smooth vector subbundle of the tangent bundle TQ of configuration space Q. For every nonnegative integer k let $C^\infty_k(D)$ be the set of all smooth functions f on D such that for each $q \in Q$ the restriction $f|D_q$ of f to the *vector space* D_q is a homogeneous function of degree k, that is, $f(r\,u_q) = r^k\, f(u_q)$ for every $r \in \mathbb{R}\backslash\{0\}$ and every $u_q \in D_q$. For fixed $q \in Q$ the function $f|D_q$ is smooth at the origin of D_q. Therefore $f|D_q$ is a homogeneous *polynomial* on D_q of degree k.

Let $C^\infty_0(D)$ be the space of smooth functions on D which are constant on the fibers of the vector bundle $\tau = \tau_Q|D : D \subseteq TQ \rightarrow Q$. Then $C^\infty_0(D) = \{\tau^* f \in C^\infty(D) \big| \, f \in C^\infty(Q)\}$. This proves

Lemma 1.8.3.37. *The map $\tau^* : C^\infty(Q) \rightarrow C^\infty_0(D)$ is an isomorphism of algebras.*

Define the space
$C^\infty_1(D) = \{f \in C^\infty(D) \big| \, f_q|D_q : D_q \subseteq T_qQ \rightarrow \mathbb{R}$ is linear for every $q \in Q\}$. Because $\mathrm{k}^\flat_q|D_q : D_q \subseteq T_qQ \rightarrow D^*_q \subseteq T^*_qQ$ is a linear isomorphism, there is a unique $Z(q) \in D_q$ such that $\mathrm{k}(q)(Z(q), u_q) = (f|D_q)(u_q)$ for every $u_q \in D_q$. Moreover, the map $q \mapsto Z(q)$ is smooth. Therefore $f = P_Z|D$, where P_Z is the momentum function corresponding to the vector field Z on Q.

Let $\mathcal{X}^\infty(Q, D)$ be the space of smooth vector fields on Q with values in the distribution D. The discussion above proves

Lemma 1.8.3.38. *The mapping*

$$\mathcal{X}^\infty(Q, D) \rightarrow C^\infty_1(D) : Z \mapsto P_Z|D$$

is an isomorphism of vector spaces.

Let $C^\infty_{\mathrm{hom}}(D)$ be the vector space of smooth homogeneous functions on D.

Lemma 1.8.3.39. *$C^\infty_{\mathrm{hom}}(D)$ is a graded almost Poisson algebra. In particular, if $f \in C^\infty_k(D)$ and $g \in C^\infty_\ell(D)$, then $\{f, g\} \in C^\infty_{k+\ell-1}(D)$.*

Proof. For any $r \in \mathbb{R} \setminus \{0\}$ let $m_r : T_qQ \rightarrow T_qQ : v_q \mapsto r\,v_q$. Then $m^t_r : T^*_qQ \rightarrow T^*_qQ : \alpha_q \mapsto r\,\alpha_q$. Let θ_Q be the canonical 1-form on T^*Q. Then for every $\alpha_q \in T^*Q$ and $w_\alpha \in T_\alpha(T^*Q)$ we get

$$\begin{aligned}\langle (m^t_r)^*\theta_Q(\alpha_q) \mid w_\alpha\rangle &= \langle \theta_Q(r\,\alpha_q) \mid T_{\alpha_q} m_r\, w_\alpha\rangle \\ &= \langle r\,\alpha_q \mid T(\tau_Q \circ m_r)w_\alpha\rangle = \langle r\,\alpha_q \mid T\tau_Q w_\alpha\rangle \text{ since } \tau_Q \circ m_r = \tau_Q \\ &= \langle r\,\alpha_q \mid T\tau_Q w_\alpha\rangle = r\,\langle \theta_Q(\alpha_q) \mid w_\alpha\rangle.\end{aligned}$$

So $(m_r^t)^*\theta_Q = r\,\theta_Q$. Therefore

$$(m_r^t)^*\omega_Q = (m_r^t)^*(-\,\mathrm{d}\theta_Q) = -\,\mathrm{d}\left((m_r^t)^*\theta_Q\right) = -\,\mathrm{d}(r\,\theta_Q) = -r\,\mathrm{d}\theta_Q = r\,\omega_Q$$

Note that the map m_r leaves the distribution D invariant and Tm_r leaves the distribution H invariant. Since $f \in C_k^\infty(D)$ if and only if $m_r^* f = r^k f$, we obtain

$$(m_r^t)^*\,\mathrm{d}f = \mathrm{d}(m_r^* f) = \mathrm{d}(r^k\, f) = r^k\,\mathrm{d}f.$$

Therefore for each $u, u' \in D$ we have

$$\begin{aligned} r^k\,\omega(Y_f(u), u') &= \langle \mathrm{d}f(m_r(u)) \mid T_{u'}m_r(u')\rangle \\ &= \omega(m_r(u))(Y_f(m_r(u))) \mid T_{u'}m_r(u')\rangle \\ &= \left((m_r^t)^*\omega\right)(u)(m_r^*Y_f(u), u'), \end{aligned}$$

that is,

$$m_r^* Y_f = r^{k-1}\,Y_f. \tag{97}$$

Consequently,

$$\begin{aligned} m_r^*(\{f,g\})(u) &= \{f,g\}(m_r(u)) = \langle \mathrm{d}g(m_r(u)) \mid Y_f(m_r(u))\rangle \\ &= r^{k-1}\,\langle \mathrm{d}g(m_r(u)) \mid T_u m_r Y_f(u)\rangle, \text{ using (97) and } f \in C_k^\infty(D) \\ &= r^{k-1}\,\langle (m_r^t)^*(\mathrm{d}g)(u) \mid Y_f(u)\rangle \\ &= r^{k+\ell-1}\,\langle \mathrm{d}g(u) \mid Y_f(u)\rangle, \text{ because } g \in C_\ell^\infty(D) \\ &= r^{k+\ell-1}\,\{f,g\}(u). \end{aligned}$$

□

1.8.4 *Momenta as coordinates*

Let U be an open subset of Q. With $\varphi^i \in C^\infty(D)$ let $\{\varphi^i\}_{i=1}^n$ be a system of coordinates on U. Let $\{Z_j\}_{j=1}^d$ be smooth vector fields on U such that $Z_j(U) \subseteq D$.

The functions $\{\tau^*\varphi^i\}_{i=1}^n$, where $\tau = \tau_Q|D$, together with the momenta $\{P_j = P_{Z_j}\}_{j=1}^d$ form a *system of coordinates* on $V = \tau^{-1}(U)$ in D if and only if for each $q \in U$ the vectors $\{Z_j(q)\}_{j=1}^d$ form a basis of D_q. The map

$$\lambda : U \times \mathbb{R}^d \to V \subseteq D : (q, c) \mapsto \sum_{j=1}^d c_j\, Z_j(q)$$

is the inverse of a trivialization of the vector bundle $\tau : V \subseteq D \rightarrow U \subseteq Q$. Substituting $v = \sum_{j=1}^{d} c_j \, Z_j(q)$ into the definition of the momentum $P_j = P_{Z_j}$ gives the relation

$$P_j(q,v) = P_{Z_j}(q,v) = \mathrm{k}(q)(Z_j(q), v) = \sum_{\ell=1}^{d} \mathrm{k}(q)(Z_j(q), Z_\ell(q))\, c_\ell \tag{98}$$

between the coordinates P_j and c_ℓ on U. Note that $c_j = P_j$ for $1 \le j \le d$ if and only if $\{Z_j(q)\}_{j=1}^{d}$ is a k-orthonormal basis of D_q for every $q \in U$.

In terms of the coordinates $\{\tau^*\varphi^1, \ldots, \tau^*\varphi^n, P_1, \ldots P_n\}$ on $V \subseteq D$ the equations of motion are

$$\dot{\varphi}^i = \sum_{j=1}^{d} (Z_j)^i \, c_j, \text{ for } 1 \le i \le n \tag{99}$$

$$\dot{P}_j = -(L_{P_j} h)|V, \text{ for } 1 \le j \le d. \tag{100}$$

Here equation (99) is the second order equation condition for φ^i, where $(Z_j)^i = \langle \mathrm{d}\varphi^i \mid Z_j \rangle$ are the components of the vector field Z_j with respect to the coordinates $\{\varphi^i\}_{i=1}^{n}$ on $U \subseteq Q$. Solving (98), the c_j's can be expressed in terms of the P_j's. Equation (100) is the momentum equation for Z_j, where the right hand side has to be expressed in terms of φ^i and P_j.

Using the fact that $h = \frac{1}{2}\,\mathrm{k}(q)(v,v) + V(q)$, where $T = \frac{1}{2}\,\mathrm{k}(q)(v,v) \in C_2^\infty(D)$ and $V \in C_0^\infty(D)$ and lemma 1.8.3.39, we obtain

$$\dot{\varphi}^i = \{h, \varphi^i\} = \{T, \varphi^i\} \in C_1^\infty(D),$$

which is a $\{\varphi^i\}_{i=1}^{n}$-dependent linear form on D and

$$\dot{P}_j = \{h, P_j\} = \{T, P_j\} + \{V, P_j\},$$

where $\{T, P_j\} \in C_2^\infty(D)$ is a $\{\varphi^i\}_{i=1}^{n}$-dependent quadratic form in $\{P_j\}_{j=1}^{d}$ and $\{V, P_j\} \in C_0^\infty(D)$ is a function of $\{\varphi^i\}_{i=1}^{n}$.

1.9 A projection principle

In this section we describe a *projection method* to obtain the distributional Hamiltonian equations of motion (63).

Equation (55) enables us to decompose $X_f|D$, the restriction to D of the Hamiltonian vector field X_f of f, into its components $X_{f,H}$ and X_{f,H^ω} in H and H^ω, respectively. In other words,

$$X_f \,|_D = X_{f,H} + X_{f,H^\omega}. \tag{101}$$

Taking equations (56), (57) and (59) into account, we see that the distributional Hamiltonian vector field Y_f of f is equal to the H-component of the Hamiltonian vector field X_f of f with respect to the decomposition (101). In other words, $Y_f = X_{f,H}$.

On the other hand, the first equality in (33) allows us to decompose $X_f\mid_D$ into its components X_{f,F^ω} and $X_{f,TD}$ in F^ω and TD, respectively.

Lemma 1.9.40. $X_{f,TD} = Y_f$.

Proof. Since $H \subseteq TD$ and $F^\omega \subseteq H^\omega$, we can intersect the decompositions $T_D(TQ) = F^\omega \oplus TD$ and $T_D(TQ) = H^\omega \oplus H$ to obtain

$$T_D(TQ) = F^\omega \oplus (H^\omega \cap TD) \oplus H. \tag{102}$$

We know that H is a symplectic distribution on D, and $F^\omega \subseteq \ker T\tau_Q$ is isotropic. Since $F^\omega \oplus (H^\omega \cap TD) = H^\omega$ and $\dim F^\omega = \dim(H^\omega \cap TD)$, it follows that F^ω and $H^\omega \cap TD$ are Lagrangian in H^ω. Using the decomposition (102) write $X_f = X_f^1 + X_f^2 + X_f^3$ where X_f^1 has values in F^ω, X_f^2 has values in $H^\omega \cap TD$, and X_f^3 has values in H. In a similar fashion, using (102) we can write

$$\mathrm{d}f = \partial_{F^\omega} f \oplus \partial_{H^\omega \cap TD} f \oplus \partial_H f.$$

Taking into account the decomposition (56) we get

$$\begin{aligned} X_f \,\lrcorner\, \omega &= (X_f^1 + X_f^2 + X_f^3) \,\lrcorner\, (\varpi_{H^\omega} \oplus \varpi_H) \\ &= (X_f^1 \,\lrcorner\, \varpi_{H^\omega}) \oplus (X_f^2 \,\lrcorner\, \varpi_{H^\omega}) \oplus (X_f^3 \,\lrcorner\, \varpi_H) \\ &= \partial_{F^\omega} f \oplus \partial_{H^\omega \cap TD} f \oplus \partial_H f. \end{aligned}$$

Since $X_f^1 \,\lrcorner\, \varpi_{H^\omega}$ annihilates F^ω, and $X_f^2 \,\lrcorner\, \varpi_{H^\omega}$ annihilates $H^\omega \cap TD$, it follows that

$$X_f^1 \,\lrcorner\, \varpi_{H^\omega} = \partial_{H^\omega \cap TD} f, \quad X_f^2 \,\lrcorner\, \varpi_{H^\omega} = \partial_{F^\omega} f, \quad \text{and} \quad X_f^3 \,\lrcorner\, \varpi_H = \partial_H f.$$

Lemma 1.6.3.18 implies that $\partial_{F^\omega} f = 0$. Hence, $X_f^2 = 0$. Therefore, $X_f^1 = X_{f,H^\omega} = X_{f,F^\omega}$, and $X_f^3 = X_{f,TD} = X_{f,H} = Y_f$. □

Let $\mathcal{P} : T_D(TQ) \to TD$ be the projection along the fibers of F^ω corresponding to the direct sum decomposition $T_D(TQ) = F^\omega \oplus TQ$. The statement of theorem 1.6.3.20 can be reformulated as the following *projection principle*.

Proposition 1.9.41. *A dynamically admissible motion of a distributional Hamiltonian system (D, H, ϖ, h) is the image under the tangent bundle projection τ_Q to Q of an integral curve of $\mathcal{P} \circ X_h\mid_D$.*

1.10 Accessible sets

An abstract structure of a Hamiltonian system with linear nonholonomic constraints is given by a quadruple (D, H, ϖ, h), where D is a manifold, (H, ϖ) is a symplectic distribution on D, and h is a smooth function on D. For each $f \in C^\infty(D)$, the distributional Hamiltonian vector field of f is defined as the unique vector field Y_f on D with values in H satisfying the equation $Y_f \lrcorner \varpi = \partial_H f$. By theorem 1.6.3.20 the evolution of our nonholonomically constrained system is given by the distributional Hamiltonian vector Y_h of the Hamiltonian h.

This abstract structure was obtained in §6.3 as the result of an analysis of dynamics of a Hamiltonian system with configuration space Q, Hamiltonian $h \in C^\infty(TQ)$, and linear nonholonomic constraints given by a distribution D on Q. However, once we have obtained the symplectic distribution (H, ϖ) on D we can forget about the linear structure of D and its embedding into TQ, and retain only the manifold structure of D.

In the following we shall need the notion of a *generalized distribution* on D, that is a linear subset H of the tangent bundle TD of D locally spanned by smooth vector fields. For $u \in D$, the number of linearly independent vector fields spanning $H_u \subseteq T_u D$ is called the *rank* of H at u. A distribution is a generalized distribution of constant rank. An *accessible* set of a generalized distribution H on D is the set of points of D that can be connected by piecewise integral curves of vector fields with values in H. Accessible sets of H are also called *reachable* sets of H or *orbits* of the family of local vector fields with values in H.

Theorem 1.10.42. (Sussmann's theorem) *Every accessible set of a generalized distribution on D is an immersed submanifold of D.*

Proof. See [112]. □

Let $L \subseteq D$ be an accessible set of H. According to theorem 10.1 it is an immersed submanifold of D. The restriction H_L of H to points in L is contained in TL. Hence, H_L is a distribution on L. Let ϖ_L be the restriction of ϖ to H_L. By construction ϖ_L is a nondegenerate 2-form on the distribution H_L. For each $f \in C^\infty(D)$, let f_L be the restriction of f to L. Clearly f_L is a smooth function on L. Let Y_{f_L} be the distributional Hamiltonian vector field of f_L on L relative to the symplectic distribution (H_L, ϖ_L). In other words, Y_{f_L} is the unique vector field on L with values in H_L such that

$$Y_{f_L} \lrcorner \varpi_L = \partial_{H_L} f_L. \tag{103}$$

Since Y_h has values in H, it follows that $Y_h(u) = Y_{h_L}(u)$ for all $u \in L$. Hence, L is preserved by the evolution of our nonholonomic Hamiltonian system. This proves

Lemma 1.10.43. *For every accessible set L of H, the nonholonomic Hamiltonian system (D, H, ϖ, h) induces on L the structure of a nonholonomic Hamiltonian system (L, H_L, ϖ_L, h_L).*

Since accessible sets of H form a partition of D, we can write

$$(D, H, \varpi, h) = \bigcup_{L \text{ a.s.} H} (L, H_L, \varpi_L, h_L). \tag{104}$$

Here the union is taken over accessible sets L of H. We say that a nonholonomic Hamiltonian system (D, H, ϖ, h) is *simple* if D is the unique accessible set of H. Since each accessible set L of H is a unique accessible set of H_L, it follows that (104) is the decomposition of (D, H, ϖ, h) into its simple components.

Accessible sets of our symplectic distribution (H, ϖ) on D are important because they provide restrictions on the evolution of the system given by integral curves of the vector field Y_h.

Theorem 1.10.44. (Stefan's theorem) *Accessible sets of a generalized distribution on D defines a structure of a smooth foliation with singularities on D.*

Proof. See [111] □

Taking into account the fact that D is a distribution on Q, the following proposition relates accessible sets of H to accessible sets of D.

Proposition 1.10.45. *Let $M \subseteq Q$ be the unique accessible set of D through $q_0 \in Q$, and let $u_0 \in D_{q_0}$. Then the restriction $D_M = \tau_Q^{-1}(M) \cap D$ of D to points of M is the accessible set of H through u_0.*

Proof. Let X be a vector field on Q with values in D. Consider the function $f \in C^\infty(D)$ defined by $f(u) = \mathrm{k}(X(\tau_Q(u)), u)$ for all $u \in D$. Let $t \mapsto u(t)$ be an integral curve of Y_f with $t \mapsto q(t) = \tau_Q(u(t))$ its projection to Q. Evaluating both sides of the equation $\dot{u}(t) \lrcorner \varpi = \mathrm{d}f(u(t))$ on vectors in

$T_{u(t)}D$ of the form $\iota_{u(t)}v$, where $v \in D_{q(t)}$, and taking into account lemma 1.5.9, we get

$$\begin{aligned} \mathrm{k}(v, T\tau_Q(\dot{u}(t))) &= \langle \dot{u}(t) \lrcorner \varpi \mid \iota_{u(t)}v \rangle = \langle \mathrm{d}f(u(t)) \mid \iota_{u(t)}v \rangle \\ &= \frac{\mathrm{d}}{\mathrm{d}s}\Big|_{s=0} f(u(t)+sv) = \frac{\mathrm{d}}{\mathrm{d}s}\Big|_{s=0} \mathrm{k}(X(q(t)), u+sv) \\ &= \mathrm{k}(X(q(t)), v), \quad \text{for all } v \in D. \end{aligned}$$

Since $T\tau_Q(\dot{u}(t))$ and $X(q(t))$ lie in D, it follows that $T\tau_Q(\dot{u}(t)) = X(q(t))$. But, $T\tau_Q(\dot{u}(t)) = \dot{q}(t)$. Therefore, $t \mapsto q(t)$ is an integral curve of X.

In theorem 1.6.3.20 we have shown that an integral curve $t \mapsto q(t)$ of a vector field on Q with values in D lifts to a curve $t \mapsto u(t)$, which is an integral curve of a vector field on D with values in H. This shows that if L is the accessible set of H through $u_0 \in D$ and $u \in L$, then the whole fiber $D_{\tau_Q}(u)$ is contained in L. Consequently, if M is the accessible set of D containing $q_0 = \tau_Q(u_0)$ then $D_M = \tau_Q^{-1}(M) \cap D \subseteq L$. On the other hand, the restriction H_{D_M} of H to points in D_M is contained in TD_M. Hence H_{D_M} is a distribution on D_M. Thus the accessible set L of H through $u_0 \in D_M$ is contained in D_M. Therefore, $L = D_M$. □

Corollary 1.10.46. *A nonholonomic system is simple if and only if Q is the unique accessible set of D.*

The upshot of the above discussion is that accessible sets provide restrictions on the evolution of a nonholonomically constrained system which are independent of external forces and depend *only* on the constraint distribution.

1.11 Constants of motion

Additional restrictions on the evolution of a nonholonomically constrained system are provided by *constants of motion.*

Theorem 1.11.47. (Nonholonomic Noether theorem) *A function $f \in C^\infty(D)$ is a constant of motion of the distributional Hamiltonian system (D, H, ϖ, h) if and only if its distributional Hamiltonian vector field Y_f preserves the Hamiltonian h.*

Proof. The dynamics on D is given by the distributional Hamiltonian vector field Y_h of the Hamiltonian h. Hence,

$$\dot{f} = \langle \mathrm{d}f \mid Y_h \rangle = \varpi(Y_f, Y_h) = -\varpi(Y_h, Y_f) = -\langle \mathrm{d}h \mid Y_f \rangle.$$

Therefore, $\dot{f} = 0$ if and only if Y_f preserves h. □

Theorem 1.11.47 is a nonholonomic counterpart of the Noether's theorem relating symmetries and conservation laws in unconstrained Hamiltonian systems. In the unconstrained case, a function $f \in C^\infty(TQ)$ is a constant of motion if and only if its Hamiltonian vector field X_f preserves the Hamiltonian. However, X_f also preserves the symplectic form ω. So it is an infinitesimal symmetry of the Hamiltonian system (T^*Q, ω, h). In the presence of constraints, the relationship between symmetries and constants of motion is more involved. The condition that the distributional Hamiltonian vector field Y_f preserves the Hamiltonian h does *not* imply that it is an infinitesimal symmetry of the nonholonomic Hamiltonian system, because it need not preserve either H or ϖ. Conservation laws corresponding to symmetries of a nonholonomically constrained system will be discussed in chapter 3.

Lemma 1.11.48. *Suppose that the nonholonomic system (D, H, ϖ, h) is simple. Then the only Casimir functions of its almost Poisson algebra $(C^\infty(D), \{\,,\,\})$ are the constant functions.*

We now consider a special case of the conservation law given in theorem 1.11.47. Let Z be a vector field on Q and let $\Phi_t : Q \to Q$ be the local 1-parameter group of local diffeomorphism of Q generated by Z. The tangent map $T\Phi_t : TQ \to TQ$ forms a local group of local diffeomorphisms of TQ generated by the tangent lift Z_{TQ} of Z to TQ. Let P_Z be a function on TQ defined by

$$P_Z(u) = \mathrm{k}(u, Z(\tau_Q(u))), \quad \text{for each } u \in TQ. \tag{105}$$

Proposition 1.11.49. *If Z is a vector field on Q with values in D such that its tangent lift Z_{TQ} preserves the Lagrangian $\ell(u) = \frac{1}{2}\,\mathrm{k}(u,u) - V(\tau_Q(u))$, then P_Z is a constant of motion of the nonholonomically constrained system with Lagrangian ℓ and constraint distribution D.*

Proof. Since Z_{TQ} preserves the Lagrangian ℓ, it follows that it preserves the kinetic energy $k(u) = \frac{1}{2}\,\mathrm{k}(u,u)$ and the potential $V(\tau_Q(u))$, separately. Hence, Z_{TQ} preserves the Hamiltonian $h(u) = \frac{1}{2}\,\mathrm{k}(u,u) + V(\tau_Q(u))$. Moreover, Z preserves the kinetic energy metric k, that is $L_Z\mathrm{k} = 0$. By lemma 1.8.1.33, Z_{TQ} is the Hamiltonian vector field of P_Z defined by equation (105).

Using decomposition (55) we can express the restriction of Z_{TQ} to D as the sum of its components in H and H^ω. In other words, $Z_{TQ}\mid_D =$

$\widetilde{Z}_H + \widetilde{Z}_{H^\omega}$. Moreover, $\widetilde{Z}_H = Y_{P_Z}$ is the distributional Hamiltonian vector field of P_Z. From the fact that Z has values in D, it follows that Z_{TQ} has values in F, and $\widetilde{Z}_{H^\omega}$ has values in $F \cap H^\omega$. Therefore by lemma 1.6.3.18 $\langle \mathrm{d}h \mid \widetilde{Z}_{H^\omega} \rangle = 0$ which implies that $\widetilde{Z}_{H^\omega}$ preserves h. Since Z_{TQ} preserves h, it follows that $Y_{P_Z} = \widetilde{Z}_H = Z_{TQ} - \widetilde{Z}_{H^\omega}$ preserves h. Then theorem 1.11.47 ensures that P_Z is a constant of motion. □

Suppose that we have k conserved functions $f_1, ..., f_k$. For each accessible set L of H and each $c = (c_a) \in \mathbb{R}^k$, consider the level set

$$L_c = \{u \in D \mid f_a(u) = c_a \quad \mathrm{a} = 1, \ldots, k\}.$$

Suppose that locally L_c is a submanifold of L. Since the functions f_a are constants of motion, the restriction of Y_h to L_c is a vector field Y_{h,L_c} on L_c with values in $H_{L_c} = H \cap TL_c$. Let ϖ_{L_c} be the restriction of ϖ to H_{L_c}, and h_{L_c} the restriction of h to L_c. We want to describe the vector field Y_{h,L_c} in terms of the data $(L_c, H_{L_c}, \varpi_{L_c}, h_{L_c})$. Equation (59) restricted to H_{L_c} implies that

$$Y_{h,L_c} \lrcorner \varpi_{L_c} = \mathrm{d}h_{L_c}. \tag{106}$$

If ϖ_{L_c} is nondegenerate, then Y_{h,L_c} is the distributional Hamiltonian vector field of h_{L_c} relative to a symplectic distribution (H_{L_c}, ϖ_{L_c}) on L_c, which we will denote by $Y_{h_{L_c}}$. Hence, we have proved

Theorem 1.11.50. *Suppose that locally L_c is a submanifold of L and that (H_{L_c}, ϖ_{L_c}) is a symplectic distribution on L_c. Then the evolution of Y_h in L_c is given by the distributional Hamiltonian vector field $Y_{h_{L_c}}$ of h_{L_c} relative to (H_{L_c}, ϖ_{L_c}).*

If the hypotheses of proposition 1.11.49 are satisfied for every accessible set L of H and for every value $c \in \mathbb{R}^d$ of the constants of motions $f_1, ..., f_d$, then the decomposition (104) of the nonholonomic Hamiltonian system (D, H, ϖ, h) admits a refinement, namely,

$$(D, H, \varpi, h) = \bigcup_{L\,\mathrm{a.s.}\,H} \; \bigcup_{c \in \mathbb{R}^d} (L_c, H_{L_c}, \varpi_{L_c}, h_{L_c}). \tag{107}$$

We can refine the above decomposition further by considering accessible sets of H_{L_c}. Let M be an accessible set H_{L_c}. By Sussmann's theorem 1.10.42, M is an immersed submanifold of L_c. The restriction H_M of H to points in M coincides with $H_{L_c} \cap TM$. Let ϖ_M be the restriction of ϖ_{L_c}

to H_M and h_M the restriction of h_{L_c} to M. Applying the decomposition (104) to $(L_c, H_{L_c}, \varpi_{L_c}, h_{L_c})$ we obtain

$$(D, H, \varpi, h) = \bigcup_{L \text{ a.s. } H} \bigcup_{c \in \mathbb{R}^d} \bigcup_{M \text{ a.s. } H_{L_c}} (M, H_M, \varpi_M, h_M). \tag{108}$$

In some examples, every distribution H_{L_c} is involutive and on every integral manifold M of H_{L_c} the 2-form ϖ_M on $H_M = TM$ gives rise to a symplectic form ω_M on M. When this happens, the distributional Hamiltonian systems (M, H_M, ϖ_M, h_M) are Hamiltonian systems (M, ω_M, h_M) and the original distributional Hamiltonian system (D, H, ϖ, h) defines a foliation of D by Hamiltonian systems, namely,

$$(D, H, \varpi, h) = \bigcup_{L \text{ a.s. } H} \bigcup_{c \in \mathbb{R}^d} \bigcup_{M \text{ a.s. } H_{L_c}} (M, \omega_M, h_M). \tag{109}$$

1.12 Notes

We review the standard approach to dynamics of systems with linear nonholonomic constraints. Our only addition is the systematic use of the Levi Civita connection of the kinetic energy metric. We note that hypothesis 2.1, which leads to the Lagrange-d'Alembert principle, has been verified experimentally by Lewis and Murray [67].

We use the name the Lagrange-d'Alembert principle on the basis of an analogy with the formalisms of d'Alembert and Lagrange. But they themselves never applied their formalisms to nonholonomically constrained systems. These applications are definitely of a later date, namely the second half of the 19th century with a lot of confusion, which was only cleared up at the end of that century.

Here is a brief account of the history. Whittaker [118] refers to Ferrers [43] as the first one who wrote down the equations of motion for a nonholonomically constrained system. But we do not understand what Ferrers meant. In §240 Routh [96] uses the Lagrange-d'Alembert principle in our form. Then Lindelöf [70] and Appell in the first edition of [4] and others used, in the case of a translation invariant nonholonomically constrained system, the *wrong* principle, namely, they take the Lagrangian on the tangent bundle of the orbit space of the translation symmetry; and then apply the Euler-Lagrange equation. Both Chaplygin [23], who refers to Lindelöf [70], and Korteweg [62], who refers to the first edition of Appell [4], correct this error. Appell corrected his mistake in the second edition of [4]. For a more detailed discussion of this point see §4 of chapter 3. From this time on the Lagrange-d'Alembert principle has been accepted as giving the "right" equations of motion.

More recently Arnol'd et al. [7] has introduced vakonomic mechanics which have nonholonomic constraints but the equations of motion are based on a variational principle. Vakonomic systems are of interest, but rolling without slipping is *not* vakonomic.

For an up to date treatment of *nonlinearly* nonholonomically constrained systems see Marle [74].

Our presentation of the distributional Hamiltonian formulation of nonholonomically constrained systems follows the theory developed in [13] and [32]. It is a special case of the partially symplectic formulation proposed by Bocharov and Vinogradov [17].

Dalsmo and van der Schaft [36] placed nonholonomically constrained systems in the context of Dirac structures. We recall the definition of a Dirac structure. The Pontryagin bundle of a manifold M is the direct sum of its tangent and cotangent bundles, that is, $TM \oplus_M T^*M$. We endow the total space of the Pontryagin bundle with a nondegenerate symmetric bilinear form of signature (dim$M$,dim$M$) given by

$$\langle (u,p),(v,q) \rangle = \langle q \mid u \rangle + \langle p \mid v \rangle$$

for every $(u,p),(v,q) \in T_mM \oplus T_m^*M$. A Dirac structure on a manifold M is a subbundle B of the $TM \oplus T^*M$ which is maximal isotropic with respect of the above bilinear form, see Courant [29]. We now show that the distributional Hamiltonian system (D, ϖ, H, h) with nonholonomic constraint distribution $D \subset TQ$, Hamiltonian h, and symplectic distribution (H, ϖ) on D is a Dirac structure. Here the symplectic form ϖ is the restriction to H of the symplectic form ω on TQ, which is obtained by pulling back the canonical symplectic form on T^*Q by the Legendre transformation $\mathrm{k}^\flat : TQ \to T^*Q$ given by the kinetic energy metric k of the system. Following Yoshimura and Marsden [119] the Dirac structure corresponding to (D, ϖ, H, h) is given by

$$\{(u,p) \in TD \oplus_D T^*D \big| \, u \in H \text{ and } p - u \,\lrcorner\, \omega \in H^0 \}.$$

Here H^0 is the annihilator of H in T^*D. The pair (Y_h, dh), where $Y_h \,\lrcorner\, \varpi = \partial_H h$, is a section of this Dirac structure.

The projection principle in the text is due to Marle [73]. However, projection principles were first used by Gibbs [47] and by Appell in the second edition [4] to derive what Pars [89] calls the Gibbs-Appell equations of motion for a nonholonomically constrained system.

Van der Schaft and Maschke [113] introduced an almost Poisson bracket to study nonholonomically constrained Hamiltonian systems. Equation (87) is the nonholonomic almost Poisson bracket analogue of the formula for the Dirac bracket given in [39]. The intrinsic relation between the almost Poisson bracket and the symplectic distribution (H, ϖ) was given by Koon and Marsden [61]. The term *almost Poisson bracket* was coined by Cantrijn, de León, and de Diego [19].

Results on accessible sets of distributions used here are taken from papers of Stefan [111] and Sussmann [112].

In [103] there is a proof of a nonholonomic version of Noether's theorem of §11. A foliation of the constraint manifold by Hamiltonian systems given by constants of motion was discovered by Kemppainen in the dynamics of a rolling disk [57].

Bloch, Krishnaprasad, Marsden and Murray establish the momentum equation in [16], but only for G-invariant vector fields which were tangent to G-orbits on configuration space Q under the hypothesis that the action of G on Q is free and proper. The version in the text is based on [105].

Chapter 2

Group actions and orbit spaces

In this chapter we treat the basic properties of the action of a Lie group on a smooth manifold, concentrating especially on the case when the action is proper. We also discuss the differential geometry of the space of orbits of a proper action using the concept of a differential space.

2.1 Group actions

A *smooth action* of a Lie group G on a smooth manifold M is a smooth mapping

$$\Phi : G \times M \to M : (g, m) \mapsto \Phi(g, m) = g \cdot m$$

such that $\Phi(g, \Phi(h, m)) = \Phi(gh, m)$ for every $g, h \in G$ and every $m \in M$. For fixed $g \in G$ let $\Phi_g : M \to M : m \mapsto \Phi(g, m)$; while for fixed $m \in M$ let $\Phi_m : G \to M : g \mapsto \Phi(g, m)$.

Example 2.1.1. Suppose that V is a smooth vector field on M. Then for each $m \in M$ there is a unique solution $\gamma_m : I_m \subseteq \mathbb{R} \to M$ of the differential equation

$$\frac{\mathrm{d}\gamma_m(t)}{\mathrm{d}t} = V(\gamma_m(t)), \quad t \in I_m,$$

with initial condition $\gamma_m(0) = m$, which is defined on a maximal open interval I_m containing 0. The set $D = \{(t, m) \in \mathbb{R} \times M \big| \, t \in I_m\}$ is an open subset of $\mathbb{R} \times M$ and the mapping

$$\varphi : D \subseteq \mathbb{R} \times M \to M : (t, m) \mapsto \gamma_m(t)$$

is the *flow* of V. The set

$$D_t = \{m \in M \big| \, (t, m) \in D\} = \{m \in M \big| \, t \in I_m\}$$

is an open subset of M. The smoothness of φ implies that

$$\varphi_t : D_t \subseteq M \to M : t \mapsto \gamma_m(t)$$

is the *flow of* V *at time* t. This flow has the *group property*, namely, if $s, t \in \mathbb{R}$, $m \in D_s$, and $\varphi_s(m) \in D_t$, then $m \in D_{s+t}$ and $\varphi_t(\varphi_s(m)) = \varphi_{t+s}(m)$. Therefore $\varphi_t : D_t \to D_{-t}$ is a diffeomorphism with inverse φ_{-t}. If for every $m \in M$, we have $I_m = \mathbb{R}$, then the flow φ (or the vector field V) is *complete* and $\varphi : \mathbb{R} \times M \to M$ defines an action of $\mathbb{R}$ on M. □

A point $m \in M$ is a *fixed point* for the G-action Φ if and only if $g \cdot m = m$ for every $g \in G$. The action Φ is *free* if and only if $g \cdot m = m$ implies $g = e$ for every $m \in M$. Here e is the identity element of G. An example of a nonfree action is given by G acting on itself by conjugation, that is,

$$\Phi : G \times G \to G : (g, h) \mapsto ghg^{-1}.$$

Indeed if $G \neq \{e\}$, then for $h \neq e$ we have $h \in G_h = \{g \in G \big| \, ghg^{-1} = h\} \neq \{e\}$.

Let $\mathfrak{g} = T_e G$ be the Lie algebra of G. For every $\xi \in \mathfrak{g}$ and every $m \in M$ let

$$\xi \cdot m = X_\xi(m) = \left. \frac{d}{dt} \right|_{t=0} \Phi_m(\exp t\, \xi)$$

be the *infinitesimal action* of $\mathfrak{g}$ on M. Then $X_\xi : M \to TM : m \mapsto X_\xi(m)$ is a smooth vector field on M, whose flow is $\varphi_t^\xi = \Phi_{\exp t\xi}$, which is complete. The infinitesimal action gives rise to a map $\mathfrak{g} \to \mathcal{X}(M) : \xi \mapsto X_\xi$, which is a homomorphism from the Lie algebra $\mathfrak{g}$ to the Lie algebra $\mathcal{X}(M)$ of smooth vector fields on M. Note that the map $T_e\Phi_m : \mathfrak{g} \to T_m M : \xi \mapsto X_\xi(m)$ is a linear mapping.

2.2 Orbit spaces

The set $G \cdot m = \{g \cdot m \in M \big| \, g \in G\}$ is called the *orbit* of the G-action Φ *through* m. The map $\Phi_m : G \to M$ induces the mapping

$$\widetilde{\Phi}_m : G/G_m \to M : g\, G_m \mapsto \Phi(g, m),$$

where $G_m = \{g \in G \big| \, \Phi_g(m) = m\}$ is the *isotropy group* at m. The mapping $\widetilde{\Phi}_m$, which is bijective onto the orbit $G{\cdot}m$, exhibits this orbit as an immersed submanifold of M. Note that tangent space $T_m(G \cdot G)$ to the G-orbit $G \cdot m$ at m is $T_e\Phi_m \mathfrak{g}$.

Two G-orbits are either equal or disjoint, which implies that they partition M. The set $\overline{M} = M/G = \{G\cdot m \big| m \in M\}$ of all G-orbits on M is called the *orbit space* of the action Φ. The map $\pi : M \to M/G = \overline{M} : m \mapsto G\cdot m$ is called the *orbit map* or *projection map*. A fiber of π is a G-orbit in M.

If we want to talk about orbits near a given one, we need a topology on $\overline{M}$. We define one by saying that a subset $\overline{U}$ in $\overline{M}$ is open if and only if the G-invariant subset $\pi^{-1}(\overline{U})$ is an open subset of M. In general, the topological space $\overline{M}$ can be quite complicated. For instance, in the case of an action of $\mathbb{R}$ defined by the flow of a complete vector field on M, orbits need not be embedded submanifolds. Moreover, the topology on $\overline{M}$ need not be Hausdorff.

In order to proceed further we make the additional assumption that the G-action Φ is *proper*, namely, the map

$$\Xi : G \times M \to M \times M : (g, m) \mapsto (m, \Phi(g, m))$$

is proper. In other words, if K is a compact subset of $M \times M$ then $\Xi^{-1}(K)$ is compact. Equivalently, if $\{m_j\}$ and $\{g_j\}$ are infinite sequences in G and M, respectively, such that as $j \to \infty$ the sequences $\{m_j\}$ and $\{g_j \cdot m_j\}$ converge in M, then there is a subsequence $\{g_{j_k}\}$ which converges in G as $k \to \infty$.

If the G-action Φ is proper, then for every $m \in M$ the isotropy group G_m is compact and hence is a Lie subgroup of G. Moreover, every G-orbit is a properly embedded submanifold of M and thus is a closed subset. Finally, the topology on the orbit space M/G is Hausdorff.

Proposition 2.2.2. *If the G-action Φ on the smooth manifold M is free and proper, then the orbit space M/G has a unique smooth manifold structure such that the projection map $\pi : M \to M/G = \overline{M}$ is a* principal fiber bundle with structure group G.

Proof. See lemma 1.11.3 in [42]. □

Let us explain this last statement a bit more. The fact that the G-orbit map $\pi : M \to \overline{M}$ is a smooth fibration implies that there exist local sections. In other words, for every $\overline{m} \in \overline{M}$ there is an open neighborhood $\overline{U}$ of $\overline{m}$ and a smooth mapping $\sigma : \overline{U} \subseteq \overline{M} \to \pi^{-1}(\overline{U})$ such that $\pi \circ \sigma = \mathrm{id}_{\overline{U}}$. The mapping

$$\tau_\sigma : \overline{U} \times G \to \pi^{-1}(\overline{U}) : (\overline{u}, g) \mapsto g \cdot \sigma(\overline{u})$$

is a diffeomorphism such that

$$\pi \circ \tau_\sigma : \overline{U} \times G \to \overline{U} : (\overline{u}, g) \mapsto \overline{u}.$$

In other words, $\tau_\sigma^{-1} : \pi^{-1}(\overline{U}) \to \overline{U} \times G$ is a local trivialization of the fibration π restricted to $\pi^{-1}(\overline{U})$. If $\sigma' : \overline{U}' \subseteq \overline{M} \to M$ is another such local section of π defined on the open set $\overline{U}'$, then for each $\overline{m} \in \overline{U} \cap \overline{U}'$ there is a unique $\rho(\overline{u}) \in G$, depending smoothly on $\overline{u}$, such that $\sigma'(\overline{u}) = \rho(\overline{u}) \cdot \sigma(\overline{u})$. Therefore

$$\tau_{\sigma'}(\overline{u}, g) = g \cdot \rho(\overline{u}) \cdot \sigma(\overline{u}) = \tau_\sigma(\overline{u}, g \cdot \rho(\overline{u})),$$

or equivalently,

$$(\tau_\sigma^{-1} \circ \tau_{\sigma'})(\overline{u}, g) = (\overline{u}, g \cdot \rho(\overline{u})).$$

A principal fiber bundle with structure group G is a smooth fibration $\nu : M \to B$ with an open covering $\{U_i\}_{i \in I}$ of B and local trivializations $\tau_i^{-1} : \nu^{-1}(U_i) \to U_i \times G$ such that

$$(\tau_i^{-1} \circ \tau_j)(u, g) = (u, g \cdot \nu_{ij}(u))$$

for smooth mappings $\nu_{ij} : U_i \cap U_j \to G$, which satisfy the cocycle conditions: $(\nu_{ij} \circ \nu_{ji})|(U_i \cap U_j) = \mathrm{id}_{U_i \cap U_j}$ and $(\nu_{ik} \circ \nu_{kj} \circ \nu_{ji})|(U_i \cap U_j \cap U_k) = \mathrm{id}_{U_i \cap U_j \cap U_k}$. The maps $\tau_i : U_i \times G \to M$ intertwine the G-action

$$G \times (U_i \times G) \to U_i \times G : (g', (u, g)) \mapsto (u, g'g)$$

with a unique G-action on M which is free and proper.

Therefore the concepts of free and proper G-action and principal fiber bundle with fiber G are equivalent.

2.3 Isotropy and orbit types

2.3.1 *Isotropy types*

For each $m \in M$ the isotropy group G_m at m is a closed subgroup of G and hence is a Lie subgroup of G. Its Lie algebra is $\mathfrak{g}_m = \{\xi \in \mathfrak{g} \,|\, X_\xi(m) = 0\}$. We have

Lemma 2.3.1.3. *The isotropy group $G_{g \cdot m}$ is conjugate to the isotropy group G_m by the element $g \in G$.*

Proof. We have $h \in G_{g \cdot m}$ if and only if $h \cdot (g \cdot m) = g \cdot m$ if and only if $(g^{-1}hg) \cdot m = m$ if and only if $g^{-1}hg \in G_m$. Therefore $g^{-1} G_{g \cdot m} g = G_m$. □

Let H be a closed subgroup of G. Then $M_H = \{m \in M \big| G_m = H\}$ is the set of points in M with *isotropy type* H.[1] As H ranges over all closed subgroups of G for which M_H is non-empty, the sets M_H partition M.

The *normalizer* $N(H)$ of H in G is $\{g \in G \big| gHg^{-1} = H\}$. In fact $N(H)$ is the largest subgroup of G which contains H as a normal subgroup. Now $N(H)$ is a closed subgroup of G and hence is a Lie group. Consequently, $N(H)/H$ is a Lie group.

The following lemma reduces the $N(H)$-action on an H-isotropy type M_H to a free action of $N(H)/H$ on M_H.

Lemma 2.3.1.4. *Let H be a closed subgroup of G such that M_H is nonempty. For every $g \in G$ we have*

$$g \in N(H) \Longleftrightarrow g \cdot M_H = M_H \Longleftrightarrow (g \cdot M_H) \cap M_H \neq \emptyset.$$

Proof. If $m \in M_H$, then $G_m = H$. From lemma 2.3.1.3 it follows that $g \cdot m \in M_H$, that is, $G_{g \cdot m} = H = G_m$ if and only if $g \in N(H)$. □

2.3.2 *Orbit types*

We say that the subgroup H' is *conjugate* to H in G if there is an element g of G such that $H' = gHg^{-1}$. If H is a closed subgroup of G then so is H'. From lemma 2.3.1.3 it follows that the isotropy type $M_{H'}$ is nonempty if and only if the isotropy type M_H is. The set of all subgroups of G which are conjugate in G to H is called the *conjugacy class* of H in G and is denoted by (H).

The set $M_{(H)} = \{m \in M \big| G_m = gHg^{-1}$, for some $g \in G\}$ is the *orbit type* of (H) in M.[2] Since $G \cdot M_H = M_{(H)}$, the orbit type $M_{(H)}$ is the smallest G-invariant subset of M which contains M_H. Being in the same orbit type defines an equivalence relation on M. This relation is coarser than the one defined by being in the same isotropy type. The orbit type $M_{(H)}$ is partitioned into isotropy types $M_{H'}$ where $H' \in (H)$, because $g \cdot M_H = M_{gHg^{-1}}$ for every $g \in G$. Therefore M is partitioned into orbit types, which in turn are partitioned into isotropy types for conjugate subgroups. The set $\overline{M_{(H)}} = M_{(H)}/G = \pi(M_{(H)})$ is the *orbit type in the orbit space* of the conjugacy class (H). Here $\pi : M \to \overline{M}$ is the G-orbit map.

[1] Also called *H-symmetry type*.

[2] The name orbit type comes from the fact that m and m' belong to $M_{(H)}$ if and only if there is a G-equivariant bijective mapping from $G \cdot m$ onto $G \cdot m'$.

Lemma 2.3.2.5. *Let H be a closed subgroup of G for which the isotropy type M_H is nonempty. Then*

1. The mapping $g \mapsto M_{gHg^{-1}}$ induces a G-equivariant bijection from $G/N(H)$ onto the collection of all isotropy types in $M_{(H)}$. Here we use the G-action on $G/N(H)$ induced by left multiplication.
2. If $\pi : M \to \overline{M}$ is the G-orbit map, then a fiber of $\pi|M_H$ is an $N(H)/H$-orbit in M_H.

2.3.3 When the action is proper

Throughout this subsection we assume that the G-action Φ is proper. Up until now we have only made *set* theoretic statements about the action of G on M.

Let $m \in M$. Then $T_e\Phi_m\mathfrak{g} = T_m(G \cdot m)$. For every $g \in G_m$ the linear transformation $T_m\Phi_g : T_mM \to T_mM$ leaves the subspace $T_m(G \cdot m)$ invariant. Hence we obtain an induced linear action $\circ$ of the compact group $H = G_m$ on the vector space $E = T_mM/T_m(G \cdot m)$. Let B be an H-invariant open subset of E. On $G \times B$ we have an action of H defined by

$$\mu : H \times (G \times B) \to G \times B : (h, (g, b)) \mapsto (gh^{-1}, h \circ b). \tag{1}$$

This action is free and proper. Therefore the orbit space $G \times_H B$ of the H action μ is a smooth manifold. Because the G action on $G \times B$ defined by

$$\nu : G \times (G \times B) \to G \times B : (g', (g, b)) \mapsto (g'g, b) \tag{2}$$

commutes with the H-action (1), it induces a G-action on $G \times_H B$.

Theorem 2.3.3.6. (tube theorem) *Let $m \in M$ and set $H = G_m$. There is a G-invariant open neighborhood U of m in M, an open H-invariant neighborhood B of the origin in E, and a diffeomorphism $\varphi : G \times_H B \to U$, which intertwines the G-action on $G \times_H B$ with the G-action on U.*

Proof. See theorem 2.4.1 in [42]. B is identified with a submanifold S of M containing m, called a *slice*, by a diffeomorphism ψ. The diffeomorphism φ is induced by the mapping $G \times B \to M : (g, b) \mapsto g \cdot \psi(b)$. □

The tube theorem is the basic result used to study local properties of the orbit space M/G.

Theorem 2.3.3.7. *Suppose that $H = G_m$ is the isotropy subgroup at m for a proper G-action on M. Then*

1. *The isotropy type M_H is a locally closed submanifold of M.*[3]
2. *The Lie group $N(H)/H$ acts smoothly, freely, and properly on M_H. There is a unique smooth manifold structure on $\overline{M_{(H)}} = \pi(M_{(H)})$ such that $\pi|M_H : M_H \to \overline{M_{(H)}}$ is a principal fiber bundle with structure group $N(H)/H$. Here $\pi : M \to \overline{M}$ is the G-orbit map.*
3. *The orbit type $M_{(H)}$ is a G-invariant, locally closed, smooth submanifold of M.*[4] *The G-action on $M_{(H)}$ induces a G-equivariant diffeomorphism from the space $(G/H)\times_{N(H)/H} M_H$ onto M_H. The map $\pi|M_{(H)} : M_{(H)} \to \overline{M_{(H)}}$ defines a smooth fibration whose fibers are G-equivariantly diffeomorphic to G/H.*

Proof. Let

$$M^H = \{m \in M \big| \, h \cdot m = m, \text{ for every } h \in H\} = \{m \in M \big| \, H \subseteq G_m\}$$

be the set of all points of M which are fixed by every element of H. Clearly $M_H \subseteq M_{(H)} \cap M^H$. Suppose that $m \in M_{(H)} \cap M^H$. Then G_m is conjugate to H in G and $H \subseteq G_m$. Because the G-action is proper, the isotropy group G_m is compact. As H is conjugate to G_m, it follows that H is also compact. Hence H has the same dimension and finite number of connected components as G_m. Since $H \subseteq G_m$, we obtain $H = G_m$, that is, $m \in M_H$. Consequently, $M_H = M_{(H)} \cap M^H$.

The theorem now follows from theorem 2.6.7 of [42], where the tube theorem is used in an essential way. □

2.3.4 *Stratification by orbit types*

A *stratification* of a smooth manifold M is a collection $\mathcal{S}$ of locally closed smooth submanifolds of M, called *strata*, having the following properties.

1. $\mathcal{S}$ is a locally finite partition of M, that is,

   a. If $S, S' \in \mathcal{S}$ and $S \neq S'$, then $S \cap S' = \emptyset$;
   b. $M = \bigcup_{S \in \mathcal{S}} S$;
   c. for each $m \in M$, there is an open neighborhood U of m in M such that $\{S \in \mathcal{S} \big| \, S \cap U \neq \emptyset\}$ is finite.

2. for each $S \in \mathcal{S}$, the closure $\mathrm{cl}(S)$ of S in M is the union of S and $\{S' \in \mathcal{S} \big| \, \dim S' < \dim S\}$.

[3] Different connected components of M_H may have different dimensions.
[4] Idem.

A geometrically well behaved stratification $\mathcal{S}$ of M satisfies Whitney's axioms A and B.[5] Such a stratification is called a *Whitney stratification*.

Theorem 2.3.4.8. *Suppose that G acts smoothly and properly on M. Then connected components of orbit types in M form a (Whitney) stratification of M.*

Proof. See theorem 2.7.4 in [42] or theorem 4.3.7 in Pflaum [91]. □

Each orbit type $M_{(H)}$ is fibered by isotropy types M_H with $H \in (H)$. For each $H \in (H)$ the codimension c of M_H in $M_{(H)}$ is $\dim G - \dim N(H)/H$. If $c > 0$ then isotropy types do *not* form a stratification of M.

In order to be able to state that orbit types in the orbit space form a Whitney stratification, we need to know that, at least locally, we can embed the G-orbit space M/G into some smooth manifold. This will be discussed in the next section.

2.4 Smooth structure on an orbit space

2.4.1 *Differential structure*

A *differential space* is a pair $(Q, C^\infty(Q))$, where Q is a topological space and $C^\infty(Q)$ is a set of continuous real valued functions having the following properties.

1. The sets $f^{-1}(I)$, with $f \in C^\infty(Q)$ and I an open interval in $\mathbb{R}$, form a subbasis for the topology of Q.
2. For every positive integer n, every $F \in C^\infty(\mathbb{R}^n)$, and every $f_1, \ldots, f_n \in C^\infty(Q)$, we have $F \circ \mathfrak{f} \in C^\infty(Q)$, where $\mathfrak{f}(q) = (f_1(q), \ldots, f_n(q)) \in \mathbb{R}^n$ for every $q \in Q$.
3. If $f : Q \mapsto \mathbb{R}$ has the property that for every $q \in Q$ there is an open neighborhood U_q of q in Q and $f_q \in C^\infty(Q)$ such that $f|U_q = f_q|U_q$, then $f \in C^\infty(Q)$.

[5]A) If i) $S, S' \in \mathcal{S}$ with $S' \subseteq \mathrm{cl}(S)$ and $S' \neq S$, ii) the sequence $\{s_j\} \subseteq S$ converges to $s' \in S$, and iii) $T_{s_j}S$ converges to a linear subspace L of $T_{s'}M$, then $T_{s'}S' \subseteq L$.

B) If $\{s_j\}$ is a sequence such as in condition A) above and $\{s'_j\} \subseteq S'$ also converges to $s' \in S'$, then each limit of the 1-dimesional subspaces $\mathbb{R}\,\lambda(s_j, s'_j)$ is contained in L. Here $\lambda : \Delta_M \subseteq M \times M \to TM$ is a diffeomorphism of an open neighborhood of the diagonal Δ_M in $M \times M$ to an open neighborhood of the zero section Z_{TM} of TM such that $\lambda(\Delta_M) = Z_{TM}$.

The set $C^\infty(Q)$ is called the *differential structure* of the *differential space* $(Q, C^\infty(Q))$.

Example 2.4.1.9. Let M be a smooth manifold with $C^\infty(M)$ its collection of smooth functions. Then $(M, C^\infty(M))$ is a differential space. □

If $(P, C^\infty(P))$ and $(Q, C^\infty(Q))$ are differential spaces, then a *smooth mapping* φ from $(P, C^\infty(P))$ to $(Q, C^\infty(Q))$ is a continuous mapping $\varphi : P \to Q$ such that $\varphi^*(C^\infty(Q)) \subseteq C^\infty(P)$. The map φ is a *diffeomorphism* from $(P, C^\infty(P))$ to $(Q, C^\infty(Q))$ if φ is a homeomorphism from P onto Q and both φ and φ^{-1} are smooth. This is equivalent to the condition that φ is a homeomorphism from P onto Q such that $\varphi^*(C^\infty(Q)) = C^\infty(P)$, because

$$\begin{aligned} C^\infty(Q) &\supseteq (\varphi^{-1})^*(C^\infty(P)) \supseteq (\varphi^{-1})^*(\varphi^*(C^\infty(Q))) \\ &= (\varphi \circ \varphi^{-1})^*(C^\infty(Q)) = C^\infty(Q). \end{aligned}$$

Differential spaces and smooth mappings form a category.

Let $(Q, C^\infty(Q))$ be a differential space and let N be a subset of Q. Define $C_i^\infty(N)$ as the set of all functions $f : N \mapsto \mathbb{R}$ with the property that for every $n \in N$ there is an open neighborhood U_n of n *in* Q and an $f_n \in C^\infty(Q)$ such that $f|(U_n \cap N) = f_n|(U_n \cap N)$. If we provide N with the topology induced from that on Q, then $(N, C_i^\infty(N))$ is a differential space, called a *differential subspace* of $(Q, C^\infty(Q))$. If U is an open subset of Q, then

$$C_i^\infty(U) = \{f|U \,\big|\, f \in C^\infty(Q)\} = C^\infty(U).$$

A differential space $(Q, C^\infty(Q))$ is a *smooth manifold* if for every $q \in Q$ there is a nonnegative integer n, an open subset U of Q containing q, and $f_1, \ldots, f_n \in C^\infty(Q)$ such that the map

$$\mathfrak{f} : U \subseteq Q \to V = f(U) \subseteq \mathbb{R}^n : q \mapsto \big(f_1(q), \ldots, f_n(q)\big)$$

is a diffeomorphism from the differential space $(U, C^\infty(U))$ onto the differential space $(V, C^\infty(V))$, seen as a subspace of the differential space $(\mathbb{R}^n, C^\infty(\mathbb{R}^n))$. In other words, $(Q, C^\infty(Q))$ is locally diffeomorphic to an open subset of $\mathbb{R}^n$ with its differential structure being given by restricting smooth functions to this open subset.

Lemma 2.4.1.10. *If the topological space Q of the differential space $(Q, C^\infty(Q))$ is Hausdorff, locally compact and paracompact, then for every open covering $\mathcal{U}$ of Q there is a partition of unity in $C^\infty(Q)$, which is subordinate to $\mathcal{U}$.*

Proof. If $q \in Q$ and U is an open neighborhood of q in Q, then it follows from property 1 of a differential structure that there is a positive integer n, an open subset W of $\mathbb{R}^n$, and $f_1, \ldots, f_n \in C^\infty(Q)$ such that $\mathfrak{f}^{-1}(W) \subset U$. There is a cutoff function $F \in C^\infty(\mathbb{R}^n)$ such that $F = 1$ on an open neighborhood of $\mathfrak{f}(q)$ in W and the support of F is a compact subset K of W. From property 2 of a differential structure it follows that $F \circ \mathfrak{f} \in C^\infty(Q)$. Also the support of $F \circ \mathfrak{f}$ is $\mathfrak{f}^{-1}(K)$, which is a closed subset of U. The set $\mathfrak{f}^{-1}(K)$ is compact if the closure of U in Q is compact. Therefore, if the topological space Q is locally compact, then there are cutoff functions in $C^\infty(Q)$.

If X is a topological space with $\mathcal{F}(X)$ a space of real valued functions on X and $\mathcal{U} = \{U_j\}_{j \in J}$ is an open covering of X, then a *partition of unity in* $\mathcal{F}(X)$ *subordinate to* $\mathcal{U}$ is a collection $\{\chi_j\}_{j \in J}$ of functions in $\mathcal{F}(X)$ having the following properties.

1. For each $j \in J$, the support of χ_j is a compact subset of some $U_j \in \mathcal{U}$.
2. The supports of the $\{\chi_j\}_{j \in J}$ form a locally finite family of compact subsets of X, whose union is X.
3. $\sum_{j \in J} \chi_j = 1$ on X.

A Hausdorff topological space X is called *paracompact* if every open covering of X has a locally finite refinement. If X is Hausdorff and *locally compact*, then it is paracompact if and only if every connected component of X is equal to the union of a countable collection of compact subsets.

Cutoff functions in $\mathcal{F}(X)$ can be used to obtain partitions of unity. □

We would like to think of $C^\infty(Q)$ as being the space of smooth functions on Q. For this to be the case, we should have a *sheaf of locally defined "smooth functions"*. Towards this goal, let U be an open subset of Q. Then we have already defined $C^\infty(U)$. Moreover, the mapping $U \to C^\infty(U)$, where U ranges over all open subsets U of Q, defines a sheaf of functions on Q. We now check to see that this sheaf behaves like the *sheaf of smooth functions on a manifold.*

Lemma 2.4.1.11. *Let N be a subset of a smooth paracompact manifold M. N is an embedded submanifold of M if and only if the identity map from the differential space $(N, C_i^\infty(N))$ into the differential space $(N, C^\infty(N))$ is a diffeomorphism of differential spaces, that is, if and only if $C_i^\infty(N) = C^\infty(N)$.*

Proof. $\Longleftarrow$. The inclusion map $\iota : N \to M$ is a smooth map of differential spaces $(N, C_i^\infty(N))$ to $(M, C^\infty(M))$. Since $C_i^\infty(N) = C^\infty(N)$ by hypothesis, it follows that ι is a smooth mapping of $(N, C^\infty(N))$ and $(M, C^\infty(M))$, where M and N are smooth manifolds. The topology on N is induced from that on M. If $n \in N$ and $\dim N = p$, then there are $\{f_j\}_{j=1}^p \subseteq C^\infty(N)$ such that $\{f_j\}_{j=1}^p$ form a system of coordinates for N on an open neighborhood of n in N. Therefore the map ι has an injective tangent at n and consequently is an embedding. Hence N is an embedded submanifold of M.

$\Longrightarrow$. Suppose that N is an embedded submanifold of M. If $f \in C^\infty(N)$, then for every $n \in N$ and any open neighborhood U_n of n in N, there is a cutoff function $\chi : M \to \mathbb{R}$ in $C^\infty(M)$ such that $\chi = 1$ on an open subset V_n of n in N whose support is a compact subset of U_n. Such a cutoff function χ exists because M is paracompact. Let $f_n = \chi \cdot f$. Then $f|V_n = f_n|V_n$. Thus $f \in C_i^\infty(N)$. Consequently, $C^\infty(N) \subseteq C_i^\infty(N)$. Conversely, suppose that $f \in C_i^\infty(N)$. Then for every $j \in N$ there is an open neighborhood U_j of j in M and $f_j \in C^\infty(M)$ such that $f|(U_j \cap N) = f_j|(U_j \cap N)$. Now $\mathcal{U} = \{U_j\}_{j \in J = N}$ is an open covering of N. We extend $\mathcal{U}$ to an open covering $\widetilde{\mathcal{U}}$ by adding the open set $U = M \setminus N$. (Since N is an embedded submanifold of M, it is a closed subset of M.) Let $\{\chi_j\}_{j \in \widetilde{J}}$ be a $C^\infty(M)$ partition of unity on M subordinate to $\widetilde{\mathcal{U}}$. Then $\{\chi_j|N\}_{j \in J}$ is a partition of unity on N subordinate to $\mathcal{U}$. So the function $f = \sum_{j \in J} \chi_j|N \cdot f_j$ lies in $C^\infty(N)$. Therefore $C_i^\infty(N) \subseteq C^\infty(N)$ and thus $C_i^\infty(N) = C^\infty(N)$. □

Because of lemma 2.4.1.11 we may call $C^\infty(Q)$ the *space of smooth functions* on the differential space $(Q, C^\infty(Q))$.

2.4.2 *The orbit space as a differential space*

Let $\overline{M} = M/G$ be the orbit space of a proper G-action on a smooth manifold M with G-orbit map $\pi : M \to \overline{M}$.

For every open subset $\overline{U}$ of $\overline{M}$ a function $\overline{f} : \overline{U} \subseteq \overline{M} \to \mathbb{R}$ is *smooth* if $\pi^*\overline{f} : U = \pi^{-1}(\overline{U}) \subseteq M \to \mathbb{R}$ is smooth. Let $C^\infty(\overline{U})$ be the space of smooth functions on $\overline{U}$.

Lemma 2.4.2.12. *For every $\overline{m} \in \overline{M}$ and every open neighborhood $\overline{U}$ of $\overline{m}$ in $\overline{m}$ there is a* cutoff function *$\overline{\chi} \in C^\infty(\overline{M})$ such that $\overline{\chi} = 1$ on an open neighborhood of $\overline{m}$ and is 0 on the complement of a compact neighborhood* *of* $\overline{m}$ *in* $\overline{U}$.

Proof. From the tube theorem 2.3.3.6 with $U = \pi^{-1}(\overline{U})$ and $H = G_m$ with $m \in \overline{m}$ it follows that the mapping $\rho : B/H \rightarrow U/G : b \circ H \rightarrow \varphi(b) \cdot G$ is a homeomorphism. Because H is a compact Lie group which acts linearly on $E = T_m M/T_m(G \cdot m)$, we may average an arbitrary inner product on E over H to obtain an H-invariant inner product β on E with norm $\| \; \|$. Because B is an H-invariant open subset of E containing the origin, there is an $\varepsilon > 0$ such that $\{\|x\| \le \epsilon\} \subseteq B$. There is a smooth function $\psi : \mathbb{R} \rightarrow \mathbb{R} : r \mapsto \psi(r)$, which is 1 in an open neighborhood of 0 in $\mathbb{R}$ and is 0 when $r \ge \varepsilon$, Then $f : H \times B \rightarrow \mathbb{R} : (h, b) \mapsto \psi(\|b\|)$ is a smooth H invariant function. Therefore f induces a smooth function $\widetilde{f} : B/H \rightarrow \mathbb{R}$, which corresponds to a function $\overline{g} = \widetilde{f} \circ \rho^{-1} \in C^\infty(\overline{U})$ with support in $\overline{U}$. Extending g by 0 outside $\overline{U}$ gives the cutoff function $\overline{\chi} \in C^\infty(\overline{M})$. □

Corollary 2.4.2.13. *Paracompactness of* M *implies that the Hausdorff space* $\overline{M}$ *is paracompact.*

Proof. To see this let $\overline{C}$ be a connected component of $\overline{M}$. Then $\overline{C} = \pi(C)$ for every connected component C of M. Because M is paracompact, C is equal to the union of a countable number of compact sets K_i. Since the G-orbit map π is continuous, $\pi(K_i)$ is compact and their union is $\overline{C}$. □

Cutoff functions can now be used to obtain a partition of unity in $C^\infty(\overline{M})$ as in lemma 2.4.1.10.

Proposition 2.4.2.14. $(M/G, C^\infty(M/G))$ *is a differential space and the* G*-orbit map* π *is a smooth mapping from* $(M, C^\infty(M))$ *to the differential space* $(M/G, C^\infty(M/G))$.

Proof. We verify that properties 1–3 defining a differential structure hold for $C^\infty(\overline{M})$.

1. If $\overline{\chi}$ is a cutoff function on $\overline{M}$ as constructed in lemma 2.4.2.12, then $\overline{\chi}^{-1}(1/2, 3/2) \subseteq \overline{U}$. This proves property 1.
2. Let $\overline{f}_1, \ldots, \overline{f}_n \in C^\infty(\overline{M})$ and let $F \in C^\infty(\mathbb{R}^n)$. Then $\overline{f}_j \circ \pi \in C^\infty(M)^G$, the space of smooth G-invariant functions on M. Hence $F \circ \bar{\mathfrak{f}} \circ \pi \in C^\infty(M)^G$, where $\bar{\mathfrak{f}}(\overline{m}) = (\overline{f}_1(\overline{m}), \ldots, \overline{f}_n(\overline{m}))$ for every $\overline{m} \in \overline{M}$. Therefore $F \circ \bar{\mathfrak{f}} \in C^\infty(\overline{M})$. This proves property 2.
3. Let $\overline{f} : \overline{M} \mapsto \mathbb{R}$. Suppose that for every $\overline{m} \in \overline{M}$ there is an open neighborhood $\overline{U}_{\overline{m}}$ of $\overline{m}$ in $\overline{M}$ and an $\overline{f}_{\overline{m}} \in C^\infty(\overline{M})$ such that $\overline{f}|\overline{U}_{\overline{m}} = \overline{f}_{\overline{m}}|\overline{U}_{\overline{m}}$. Then $\overline{f} \circ \pi$ is G-invariant and is equal to the smooth function $\overline{f}_{\overline{m}} \circ \pi$

on $U_m = \pi^{-1}(\overline{U}_m)$. Here $m \in \overline{m}$. Because the $\{U_m\}_{m \in M}$ form an open covering of M, it follows that $\overline{f} \circ \pi \in C^\infty(M)$. Therefore $\overline{f} \circ \pi \in C^\infty(M)^G$, which implies that $\overline{f} \in C^\infty(\overline{M})$. This proves property 3.

Thus $C^\infty(\overline{M})$ is a differential structure.

The G-orbit map $\pi : M \to \overline{M}$ is smooth because $\pi^*(C^\infty(\overline{M})) = C^\infty(M)^G \subseteq C^\infty(M)$. □

Corollary 2.4.2.15. *$C_i^\infty(\overline{U}) = C^\infty(\overline{U})$, for any open subset $\overline{U}$ of $\overline{M}$.*

Proof. Let $\overline{f} : \overline{U} \mapsto \mathbb{R}$. Then locally $\overline{f}$ agrees with an element of $C^\infty(\overline{M})$. Using the argument which proved property 3 in the proof of the proposition, it follows that $\overline{f} \circ \pi \in C^\infty(U)^G$, where $U = \pi^{-1}(\overline{U})$. Hence $\overline{f} \in C^\infty(\overline{U})$. Conversely, suppose that $\overline{f} : \overline{U} \to \mathbb{R}$ such that $\overline{f} \circ \pi \in C^\infty(U)^G$. Using lemma 2.4.2.12 we know that for every $\overline{m} \in \overline{U}$ there is a cutoff function $\overline{\chi}_{\overline{m}} \in C^\infty(\overline{M})$ whose support is a compact subset $\overline{K}$ of $\overline{U}$ and which is 1 in an open neighborhood of $\overline{m}$ in $\overline{M}$. Define $\overline{f}_{\overline{m}} : \overline{M} \to \mathbb{R}$ by $\overline{f}_{\overline{m}} = \overline{\chi}_{\overline{m}} \cdot \overline{f}$ on $\overline{U}$ and 0 on $\overline{M} \setminus \overline{U}$. Then $\overline{f}_{\overline{m}} = 0$ on $\overline{M} \setminus \overline{K} = \overline{W}$. Therefore $U = \pi^{-1}(\overline{U})$ and $W = \pi^{-1}(\overline{W})$ are open subsets of $M = U \cup W$. On U we have $\overline{f}_{\overline{m}} \circ \pi = (\overline{\chi}_{\overline{m}} \circ \pi) \cdot (\overline{f} \circ \pi)$, which is smooth and G-invariant; while on W we have $\overline{f}_{\overline{m}} \circ \pi = 0$, which is also smooth and G-invariant. Therefore $\overline{f}_{\overline{m}} \in C^\infty(\overline{M})$. But $\overline{f} = \overline{f}_{\overline{m}}$ in an open neighborhood of $\overline{m} \in \overline{M}$. Therefore $\overline{f} \in C_i^\infty(\overline{U})$. Consequently, $C^\infty(\overline{U}) = C_i^\infty(\overline{U})$. □

Using the notation of the tube theorem 2.3.3.6 we have

Lemma 2.4.2.16. *The differential spaces $(B/H, C^\infty(B/H))$ and $(U/G, C^\infty(U/G))$ are diffeomorphic.*

Proof. On $G \times B$ we have the G-action ν (2) with G-orbit map π_ν. Since every G-orbit on $G \times B$ intersects $\{e\} \times B$ exactly once, the G-orbit space $(G \times B)/G$ is diffeomorphic to B. In particular, the map

$$i : B \to (G \times B)/G : b \mapsto \pi_\nu(e, b) \tag{3}$$

is a diffeomorphism, whose inverse is given by the smooth map $\widetilde{\pi} : (G \times B)/G \to B$, induced from the G-invariant map $\pi : G \times B \to B : (g, b) \mapsto b$.

On $G \times B$ we have an H-action μ (1) whose orbit space is $G \times_H B$. In addition, we have an action of $G \times H$ given by

$$\lambda : (G \times H) \times (G \times B) \to G \times B : ((g', h), (g, b)) \mapsto (g'^{-1}, h \circ b).$$

Because the actions μ and ν commute, we have an induced H-action on the orbit space $(G \times B)/G$ and an induced G-action on $G \times_H B$, whose orbit spaces $((G \times B)/G)/H$ and $(G \times_H B)/G$ are equal to the orbit space $(G \times B)/(G \times H)$ of the action λ.

Let $\varphi : G \times_H B \to U$ be the G-equivariant diffeomorphism given by the tube theorem 2.3.3.6. Then φ induces the homeomorphism $\widetilde{\varphi} : (G \times_H B)/G = (G \times B)/(G \times H) \to U/G$ and the isomorphism $\varphi^* : C^\infty(U) \to C^\infty(G \times B)^H$, which restricts to the isomorphism $\widetilde{\varphi}^* : C^\infty(U)^G \to C^\infty(G \times B)^{G \times H}$. The diffeomorphism i (3) induces the homeomorphism

$$\widetilde{i} : B/H \to ((G \times B)/G)/H = (G \times B)/(G \times H)$$

and the isomorphism $i^* : C^\infty(G \times B)^G \to C^\infty(B)$, which restricts to the isomorphism

$$\widetilde{i^*} : C^\infty(G \times B)^{G \times H} \to C^\infty(B)^H = C^\infty(B/H).$$

Therefore $\widetilde{\varphi} \circ \widetilde{i} : B/H \to U/G$ is a homeomorphism and $\widetilde{i^*} \circ \widetilde{\varphi}^* : C^\infty(U/G) \to C^\infty(B/H)$ is an isomorphism. In other words, the differential spaces $(B/H, C^\infty(B/H))$ and $(U/G, C^\infty(U/G))$ are diffeomorphic. □

2.5 Subcartesian spaces

In this section we will show that the orbit space of a proper group action is a subcartesian space.

A differential space $(Q, C^\infty(Q))$ is *subcartesian* if Q is a Hausdorff topological space and $(Q, C^\infty(Q))$ is locally diffeomorphic to $(N, C_i^\infty(N))$, where N is a subset of $\mathbb{R}^n$. In other words, if for each $q \in Q$ there is an open neighborhood U of q in Q, a nonnegative integer n, a subset N of $\mathbb{R}^n$, and a diffeomorphism ψ from $(U, C^\infty(U))$ onto $(N, C_i^\infty(N))$.

A subcartesian differential space $(Q, C^\infty(Q))$ is *locally compact* if and only if for every $q \in Q$ there is an open neighborhood U of Q containing q and a diffeomorphism $\varphi : U \subseteq Q \to V \subseteq \mathbb{R}^n$, which maps U onto a locally closed subset V.

Using the notation of the tube theorem 2.3.3.6, we now investigate the orbit space of the linear action of the compact Lie group $H = G_m$ on the vector space $E = T_mM/T_m(G \cdot m)$. From classical invariant theory we know that the algebra $P(E)^H$ of *H-invariant polynomial functions* on E

is finitely generated, that is, there is a positive integer n and polynomials $p_1, \ldots, p_n \in P(E)$ such that every $p \in P(E)^H$ can be written as $p = F(p_1, \ldots, p_n)$, where F is a polynomial on $\mathbb{R}^n$. Because the action of H on E is linear, we can choose p_i to be homogeneous of degree $d_i > 0$. We may also suppose that n is minimal. We then say that $\{p_i\}_{i=1}^n$ is a *Hilbert basis* of $P(E)^H$ and that

$$\sigma : E \to \mathbb{R}^n : x \mapsto \big(p_1(x), \ldots, p_n(x)\big) \tag{4}$$

is the *Hilbert map* corresponding to the given Hilbert basis. Because there can be nontrivial relations among the generators p_i, neither the Hilbert basis of E nor the Hilbert map is unique. Since elements of $P(E)^H$ separate H-orbits on E, the Hilbert map σ induces a continuous bijective map

$$\widetilde{\sigma} : E/H \to \Sigma = \sigma(E) \subseteq \mathbb{R}^n. \tag{5}$$

The *Tarski-Seidenberg theorem* states that the image of a semialgebraic set under a polynomial mapping is *semialgebraic*. Therefore Σ is a semialgebraic subset of $\mathbb{R}^n$. Because $\sigma(t \cdot x) = (t^{d_1}p_1(x), \ldots, t^{d_n}p_n(x))$ the set Σ is *quasi homogeneous* in the sense that if $y \in \Sigma$ then $t \cdot y \in \Sigma$. In particular, Σ is contactible to the origin in $\mathbb{R}^n$. Averaging an arbitrary inner product on E over the orbits of the H-action, we obtain an H-invariant inner product β on E. Since the function $E \to \mathbb{R} : x \mapsto \beta(x,x)$ is H-invariant, there is a polynomial $P : \mathbb{R}^n \to \mathbb{R}$ such that $\beta(x,x) = P(\sigma(x))$ for every $x \in E$. Consequently, the Hilbert mapping σ (4) is proper. This implies that Σ is a closed subset of $\mathbb{R}^n$.

Because the induced mapping $\widetilde{\sigma}$ (5) is continuous, bijective, and proper, it is a homeomorphism from the locally compact Hausdorff space E/H onto Σ. Thus we have proved

Lemma 2.5.17. *The orbit space E/H of the H-action on E is homeomorphic to the image Σ of the Hilbert map σ.*

Shrinking B (and thereby also U), if necessary, we may assume that $B = \{x \in E \,|\, \beta(x,x) < c\}$ for some $c > 0$. Then the map $\widetilde{\sigma}$ (5) is a homeomorphism from B/H onto $\sigma(B) = \{y \in \Sigma = \sigma(E) \,|\, P(y) < c\}$. Here P is an H-invariant polynomial which expresses β in terms of the invariant polynomials $\{p_i\}_{i=1}^n$ on E. Note that $\sigma(B)$ is an open semialgebraic subset of the closed semialgebraic subset Σ of $\mathbb{R}^n$.

Proposition 2.5.18. *Let $\phi : B/H \to U/G$ be the diffeomorphism given by lemma* 2.4.2.16. *Then*

$$\widetilde{\sigma} \circ \phi^{-1} : (U/G, C^\infty(U/G)) \to (\sigma(B), C^\infty(\sigma(B)))$$

is a diffeomorphism *of differential spaces.*

Proof. Let $f \in C^\infty(\sigma(B))$ and $b \in B$. Then there is an open neighborhood U of $\sigma(b)$ in $\mathbb{R}^n$ and a $g \in C^\infty(\mathbb{R}^n)$ such that $f = g$ on $\sigma(B) \cap U$. Therefore $f \circ \sigma = g \circ \sigma$ is smooth on the open neighborhood $\sigma^{-1}(U)$ of b in B. Because this holds for every $b \in B$, we get $f \circ \sigma \in C^\infty(B)$. Since $f \circ \sigma$ is H-invariant, it follows that $\sigma^*(C^\infty(\sigma(B))) \subseteq C^\infty(B)^H$. Conversely, the theorem of Schwarz [98] states that $C^\infty(B)^H \subseteq \sigma^*(C^\infty(\mathbb{R}^n))$. Let $i : \sigma(B) \to \mathbb{R}^n$ be the inclusion mapping. Then $i^* : C^\infty(\mathbb{R}^n) \to C^\infty(\sigma(B))$ is surjective, that is, $i^*(C^\infty(\mathbb{R}^n)) = C^\infty(\sigma(B))$. Since $\sigma = i \circ \sigma$, we obtain

$$C^\infty(B)^H \subseteq \sigma^*(C^\infty(\mathbb{R}^n)) = \sigma^* \circ i^*(C^\infty(\mathbb{R}^n)) = \sigma^*(C^\infty(\sigma(B))).$$

Therefore $C^\infty(B)^H = \sigma^*(C^\infty(\sigma(B)))$. Since the mapping $\widetilde{\sigma} : B/H \to \sigma(B)$ is a homeomorphism by lemma 2.5.17, we deduce that it is a diffeomorphism from $(B/H, C^\infty(B/H))$ onto $(\sigma(B), C^\infty(\sigma(B)))$. The proposition follows using lemma 2.4.2.16. □

Because the open sets $\{U/G\}$ form an open covering of the orbit space M/G we have proved

Corollary 2.5.19. *For a proper action of a Lie group G on a smooth manifold M, the differential space $(M/G, C^\infty(M/G) = C^\infty(M)^G)$ is a subcartesian space. More precisely, M/G has a covering by open subsets each of which is diffeomorphic as a differential space to an open subset of a closed semialgebraic set.*

2.6 Stratification of the orbit space by orbit types

In the section we investigate the orbit types $\overline{M_{(H)}} = M_{(H)}/G$ in the orbit space $\overline{M} = M/G$ of a proper action of G on a manifold M.

2.6.1 *Orbit types in an orbit space*

Let M_H be the H-isotropy type where $H = G_m$ for some $m \in M$. Then

$$\overline{M_H} = \pi(M_H) = M_{(H)}/G = \pi(M_{(H)}) = \overline{M_{(H)}}.$$

Moreover, there is a smooth manifold structure on $\overline{M_{(H)}}$ such that $\pi|M_H : M_H \to \overline{M_{(H)}}$ is a principal $N(H)/H$-bundle, see theorem 2.3.3.7.

Using the notation of the tube theorem 2.3.3.6, consider the G-action

$$G \times (G \times_H E) \to G \times_H E : \big(g', \rho(g,x)\big) \mapsto \rho(g'g, x),$$

where $\rho : G \times E \to G\times_H E$ is the orbit map of the H-action

$$H \times (G \times E) \to G \times E : (h, (g, x)) \mapsto (gh^{-1}, h \circ x).$$

(See the second paragraph in §3.3 for the definition of the action $\circ$.) Then $G_{\rho(g,x)} = gH_x g^{-1}$. To see this suppose that $g' \in G_{\rho(g,x)}$. Then $g' \cdot \rho(g, x) = \rho(g, x)$ or $\rho(g'g, x) = \rho(g, x)$. Therefore there is an $h \in H$ such that $g'g = gh^{-1}$ and $h \circ x = x$. In other words, we have $g' \in gH_x g^{-1}$. So $G_{\rho(g,x)} \subseteq gH_x g^{-1}$. Conversely, suppose that $g' \in gH_x g^{-1}$. Then for some $h^{-1} \in H_x$, we have $g'g = gh^{-1}$. Therefore g' and $h \circ x = x$. This implies that $g' \cdot \rho(g, x) = \rho(g, x)$, that is, $g' \in G_{\rho(g,x)}$. So $gH_x g^{-1} \subseteq G_{\rho(g,x)}$. Therefore $G_{\rho(g,x)} = gH_x g^{-1}$.

Let $E^H = \{x \in E \big|, h \circ x = x \text{ for every } h \in H\} = \{x \in E \big| \, H_x = H\}$.

Lemma 2.6.1.20. *$(G\times_H E)_{(H)}$ is isomorphic to E^H.*

Proof. Suppose that $x \in E^H$. Then $H = H_x$. Therefore for each $g \in G$, we have $G_{\rho(g,x)} = gH_x g^{-1} = gHg^{-1}$. In other words, $\rho(g, x) \in (G\times_H E)_{(H)}$. Conversely, if $\rho(g, x) \in (G\times_H E)_{(H)}$, then $G_{\rho(g,x)} \in (H)$. But $G_{\rho(g,x)} = gH_x g^{-1}$, so $H_x \in (H)$. However, $H_x \subseteq H$ and has the same dimension and number of connected components as H, since H_x is conjugate to H in G. Therefore $H_x = H$, that is, $x \in E^H$. □

The fact that $\pi(M_H) = \overline{M_{(H)}}$ in combination with lemma 2.4.2.16 gives $\overline{U_{(H)}} = U_{(H)}/G = \phi((B \cap E^H)/H)$. Here ϕ is the diffeomorphism given by lemma 2.4.2.16 and $(B \cap E^H)/H$ is the image of $B \cap E^H$ under the H orbit map on B.

Because H acts linearly on E, it follows that E^H is a linear subspace of E. Let F be the orthogonal complement of E with respect to the H-invariant inner product β. Then F is H-invariant and $F \cap E^H = \{0\}$. The mapping

$$E^H \times F \to E : (x, y) \mapsto x + y$$

is a linear isomorphism, which is H-equivariant if we let H act on $E^H \times F$ by

$$H \times (E^H \times F) \to E^H \times F : (h, (x, y)) \mapsto (x, h \circ y).$$

If $x = (x_1, \ldots, x_\ell)$ is a coordinate system on E^H and $y = (y_1, \ldots, y_r)$ is one on F, then every polynomial $p \in P(E^H \times F)$ can be written uniquely as $p(x, y) = \sum_\alpha x^\alpha \, q_\alpha(y)$, where $\alpha = (\alpha_i, \ldots, \alpha_\ell)$, $x^\alpha = x_1^{\alpha_1} \cdots x_\ell^{\alpha_\ell}$ and $q_\alpha(y)$ is a polynomial in $y_1, \ldots, y_r$. Since $p \in P(E^H \times F)^H \iff q_\alpha \in P(F)^H$ for every α, a Hilbert basis of $P(E)^H$, which is isomorphic to $P(E^H \times F)^H$, is given by $x_1, \ldots, x_\ell, q_1, \ldots, q_m$, where $\{q_j\}_{j=1}^m$ is a Hilbert basis of $P(F)^H$.

The image of $B \cap E^H$ under the Hilbert map

$$\sigma : E = E^H \times F \to \mathbb{R}^n = \mathbb{R}^\ell \times \mathbb{R}^m : (x, y) \mapsto (x, q_1(y), \ldots, q_m(y)) \qquad (6)$$

is an open subset of $\mathbb{R}^\ell \times \{0\}$, which is a smooth ℓ-dimensional submanifold. In view of the fact that the differential spaces $\big((B \cap E^H)/H, C^\infty(B \cap E^H)^H\big)$ and $(\sigma(B \cap E^H), C^\infty(\sigma(B \cap E^H)))$ are diffeomorphic by proposition 2.5.18 and $(\sigma(B \cap E^H), C^\infty(\sigma(B \cap E^H)))$ is diffeomorphic to $(U_{(H)}/G, C^\infty(U_{(H)}/G))$ by lemma 2.4.2.16, we have proved

Proposition 2.6.1.21. *Each connected component of each orbit type $\overline{M_{(H)}}$ in the G-orbit space $\overline{M}$ is a smooth manifold, when regarded as a differential subspace of the differential space $(\overline{M}, C^\infty(\overline{M}))$. In other words, $(\overline{M_{(H)}}, C^\infty_i(\overline{M_{(H)}}))$ is a smooth manifold.*

2.6.2 *Stratification of an orbit space*

The orbit types for the action of H on $E = E^H \times F$ are of the form $E^H \times R$, where R is an orbit type for the linear H-action on F. Furthermore, the $\mathbb{R}_{>0}$-action on E of multiplication by $t > 0$ commutes with the linear action of H on F. Therefore $t \cdot R = R$. Let S^{r-1} be the unit sphere in F with respect to the H-invariant inner product β. Then $R \to R \cap S^{r-1}$ is a bijective map from all orbit types $R \neq \{0\}$ to orbit types of the induced H-action on S^{r-1}. Using induction on the dimension of S^{r-1}, it follows that there are only finitely many H orbit types on S^{r-1}. Consequently, there are only finitely many H-orbit types for the action of H on E.

From the quasihomogenity of the Hilbert map σ (6), it follows that $\sigma(E)$ is invariant under the transformation

$$(x_1, \ldots, x_\ell, q_1, \ldots, q_m) \mapsto (x_1, \ldots, x_\ell, t^{d_1} q_1, \ldots, t^{d_m} q_m),$$

where $d_j = \deg q_j$. Note that when $(q, \ldots, q_m) \neq (0, \ldots, 0)$ then

$$(x_1, \ldots, x_\ell, t^{d_1} q_1, \ldots, t^{d_m} q_m) \neq (x_1, \ldots, x_\ell, s^{d_1} q_1, \ldots, s^{d_m} q_m),$$

when $s > 0$ and $s \neq t$. Therefore we have proved

Lemma 2.6.2.22. *Each orbit type in $\sigma(E)$, which is different from $\mathbb{R}^\ell \times \{0\}$ is equal to a product of $\mathbb{R}^\ell$ with a submanifold of $\mathbb{R}^m$ of dimension greater than or equal to 1.*

Therefore each connected component of an orbit type in the orbit space near a given connected component of a given orbit type has dimension greater than the dimension of the given connected component of the given orbit type. This proves

Proposition 2.6.2.23. *Connected components of orbit types in the orbit space $\overline{M}$, when viewed as differential subspaces of $(\overline{M}, C^\infty(\overline{M}))$ are smooth manifolds. These manifolds define a stratification $\mathcal{S}$ of $\overline{M}$, called the* orbit type stratification of the orbit space.

2.6.3 *Minimality of $\mathcal{S}$*

In this subsection we show that the orbit type stratification $\mathcal{S}$ of the orbit space is *minimal* when stratifications are partially ordered by inclusion of the strata.

We begin with some observations. Let $q \in P(F)^H$ be a homogeneous polynomial of degree 1. Then $q \in F^*$, which shows that $\beta^\sharp(q) \in F$. (The inner product β on F induces a bijective linear map $\beta^\flat : F \to F^*$ given by $\beta(y)y' = \beta(y, y')$. The inverse of $\beta^\flat$ is $\beta^\sharp$.) Since β is H-invariant, it follows that $\beta^\sharp(q) \in F^H = \{0\}$. Therefore $q = 0$. Consequently, $d_j = \deg q_j$, where $\{q_j\}_{j=1}^m$ form a Hilbert basis of $P(F)^H$, is greater than or equal to 2. Because the polynomial $F \to \mathbb{R} : y \mapsto \beta(y, y)$ is H-invariant and of degree 2, we may choose $q_1(y) = \beta(y, y)$ for every $y \in F$.

Let $S = \{y \in F \,|\, q_1(y) = 1\}$ be the unit sphere in F with respect to the inner product β. For each j with $2 \le j \le m$ let C_j be the maximum of $|q_j(y)|$ with $y \in S$. When $y \in F \setminus \{0\}$, let $t = q_1(y)^{1/2}$. Then $t^{-1}y \in S$. Since

$$q_j(y) = q_j(t(t^{-1}y)) = t^{d_j}\, q_j(t^{-1}y)$$

and $|q_j(t^{-1}y)| \le C_j$, we obtain $|q_j(y)| \le q_1(y)^{d_j/2}\, C_j$, which also holds when $y = 0$. Therefore $\sigma(E)$ is contained in

$$\{(x, q) \in \mathbb{R}^\ell \times \mathbb{R}^m \,|\, q_1 \ge 0 \text{ and } |q_j| \le C_j\, q_1^{d_j/2} \text{ for every } 2 \le j \le m\}. \quad (7)$$

Here σ is the Hilbert map given in (6).

Lemma 2.6.3.24. *Let I be an open interval in $\mathbb{R}$ containing 0 and let $\gamma : I \to \sigma(E) \subseteq \mathbb{R}^\ell \times \mathbb{R}^m : t \mapsto \gamma(t) = (x(t), q(t))$. Suppose that $q_1(0) = 0$ and that $q : I \to \mathbb{R}^m : t \mapsto q(t)$ is differentiable at $t = 0$. Then $q'(0) = 0$.*

Proof. From the fact that $\gamma(I) \subseteq \sigma(E)$ and the inequality $|q_j| \le C_j\, q_1^{d_j/2}$ (7), it follows that $q_1(t) \ge 0$ for every $t \in I$. Because $q_1(0) = 0$, we see that the function $t \mapsto q_1(t)$ attains its minimum value at 0. Therefore $q_1'(0) = 0$, that is, $\frac{q_1(t)}{t} \to 0$ as $t \to 0$. For each j where $2 \le j \le m$, the fact that $\gamma(I) \subseteq \sigma(E)$ and inequality (7) imply

$$\frac{|q_j(t)|}{|t|} \le C_j\, \frac{q_1(t)^{d_j/2}}{|t|} = C_j \left| \frac{q_1(t)}{t} \right|^{d_j/2} |t|^{d_j/2-1}. \tag{8}$$

The right hand side of (8) converges to 0 as $t \to 0$, because $\frac{q_1(t)}{t} \to 0$ as $t \to 0$ and $\frac{1}{2}\, d_j - 1 \ge 0$. Therefore $q_j'(0) = 0$ for every $2 \le j \le m$. □

From lemma 2.6.3.24 we see that if N is a C^1 submanifold of $\mathbb{R}^n = \mathbb{R}^\ell \times \mathbb{R}^m$ such that $N \subseteq \sigma(E)$ and $0 \in N$, then $T_0N \subseteq \mathbb{R}^\ell \times \{0\}$. In particular, $\dim N \le \ell$. From lemma 2.6.2.22 we know that all the orbit type strata in $\sigma(E)$ different from $\mathbb{R}^\ell \times \{0\}$ have dimension strictly greater than ℓ. Therefore no union of $\mathbb{R}^\ell \times \{0\}$ with different strata in $\sigma(E)$ can be a C^1 manifold through the origin. This proves

Proposition 2.6.3.25. *The orbit type stratification of the orbit space M/G of a proper G-action on M is* minimal, *that is, no union of different strata can be a connected smooth manifold in the differential space $(M/G, C^\infty(M/G))$.*

Corollary 2.6.3.26. *If M/G is connected, then the differential space $(M/G,\, C^\infty(M/G))$ is a smooth manifold if and only if there is exactly one orbit type.*

Given a semialgebraic subset, it has a *primary stratification* given by iteratively forming the semialgebraic set of singular points of the preceding semialgebraic variety.

Proposition 2.6.3.27. *In the local model of $\sigma(B) \subseteq \mathbb{R}^n$ of the orbit space U/G given in lemma* 2.4.2.16, *the orbit type stratification of B/H coincides with the primary stratification of the semialgebraic set $\sigma(B)$.*

Proof. See theorem A in Bierstone [14]. □

2.7 Derivations and vector fields on a differential space

Let $\mathcal{A}$ be an algebra over $\mathbb{R}$ with multiplication $\cdot$. A *derivation* of $\mathcal{A}$ is a linear mapping δ of $\mathcal{A}$ into itself such that Leibniz' rule holds, namely,

$\delta(f \cdot g) = (\delta f) \cdot g + f \cdot (\delta g)$ for every $f, g \in \mathcal{A}$. We denote the set of all derivations of $\mathcal{A}$ by $\mathrm{Der}(\mathcal{A})$. Note that $\mathrm{Der}(\mathcal{A})$ is a Lie algebra with bracket

$$[\delta, \delta'] = \delta \circ \delta' - \delta' \circ \delta \text{ for every } \delta, \delta' \in \mathrm{Der}\,\mathcal{A}.$$

Example 2.7.28. Let Q be a smooth manifold with $C^\infty(Q)$ its space of smooth functions. If V is a smooth vector field on Q, then for every $f \in C^\infty(Q)$ the *Lie derivative* $L_V f$ of f with respect to V, namely, $L_V f : Q \to \mathbb{R} : q \mapsto \langle \mathrm{d}f(q) | V(q) \rangle$, is a smooth function on Q. Moreover, the linear mapping $L_V : C^\infty(Q) \to C^\infty(Q)$ is a derivation.[6] □

Let $(Q, C^\infty(Q))$ be a differential space, which is not necessarily a smooth manifold, and let $\delta \in \mathrm{Der}(C^\infty(Q))$. An *integral curve* of δ is a smooth mapping $\gamma : I \subseteq \mathbb{R} \to Q$, where I is an interval, such that

$$\frac{\mathrm{d}f(\gamma(t))}{\mathrm{d}t} = \delta(f)(\gamma(t)) \text{ for every } f \in C^\infty(Q) \text{ and every } t \in I.$$

Proposition 2.7.29. *Let $(Q, C^\infty(Q))$ be a locally compact subcartesian differential space and let $\delta \in \mathrm{Der}(C^\infty(Q))$. Suppose that for every $q \in Q$, there is an* open *interval $I \subseteq \mathbb{R}$ containing* 0 *and an integral curve $\gamma_q : I \subseteq \mathbb{R} \to Q$ of δ such that $\gamma_q(0) = q$. Then*

1. *For each $q \in Q$ there is a unique integral curve $\gamma : I_q \to Q$ of δ, which is defined on a maximal open interval I_q in $\mathbb{R}$ containing* 0 *such that $\gamma(0) = q$.*
2. *The set $D = \{(t, q) \in R \times Q \,\big|\, t \in I_q\}$ is an open subset of $\mathbb{R} \times Q$ and the map $\varphi : D \subseteq \mathbb{R} \times Q \to Q : (t, q) \mapsto \gamma_q(t)$ is smooth.*
3. *For each $t \in \mathbb{R}$, the set $D_t = \{q \in Q \,\big|\, t \in I_q\}$ is an open subset of Q and the mapping $\varphi_t : D_t \to Q : q \mapsto \gamma_q(t)$ is smooth. φ_t is called the* flow *of δ* after time *t.*
4. *If $s, t \in R$, $q \in D_s$, $\varphi_s(q) \in D_t$, then $q \in D_{t+s}$ and $\varphi_t(\varphi_s(q)) = \varphi_{t+s}(q)$. Therefore $\varphi_t : D_t \to D_{-t}$ is a diffeomorphism with inverse φ_{-t}.*

Proof. Using the fact that Q is locally compact, we may identify a suitable open neighborhood of q in Q with a locally closed subset V of $\mathbb{R}^n$. Let $\{x_i\}_{i=1}^n$ be coordinate functions on $\mathbb{R}^n$. There is an open neighborhood U_i of q in $\mathbb{R}^n$ and $\delta_i \in \mathrm{Der}(C^\infty(U_i))$ such that $\delta x_i = \delta_i | U_i$. Let $U = \cap_{i=1}^n U_i$ and let $\widetilde{\delta} : U \to \mathbb{R}^n$ be a smooth vector field on U such that $L_{\widetilde{\delta}} x_i = \delta_i | U$. Shrinking U if necessary, we may assume that $V \cap U$ is a closed subset of U.

[6] In many texts on manifolds, smooth vector fields are *defined* as derivations on $C^\infty(Q)$, after which tangent spaces are introduced. It is then shown that derivations correspond to smooth sections of the tangent bundle.

Let $\gamma : I \to V \cap U$ be an integral curve of δ. Then

$$\frac{\mathrm{d}\gamma_i(t)}{\mathrm{d}t} = \frac{\mathrm{d}x_i(\gamma(t))}{\mathrm{d}t} = \delta(x_i)(\gamma(t)) = \delta_i(\gamma(t))$$

shows that γ is an integral curve of the derivation $L_{\widetilde{\delta}}$, thought of as a smooth vector field on $\mathbb{R}^n$. Therefore the local existence and uniqueness of integral curves of the smooth vector field implies the existence and uniqueness of smooth integral curves of $\widetilde{\delta}$ with prescribed initial value. Consequently, for each $q \in V \cap U$ there is a unique integral curve $\gamma : I_q \to V \cap U$ of the derivation δ on a maximal open interval I_q in $\mathbb{R}$ containing 0 with $\gamma(0) = q$.

Let $\widetilde{\varphi}_t$ be the flow after time t of the vector field $\widetilde{\delta}$ on U. Let $\widetilde{I}$ be the maximal domain of definition of the integral curve $t \mapsto \widetilde{\gamma}(t) = \widetilde{\varphi}_t(q)$ of $\widetilde{\delta}$ with $\widetilde{\gamma}(0) = q$. We will now show that $I_q = \widetilde{I}$. Now $\widetilde{\gamma}|I_q = \gamma$, so $I_q \subseteq \widetilde{I}$. Suppose that $s = \sup I_q \in \widetilde{I}$ and let $r = \widetilde{\gamma}(s)$. Then

$$r = \lim_{t \uparrow s} \widetilde{\gamma}(t) = \lim_{t \uparrow s} \gamma(t) \in V \cap U,$$

because $\gamma(t) \in V \cap U$ for every $t \in I_q$ and $V \cap U$ is closed in U. The hypothesis that every integral curve of δ is defined on an open interval, implies that there is an open interval J in $\mathbb{R}$ containing 0 and an integral curve $\widehat{\gamma} : J \to V \cap U$ of δ such that $\widehat{\gamma}(0) = r$. Therefore

$$\widehat{\gamma}(t-s) = \varphi_{t-s}(r) = \varphi_{t-s}(\varphi_s(q)) = \widetilde{\gamma}(t)$$

for all $t \in I_q \cap (s+J)$. From uniqueness it follows that $\widetilde{\gamma}$ and $t \mapsto \widehat{\gamma}(t-s)$ piece together to form an integral curve of δ, which is defined on $I_q \cap (s+J)$. This contradicts the maximality of I_q. A similar argument shows that if $\inf I_q \in \widetilde{I}$, then we obtain a contradiction. Therefore $I_q = \widetilde{I}$. This proves assertion 1.

Let $\widetilde{D} \subseteq \mathbb{R} \times U$ be the domain of definition of the flow $\widetilde{\varphi} : \widetilde{D} \to U$ of the smooth vector field $\widetilde{\delta}$ on U. Let $\varphi : D \to V \cap U$ be as defined in assertion 2 of the proposition with Q replaced by $V \cap U$. The argument of the preceding paragraph shows that $D = \widetilde{D} \cap (\mathbb{R} \times (V \cap U))$ and that $\varphi = \widetilde{\varphi}|D$. Because $\widetilde{D}$ is an open subset of $\mathbb{R} \times U$ and the flow $\widetilde{\varphi}$ is smooth, it follows that D is an open subset of $\mathbb{R} \times (V \cap U)$ and that the mapping $\varphi : D \to V \cap U$ is smooth as a map of differential spaces. Because the preceding two properties are local, we have shown that the set D, given in statement 2 of the proposition, is an open subset of $\mathbb{R} \times Q$ and that the mapping $\varphi : D \to Q$ is smooth. This proves assertion 2.

The other assertions follow in the same way as for smooth vector fields on a smooth manifold. □

Example 2.7.30. Consider the set

$$S = \{(x_1, x_2) \in \mathbb{R}^2 \mid x_1^2 + (x_2 - 1)^2 < 1 \text{ or } x_2 = 0\}.$$

The vector field $\widetilde{X} = \frac{\partial}{\partial x_1}$ on $\mathbb{R}^2$ restricts to a derivation L_X of $C^\infty(S)$. For every $x = (x_1, x_2) \in S$, $\varphi_t^{\widetilde{X}}(x) = (x_1 + t, x_2)$ for all $t \in \mathbb{R}$. Its restriction to S induces φ_t^X given by $\varphi_t^X(x_1, x_2) = (x_1 + t, x_2)$ for $t \in (-x_1 - \sqrt{1 - (x_2 - 1)^2}, -x_1 + \sqrt{1 - (x_2 - 1)^2})$ if $x_2 > 0$, and for $t \in \mathbb{R}$ if $x_2 = 0$. Hence, all integral curves of X have open domains. Nevertheless, φ_t^X fails to be a local one-parameter group of local diffeomorphisms of S. The proof that S is not a locally closed subset of $\mathbb{R}^2$ is left to the reader. □

Proposition 2.7.29 motivates the following definition. Let $(Q, C^\infty(Q))$ be a locally compact subcartesian space and let $\delta \in \mathrm{Der}(C^\infty(Q))$. We call δ a *vector field* on $(Q, C^\infty(Q))$ if and only if for every $q \in Q$ there is an *open* interval I in $\mathbb{R}$ containing 0 and an integral curve $\gamma : I \subseteq \mathbb{R} \to Q$ of δ such that $\gamma(0) = q$. The collection of all vector fields on Q we denote by $\mathcal{X}(Q, C^\infty(Q))$. We will refer to $\mathcal{X}(Q, C^\infty(Q))$ as the space of all *smooth vector fields* on the differential space $(Q, C^\infty(Q))$.

We conclude this subsection with the statement of a generalization of Sussmann's theorem to subcartesian spaces $(S, C^\infty(S))$, see theorem 1.10.42 of chapter 1. Let $\mathcal{F}$ be a family of vector fields on a subcartesian space S. We say that the family $\mathcal{F}$ is *locally complete* if, for every $X, Y \in \mathcal{F}$, $t \in \mathbb{R}$ and $x \in S$, for which $(\varphi_t^X)_* Y(x)$ is defined, there exists an open neighborhood U of x and $Z \in \mathcal{F}$ such that $(\varphi_t^X)_* Y \mid U = Z|U$. For example, a family consisting of a single vector field X is locally complete because $(\varphi_t^X)_* X(x) = X(x)$ at all points $x \in S$ for which $\varphi_t^X(x)$ is defined. For $x \in S$, we define the orbit through x of a family $\mathcal{F}$ of vector fields on S to be the set of points in S which can be connected to x by piecewise integral curves of vector fields in $\mathcal{F}$.

Theorem 2.7.31. *Let $\mathcal{F}$ be a* locally complete family of vector fields on a subcartesian space S. Each orbit *O of $\mathcal{F}$ admits a unique manifold structure such that the inclusion map $\iota_O : O \hookrightarrow S$ is smooth. Moreover, smooth functions on O are locally pull backs of smooth functions on S.*

Outline of proof. For each $x \in S$, let $D_{\mathcal{F}_x} = \{X(x) \mid X \in \mathcal{F}\}$. In other words, $D_{\mathcal{F}_x}$ is the subspace of the space of derivations of $C^\infty(S)$ consisting of values at x of vector fields $X \in \mathcal{F}$. Since the family $\mathcal{F}$ is locally complete, it follows that $\dim D_{\mathcal{F}_x}$ is constant along orbits of $\mathcal{F}$.

Let O be the orbit of $\mathcal{F}$ through $z \in S$ and let $m = \dim D_{\mathcal{F}_z}$. There exist m vector fields $X^1, ..., X^m \in \mathcal{F}$ which are linearly independent in a neighborhood V of z in S. Since S is subcartesian, without loss of generality, we may assume that V is a subset of $\mathbb{R}^n$. Then, there exists a neighborhood U of z in $\mathbb{R}^n$, and linearly independent vector fields $\tilde{X}^1, ..., \tilde{X}^m$ on U such that, $X^i(x) = \tilde{X}^i(x)$ for each $i = 1, ..., m$ and each $x \in V \cap U$. Consider the map

$$\tilde{\rho} : T = (t_1, ..., t_m) \mapsto (\varphi_{t_m}^{\tilde{X}^m} \circ \cdots \circ \varphi_{t_1}^{\tilde{X}^1})(x) \in \mathbb{R}^n$$

defined in a neighborhood of zero in $\mathbb{R}^n$. Since the vector fields $\tilde{X}^1, ..., \tilde{X}^m$ are linearly independent on U, there is a neighborhood Ω of zero in $\mathbb{R}^m$ such that $\tilde{\rho}$ is a diffeomorphism of Ω onto its image $W = \tilde{\rho}(\Omega) \subset \mathbb{R}^n$.

But $x \in S$ and the vector fields $\tilde{X}^1, ..., \tilde{X}^m$ are extensions to U of vector fields $X^1, ..., X^m$ on S restricted to $V \cap U$. Moreover, for each $i = 1, \ldots, m$ the flow $\varphi_{t_i}^{X^i}$ of X^i is a local one parameter group of local diffeomorphisms of S. Hence, $W = \tilde{\rho}(\Omega)$ is in S, and it is an open subset of the orbit O through x.

Since W is a submanifold of $\mathbb{R}^n$, a function $f : W \to \mathbb{R}$ is smooth if it locally coincides with restrictions to W of smooth functions on $\mathbb{R}^n$. Let $\iota_V : V \hookrightarrow \mathbb{R}^n$, $\iota_W : W \hookrightarrow \mathbb{R}^n$ and $\iota_{WV} : W \hookrightarrow V$. Then $\iota_W = \iota_{WV} \circ \iota_V$. Moreover, smooth functions on V are locally restrictions to V of smooth functions on $\mathbb{R}^n$. Hence, smooth functions on W are locally restrictions to W of smooth functions on V. But V is an open subset of S. Hence, smooth functions on V are locally restrictions to V of smooth functions on S. Therefore, smooth functions on W are locally restrictions of smooth functions on S.

Analogous reasoning for every $x \in O$ gives a covering of O by smooth manifolds W_x. One needs to show that the differential structures are compatible on overlaps and that they are independent of the choice of vector fields $X^1, ..., X^m \in \mathcal{F}$ which are linearly independent in W_x. For details see [106]. □

Let $\mathcal{X}(S, C^\infty(S))$ denote the family of all vector fields on a subcartesian space $(S, C^\infty(S))$. It is easy to see that the family $\mathcal{X}(S, C^\infty(S))$ is locally complete. Hence, we obtain the following

Corollary 2.7.32. *For each point x in a subcartesian space $(S, C^\infty(S))$, the orbit O of the family $\mathcal{X}(S, C^\infty(S))$ of all vector fields on $(S, C^\infty(S))$ is a manifold and the inclusion map $O \hookrightarrow S$ is smooth.*

Proof. See [106]. □

2.8 Vector fields on a stratified differential space

Let $(Q, C^\infty(Q))$ be a differential space. A *stratification* $\mathcal{S}$ of $(Q, C^\infty(Q))$ is a collection of differential subspaces, which are smooth submanifolds of $(Q, C^\infty(Q))$ such that

1. $\mathcal{S}$ is a locally finite partition of Q;
2. for each $S \in \mathcal{S}$ the closure of S in Q is the union of S and $\{S' \in \mathcal{S} \big| \dim S' < \dim S\}$.

Denote a differential space $(Q, C^\infty(Q))$ with the stratification $\mathcal{S}$ by $(Q, \mathcal{S}, C^\infty(Q))$.

A *stratified vector field* W on $(Q, \mathcal{S}, C^\infty(Q))$ is a map which assigns to each stratum S of the stratification $\mathcal{S}$ a smooth vector field W_S on S. If $f \in C^\infty(Q)$ and $S \in \mathcal{S}$, then $f|S \in C^\infty(S)$. Therefore $L_{W_S}(f|S) \in C^\infty(S)$.

Lemma 2.8.33. *For every $f \in C^\infty(Q)$ the function $\partial_W f$, which assigns to each $S \in \mathcal{S}$ the smooth function $L_{W_S}(f|S)$ in $C^\infty(S)$, is a smooth function on Q, which lies in* $\mathrm{Der}(C^\infty(Q))$.

Proof. Let $q \in Q$. Because $f \in C^\infty(Q)$, for every open neighborhood U of q in Q, we have $f|U \in C^\infty(U)$. Since the stratification $\mathcal{S}$ is a locally finite partition of Q, there are a finite number of strata $\{S_j\}_{j\in J}$ such that $S_j \cap U \neq \emptyset$. Let $F = \sum_{j\in J} L_{W_{S_j}}(f|(U \cap S_j))$. Then $F \in C^\infty(U)$ because $L_{W_{S_j}}(f|(U \cap S_j)) \in C^\infty(U \cap S_j) \subseteq C^\infty(U)$, since S_j is an embedded submanifold of $(Q, C^\infty(Q))$. From

$$(\partial_W f)|(U \cap S_j) = L_{W_{S_j}}(f|(U \cap S_j)) = F|(U \cap S_j)$$

it follows that $L_W f \in C_i^\infty(U) = C^\infty(U)$. Therefore $\partial_W f \in C^\infty(Q)$. Clearly $\partial_W f$ is a derivation on $C^\infty(Q)$. □

A stratified vector field W on $(Q, \mathcal{S}, C^\infty(Q))$ is *smooth* if and only if $\partial_W f \in C^\infty(Q)$ for every $f \in C^\infty(Q)$. Let $\mathcal{X}^\infty(Q, \mathcal{S})$ be the *set of all smooth stratified vector fields* on $(Q, \mathcal{S}, C^\infty(Q))$.

Proposition 2.8.34. *The map $\partial : \mathcal{X}^\infty(Q, \mathcal{S}) \to \mathrm{Der}(C^\infty(Q)) : W \mapsto \partial_W$ is an injective homomorphism of Lie algebras. Moreover,*

$$\partial(\mathcal{X}^\infty(Q, \mathcal{S})) \subseteq \mathcal{X}(Q, C^\infty(Q)), \tag{9}$$

if the differential space $(Q, C^\infty(Q))$ is locally compact and is subcartesian.

Proof. First we show that ∂ is injective. Let $W \in \mathcal{X}^{\infty}(Q, \mathcal{S})$ and suppose that $\partial_W = 0$. Let $S \in \mathcal{S}$, $f \in C^{\infty}(S)$, and $q \in S$. Then there is an open neighborhood U of q in Q and $g \in C^{\infty}(Q)$ such that $f|(S \cap U) = g|(S \cap U)$. Because $\partial_W g = 0$, we obtain $L_{W_S}(f|(S \cap U)) = 0$. Since this holds for every $q \in S$, we see that $L_{W_S} f = 0$. Because $L_{W_S} : \mathcal{X}^{\infty}(S) \to \mathrm{Der}(C^{\infty}(S))$ is an isomorphism for smooth manifolds, it follows that $W_S = 0$. Since this holds for every $S \in \mathcal{S}$, we deduce that $W = 0$.

We now show that (9) holds. Let $W \in \mathcal{X}^{\infty}(Q, \mathcal{S})$ and $q \in Q$. Then there is an $S \in \mathcal{S}$ with $q \in S$. The smooth vector field W_S on S has a smooth integral curve $\gamma : I \to S$, where I is an open interval in $\mathbb{R}$ containing 0 and $\gamma(0) = q$. For any $f \in C^{\infty}(Q)$, we know that $f|S$ is a smooth function on the smooth manifold S. Therefore

$$\begin{aligned}\frac{\mathrm{d}f(\gamma(t))}{\mathrm{d}t} &= \langle \mathrm{d}(f|S)(\gamma(t)) \mid \gamma'(t)\rangle = \langle \mathrm{d}(f|S)(\gamma(t)) \mid W_S(\gamma(t))\rangle \\ &= L_{W_S}(f|S)(\gamma(t)) = \partial_W f(\gamma(t)).\end{aligned}$$

So γ is an integral curve of the derivation ∂_W with $\gamma(0) = q$. Because this holds for every $q \in Q$, the derivation ∂_W is a smooth vector field on $(Q, C^{\infty}(Q))$, that is, $\partial_W \in \mathcal{X}(Q, C^{\infty}(Q))$. □

2.9 Vector fields on an orbit space

Let G be a Lie group which acts smoothly and properly on a smooth manifold M. Then $(M/G, C^{\infty}(M/G))$ is a differential space, which has a stratification $\mathcal{S}$ by orbit types.

Let $V \in \mathcal{X}(M)^G$, that is, V is a smooth G-invariant vector field on M. Then $V_{(H)} = V|M_{(H)}$ is a smooth vector field on the orbit type $M_{(H)}$. Let $\overline{V_{(H)}} = \pi_* V_{(H)}$, where $\pi : M \to \overline{M} = M/G$ is the G-orbit map. Then $\overline{V_{(H)}}$ is a smooth vector field on the orbit type $\overline{M_{(H)}} = \pi(M_{(H)})$ in the orbit space $\overline{M}$. The map which assigns to the orbit type $\overline{M_{(H)}}$ the smooth vector field $\overline{V_{(H)}}$ defines a smooth stratified vector field $\overline{V}$ on $(\overline{M}, \mathcal{S}, C^{\infty}(\overline{M}))$ such that $\pi_* V = \overline{V}$. Thus we have proved

Proposition 2.9.35. *We have*

$$\pi_*(\mathcal{X}(M)^G) \subseteq \mathcal{X}^{\infty}(\overline{M}, \mathcal{S}). \tag{10}$$

Schwarz [99] has shown that equality holds in (10).

Proposition 2.9.36. *Using the map* $\partial : \mathcal{X}^\infty(\overline{M}, \mathcal{S}) \to \mathrm{Der}(C^\infty(\overline{M}))$ *defined in §8, we have*

$$\partial\mathcal{X}^\infty(\overline{M}, \mathcal{S}) = \mathcal{X}(\overline{M}, C^\infty(\overline{M})). \tag{11}$$

Proof. We begin by observing that the orbit space $\overline{M}$ is locally compact, since M is and the orbit map π is continuous, open and surjective. In addition, the differential space $(\overline{M}, C^\infty(\overline{M}))$ is subcartesian.

Let δ be a smooth vector field on $\overline{M}$. Let $\gamma : I \to \overline{M} : t \mapsto \gamma(t)$ which starts at $\overline{m}$. For any point $\overline{m} \in \overline{M}$, we use the diffeomorphism $\varphi = \widetilde{\sigma} \circ \phi^{-1}$ to identify an open neighborhood $\overline{U}$ of $\overline{m}$ in $\overline{M}$ with the image $\sigma(B) \subseteq \mathbb{R}^n$ of the Hilbert map, as in the proof of proposition 2.5.18. We also use the decomposition $\mathbb{R}^n = \mathbb{R}^\ell \times \mathbb{R}^m$ so that the orbit types in $\sigma(E)$ are of the form $\mathbb{R}^\ell \times R$, where R is an orbit type for the action of H on F. We may assume that $\widetilde{\sigma}(\overline{m}) = (0, 0) \in \mathbb{R}^\ell \times \mathbb{R}^m$ and that its orbit type corresponds locally to $\mathbb{R}^\ell \times \{0\}$.

Let $V(x, q) = (\dot{x}, \dot{q})$ be a smooth vector field in an open neighborhood of 0 in $\mathbb{R}^n$ induced by the smooth vector field δ. Because $\gamma(t) \in \sigma(E)$ for all $t \in I$, from lemma 2.6.3.24 it follows that $q'(0) = 0$. Therefore $\dot{q} = 0$ when $q = 0$. In other words, V is tangent to $S = \mathbb{R}^\ell \times \{0\}$ at the origin. This implies that $V|S$ near the origin is a smooth vector field on S. Therefore the integral curves of δ, which start on S remain on S.

Let f be a smooth function on an open neighborhood of the origin in $\mathbb{R}^n$. For any integral curve γ of δ, which lies on S, we have

$$\delta(f)(\gamma(t)) = \frac{\mathrm{d}f(\gamma(t))}{\mathrm{d}t} = L_W f(\gamma(t)).$$

This implies that $\delta(f)|S = L_W(f|S)$. Thus, for every $S \in \mathcal{S}$ and every $\overline{m} \in S$, there is an open neighborhood $\overline{U}$ of $\overline{m}$ in $\overline{M}$ and a smooth vector field $W_{S\cap U}$ such that

$$\delta(f)|(S \cap U) = L_{W_{S\cap U}}(f|(S \cap U)) \text{ for every } f \in C^\infty(\overline{U}). \tag{12}$$

Because equation (12) determines $W_{S\cap U}$ uniquely, the vector fields $W_{S\cap U}$ patch together to form a smooth vector field W_S on S, where

$$\delta(f)|S = L_{W_S}(f|S) \tag{13}$$

holds for every $f \in C^\infty(\overline{M})$. Because (13) holds for every $S \in \mathcal{S}$, we see that $S \mapsto W_S$ defines a smooth stratified vector field $\widetilde{W}$ on $\overline{M}$, that is, $\widetilde{W} \in \mathcal{X}^\infty(\overline{M}, \mathcal{S})$. Thus $\delta = \partial_{\widetilde{W}}$, which implies $\mathcal{X}(\overline{M}, C^\infty(\overline{M})) \subseteq \partial\mathcal{X}^\infty(\overline{M}, \mathcal{S})$. The inclusion $\partial\mathcal{X}^\infty(\overline{M}, \mathcal{S}) \subseteq \mathcal{X}(\overline{M}, C^\infty(\overline{M}))$ follows from proposition 2.8.34 with $(Q, C^\infty(Q)) = (\overline{M}, C^\infty(\overline{M}))$. □

In view of propositions 2.9.35 and 2.9.36, we call the set of smooth vector fields on $\overline{M}$ the space $\mathcal{X}(\overline{M}, C^\infty(\overline{M}))$ of *smooth vector fields* on the G orbit space $\overline{M}$. These vector fields depend only on the differential structure $C^\infty(\overline{M})$ of $\overline{M}$; whereas the space of induced vector fields $\pi_*(\mathcal{X}^\infty(M)^G)$ depends on the manifold structure of M, and the space $\mathcal{X}^\infty(\overline{M}, \mathcal{S})$ of stratified vector fields on $\overline{M}$ depends on the stratification $\mathcal{S}$ of $\overline{M}$ by orbit types.

Examples 2.9.37.

1. Let $M = \mathbb{R}$ and $G = \mathbb{Z}_2 = \{\pm 1\}$. Then $p : \mathbb{R} \to \mathbb{R}_{\geq 0} \subseteq \mathbb{R} : x \mapsto x^2$ is a $\mathbb{Z}_2$-invariant polynomial, which freely generates $C^\infty(\mathbb{R})^{\mathbb{Z}_2}$. In other words, every smooth even function is a smooth function of p. Therefore $p^* : C^\infty(\mathbb{R}_{\geq 0}) \to C^\infty(\mathbb{R})^{\mathbb{Z}_2}$ is an isomorphism, which gives rise to a diffeomorphism of the differential space $(\mathbb{R}/\mathbb{Z}_2, C^\infty(\mathbb{R}/\mathbb{Z}_2))$ onto the differential space $(\mathbb{R}_{\geq 0}, C^\infty(\mathbb{R}_{\geq 0}))$. A derivation on $C^\infty(\mathbb{R}_{\geq 0})$ is a smooth vector field W on $\mathbb{R}$. Now L_W is a smooth vector field on $C^\infty(\mathbb{R}_{\geq 0})$ if and only if the flow φ_t of W maps 0 into $\mathbb{R}_{\geq 0}$ for every t in an open neighborhood of 0 in $\mathbb{R}$. This holds if and only if $W(0) = 0$. Therefore *not* every derivation of $C^\infty(\mathbb{R}/\mathbb{Z}_2)$ is a smooth vector field on $\mathbb{R}/\mathbb{Z}_2$.

2. Let $M = \mathbb{C}$ and $G = S^1 = \{z \in C \,\big|\, |z| = 1\}$. Suppose that S^1 acts on $\mathbb{C}$ by multiplication. Then $C^\infty(\mathbb{C})^{S^1}$ is generated freely by the real valued function $p : \mathbb{C} \to \mathbb{R}_{\geq 0} \subseteq \mathbb{R} : z \mapsto z\,\overline{z}$. Thus $p^* : C^\infty(\mathbb{R}_{\geq 0}) \to C^\infty(\mathbb{C}/S^1)$ is an isomorphism. So again not every derivation of $C^\infty(\mathbb{C}/S^1)$ is a vector field on $\mathbb{C}/S^1$. The inclusion map $i : \mathbb{R} \to \mathbb{C}$ induces a diffeomorphism of the differential space $(\mathbb{C}/S^1, C^\infty(\mathbb{C}/S^1))$ onto the differential space $(\mathbb{R}/\mathbb{Z}_2, C^\infty(\mathbb{R}/\mathbb{Z}_2))$. □

2.10 Tangent objects to an orbit space

In the preceding section we have introduced the notion of a smooth vector field on the orbit space $\overline{M}$ of a smooth proper action of a Lie group G on a smooth manifold M. We did not define what we meant by being "tangent" to $\overline{M}$ at $\overline{m} \in \overline{M}$, so we cannot view our smooth vector fields as being "tangent" to $\overline{M}$. Below we discuss various ways of dealing with this.

2.10.1 *Stratified tangent bundle*

Let $\mathcal{S}$ be the orbit type stratification of the orbit space $\overline{M}$, which due to its local model as an open subset of the semialgebraic variety $\sigma(E)$, is a

Whitney stratification. Let $T^{\mathcal{S}}\overline{M}$ be $\coprod_{S\in\mathcal{S}} TS$, the disjoint union of the tangent bundle TS of the strata S of $\mathcal{S}$. Then $T^{\mathcal{S}}\overline{M}$ is the *stratified tangent bundle* of $\overline{M}$. We can give $T^{\mathcal{S}}\overline{M}$ the structure of a differential space so that $\mathcal{T} = \{TS\}_{S\in\mathcal{S}}$ forms a stratification of $T^{\mathcal{S}}\overline{M}$ and the projection map $\pi^{\mathcal{S}} : T^{\mathcal{S}}\overline{M} \to \overline{M}$ is a smooth map of stratified differential spaces. In addition, the space $\mathcal{X}^{\infty}(\overline{M}, \mathcal{S})$ of smooth stratified vector fields on $\overline{M}$ is equal to the space of smooth sections of $\pi^{\mathcal{S}} : T^{\mathcal{S}}\overline{M} \to \overline{M}$, that is, the space of smooth mappings $\sigma : \overline{M} \to T^{\mathcal{S}}\overline{M}$ such that $\pi^{\mathcal{S}} \circ \sigma = \mathrm{id}_{\overline{M}}$.

2.10.2 *Zariski tangent bundle*

Let $(Q, C^{\infty}(Q))$ be a differential space. For $q \in Q$, let $\mathcal{M}_q = \{f \in C^{\infty}(Q) \,|\, f(q) = 0\}$. Then $\mathcal{M}_q$ is a maximal ideal in the algebra $C^{\infty}(Q)$. Following ideas in algebraic geometry, we call $\mathcal{M}_q/\mathcal{M}_q^2$ the *Zariski cotangent space* to Q at q and its topological dual $\big(\mathcal{M}_q/\mathcal{M}_q^2\big)^*$ the *Zariski tangent space* $T_q^Z Q$ to Q at q.

If S is a smooth submanifold of Q, then for each $s \in S$ with $v_s \in T_sS$ and for each $f \in C^{\infty}(Q)$ the function $\widetilde{\partial}_{v_s} : C^{\infty}(Q) \to \mathbb{R} : f \mapsto \langle \mathrm{d}(f|S)(s) \mid v_s\rangle$ gives rise to the mapping $\partial : T_sS \to T_s^Z Q : v_s \to \widetilde{\partial}_{v_s}$, which is injective and therefore can be used to identify T_sS with a linear subspace of $T_s^Z Q$. In this way the stratified tangent bundle $T^{\mathcal{S}}Q$ of Q is contained in the *Zariski tangent bundle* $T^Z Q = \coprod_{q\in Q} T_q^Z Q$ of Q.

When $Q = \overline{M}$, using the local model of $\overline{M}$ as an open subset of a semialgebraic set $\sigma(E)$, then $T_0^Z(\sigma(E))$ is isomorphic to $T_0^Z\mathbb{R}^n$, which in turn is isomorphic to $\mathbb{R}^n$. Therefore the dimension of the Zariski tangent space to $\overline{M}$ at $\overline{m}$ is equal to the number of generators in a Hilbert basis of the algebra of G_m-invariant polynomials on $E = T_mM/T_m(G\cdot m)$.

2.10.3 *Tangent cone*

The *tangent cone* $T_0\mathcal{C}$ in $T_0^Z\sigma(E)$ is the set of all limits of sequences $\{\tau_j\,\sigma(x_j)\} \in \sigma(E)$, where $\tau_j \in \mathbb{R}_{>0}$ and $\{\sigma(x_j)\} \to 0$. The conic structure of $T_0\mathcal{C}$ is given by multiplication of a vector in $\mathbb{R}^n$ by a positive real number. Because the Hilbert map $\sigma : E \to \mathbb{R}^n$ is not homogeneous, this conic structure is *different* from the conic structure on $\sigma(E)$, which is induced by the Hilbert map σ from the H-invariant conic structure on E defined by multiplication of vectors in E by positive real numbers.

2.10.4 *Tangent wedge*

Throughout this subsection we suppose that the action of G on M is proper. Let O be the G-orbit $G \cdot m$ through $m \in M$. The *normal bundle* NO of O is $\coprod_{m' \in O} T_{m'}M/T_{m'}(G \cdot m)$, namely, the disjoint union of the vector spaces $E_{m'} = T_{m'}M/T_{m'}(G \cdot m)$ for every $m' \in O$. The normal bundle NO is a smooth vector bundle over O. On NO there is an action of G induced by the tangent of the G-action on M. The orbit space NO/G is called the *tangent wedge* $W_{\overline{m}}$ at $\overline{m} = O \in \overline{M}$. Let $C^\infty(W_{\overline{m}}) = C^\infty(NO/G) = C^\infty(NO)^G$. Then $(W_{\overline{m}}, C^\infty(W_{\overline{m}}))$ is a differential space. The inclusion map $E_m \rightarrow NO$ leads to an identification of the tangent wedge $W_{\overline{m}}$ at $\overline{m}$ with E/H, where $E = E_m$ and $H = G_m$. The discussion in §6.1 shows that $E/H = E^H \times F/H$, where E^H corresponds to the tangent space at $\overline{m}$ of the orbit type $\overline{M_{(H)}}$ in $\overline{M}$ and F is a vector space isomorphic to E/E^H. The tangent wedge at $\overline{m}$ is a wedge over the cone F/H. The conic structure on F/H is the one induced by the Hilbert map σ from the H-invariant conic structure on F.

Proposition 2.10.4.38. *The tangent wedge* $(W_{\overline{m}}, C^\infty(W_{\overline{m}}))$ *at* $\overline{m} \in \overline{M}$ *is* diffeomorphic *as a differential space to an open neighborhood of* $\overline{m}$ *in* $(\overline{M}, C^\infty(\overline{M}))$.

Proof. This follows from proposition 2.5.18. □

We say that a *stratified differential space* $(Q, \mathcal{S}, C^\infty(Q))$ is *smoothly locally trivial* if for every $q \in Q$ there is an open neighborhood U of q in Q, a stratified differential space $(P, \mathcal{T}, C^\infty(P))$ with a distinguished point $\mathfrak{o}$, and a map $\varphi : U \rightarrow (S \cap U) \times P$, where $S \in \mathcal{S}$ with $q \in S$ such that

1. $\varphi^* : C^\infty((S \cap U) \times P) \rightarrow C^\infty(U)$ is bijective.
2. $\varphi|((S \cap U) \times \{\mathfrak{o}\})$ is a diffeomorphism with $\varphi(s, \mathfrak{o}) = (s, \mathfrak{o})$ for every $s \in S$.

From proposition 2.10.4.38 we obtain

Corollary 2.10.4.39. *The stratified differential space* $(\overline{M}, \mathcal{S}, C^\infty(\overline{M}))$, *where* $\mathcal{S}$ *is the orbit type stratification, is smoothly locally trivial and thus is subcartesian.*

Proof. For a proof of the last assertion see [106]. □

Lemma 2.10.4.40. *Let* $(\overline{M}, \mathcal{S}, C^\infty(\overline{M}))$ *be a smoothly locally trivial stratified subcartesian space and* X_S *a smooth vector field on a stratum* S *of* $\mathcal{S}$.

For each $s \in S$ there exists a neighborhood U of s and a vector field X on $\overline{M}$ such that the restriction of X_S to U coincides with the restriction of X to U.

Proof. See [106]. □

Theorem 2.10.4.41. *Each stratum of the stratified differential space $(\overline{M}, \mathcal{S}, C^\infty(\overline{M}))$, where $\mathcal{S}$ is the orbit type stratification, is an orbit of the family $\mathcal{X}(\overline{M}, C^\infty(\overline{M}))$ of all vector fields on $\overline{M}$.*

Proof. By corollary 2.10.4.39, $(\overline{M}, \mathcal{S}, C^\infty(\overline{M}))$ is smoothly locally trivial. Lemma 2.10.4.40 implies that every stratum S of $\mathcal{S}$ is contained in an orbit O of the family $\mathcal{X}(\overline{M}, C^\infty(\overline{M}))$ of all vector fields on $\overline{M}$. By corollary 2.7.32 the orbit O of $\mathcal{X}(\overline{M}, C^\infty(\overline{M}))$ is a smooth manifold. However, proposition 2.6.3.25 states that the stratification S by orbit type is minimal, that is, no union of different strata can be a connected smooth manifold in the differential space $(\overline{M}, \mathcal{S}, C^\infty(\overline{M}))$. Hence, $S = O$, which completes the proof. □

2.11 Notes

The text in this chapter follows that of the survey article of Duistermaat [41]. The results about the flow of a vector field are standard and may be found in Coddington and Levinson [27]. The notion of a proper action is due to Palais [88]. The proofs of the standard facts about proper actions not given in the text, namely, proposition 2.2.2, theorems 2.3.3.6, 2.3.3.7, and 2.3.4.8, may be found in Duistermaat and Kolk [42]. The equivalence of a free and proper action and a principal G-bundle is due to H. Cartan [20]. The facts about paracompactness and partitions of unity used to prove lemma 2.4.1.11 come from Dieudonne [37], [38]. The notion of a differential space originates with Sikorski [100], [101]. Proposition 2.4.2.14 showing that the orbit space of a proper action is a differential space comes from Cushman and Śniatycki [33], as does lemma 2.4.2.16. The fact that the sheaf of smooth functions on a locally compact and paracompact differential space is fine follows from lemma 2.4.1.10, see Gunning [50]. The notion of a subcartesian space is due to Aronszjan [9], but its application to studying the orbit type stratification of an orbit space of a proper group action is due to Śniatycki [106]. Facts about Hilbert maps and stratifications may be found in Pflaum [91]. The Tarski-Seidenberg theorem is proved in Friedman [46] on pp. 225–235. For a discussion of the inequalities appearing in the image of the Hilbert map we refer the reader to Procesi and Schwarz [93]. The approach to proving the minimality of the orbit type stratification of the orbit space is due to Bierstone [15], who in [14] proved proposition 2.6.3.27 relating the stratification by orbit types with the primary stratification of the semialgebraic variety given by the image of the Hilbert map. The result that strata of the orbit type stratification are orbits of the family of all vector fields is due to Lusala and Śniatycki [71]. The notion of smooth stratified tangent bundle and smooth stratified vector field is taken from theorems 2.1.2, 2.2.6, and 2.2.9 of Pflaum [91]. The fact that the dimension of the Zariski tangent space at a point $\overline{m}$ of the orbit space $\overline{M}$ of a proper action G on a smooth manifold M is equal to the number of generators in a Hilbert basis of the algebra of G_m-invariant polynomials on $T_mM/T_m(G \cdot m)$ is due to Mather [78]. The definition of tangent cone is due to Whitney [117]. The definition of the tangent wedge $W_{\overline{m}}$ to $\overline{M}$ at $\overline{m}$ is due to Cushman and Śniatycki [33] as is the first proof of proposition 2.10.4.38 that the tangent wedge $(W_{\overline{m}}, C^\infty(W_{\overline{m}}))$ is diffeomorphic to the orbit space $(\overline{M}, C^\infty(\overline{M}))$ near $\overline{m}$.

Chapter 3

Symmetry and reduction

3.1 Dynamical systems with symmetry

In this section we discuss symmetry and reduction for a general dynamical system. These results will be used in the discussion of symmetry and reduction of nonholonomically constrained systems given in subsequent sections.

3.1.1 *Invariant vector fields*

A *smooth dynamical system* is a pair (M, V), where M is a smooth manifold and V is a smooth vector field on M. Let Φ be a smooth action of a Lie group G on a smooth manifold M. We say that G is a *symmetry* of (M, V) if the G-action Φ leaves V *invariant*, that is, if $T_m\Phi_g(V(m)) = V(\Phi_g(m))$ for every $(g, m) \in G \times M$.

Lemma 3.1.1.1. *V is a smooth G-invariant vector field if and only if its flow φ_t commutes with the action of G. More formally, $I_{g\cdot m} = I_m$, the maximum time interval in $\mathbb{R}$ containing 0, where the integral curve $t \mapsto \varphi_t(m)$ of V starting at m is defined, and*

$$\Phi_g(\varphi_t(m)) = \varphi_t(\Phi_g(m)), \text{ for every } (g, m) \in G \times M \text{ and } t \in I_m. \quad (1)$$

Proof. Suppose that V is a smooth G-invariant vector field on M, then

$$\frac{\mathrm{d}}{\mathrm{d}t}(\Phi_g(\varphi_t(m))) = T_m\Phi_g\, V(\varphi_t(m)) = V(\Phi_g(\varphi_t(m)))$$

and $\Phi_g(\varphi_0(m)) = \Phi_g(m)$. Therefore $\gamma : I_m \to M : t \mapsto \Phi_g(\varphi_t(m))$ is an integral curve of V, that is, $\frac{\mathrm{d}\,\gamma(t)}{\mathrm{d}\,t} = V(\gamma(t))$, starting at $\Phi_g(m)$ as is $\widetilde{\gamma} : I_m \to M : t \mapsto \varphi_t(\Phi_g(m))$. Therefore by uniqueness of integral curves starting at the same point, we obtain $I_{g\cdot m} = I_m$ and that (1) holds. The converse follows by differentiating (1). □

3.1.2 *Reduction of symmetry*

In this section we show how to reduce the symmetry of the G-invariant vector field V on a smooth manifold M.

Our first observation about a G-invariant vector field V is that equation (1) implies that for $t \in I_m$ the flow φ_t of V maps the G-orbit $G \cdot m$ through m onto the G-orbit $G \cdot \varphi_t(m)$ through $\varphi_t(m)$ in a G-equivariant fashion. Therefore φ_t induces a map $\overline{\varphi}_t : \pi(D_t) \to \overline{M} = M/G$ such that

$$\overline{\varphi}_t \circ \pi = \pi \circ \varphi_t \text{ on } D_t. \tag{2}$$

Here $\pi : M \to \overline{M} = M/G$ is the G-orbit map and $D_t = \{m \in M \big| \, t \in I_m\}$. Since D_t is an open G-invariant subset of M, the domain $\pi(D_t)$ of definition of $\overline{\varphi}_t$ is an open subset of the G-orbit space $\overline{M}$. In addition, the group property (see §1 of chapter 2) holds for $\overline{\varphi}_t$, namely, if $s, t \in \mathbb{R}$, $\overline{m} \in \pi(D_s)$, and $\overline{\varphi}_s(\overline{m}) \in \pi(D_t)$, then $\overline{m} \in \pi(D_{t+s})$ and $\overline{\varphi}_t(\overline{\varphi}_s(\overline{m})) = \overline{\varphi}_{t+s}(\overline{m})$. Thus $\overline{\varphi}_t$ is a continuous flow on $\overline{M}$, which we call the *reduced flow* or *reduced dynamical system*. If the flow φ_t of V is complete, then so is the *reduced flow* $\overline{\varphi}_t$. Then $t \mapsto \overline{\varphi}_t$ is a *one parameter group* of homeomorphisms of $\overline{M}$.

3.1.3 *Reduction for a free and proper G-action*

If G acts freely and properly on M, then the orbit space $\overline{M}$ is a smooth manifold, and the reduced flow $\overline{\varphi}_t$ is a smooth flow of a reduced smooth vector field $\overline{V}$ on $\overline{M}$ defined by $\overline{V}(\overline{m}) = \frac{\mathrm{d}}{\mathrm{d}t}\Big|_{t=0} \overline{\varphi}_t(\overline{m})$. Note that we have

$$T_m\pi(V(m)) = \overline{V}(\pi(m)), \text{ for every } m \in M.$$

Because the orbit space $\overline{M}$ has smaller dimension than M, if $\dim G > 0$, there is a chance that the reduced dynamical system is simpler than the original one.

3.1.4 *Reduction of a nonfree, proper G-action*

Assume that the action of G on M is proper, but is not necessarily free. In this case, the *orbit space* $\overline{M}$ is a subcartesian differential space endowed with the *orbit type stratification* $\mathcal{S}$. The ring $C^\infty(\overline{M})$ of smooth functions on $\overline{M}$ is

$$\{\overline{f} : \overline{M} \to \mathbb{R} \mid \pi^*\overline{f} \in C^\infty(M)\},$$

where $\pi : M \to \overline{M}$ is the G-orbit map, which is isomorphic to the ring $C^\infty(M)^G$ of G-invariant functions in $C^\infty(M)$.

Since the vector field V is G-invariant, it is a derivation of $C^\infty(M)$ which preserves $C^\infty(M)^G$. Hence it induces a derivation $\overline{V}$ of $C^\infty(\overline{M})$ such that, $\pi^*(\overline{V}(\overline{f})\ ) = V(\pi^*\overline{f})$ for every $\overline{f} \in C^\infty(\overline{M})$.

Proposition 3.1.4.2. *The* reduced flow $\overline{\varphi}_t$, *defined by equation* (2), *is the local one parameter group of local diffeomorphisms of* $\overline{M}$ *generated by* $\overline{V}$.

Proof. Let φ_t be the flow of V and $\overline{\varphi}_t$ the reduced flow on $\overline{M}$. We have shown above that $\overline{\varphi}_t$ is a local one parameter group of local homeomorphisms of $\overline{M}$. For each $\overline{f} \in C^\infty(\overline{M})$, the argument above implies that $\pi^*(\overline{\varphi}_t^*\overline{f}) = \varphi_t^*(\pi^*\overline{f}) \in C^\infty(M)$. Hence, $\overline{\varphi}_t$ is a local one parameter group of local diffeomorphisms of $\overline{M}$. Moreover,

$$\pi^*\frac{\mathrm{d}}{\mathrm{d}t}(\overline{\varphi}_t^*\overline{f}) = \frac{\mathrm{d}}{\mathrm{d}t}\varphi_t^*\pi^*(\overline{f}) = V(\varphi_t^*\pi^*(\overline{f})) = V(\pi^*\varphi_t^*(\overline{f})) = \pi^*(\overline{V}(\overline{\varphi}_t^*(\overline{f}))).$$

Hence, $\frac{\mathrm{d}}{\mathrm{d}t}(\overline{\varphi}_t^*\overline{f}) = \overline{V}(\overline{\varphi}_t^*(\overline{f}))$ for every $\overline{f} \in C^\infty(\overline{M})$. This implies that $\overline{V}$ is a vector field on $\overline{M}$ and $\overline{\varphi}_t$ is the local one parameter local group of diffeomorphisms of $\overline{M}$ generated by $\overline{V}$. □

Since $\overline{V}$ is a vector field on $\overline{M}$, it is tangent to strata of the orbit type stratification $\mathcal{S}$. For every $S \in \mathcal{S}$, the restriction of $\overline{V}$ to S is a vector field V_S on S. Morever, the reduced flow $\overline{\varphi}_t$ preserves S and it coincides on S with the flow of V_S.

We summarize the above discussion with

Theorem 3.1.4.3. (singular reduction for dynamical systems) *Let V be a vector field on M and G a symmetry group of V which acts properly on M. Then, the space $\overline{M} = M/G$ of G-orbits on M is a subcartesian space stratified by* reduced dynamical systems (S, V_S). *The orbit map $\pi : M \to \overline{M}$ maps each trajectory of the original dynamical system to a trajectory of one of the reduced dynamical systems (S, V_S).*

We make the following observation.

Proposition 3.1.4.4. *Each path component of an isotropy type of the G-action on M is invariant under the flow of a G-invariant vector field V.*

Proof. Let I_m be the maximal domain of definition of the integral curve $t \mapsto \varphi_t(m)$ of V starting at m. Then for every $t \in I_m$ we have

$$\begin{aligned} g \in G_m &\iff \Phi_g(m) = m \iff \Phi_g(\varphi_t(m)) = \varphi_t(\Phi_g(m)) = \varphi_t(m) \\ &\iff g \in G_{\varphi_t(m)}. \end{aligned}$$

In other words, $G_m = G_{\varphi_t(m)}$. Thus the flow of V leaves each isotropy type M_{G_m} invariant. Because the integral curve $t \mapsto \varphi_t(m)$ is continuous, it follows that every path component of each isotropy type is left invariant by the flow of V. □

Using lemma 2.3.1.4, we can replace the action of G on the path component N of the isotropy type M_H, where $H = G_m$ for some $m \in M$, by the free action of $N(H)/H$ on N. Since the G-action on M is proper, it is a proper action on N. Therefore the action of $N(H)$ on N is proper, since $N(H)$ is a Lie subgroup of G. Consequently, the action of $N(H)/H$ on N is free and proper.

Remark 3.1.4.5. Proposition 3.1.4.4 can be viewed as a *conservation law* which is a consequence of the non-freeness of the G-action.

Corollary 3.1.4.6. *If $\{m\}$ is an isolated path component of the isotropy type M_H, then m is an equilibrium point of V, that is, $V(m) = 0$.*

3.2 Nonholonomic singular reduction for a proper action

In this section we define the notion of symmetry of a nonholonomically constrained system and then show how to reduce it, when the action of the symmetry group is proper.

Let $\Phi : G \times D \to D$ be a smooth action of a Lie group G on a smooth manifold D. We say that G is a *symmetry of the distributional Hamiltonian system* (D, H, ϖ, h) if and only if

1. the distribution H on D is G-invariant, that is for every $(g, u) \in G \times D$

$$H_{g \cdot u} = T_u\Phi_g \, H_u.$$

2. the symplectic form ϖ on H is G-invariant, that is,

$$\varpi_{g\cdot u}(T_u\Phi_g v_u, T_u\Phi_g w_u) = \varpi_u(v_u, w_u),$$

for every $(g, u) \in G \times D$ and every $v_u, w_u \in H_u$.
3. h is a G-invariant smooth function on D.

In treating reduction of a nonholonomic system with a symmetry, we will begin with the almost Poisson bracket formulation of §7 of chapter 1. Recall that on $C^\infty(D)$ we have an *almost Poisson bracket* $\{\,,\}$ defined by $\{f, k\} = -L_{Y_f} k$ for every $f, k \in C^\infty(D)$. Here Y_f is the almost Hamiltonian vector field on D with values in H defined by $Y_f \mathbin{\lrcorner} \varpi = \partial_H f$, see §6.3 of chapter 1.

Let $C^\infty(D)^G$ be the space of smooth G-invariant functions on D.

Lemma 3.2.7. *For every* $f \in C^\infty(D)^G$, *the* almost Hamiltonian vector field Y_f *is* G-invariant.

Proof. For every $(g,u) \in G \times D$ we have

$$\begin{aligned}
&\varpi_{g\cdot u}(T_u\Phi_g\, Y_f(u), T_u\Phi_g\, v_u)\\
&\quad = (\Phi_g^*\varpi)_u(Y_f(u), v_u) = \varpi_u(Y_f(u), v_u), \text{ since } \varpi \text{ is } G\text{-invariant}\\
&\quad = \langle \mathrm{d}f(u) \mid v_u\rangle = \langle \mathrm{d}(\Phi_g^* f)(u) \mid v_u\rangle = \langle \Phi_g^*(\mathrm{d}f)(u) \mid v_u\rangle\\
&\quad = \langle \mathrm{d}f(\Phi_g(u)) \mid T_u\Phi_g v_u\rangle = \varpi_{g\cdot u}(Y_f(\Phi_g(u)), T_u\Phi_g v_u). \qquad (3)
\end{aligned}$$

But $T_u\Phi_g : T_uH_u \to T_uH_{g\cdot u}$ is bijective, since H is G-invariant. Moreover, $\varpi_{g\cdot u}$ is nondegenerate. Therefore (3) implies $Y_f(\Phi_g(u)) = T_u\Phi_g Y_f(u)$, that is, Y_f is G-invariant. □

Corollary 3.2.8. *The space* $(C^\infty(D)^G, \{\,,\,\})$ *is an* almost Poisson subalgebra *of* $(C^\infty(D), \{\,,\,\})$.

Proof. We need only show that if $f, k \in C^\infty(D)^G$, then $\{f,k\} \in C^\infty(D)^G$. By definition of the almost Poisson bracket $\{\,,\,\}$ on $C^\infty(D)$ we have

$$\{f,k\}(u) = -L_{Y_f}k(u) = -\langle \mathrm{d}k(u) \mid Y_f(u)\rangle,$$

for every $u \in D$. Now

$$\begin{aligned}
\langle \mathrm{d}(\Phi_g^*k)(u) \mid \Phi_g^*(Y_f)(u)\rangle &= \langle {T_u\Phi_g}^t(\mathrm{d}k(\Phi_g(u))) \mid (T_u\Phi_g)^{-1}Y_f(\Phi_g(u))\rangle\\
&= \langle \mathrm{d}k(\Phi_g(u)) \mid Y_f(\Phi_g(u))\rangle = -\{f,k\}(\Phi_g(u)). \qquad (4)
\end{aligned}$$

But for every $u \in D$ and every $v_u \in H_u$

$$\begin{aligned}
\varpi_u(Y_{\Phi_g^*f}(u), v_u) &= \langle \mathrm{d}(\Phi_g^*f)(u) \mid v_u\rangle = \langle \Phi_g^*(\mathrm{d}f)(u) \mid v_u\rangle\\
&= \langle \mathrm{d}f(\Phi_g(u)) \mid T_u\Phi_g\, v_u\rangle = \varpi_{g\cdot u}(Y_f(\Phi_g(u)), T_u\Phi_g v_u)\\
&= (\Phi_g^*\varpi)_u((T_u\Phi_g)^{-1}Y_f(\Phi_g(u)), v_u)\\
&= \varpi_u(\Phi_g^*Y_f(u), v_u),
\end{aligned}$$

since ϖ_u is G-invariant. Therefore, $Y_{\Phi_g^*f} = \Phi_g^*Y_f$, since ϖ_u is nondegenerate on H_u. Thus we can rewrite (4) as

$$\begin{aligned}
\{f,k\}(\Phi_g(u)) &= -\langle \mathrm{d}(\Phi_g^*k)(u) \mid Y_{\Phi_g^*f}(u)\rangle\\
&= -\langle \mathrm{d}k(u) \mid Y_f(u)\rangle, \text{ since } f,k \in C^\infty(D)^G\\
&= \{f,k\}(u).
\end{aligned}$$

□

We denote by $\overline{D}$ the space of G-orbits on D and by $\rho : D \to \overline{D}$ the orbit map. We know from the discussion in §4 of chapter 2 that $(\overline{D}, C^\infty(\overline{D}))$ is a subcartesian differential space, where

$$C^\infty(\overline{D}) = \{\overline{f} : C^\infty(\overline{D}) \to \mathbb{R} \mid \rho^*\overline{f} \in C^\infty(D)\}.$$

The map $\rho^* : C^\infty(\overline{D}) \to C^\infty(D)$ defines an associative algebra isomorphism of $C^\infty(\overline{D})$ onto $C^\infty(D)^G$. We can use this isomorphism to define in $C^\infty(\overline{D})$ the structure of almost Poisson algebra. For each $\overline{f}, \overline{k} \in C^\infty(\overline{D})$, the bracket $\{\overline{f}, \overline{k}\}$ is uniquely defined by the condition

$$\rho^*\{\overline{f}, \overline{k}\} = \{\rho^* f, \rho^* k\}.$$

Proposition 3.2.9. *$(C^\infty(\overline{D}), \{\,,\,\})$ is an almost Poisson algebra.*

Proof. We only need to verify

$$\{\overline{f}, \overline{g} \cdot \overline{k}\} = \{\overline{f}, \overline{g}\} \cdot \overline{k} + \overline{g} \cdot \{\overline{f}, \overline{k}\}. \tag{5}$$

This we do as follows.

$$\begin{aligned}\rho^*\{\overline{f}, \overline{g} \cdot \overline{k}\} &= \{\rho^*\overline{f}, \rho^*\overline{g} \cdot \rho^*\overline{k}\} \\ &= \{\rho^*\overline{f}, \rho^*\overline{g}\} \cdot \rho^*\overline{k} + \rho^*\overline{g} \cdot \{\rho^*\overline{f}, \rho^*\overline{k}\}, \\ &\qquad \text{since Leibniz' rule holds on } (C^\infty(D), \{\,,\,\}) \\ &= \rho^*(\{\overline{f}, \overline{g}\} \cdot \overline{k} + \overline{g} \cdot \{\overline{f}, \overline{k}\}).\end{aligned}$$

Equation (5) follows because ρ^* an injective map from $C^\infty(\overline{D})$ into $C^\infty(D)$, since ρ is surjective. □

We call $(C^\infty(\overline{D}), \{\,,\,\})$ the *reduced almost Poisson algebra* on $C^\infty(\overline{D})$ with almost Poisson structure $\{\,,\,\}$. Information about the geometry of the orbit space $\overline{D}$ is encoded in the structure of the reduced almost Poisson algebra $(C^\infty(\overline{D}), \{\,,\,\})$. The remainder of this section is devoted to retrieving this information and formulating it in explicitly geometric terms.

Since the action of G on D is proper, we know from the discussion of §6 of chapter 2 that $\overline{D} = D/G$ has the orbit type stratification $\mathcal{S}$. It follows from theorem 2.10.4.41 of chapter 2 that each stratum of $\mathcal{S}$ is an orbit of the family $\mathcal{X}(\overline{D}, C^\infty(\overline{D}))$ of all vector fields on $\overline{D}$. Thus, the stratification $\mathcal{S}$ is determined by derivations of $C^\infty(\overline{D})$ which generate local one-parameter groups of diffeomorphisms of $\overline{D}$. On the other hand, the notion of a diffeomorphism of $\overline{D}$ is defined in terms of its differential structure $C^\infty(\overline{D})$. Hence, we know how to obtain the orbit type stratification of $\overline{D}$ from its differential structure $C^\infty(\overline{D})$.

Since the almost Poisson bracket on $\overline{D}$ satisfies *Leibniz' rule*, it assigns to each $\overline{f} \in C^\infty(\overline{D})$ a derivation $P_{\overline{f}}$ of $C^\infty(\overline{D})$ such that

$$P_{\overline{f}}\,\overline{k} = \{\overline{f}, \overline{k}\}. \tag{6}$$

Lemma 3.2.10. *For each $\overline{f} \in C^\infty(\overline{D})$, the derivation $P_{\overline{f}}$ of $C^\infty(\overline{D})$ is a vector field on $\overline{D}$.*

Proof. The pull back $\rho^*\overline{f}$ is a function f which lies in $C^\infty(D)^G$. The almost Hamiltonian vector field Y_f associated to f generates a local one-parameter group of local diffeomorphism φ_t of D. Since f is G-invariant, Y_f is G-invariant and φ_t commutes with the action of G. Hence, φ_t gives rise to a local one-parameter group of local diffeomorphisms of $\overline{D}$ such that $\overline{\varphi}_t \circ \rho = \rho \circ \varphi_t$. For every $\overline{k} \in C^\infty(\overline{D})$ and $\overline{u} \in \overline{D}$, let $k = \rho^*\overline{k}$ and u be an element of $\rho^{-1}(\overline{u})$. Then,

$$\begin{aligned} \frac{\mathrm{d}}{\mathrm{d}t}\overline{k}(\overline{\varphi}_t(\overline{u})) &= \frac{\mathrm{d}}{\mathrm{d}t}\overline{k}(\overline{\varphi}_t(\rho(u))) = \frac{\mathrm{d}}{\mathrm{d}t}\overline{k}(\rho \circ \varphi_t(u)) = \frac{\mathrm{d}}{\mathrm{d}t}k(\varphi_t(u)) \\ &= (Y_f k)(\varphi_t(u)) = \{f, k\}(\varphi_t(u)) = \{\overline{f}, \overline{k}\}(\rho(\varphi_t(u)) \\ &= \{\overline{f}, \overline{k}\}(\overline{\varphi}_t(\rho(u)) = (Y_{\overline{f}}\overline{k})(\overline{\varphi}_t(\overline{u})). \end{aligned}$$

Hence, $t \mapsto \overline{\varphi}_t(\overline{u})$ is an integral curve of $P_{\overline{f}}$ through $\overline{u}$. This implies $\overline{\varphi}_t$ is a local one parameter group of local diffeomorphisms of $\overline{D}$ generated by $Y_{\overline{f}}$. Hence $P_{\overline{f}}$ is a vector field on $\overline{D}$. □

The vector field $P_{\bar{f}}$ on $\bar{D}$ is called the *almost Poisson vector field* corresponding to $\bar{f} \in C^\infty(\overline{D})$

Proposition 3.2.11. *Let S be a stratum of the orbit type stratification $\mathcal{S}$ of $\overline{D}$. The ring $C^\infty(S)$ of smooth functions on S inherits an almost Poisson algebra structure from the reduced almost Poisson algebra $(C^\infty(\overline{D}), \{\,,\,\})$.*

Since a stratum S of $\mathcal{S}$ is an orbit of the familly $\mathcal{X}(\overline{D}, C^\infty(\overline{D}))$ of all vector fields on $\overline{D}$, it follows from lemma 3.2.10 that, for every $\overline{f} \in C^\infty(\overline{D})$, the vector field $P_{\overline{f}}$ is tangent to S. Hence, for every $\overline{u}$ in S and $\overline{k} \in C^\infty(\overline{D})$,

$$\{\overline{f}, \overline{k}\}(\overline{u}) = (P_{\overline{f}}\overline{k})(\overline{u})$$

depends on $\overline{k}$ only through the restriction of $\overline{k}$ to S. On the other hand, $\{\overline{f}, \overline{k}\}(\overline{u}) = -\{\overline{k}, \overline{f}\}(\overline{u})$ and by the same argument as above it depends on $\overline{f}$ only through the restriction of $\overline{f}$ to S. Hence, the restriction to S of the almost Poisson bracket on $\overline{D}$ is a well defined bracket on the space of restrictions to S of functions in $C^\infty(\overline{D})$.

By lemma 2.4.1.11 of chapter 2 the ring $C^\infty(S)$ of smooth functions on S given by the manifold structure of S is generated by restrictions to S of smooth functions on $\overline{D}$. That is, for each $f_S \in C^\infty(S)$ and each point $\overline{u} \in S$, there is a neighborhood U of $\overline{u}$ in S and a function $\overline{f} \in C^\infty(\overline{D})$ such that $f_S \mid U = \overline{f} \mid U$. This implies that the bracket defined on the space of restrictions to S of smooth functions on $\overline{D}$ extends to a bracket defined on $C^\infty(S)$ which we denote also by $\{\ ,\ \}$.

Clearly, the bracket $\{\ ,\ \}$ on $C^\infty(S)$ is bilinear, skew symmetric, and satisfies Leibniz' rule. Hence, $(C^\infty(S), \{\ ,\ \})$ is an almost Poisson algebra. □

For each $f_S \in C^\infty(S)$, we denote by P_{f_S} the almost Poisson vector field of f_S given by $P_{f_S} k_S = \{f_S, k_S\}$ for every $k_S \in C^\infty(S)$. Let H_S be a generalized distribution spanned by the family

$$\mathcal{P}_S = \{P_{f_S} \mid f_S \in C^\infty(S)\}$$

of Poisson vector fields on S. Since the family $\mathcal{P}_S$ of almost Poisson vector fields on S is closed under addition and multiplication by numbers, it follows that

$$H_S = \{P_{f_S}(s) \mid s \in S \text{ and } f_S \in C^\infty(S)\}. \tag{7}$$

Proposition 3.2.12. *The almost Poisson bracket on $C^\infty(S)$ gives rise to a unique antisymmetric bilinear form ϖ_S on H_S such that*

$$\varpi_S(P_{f_S}, P_{k_S}) = \{k_S, f_S\} \tag{8}$$

for every $f_S, k_S \in C^\infty(S)$.

Proof. For each $s \in S$,

$$\varpi_S(P_{f_S}(s), P_{k_S}(s)) = \{k_S, f_S\}(s). \tag{9}$$

Suppose that f_S and f'_S in $C^\infty(S)$ are such that $\check{P}_{f_S}(s) = P_{f'_S}(s)$. Let $\overline{f}$ and $\overline{f}'$ be functions in $C^\infty(\overline{D})$ such that their restrictions to S are f_S and f'_S, respectively. Then P_{f_S} and $P_{f'_S}$ are the restriction to S of $P_{\overline{f}}$ and $P_{\overline{f}'}$, respectively. Moreover, if $\overline{k} \in C^\infty(\overline{D})$ is an extension of k_S, then

$$\begin{aligned}\{k_S, f_S\}(s) &= \{\overline{k}, \overline{f}\}(s) = -(P_{\overline{f}}\overline{k})(s) = -P_{\overline{f}}(s)\overline{k} = -P_{f_S}(s)k = -P_{f'_S}(s)k \\ &= -P_{\overline{f}'}(s)\overline{k} = -(P_{\overline{f}'}\overline{k})(s) = \{\overline{k}, \overline{f}'\}(s) = \{k_S, f'_S\}(s).\end{aligned}$$

Hence, the right hand side of equation (9) depends on f_S only through $P_{f_S}(s)$. Similarly, one can show that the right hand side of equation (9) depends on k_S only through $P_{k_S}(s)$. Hence, $\varpi_S(P_{f_S}(s), P_{k_S}(s))$ is well defined by equation (9).

Bilinearity and antisymmetry of $\varpi_S(s)$ follow from bilinearity and antisymmetry of the almost Poisson bracket. □

Proposition 3.2.13. *The antisymmetric bilinear form ϖ_S on H_S is nondegenerate, so that (H_S, ϖ_S) is a symplectic distribution on S.*

Proof. See the argument leading to proposition 1.7.1.26 of chapter 1. □

For each $f_S \in C^\infty(S)$, we denote by Y_{f_S} the almost Hamiltonian vector field on S defined in terms of the symplectic distribution (H_S, ϖ_S). In other words, Y_S is the unique vector field on S with values in H_S such that $Y_S \lrcorner \varpi_S = \partial_{H_S} f_S$, where ∂_{H_S} denotes the restriction of d to H_S. Equation (8) gives

$$\varpi_S(P_{f_S}, P_{k_S}) = \{k_S, f_S\} = P_{k_S} f_S = \langle df_S \mid P_{k_S}\rangle.$$

Since the vector fields P_{k_S}, for $k \in C^\infty(S)$, span H_S and ϖ_S is nondegenerate, it follows that $P_{f_S} \lrcorner \varpi_S = \partial_{H_S} f_S$. Thus, we have proved

Proposition 3.2.14. *For each $f_S \in C^\infty(S)$, the almost Poisson vector field P_{f_S} coincides with the almost Hamiltonian vector field $P_{f_S} = Y_{f_S}$.*

By definition the Hamiltonian h of the original distributional Hamiltonian system (D, H, ϖ, h) is G-invariant. Hence, there exists a unique $\overline{h} \in C^\infty(\overline{D})$ such that $h = \rho^*\overline{h}$. The pull back of $\overline{h}$ to a stratum S of $\overline{D}$ is a function $h_S \in C^\infty(S)$. Thus, each stratum S of $\overline{D}$ carries the structure (S, H_S, ϖ_S, h_S) of a distributional Hamiltonian system.

Motions of the distributional Hamiltonian system (D, H, ϖ, h) are given by integral curves of the almost Hamiltonian vector field Y_h of h. Let $c : (a, b) \to D$ be an integral curve of Y_h. It follows from the discussion in §7 of chapter 2 that $\overline{c} = \rho \circ c : (a, b) \to \overline{D}$ is an integral curve of the vector field $\overline{Y}_h = \rho_* Y_h$. This implies that $\overline{c} : (a, b) \to \overline{D}$ is contained in a stratum S of $\overline{D}$. Moreover, for each $\overline{f} \in C^\infty(\overline{D})$,

$$\rho^*(\overline{Y}_h \overline{f}) = Y_h \rho^* \overline{f} = \varpi(Y_{\rho^*\overline{f}}, Y_h) = \{h, \rho^*\overline{f}\} = \rho^*\{\overline{h}, \overline{f}\} = \rho^* P_{\overline{h}} \overline{f}.$$

This implies that $\overline{Y}_h = P_{\overline{h}}$. Since the restriction of $P_{\overline{h}}$ to the stratum S of $\overline{D}$ is equal to the almost Hamiltonian vector field Y_{h_S} of the restriction h_S of $\overline{h}$ to S. This implies that, the curve $\overline{c} = \rho \circ c$ is an integral curve of the almost Hamiltonian vector field Y_{h_S}. We summarize the above discussion in

Theorem 3.2.15. (singular nonholonomic reduction) *Given a distributional Hamiltonian system (D, H, ϖ, h) with a symmetry group G which acts properly on D, the space $\overline{D} = D/G$ of G-orbits on D is a subcartesian differential space stratified by* reduced distributional Hamiltonian systems (S, H_S, ϖ_S, h_S). *The orbit map $\rho : D \to \overline{D}$ sends each trajectory of the original distributional Hamiltonian system (D, H, ϖ, h) to a trajectory of one of the reduced distributional Hamiltonian systems (S, H_S, ϖ_S, h_S).*

Remark 3.2.16. *Singular Hamiltonian reduction.* Singular Hamiltonian reduction is a special case of singular nonholonomic reduction. A Hamiltonian system (P, ω, h) can be considered as an integrable distributional Hamiltonian system (P, H, ϖ, h), where $H = TP$ and $\varpi = \omega$. If G is a symmetry group of (P, ω, h) acting properly on P, then it is a symmetry group of the distributional Hamltonian system. The orbit space $\overline{P} = P/G$ is a subcartesian space stratified by reduced distributional Hamiltonian systems (S, H_S, ϖ_S, h_S). In this case, for each stratum S, the distribution H_S is involutive and, for each integral manifold M of S, the form ϖ_S restricted to TM gives a symplectic form ω_M on M. Thus, each stratum S of $\overline{P}$ is singularly foliated by reduced Hamiltonian systems (M, ω_M, h_M), where h_M is the pull back of $\overline{h}$ to M.

Remark 3.2.17. *Singular symplectic reduction.* Let G be a Lie group, which acts properly and symplectically on the smooth symplectic manifold (M, ω). The algebra of smooth functions on (M, ω) is $C^\infty(M)$. We now describe the process of *singular symplectic reduction.*

We start by noting that the space $P = M/G$ of G orbits on M is a differential space with differential structure $C^\infty(P) = \{h : P \to \mathbb{R} \bigm| \pi^* h \in C^\infty(M)\}$. Here $\pi : M \to P$ is the G-orbit map. Since the action of G on M preserves the symplectic form ω on M, the induced action on $C^\infty(M)$ preserves the Poisson bracket. Hence $C^\infty(M)^G$ is a Poisson subagebra of $C^\infty(M)$. Using the ring isomorphism $\pi^* : C^\infty(P) \to C^\infty(M)^G$ we can pull back the Poisson bracket on $C^\infty(M)^G$ to a Poisson bracket on $C^\infty(P)$, namely, for each $h_1, h_2 \in C^\infty(P)$ their Poisson bracket $\{h_1, h_2\}$ is given by $\pi^*\{h_1, h_2\} = \{\pi^* h_1, \pi^* h_2\}$. For each $h \in C^\infty(P)$ define the Poisson derivation Y_h corresponding to h by $Y_h(h') = \{h, h'\}$ for every $h' \in C^\infty(P)$. Each Poisson derivation is a vector field on P in the sense that its flow at a fixed time is a local one parameter group of local diffeomorphisms of P.

Let $\mathfrak{P}(P) \subseteq \mathcal{X}(P)$ be the collection of all Poisson derivations of $C^\infty(P)$. Here $\mathcal{X}(P)$ is the set of all vector fields on P.

Recall that P is a stratified space $\bigcup_{p \in P} P_p$ where the stratum P_p through the point p of P is an orbit of the family $\mathcal{X}(P)$ of all vector fields on P. The space $C^\infty(P_p)$ of smooth functions on P_p is generated by the restriction to P_p of smooth functions on P. The restriction to P_p of vector fields on P are tangent to P_p, because P_p is an orbit of $\mathcal{X}(P)$. Hence $C^\infty(P_p)$ inherits the structure of a Poisson algebra from $C^\infty(P)$. Thus P_p is a Poisson manifold. The space $\mathfrak{P}(P_p)$ of all Poisson vector fields on P_p coincides with the space of restrictions to P_p of Poisson vector fields on P. The orbit through $p \in P_p$ of the family $\mathfrak{P}(P_p)$ of Poisson vector fields on P_p is the symplectic leaf $(P_p^{\mathrm{symp}}, \omega_p)$ through p. Using the Poisson structure on $C^\infty(P)$, the symplectic form ω_p on P_p^{symp} is defined as follows: $\omega_p(Y_{h_1}(p), Y_{h_2}(p)) = \{h_1, h_2\}(p)$, for every $h_1, h_2 \in C^\infty(P)$. See [107]. Note that the symplectic form ω_p on the symplectic leaf P_p^{symp} is completely determined by the Poisson structure on $C^\infty(P)$.

Let R be a subcartesian differential space such that its differential structure $C^\infty(R)$ is a Poisson algebra. Then the orbits of the family $\mathcal{X}(R)$ of vector fields on R are Poisson manifolds, which are singularly foliated by symplectic manifolds. Suppose that $\varphi : (P, C^\infty(P)) \to (R, C^\infty(R))$ is a diffeomorphism of differential spaces such that $\varphi^* : C^\infty(R) \to C^\infty(P)$ is an isomorphism of Poisson algebras. Then for each $p \in P$ the restriction of φ to the symplectic leaf P_p^{symp} through p in P is a symplectomorphism of $(P_p^{\mathrm{symp}}, \omega_p)$ onto the symplectic leaf of R through $r = \varphi(p)$.

Remark 3.2.18. *Singular reduction by stages.* Suppose that the Lie group G acts properly and symplectically on a smooth symplectic manifold (M, ω).

If G has a normal subgroup H, then we may first reduce by the action of H, obtaining the space $Q = M/H$ of H-orbits on M. The space Q is a stratified subcartesian differential space with differential structure $C^\infty(Q) = \{h_Q : Q \to \mathbb{R} \,|\, \rho^* h_Q \in C^\infty(M)\}$, where $\rho : M \to Q$ is the H-orbit map. $C^\infty(Q)$ has a Poisson algebra structure and for each $q \in Q$ the stratum Q_q through q is a smooth Poisson manifold, which is singularly foliated by symplectic leaves.

Now consider the action of the quotient group $L = G/H$ on $Q = M/H$. Since M is locally compact and the G-action is proper, it follows that the action of L on Q is proper. In addition the space $R = Q/L$ of L-orbits on Q is a locally compact, subcartesian differential space with respect to the differential structure $C^\infty(R) = \{h_R : R \to \mathbb{R} \,\big|\, \sigma^* h_R \in C^\infty(Q)\}$. Here $\sigma : Q \to R$ is the L-orbit map. See [72]. The definition of $C^\infty(R)$ implies that $C^\infty(R) = \{h_R : R \to \mathbb{R} \,\big|\, (\sigma \circ \rho)^* \in C^\infty(M)^G\}$. Hence $(\sigma \circ \rho)^* : C^\infty(R) \to C^\infty(M)^G$ is a ring homomorphism, which we can use to pull back the Poisson algebra structure on $C^\infty(M)^G$ to $C^\infty(R)$.

Proposition 3.2.19. *There is a unique isomorphism $\varphi : (P, C^\infty(P)) \to (R, C^\infty(R))$ of differential spaces such that $\varphi \circ \pi = \sigma \circ \rho$. Moreover, $\varphi^* : C^\infty(R) \to C^\infty(P)$ is an isomorphism of Poisson algebras.*

Proof. Both the maps $\pi : M \to P$ and $\sigma \circ \rho : M \to R$ are surjective. Let $x_0 \in M$, let $p_0 = \pi(x_0)$, $q_0 = \rho(x_0)$, and $r_0 = \sigma(q_0)$ be points on P, Q, and R, respectively. The fiber $\pi^{-1}(p)$ is the orbit Gx_0 of G through x_0. The action of L on Q associates to each element $gH \in G/H$ in L and each $q \in Q$ the point $(gH)q = \rho(gx)$ for any $x \in \rho^{-1}(q)$. Hence

$$(\sigma \circ \rho)^{-1}(r_0) = \rho^{-1}\big(\sigma^{-1}(r_0)\big) = \rho^{-1}(Lq_0) = \rho^{-1}(\{\rho(gx_0) \,\big|\, g \in G\}) = Gx_0.$$

Consequently, $\pi^{-1}(\pi(x_0)) = (\sigma \circ \rho)^{-1}(\sigma(\rho(x_0)))$ for every $x_0 \in M$. Hence there is a unique bijection $\varphi : P \to R$ such that $\varphi(\pi(x)) = \sigma(\rho(x))$ for every $x \in M$.

To show that φ is smooth, note that $\varphi \circ \pi = \sigma \circ \rho$ implies that $(\sigma \circ \rho)^* h_R \in C^\infty(M)^G$ for every $h_R \in C^\infty(R)$. Therefore there is a unique $h_P \in C^\infty(P)$ such that $(\sigma \circ \rho)^* = \pi^* h_P$. But $\varphi \circ \pi = \sigma \circ \rho$ implies that $h_P = \varphi^* h_R$. Moreover, every $h_P \in C^\infty(P)$ is of the form $\pi^* f$ for some $f \in C^\infty(M)^G$. On the other hand $f = (\sigma \circ \rho)^* h_R$ for some $h_R \in C^\infty(R)$. Hence φ^* maps $C^\infty(R)$ onto $C^\infty(P)$, which implies that $\varphi : (P, C^\infty(P)) \to (R, C^\infty(R))$ is a diffeomorphism.

The Poisson algebra structures on $C^\infty(P)$ and $C^\infty(R)$ are induced by the Poisson algebra structure on $C^\infty(M)^G$ and the ring homomorphisms $\pi^* : C^\infty(P) \to C^\infty(M)^G$ and $(\sigma \circ \rho)^* : C^\infty(R) \to C^\infty(M)^G$. This implies that $\varphi^* : C^\infty(R) \to C^\infty(P)$ is an isomorphism of Poisson algebras. □

Taking into account remark 3.2.18 on singular symplectic reduction we have proved

Corollary 3.2.20. *For every $x \in M$, the restriction of the map φ to the symplectic leaf $P^{\mathrm{symp}}_{\pi(x)}$ of the stratum $P_{\pi(x)}$ of P is a symplectomorphism of $(P^{\mathrm{symp}}_{\pi(x)}, \omega_{\pi(x)})$ onto the symplectic leaf through $r = \varphi(\pi(x)) = \sigma(\rho(x))$ of the orbit through r of the family $\mathcal{X}(R)$ of all vector fields on R.*

The above results can be informally stated that for a proper action of a Lie group on a symplectic manifold that Poisson reduction by stages proposition 3.2.19 implies symplectic reduction by stages corollary 3.2.20.

Corollary 3.2.21. *Suppose that the reduced space has one stratum, that is, $S = \overline{D}$, and that the reduced distributional system $(\overline{D}, H_{\overline{D}}, \varpi_{\overline{D}}, h_{\overline{D}})$ is simple, that is, $\overline{D}$ is the only accessible set of $H_{\overline{D}}$. Then the only Casimir functions of the reduced Poisson algebra $(C^{\infty}(\overline{D}), \{\ ,\ \})$ are the constant functions.*

Remark 3.2.22. The conclusion of corollary 3.2.21 is in contrast with G-invariant Hamiltonian systems, where the momentum functions associated to the symmetries are nonconstant Casimirs for the reduced Poisson algebra.

3.3 Nonholonomic reduction for a free and proper action

Suppose that $\Phi : G \times D \to D$ is a free and proper action, which is a symmetry of the distributional Hamiltonian system (D, H, ϖ, h). In this section we relate the construction of the reduced distributional Hamiltonian system given above to regular nonholonomic reduction.

Because the G-action is free and proper, the space $\overline{D}$ of G-orbits on D is a smooth manifold and the G-orbit map $\rho : D \to \overline{D}$ exhibits D as a principal G-bundle. In other words, the orbit type stratification $\mathcal{S}$ of $\overline{D}$ has only one stratum $S = \overline{D}$. Hence, the reduced distributional system (S, H_S, ϖ_S, h_S) will be denoted by $(\overline{D}, \overline{H}, \overline{\varpi}, \overline{h})$.

For each $u \in D$ let

$$U_u = \operatorname{span}\{Y_f(u) \in H_u \,\big|\, f \in C^{\infty}(D)^G\}. \tag{10}$$

From lemma 1.7.1.24 of chapter 1 it follows that $u \mapsto U_u$ is a *generalized distribution* on D.

Proposition 3.3.23. *Let $(\ker T_u\rho)^{\circ} = \{\alpha_u \in T_u^*D \,\big|\, \alpha_u | \ker T_u\rho = 0\}$. Then $u \mapsto (\ker T_u\rho)^{\circ}$ is a distribution on D and*

$$(\ker T_u\rho)^{\circ} = \operatorname{span}\{\mathrm{d}f(u) \in T_u^*D \,\big|\, f \in C^{\infty}(D)^G\}. \tag{11}$$

Proof. Since $\rho : D \to \overline{D}$ is a principal G-bundle, $\dim \ker T_u\rho = \dim \mathfrak{g}$, where $\mathfrak{g}$ is the Lie algebra of G. Therefore $\dim(\ker T_u\rho)^{\circ} = \dim D - \dim \mathfrak{g}$. Clearly, $u \mapsto (\ker T_u\rho)^{\circ}$ is smooth and therefore defines a distribution on D.

Let $f \in C^\infty(D)^G$. For every $u \in D$ and every $\xi \in \mathfrak{g}$ we have

$$\begin{aligned}\langle \mathrm{d}f(u) \mid X^\xi(u)\rangle &= (L_{X^\xi} f)(u) = \frac{\mathrm{d}}{\mathrm{d}t}\Big|_{t=0} f(\Phi_u(\exp t\xi)) \\ &= \frac{\mathrm{d}}{\mathrm{d}t}\Big|_{t=0} f(u), \text{ since } f \in C^\infty(D)^G \\ &= 0.\end{aligned}$$

Therefore $\mathrm{d}f(u)|\ker T_u\rho = 0$, since $\ker T_u\rho = \mathrm{span}\{X^\xi(u) \,\big|\, \xi \in \mathfrak{g}\}$. So $\mathrm{d}f(u) \in (\ker T_u\rho)^\circ$. Consequently,

$$\mathrm{span}\{\mathrm{d}f(u) \in T_u^*D \,\big|\, f \in C^\infty(D)\} \subseteq (\ker T_u\rho)^\circ.$$

Let $\overline{u} = \rho(u)$. Since $\overline{D}$ is a smooth manifold,

$$T_{\overline{u}}^*\overline{D} = \mathrm{span}\{\mathrm{d}\overline{f}(\overline{u}) \,\big|\, \overline{f} \in C^\infty(\overline{D})\}.$$

Because $T_u\rho : T_uD \to T_{\overline{u}}\overline{D}$ is surjective, its transpose $(T_u\rho)^t : T_{\overline{u}}^*\overline{D} \to T_u^*D$ is injective. But then

$$\begin{aligned}(T_u\rho)^t(T_{\overline{u}}^*\overline{D}) &= \mathrm{span}\{(T_u\rho)^t(\mathrm{d}\overline{f}(\overline{u})) \,\big|\, \overline{f} \in C^\infty(\overline{D})\} \\ &= \mathrm{span}\{\mathrm{d}(\rho^*\overline{f})(u) \,\big|\, \overline{f} \in C^\infty(\overline{D})\} \\ &= \mathrm{span}\{\mathrm{d}f(u) \,\big|\, f \in C^\infty(D)^G\}. \qquad (12)\end{aligned}$$

Now

$$\begin{aligned}\dim(\ker T_u\rho)^\circ &= \dim T_u^*D - \dim \mathfrak{g} \\ &= \dim T_{\overline{u}}^*\overline{D} \le \dim \mathrm{span}\{\mathrm{d}f(u) \,\big|\, f \in C^\infty(D)^G\}.\end{aligned}$$

The last inequality above follows from (12). Therefore (11) holds. □

Corollary 3.3.24. *The map $u \mapsto U_u$ (10) is a distribution on D.*

Proof. Since the linear map $\varpi_u^\flat$ is invertible we have

$$\begin{aligned}\dim D - \dim \mathfrak{g} &= \dim \varpi_u^\flat\big(\mathrm{span}\{\mathrm{d}f(u) \in T_u^*D \,\big|\, f \in C^\infty(D)^G\}\big) \\ &= \dim \mathrm{span}\{\varpi_u^\flat(\mathrm{d}f(u))|H_u \,\big|\, f \in C^\infty(D)^G\} \\ &= \dim \mathrm{span}\{Y_f(u) \in H_u \,\big|\, f \in C^\infty(D)^G\} \\ &= \dim U_u.\end{aligned}$$

Therefore $u \mapsto U_u$ is a distribution. □

An alternative characterization of U is given in the following

Proposition 3.3.25. *For each $u \in D$ we have*

$$U_u = \{w_u \in H_u \,\big|\, \varpi_u(v_u, w_u) = 0 \text{ for every } v_u \in H_u \cap \ker T_u\rho\}. \qquad (13)$$

Proof. We compute.

$$\begin{aligned}\varpi_u^\flat(U_u) &= \varpi_u^\flat(\mathrm{span}\{Y_f(u)\in H_u \,\big|\, f\in C^\infty(D)^G\}), \text{ by definition of } U_u\\ &= \mathrm{span}\{\mathrm{d}f(u)|H_u \,\big|\, f\in C^\infty(D)^G\},\\ &\qquad\qquad \text{by definition of almost Hamiltonian vector field.}\\ &= \mathrm{span}\{\alpha_u\in H_u^* \,\big|\, \alpha_u|\ker T_u\rho = 0\}, \text{ by (11)}\\ &= \varpi_u^\flat(\mathrm{span}\{w_u\in H_u \,\big|\, \varpi_u(w_u,v_u)=0, \text{ for every } v_u\in H_u\cap\ker T_u\rho\}),\end{aligned}$$

since $\alpha_u = \varpi_u^\flat(w_u)$ for a unique $w_u \in H_u$. Because $\varpi^\flat$ is invertible, equation (13) holds. □

Let $\overline{H} = T\rho(U)$. Since, for each $f \in C^\infty(D)$, the vector field Y_f is G-invariant, it follows that it projects to a vector field $\overline{Y_f}$ on $\overline{D}$. Hence, $\overline{H}$ is a generalized distribution on $\overline{D}$. In §3.3 of chapter 5 we show that the *reduced generalized distribution* for Carathéodory's sleigh is *not* a distribution.

For each $u \in D$ the linear map $T_u\rho|U_u : U_u \subseteq H_u \to \overline{H}_{\overline{u}}$ is surjective with kernel equal to $U_u \cap (H_u \cap \ker T_u\rho) = U_u \cap U_u^{\varpi_u}$. Therefore $T_u\rho|U_u$ induces a bijective linear map $\overline{\rho}_u : U_u/(U_u \cap U_u^{\varpi_u}) \to \overline{H}_{\overline{u}}$. The symplectic form ϖ_u on H_u induces a nondegenerate skew symmetric bilinear form $\widehat{\varpi}_u$ on $U_u/(U_u \cap U_u^{\varpi_u})$ defined by

$$\widehat{\varpi}_u(v_u + U_u \cap U_u^{\varpi_u}, v_u' + U_u \cap U_u^{\varpi_u}) = \varpi_u(v_u, v_u'),$$

for every $v_u, v_u' \in U_u$. To see this suppose that $0 = \widehat{\varpi}_u(v_u + U_u \cap U_u^{\varpi_u}, v_u' + U_u \cap U_u^{\varpi_u})$ for every $v_u' \in U_u$. Then $0 = \varpi_u(v_u, v_u')$ for every $v_u' \in U_u$. This implies that $v_u \in U_u^{\varpi_u}$. But $v_u \in U_u$ by hypothesis. So $v_u \in U_u \cap U_u^{\varpi_u}$. Therefore $v_u + U_u \cap U_u^{\varpi_u} = 0 \in U_u/(U_u \cap U_u^{\varpi_u})$. In other words, $\widehat{\varpi}_u$ is nondegenerate. On $\overline{H}_{\overline{u}}$ define $\overline{\varpi}_{\overline{u}}$ by $\overline{\varpi}_{\overline{u}} = ((\overline{\rho}_u)^{-1})^*\widehat{\varpi}_u$. Since $\overline{\rho}_u$ and $\widehat{\varpi}_u^\flat$ are invertible, so is $\overline{\varpi}_{\overline{u}}^\flat$. Therefore $\overline{\varpi}_{\overline{u}}$ is nondegenerate. This proves

Proposition 3.3.26. *For each $\overline{u} \in \overline{D}$ the skew symmetric bilinear form $\overline{\varpi}_{\overline{u}}$ on $\overline{H}_{\overline{u}}$ is nondegenerate.*

For each $\overline{f} \in C^\infty(\overline{D})$, we denote by $Y_{\overline{f}}$ the *almost Hamiltonian vector field* corresponding to the symplectic distribution $(\overline{H}, \overline{\varpi})$ on $\overline{D}$. In other words, $Y_{\overline{f}}$ is the unique vector field on $\overline{D}$ with values in $\overline{H}$ such that

$$Y_{\overline{f}} \,\lrcorner\, \overline{\varpi} = \partial_{\overline{H}}\overline{f}.$$

Proposition 3.3.27. *For each $f \in C^\infty(D)^G$ the push forward $\overline{Y_f}$ of the almost Hamiltonian vector field Y_f on D by the orbit map $\rho : D \to \overline{D}$ is equal to the almost Hamiltonian vector field $Y_{\overline{f}}$ where $\rho^*\overline{f} = f$.*

Proof. Since $f = \rho^* \overline{f}$ is G-invariant, it follows that Y_f has values in the distribution U on D defined by equation (10). Hence, its push forward $\overline{Y}_f$ to $\overline{D}$ has values in $\overline{H} = T\rho(U)$. Given $u \in D$ and $w \in U_u$, let $\overline{u} = \rho(u)$ and $\overline{w} = T\rho(w) \in \overline{H}_{\overline{u}}$. Then,

$$\langle Y_{\overline{f}}(\overline{u}) \mathbin{\lrcorner} \overline{\varpi} \mid \overline{w} \rangle = \langle \partial_{\overline{H}} \overline{f} \mid \overline{w} \rangle = \langle \mathrm{d}\overline{f} \mid \overline{w} \rangle = \langle \mathrm{d}f \mid w \rangle.$$

On the other hand, $Y_f \mathbin{\lrcorner} \varpi = \partial_H f$ implies

$$\begin{aligned}
\langle \mathrm{d}f \mid w \rangle &= \langle \partial_H f \mid w \rangle = \langle Y_f(u) \mathbin{\lrcorner} \varpi \mid w \rangle \\
&= \varpi(Y_f(u), w) = \overline{\varpi}(T\rho(Y_f(u)) \mid T\rho(w)) = \overline{\varpi}(\overline{Y_f}(\overline{u}), \overline{w}) \\
&= \langle \overline{Y_f}(\overline{u}) \mathbin{\lrcorner} \overline{\varpi} \mid \overline{w} \rangle.
\end{aligned}$$

Hence, $Y_{\overline{f}}(\overline{u}) \mathbin{\lrcorner} \overline{\varpi} = \overline{Y_f}(\overline{u}) \mathbin{\lrcorner} \overline{\varpi}$ for every $\overline{u} \in \overline{D}$. Since, $\overline{\varpi}$ is non-degenerate, it follows that $Y_{\overline{f}} = \overline{Y_f}$. □

The Hamiltonian h of our distributional Hamiltonian system (D, H, ϖ, h) is G-invariant and it pushes forward to a function $\overline{h} \in C^\infty(\overline{D})$. Hence, we obtain a *reduced distributional Hamiltonian system* $(\overline{D}, \overline{H}, \overline{\varpi}, \overline{h})$.

On the other hand, singular reduction applied to our case leads to a distributional Hamiltonian system (S, H_S, ϖ_S, h_S) on each stratum of $\overline{D}$. Since the action of G on D is free and proper, the orbit space $\overline{D}$ is a manifold. It means that the stratification $\mathcal{S}$ of $\overline{D}$ by orbit type has only one stratum $S = \overline{D}$. We are going to show that $H_S = \overline{H}$, $\varpi_S = \overline{\varpi}$ and $h_S = \overline{h}$.

Lemma 3.3.28. *For each $f \in C^\infty(D)^G$ the push forward $\overline{Y_f}$ of the almost Hamiltonian vector field Y_f on D by the orbit map $\rho : D \to \overline{D}$ is equal to the almost Poisson vector field $P_{\overline{f}}$ where $\rho^* \overline{f} = f$.*

Proof. We compute. For every $u \in D$ and every $\overline{k} \in C^\infty(\overline{D})$ we have

$$\begin{aligned}
\rho^*(L_{\overline{Y_f}} \overline{k})(u) &= \langle \mathrm{d}\overline{k}(\overline{u}) \mid \overline{Y}_f(\overline{u}) \rangle, \text{ where } \overline{u} = \rho(u) \\
&= \langle \mathrm{d}\overline{k}(\overline{u}) \mid T_u \rho Y_f(u) \rangle, \\
&= \langle (T_u \rho)^t \, \mathrm{d}\overline{k}(\rho(u)) \mid Y_f(u) \rangle = \langle \rho^*(\mathrm{d}\overline{k})(u) \mid Y_f(u) \rangle \\
&= \langle \mathrm{d}(\rho^* \overline{k})(u) \mid Y_f(u) \rangle = \langle \mathrm{d}k(u) \mid Y_f(u) \rangle, \text{ where } \rho^* \overline{k} = k \\
&= \{f, k\}(u) = \{\rho^* \overline{f}, \rho^* \overline{k}\}(u) = \rho^* \{\overline{f}, \overline{k}\}(u) = \rho^*(L_{P_{\overline{f}}} \overline{k})(u).
\end{aligned}$$

But $\rho^* : C^\infty(\overline{D}) \to C^\infty(D)^G : \overline{f} \mapsto f \circ \rho$ is injective, since $\rho : D \to \overline{D}$ is surjective. Therefore $L_{P_{\overline{f}}} \overline{k} = L_{Y_{\overline{f}}} \overline{k}$ for every $\overline{k} \in C^\infty(\overline{D})$, that is, the vector fields $P_{\overline{f}}$ and $Y_{\overline{f}}$ on $\overline{D}$ are equal. □

Equation (7) yields

$$\begin{aligned} H_S &= \{P_{f_S}(s) \mid s \in S \text{ and } f_S \in C^\infty(S)\} \\ &= \{P_{\overline{f}}(\overline{u}) \mid \overline{u} \in \overline{D} \text{ and } \overline{f} \in C^\infty(\overline{D})\} \quad \text{since } S = \overline{D} \\ &= \{\overline{Y_f}(\overline{u}) \mid \overline{u} \in \overline{D} \text{ and } f \in C^\infty(D)\} \quad \text{by proposition 3.2.12} \\ &= \{T\rho(Y_f(u)) \mid u \in D \text{ and } f \in C^\infty(D)^G\} \\ &= T\rho(U_u) = \overline{H}_u. \end{aligned}$$

Moreover, since $S = \overline{D}$, we can write $\overline{f}$ for f_S in $C^\infty(S)$. Proposition 3.2.11 and proposition 3.2.12 give $\overline{\varpi}_S(P_{f_S}, P_{k_S}) = \overline{\varpi}_S(P_{\overline{f}}, P_{\overline{k}}) = \overline{\varpi}_S(Y_{\overline{f}}, Y_{\overline{k}})$. On the other hand, equation (8) yields $\varpi_S(P_{f_S}, P_{k_S}) = \{k_S, f_S\} = \{\overline{k}, \overline{f}\} = P_{\overline{k}}\overline{f} = Y_{\overline{k}}\overline{f} = \overline{\varpi}(Y_{\overline{f}}, Y_{\overline{k}})$. Hence, $\varpi_S = \overline{\varpi}$.

We summarize our results with

Theorem 3.3.29. (regular nonholonomic reduction) *Given a distributional Hamiltonian system (D, H, ϖ, h) which has a symmetry group G that acts freely and properly on D, there is a* reduced distributional Hamiltonian system *$(\overline{D}, \overline{H}, \overline{\varpi}, \overline{h})$ where $(\overline{H}, \overline{\varpi})$ is a generalized symplectic distribution on the G-orbit space $\overline{D}$ and $\rho^*\overline{h} = h$. Here $\rho : D \to \overline{D}$ is the G-orbit map. The reduced distributional Hamiltonian system $(\overline{D}, \overline{H}, \overline{\varpi}, \overline{h})$ coincides with the reduced distributional Hamiltonian system obtained by singular reduction.*

3.4 Chaplygin systems

Let $\widetilde{\Phi}$ be a smooth free and proper action of a Lie group G on a smooth manifold Q with orbit map $\pi : Q \to Q/G = R$.

Suppose that we have a nonholonomically constrained dynamical system (D, ℓ) with constraint distribution D on Q and Lagrangian ℓ on TQ. Suppose that this system has the following properties.

1. The G-action $\Phi : G \times TQ \to TQ : (g, (q, \dot{q})) \mapsto T_q\widetilde{\Phi}_g\dot{q}$ leaves the distribution D invariant.
2. For each $q \in Q$, we have $T_qQ = D_q \oplus \mathfrak{g} \cdot q$, where $\mathfrak{g} \cdot q = \operatorname{span}\{X_\xi(q) = \frac{\mathrm{d}}{\mathrm{d}t}\big|_{t=0} \widetilde{\Phi}_q(\exp t\,\xi) \in T_qQ \,\big|\, \xi \in \mathfrak{g}\}$.
3. The Lagrangian $\ell : TQ \to \mathbb{R}$ is invariant under the G-action $\widetilde{\Phi}$.

We call (D, ℓ, G) a *Chaplygin system*.

Because the G-action $\widetilde{\Phi}$ is free and proper, the G-orbit map $\pi : Q \to R$ defines a principal G-bundle. The lifted G-action Φ on TQ is proper, as is its restriction to the G-invariant distribution D, which is a closed subset of TQ because it is a submanifold. The restricted action $\Phi|D : G \times D \to D$ is free. Thus $T\pi|D : D \to TR$ is a smooth fibration.

On Q we have two smooth distributions: ver and hor with $\mathrm{ver}_q = \mathfrak{g} \cdot q = \ker T_q\pi$ and $\mathrm{hor}_q = D_q$. By property 2 we have $T_qQ = \mathrm{ver}_q \oplus \mathrm{hor}_q$. Moreover $T_q\pi : \mathrm{hor}_q \to T_{\pi(q)}R$ is a bijective linear map. Therefore we have an Ehresmann connection on Q with horizontal distribution D, which for short we call the connection D. From properties 1 and 3 it follows that the connection D is G-invariant. For each $w_r \in T_rR$ with $r = \pi(q)$ there is a unique $v_q \in D_q$ such that $T_q\pi\, v_q = w_r$. The vector $v_q = \mathrm{lift}_q\, w_r$ is called the *horizontal lift* of w_r with respect to the connection D.

Lemma 3.4.30. *The G-invariant map $T\pi|D : D \to TR$ induces a diffeomorphism $\psi : D/G \to TR$.*

Proof. Let $q' \in Q$. Then $q' \in G \cdot q$ if and only if there is a unique $g \in G$ such that $q' = g \cdot q$, since the G action $\widetilde{\Phi}$ on Q is free. Because D is G-invariant, for every $w_r \in T_rR$ with $r = \pi(q) = \pi(q')$ we have

$$\mathrm{lift}_{q'}\, w_r = T_q\widetilde{\Phi}_g(\mathrm{lift}_q\, w_r).$$

Therefore each fiber of $T\pi|D$ is a G-orbit on D. Consequently, the map $T\pi|D : D \to TR$ induces a diffeomorphism $\psi : D/G \to TR$. □

In the rest of this section we identify the orbit space D/G with TR.

Because the Lagrangian $\ell : TQ \to \mathbb{R}$ is Φ-invariant, there is a unique function $\overline{\ell} : TR \to \mathbb{R}$ such that $\ell|D = (T\pi|D)^*\overline{\ell}$. We call $\overline{\ell}$ the *reduced Lagrangian* on TR. In fact, for every $w_r \in T_rR$ with $r = \pi(q)$ we have

$$\overline{\ell}(r, w_r) = \ell(q, \mathrm{lift}_q\, w_r).$$

For each $(q, \dot{q}) \in TQ$, the Lagrange derivative $\delta\ell = \delta(q, \dot{q})$ of ℓ at $(q, \dot{q})$ is a linear function on T_qQ. The Lagrange-d'Alembert principle in §3.2 of chapter 1 states that $\delta\ell|D_q = 0$. From properties 1 and 3 of a Chaplygin system it follows that $\delta\ell|D$ is a G-invariant function on D. Therefore there is a unique mapping $\overline{\delta\ell} : TR \to T^*R$ such that

$$\langle \overline{\delta\ell}(r, \dot{r}) \mid w_r \rangle = \langle \delta\ell(q, \dot{q}) \mid v_q \rangle,$$

where $(r, \dot{r}) = T(T\pi)(q, \dot{q})$, $w_r = T_q\pi\, v_q$, and $v_q \in D_q$. The last two of the preceding conditions just say that $v_q = \mathrm{lift}_q\, w_r$. We call $\overline{\delta\ell}$ the *reduced Lagrange derivative* of ℓ.

We now compare the reduced Lagrange derivative $\overline{\delta\ell}$ with the Lagrange derivative $\delta\overline{\ell}$ of the reduced Lagrangian. To do this we use a trivialization of TR. Suppose that we have an isomorphism $\overline{\lambda} : R \times \mathbb{R}^p \to TR$ of vector bundles. Then the map

$$\lambda : Q \times (\mathbb{R}^p \times \mathfrak{g}) \to TQ : (q, (c, \xi)) \mapsto \mathrm{lift}_q \, \overline{\lambda}(\pi(q), c) + X_\xi(q) \tag{14}$$

is an isomorphism of vector bundles. Using the notation of proposition 1.4.6 of chapter 1, we see that $\widetilde{c}^{\,Q}_q = (c, 0)^Q_q = \mathrm{lift}_q \, \overline{c}^{\,R}_{\pi(q)}$ lies in D_q. Therefore $\lambda|(Q \times (\mathbb{R}^p \times \{0\}))$ is the inverse of a trivialization of D, thought of as a vector subbundle of TQ.

Let $\gamma : \mathbb{R} \to Q$ be a smooth curve. Define a smooth curve $\mathbb{R} \to \mathbb{R}^p \times \{0\} : t \mapsto \widetilde{c}(t) = (c(t), 0)$ by requiring that $\gamma'(t) = \lambda(\gamma(t), \widetilde{c}(t))$ lie in $D_{\gamma(t)}$ for every $t \in \mathbb{R}$. Applying the formula for the Lagrange derivative in a trivialization from proposition 1.4.6 of chapter 1 to $\mathcal{L} = \ell \circ \lambda$ we obtain

$$\begin{aligned}
&\langle \delta\mathcal{L}(\gamma(t), \widetilde{c}(t)) \mid \widetilde{a} \rangle \\
&= \langle \frac{\mathrm{d}}{\mathrm{d}t} \frac{\partial \mathcal{L}}{\partial \widetilde{c}}(\gamma(t), \widetilde{c}) \Big|_{\widetilde{c} = \widetilde{c}(t)} \mid \widetilde{a} \rangle - \langle \frac{\partial \mathcal{L}}{\partial q}(q, \widetilde{c}(t)) \Big|_{q = \gamma(t)} \mid \widetilde{a}^{\,Q}_{\gamma(t)} \rangle \\
&\qquad - \langle \frac{\partial \ell}{\partial \dot{q}}(\gamma(t), \dot{q}) \Big|_{\dot{q} = \widetilde{c}^{\,Q}_{\gamma(t)}} \mid \left[\widetilde{c}(t)^Q, \widetilde{a}^{\,Q} \right]_{\gamma(t)} \rangle;
\end{aligned} \tag{15}$$

while applying the same formula to $\overline{\mathcal{L}} = \overline{\ell} \circ \overline{\lambda}$ gives

$$\begin{aligned}
&\langle \delta\overline{\mathcal{L}}(\overline{\gamma}(t), \overline{c}(t)) \mid \overline{a} \rangle \\
&= \langle \frac{\mathrm{d}}{\mathrm{d}t} \frac{\partial \overline{\mathcal{L}}}{\partial \overline{c}}(\overline{\gamma}(t), \overline{c}) \Big|_{\overline{c} = \overline{c}(t)} \mid \overline{a} \rangle - \langle \frac{\partial \mathcal{L}}{\partial r}(r, \overline{c}(t)) \Big|_{r = \overline{\gamma}(t)} \mid \overline{a}^{\,R}_{\overline{\gamma}(t)} \rangle \\
&\qquad - \langle \frac{\partial \overline{\ell}}{\partial \dot{r}}(\overline{\gamma}(t), \dot{r}) \Big|_{\dot{r} = \overline{c}^{\,Q}_{\overline{\gamma}(t)}} \mid \left[\overline{c}(t)^R, \overline{a}^{\,R} \right]_{\overline{\gamma}(t)} \rangle.
\end{aligned} \tag{16}$$

Here $\overline{\gamma}(t) = \pi(\gamma(t))$, $\overline{c}(t) = T_{\gamma(t)}\pi \, \widetilde{c}(t)$, and $\overline{a}^{\,R}_{\overline{\gamma}(t)} = T_{\gamma(t)}\pi \, \widetilde{a}^{\,R}_{\gamma(t)}$. Since

$$\begin{aligned}
\mathcal{L}(q, \widetilde{c}) &= \ell(q, \widetilde{c}^{\,Q}_q) = \ell(q, \mathrm{lift}_q \, \overline{c}^{\,R}_r), \text{ where } r = \pi(q) \\
&= \overline{\ell}(r, \overline{c}^{\,R}_r) = \overline{\mathcal{L}}(r, \overline{c}),
\end{aligned}$$

the first two terms in (15) and (16) are equal.

The tangent vector $\left[\widetilde{c}(t)^Q, \widetilde{a}^{\,Q} \right]_{\gamma(t)}$ to Q at $\gamma(t)$ does *not* lie in $D_{\gamma(t)}$, even though the vector fields $\widetilde{c}(t)^Q = \mathrm{lift}_{\gamma(t)} \, \overline{c}^{\,R}_{\overline{\gamma}(t)}$ and $\widetilde{a}^{\,Q} = \mathrm{lift}_{\gamma(t)} \, \overline{a}^{\,R}_{\overline{\gamma}(t)}$ on Q

have values in $D_{\gamma(t)}$. Because $(T\pi|D)_* \widetilde{c}(t)^Q = \overline{c}(t)^R$ and $(T\pi|D)_* \widetilde{a}^Q = \overline{a}^R$, it follows that $(T\pi|D)_* \left[\widetilde{c}(t)^Q, \widetilde{a}^Q\right] = \left[\overline{c}(t)^R, \overline{a}^R\right]$. Therefore

$$\left[\widetilde{c}(t)^Q, \widetilde{a}^Q\right]_{\gamma(t)} = \mathrm{lift}_{\gamma(t)} \left[\overline{c}(t)^R, \overline{a}^R\right]_{\overline{\gamma}(t)} + X_\xi(\gamma(t)), \tag{17}$$

for some $\xi \in \mathfrak{g}$.

We now determine ξ. First recall that a $\mathfrak{g}$-*valued connection* 1-*form* Θ corresponding to the connection D on the G-principal bundle $\pi : Q \to R = Q/G$ is a linear map $\Theta(q) : T_qQ \to \mathfrak{g}$, which depends smoothly on $q \in Q$, such that

1. $\ker \Theta(q) = D_q$, for every $q \in Q$;
2. $\langle \Theta(q) \mid X_\xi(q) \rangle = \xi$, for every $\xi \in \mathfrak{g}$.

Applying $\Theta_{\gamma(t)}$ to both sides of (17) and using the fact that the *horizontal lift* of $\left[\overline{c}(t)^R, \overline{a}^R\right]_{\overline{\gamma}(t)}$ lies in $D_{\gamma(t)} = \ker \Theta_{\gamma(t)}$, we obtain

$$\xi = \langle \Theta_{\gamma(t)} \mid \left[\widetilde{c}(t)^Q, \widetilde{a}^Q\right]_{\gamma(t)} \rangle = -(\mathrm{d}\Theta)_{\gamma(t)}(\widetilde{c}(t)^Q, \widetilde{a}^Q), \tag{18}$$

because $\widetilde{c}(t)^Q, \widetilde{a}^Q \in D_{\gamma(t)} = \ker \Theta_{\gamma(t)}$ and

$$\mathrm{d}\Theta(X,Y) = X \,\lrcorner\, \mathrm{d}(Y \,\lrcorner\, \Theta) - Y \,\lrcorner\, \mathrm{d}(X \,\lrcorner\, \Theta) - [X,Y] \,\lrcorner\, \Theta$$

for every vector field X, Y on Q with values in D. The $\mathfrak{g}$-valued 2-form $\mathrm{d}\Theta$ applied to a pair of horizontal vector fields on Q is called the *curvature* 2-*form* Ω of the connection D. Therefore we may rewrite (18) as $\xi = -\Omega_{\gamma(t)}(\widetilde{c}(t)^Q, \widetilde{a}^Q)$. Because

$$\begin{aligned} &\langle \frac{\partial \ell}{\partial \dot{q}}(\gamma(t), \widetilde{c})\Big|_{\widetilde{c}=\widetilde{c}(t)^Q_{\gamma(t)}} \mid \mathrm{lift}_{\gamma(t)} \left[\overline{c}(t)^R, \overline{a}^R\right]_{\overline{\gamma}(t)} \rangle \\ &\qquad = \langle \frac{\partial \overline{\ell}}{\partial \dot{r}}(\overline{\gamma}(t), \overline{c})\Big|_{\overline{c}=\overline{c}(t)} \mid \left[\overline{c}(t)^R, \overline{a}^R\right]_{\overline{\gamma}(t)} \rangle, \end{aligned}$$

we may rewrite (15) as

$$\begin{aligned} &\langle \overline{\delta \ell}(\overline{\gamma}(t), \overline{c}(t)) \mid \overline{a}^R \rangle = \langle \delta \overline{\ell}(\overline{\gamma}(t), \overline{c}(t)) \mid \overline{a}^R \rangle \\ &+ \langle \frac{\partial \ell}{\partial \dot{q}}(\gamma(t), \widetilde{c})\Big|_{\widetilde{c}=\widetilde{c}(t)^Q_{\gamma(t)}} \mid \Big(0, \Omega_{\gamma(t)}\big(\mathrm{lift}_{\gamma(t)}\overline{c}(t)^R, \mathrm{lift}_{\gamma(t)}\overline{a}^R\big)\Big)^Q \rangle, \end{aligned} \tag{19}$$

where $\dot{q} = \widetilde{c}(t)^Q = \mathrm{lift}_{\gamma(t)}\overline{c}(t)^R$. Since the correction term in (19) is a homogeneous quadratic polynomial in $\overline{c}^R$, the vector fields on R defined by $\overline{\delta \ell} = 0$ and $\delta \overline{\ell} = 0$ have the same zeroes and linearizations at each zero.

The correction term in (19) may be written as

$$\langle \frac{\partial \ell}{\partial \dot{q}}(\gamma(t), \widetilde{c})\Big|_{\widetilde{c}=\widetilde{c}(t)^Q_{\gamma(t)}} \mid \big(0, \Omega_{\gamma(t)}(\widetilde{c}(t), \widetilde{a})\big)^Q \rangle, \tag{20}$$

which is the momentum applied to the *curvature*.

Proposition 3.4.31. *Let (D, ℓ, G) be a Chaplygin system. If the distribution D is integrable, then $\overline{\delta\ell} = \delta\overline{\ell}$.*

Proof. Since D is integrable, it follows that $\Omega = 0$. The proposition follows because the correction term in (19) vanishes since it vanishes in (20). □

Counterexample 3.4.32. Chaplygin claimed[1] that the converse to proposition 3.4.31 is valid. We give the following counterexample.

Suppose that G acts freely and properly on a smooth manifold Q. Then there is a G-invariant Riemannian metric k on Q. Let $\pi : Q \to R = Q/G$ be the G-orbit map. Then $q \mapsto \mathrm{ver}_q = \mathfrak{g} \cdot q = \{X_\xi(q) \in T_qQ \,|\, \xi \in \mathfrak{g}\}$ defines a vertical distribution ver on the bundle π, because $\mathrm{ver}_q = \ker T_q\pi$. Let $D_q = \mathrm{ver}_q^{\perp}$ be the k_q orthogonal complement to ver_q in T_qQ. Then $q \mapsto D_q$ defines a horizontal distribution D on Q, which is invariant under the G-action $\widetilde{\Phi} : G \times TQ \to TQ : (q, \dot{q}) \mapsto T_q\Phi_g\dot{q}$, so that $T_qQ = \mathrm{ver}_q \oplus D_q$. More concretely, let $\overline{\mathrm{k}}$ be a Riemannian metric on R and let $\mathrm{k}_{\mathfrak{g}}$ be any inner product of $\mathfrak{g}$. Let D be a G-invariant distribution on Q which is complementary to the distribution ver and for which $T\pi|D : D \to TR$ is a bundle isomorphism. In other words, D is a connection on the bundle $\pi : Q \to R = Q/G$. For any $w_r, w_r' \in T_rR$ and any $\xi, \xi' \in \mathfrak{g}$ let

$$\mathrm{k}(q)(\mathrm{lift}_q w_r + X_\xi(q), \mathrm{lift}_q w_r' + X_{\xi'}(q)) = \overline{\mathrm{k}}(r)(w_r, w_r') + \mathrm{k}_{\mathfrak{g}}(\xi, \xi'),$$

where $r = \pi(q)$ is a Riemannian metric on Q which is G-invariant. Moreover, $D_q^{\perp} = \mathrm{ver}_q = \mathfrak{g} \cdot q$ for every $q \in Q$. Let $\ell : TQ \to \mathbb{R} : \dot{q} \mapsto \frac{1}{2}\mathrm{k}(q)(\dot{q}, \dot{q}) = \mathrm{k}^\flat(q)(\dot{q})$ be the kinetic energy associated to the metric k. Then (D, ℓ, G) is a Chaplygin system. Since $\frac{\partial \ell}{\partial \dot{q}} = \mathrm{k}^\flat(q)(\dot{q})$, the correction term in (19) equals $\mathrm{k}(\gamma(t))\big(\widetilde{c}(t)^Q_{\gamma(t)}, \Omega_{\gamma(t)}(\widetilde{c}(t), \widetilde{a}^Q_{\gamma(t)})\big) = 0$, because $(\widetilde{c}(t), \widetilde{a}^Q_{\gamma(t)}) \in D_{\gamma(t)}$ while $\Omega_{\gamma(t)}(\widetilde{c}(t), \widetilde{a}^Q_{\gamma(t)}) \in \mathfrak{g} = \mathrm{ver}_{\gamma(t)}$. From the fact that in general D is *not* integrable, it follows that $\Omega \neq 0$, in general. □

[1] See equation (7) in Chaplygin [23].

3.5 Orbit types and reduction

Let (D, H, ϖ, h) be a distributional Hamiltonian system with a symmetry G which acts properly on the smooth manifold D. In this section we want to describe the reduced almost Poisson algebra $(C^\infty(\overline{D}), \{\,,\,\})$ in terms of the reduced almost Hamiltonian systems on orbit types in $\overline{D}$.

From lemma 1.7.1.23 of chapter 1, it follows that the distribution H on D is given by $u \mapsto H_u = \mathrm{span}\{Y_f(u) \big| f \in C^\infty(D)\}$. Let K be the isotropy group G_u of the G-action Φ on D. The K-symmetry type[2] $D_K = \{v \in D \big| G_v = K\}$ is a smooth submanifold of D. Since the distribution H on D is G-invariant, $T_u\Phi_g\, H_u = H_{g\cdot u}$ for every $(g, u) \in G \times D$. Therefore, $T_u\Phi_k\, H_u = H_u$ for every $k \in K$. This implies that $T_uD_K = \bigcap_{k\in K}(\ker T_u\Phi_k - \mathrm{id})$ and that K acts trivially on D_K.

Lemma 3.5.33. *For each $u \in D$, the subspace $H_u^K = T_uD_K \cap H_u$ of (H_u, ϖ_u) is symplectic.*

Proof. For each $k \in K$, we know that $A_k = T_u\Phi_k$ is a linear map of H_u into itself. Moreover, $K \to \mathrm{Gl}\,(H_u, \mathbb{R}) : k \mapsto A_k$ is a homomorphism of groups. Because K is compact and H_u^K is invariant under A_k for every $k \in K$, there is a subspace F of H_u, which is invariant under A_k for every $k \in K$ such that $H_u = H_u^K \oplus F$. From the fact that ϖ_u is Φ_g-invariant for every $g \in G$, it follows that $\Phi_k^*\varpi_u = \varpi_u$ for every $k \in K$. For $k \in K$, $w_u \in H_u$ and $v_u \in H_u^K$ we have

$$\begin{aligned}\varpi_u(v_u, A_kw_u - w_u) &= \varpi_u(v_u, A_kw_u) - \varpi(v_u, w_u)\\ &= \varpi_u(A_k^{-1}v_u, w_u) - \varpi_u(v_u, w_u) = \varpi_u((A_{k^{-1}} - \mathrm{id})v_u, w_u) = 0.\end{aligned}$$

Therefore $H_u^K \subseteq ((A_k - \mathrm{id})H_u)^{\varpi_u}$. Conversely, if $w_u \in ((A_k - \mathrm{id})H_u)^{\varpi_u}$, then for every $v_u \in H_u$ we have

$$0 = \varpi_u(w_u, (A_k - \mathrm{id})v_u) = \varpi_u((A_{k^{-1}} - \mathrm{id})w_u, v_u).$$

Because ϖ_u is nondegenerate on H_u, we conclude that $w_u \in \ker(A_{k^{-1}} - \mathrm{id})$. Since this conclusion holds for every $k \in K$, we obtain $w_u \in H_u^K$. Consequently, $H_u^K = ((A_k - \mathrm{id})H_u)^{\varpi_u}$ for every $k \in K$. But

$$(A_k - \mathrm{id})H_u = (A_k - \mathrm{id})H_u^K + (A_k - \mathrm{id})F = (A_k - \mathrm{id})F \subseteq F.$$

So $F^{\varpi_u} \subseteq ((A_k - \mathrm{id})H_u)^{\varpi_u} = H_u^K$. This implies that $\{0\} \subseteq F \cap F^{\varpi_u} \subseteq F \cap H_u^K = \{0\}$. Therefore $\varpi_u|F$ is nondegenerate, which implies that $\varpi_u|H_u^K$ is nondegenerate, since $\varpi_u|H_u$ is nondegenerate. □

[2] Also called the K-isotropy type

Lemma 3.5.34. *If $f \in C^\infty(D)^G$, then $Y_f(u) \in H_u^K$ for every $u \in D$. Moreover, the condition $\varpi_u(Y_f(u), v) = \langle \mathrm{d}f(u)|v\rangle$ for every $v \in H_u^K$, determines $Y_f(u)$ uniquely.*

Proof. Because $\Phi_g^* Y_f = Y_{\Phi_g^* f}$ for every $g \in G$ and $f \in C^\infty(D)^G$, it follows that the almost Hamiltonian vector field Y_f on D is G-invariant, that is, $T_u\Phi_g\, Y_f(u) = Y_f(\Phi_g(u))$ for every $(g,u) \in G\times U$. Therefore for every $k \in K$ we have $Y_f(u) \in \ker(T_u\Phi_k - \mathrm{id}) \cap H_u$. In other words, $Y_f(u) \in H_u^K \cap H_u$. For every $v \in H_u$ we have $\varpi_u(Y_f(u), v) = \langle \mathrm{d}f(u) \mid v\rangle$. But $\varpi_u|(H_u^K \cap H_u)$ is nondegenerate and $Y_f(u) \in H_u^K \cap H_u$. Therefore $\varpi_u(Y_f(u), v) = \langle \mathrm{d}f(u) \mid v\rangle$ for every $v \in H_u^K \cap H_u$ determines $Y_f(u)$ uniquely. □

Corollary 3.5.35. *Let $f, k \in C^\infty(D)^G$ and $u \in D$. Then $\{f|D_K, k|D_K\} = \{f,k\}|D_K$, where $K = G_u$.*

Proof. By definition of the almost Poisson bracket $\{\,,\,\}$ on $C^\infty(D)^G$, for each $u \in D_K$ we have

$$\begin{aligned}
\{f,k\}(u) &= -\langle \mathrm{d}k(u) \mid Y_f(u)\rangle \\
&= -\langle \mathrm{d}k(u)|(H_u \cap H_u^K) \mid Y_f(u)|(H_u \cap H_u^K)\rangle, \\
&\qquad\qquad \text{since } Y_f(u) \in H_u \cap H_u^K \\
&= -\langle \mathrm{d}(k|D_K)(u) \mid Y_{f|D_K}(u)\rangle = \{f|D_K, k|D_K\}(u).
\end{aligned}$$

□

Let $M = M_K$ be a connected component of the symmetry type D_K. Then M is a smooth submanifold of D with smooth distribution $H_M = \{H_u \big| u \in M\}$ and smooth symplectic form $\varpi|M$. The proper G-action Φ on D induces a free and proper action of $G_K = N(K)/K$ on D_K, since K acts trivially on D_K. Here $N(K)$ is the normalizer of K in G. Let $G_M = \{g \in G_K \big| g \cdot m \subset M \text{ for every } m \in M\}$. Then G_M is a closed subgroup of G_K which acts freely and properly on M. Therefore G_M is a symmetry group of the distributional Hamiltonian system $(M, H_M, \varpi_M, h|M)$.

Lemma 3.5.36. *Let $f_M : M \to \mathbb{R}$ be a smooth G_M-invariant function on M. Then there is a smooth G-invariant function $f_D : D \to \mathbb{R}$ such that $f|M = f_M$.*

Proof. Let $u \in M$ with $G_u = K$. Let S_u be a slice to the G_K action through u such that $V = (G_K \cdot u) \times S_u$ is a G_K-invariant open neighborhood of u in D_K. Since M is a smooth submanifold of D_K, there is an open neighborhood of u in D_K of the form $W \times U$, where W is a connected open neighborhood of u in M. By hypothesis $f_M \in C^\infty(M)^{G_M}$. Let $f_{S_u \cap W} = f_M|(S_u \cap W)$. Extend $f_{S_u \cap W}$ to a smooth function f_{S_u} on S_u, which is constant on the fibers of the projection map $(S_u \cap W) \times (S_u \times U) \to S_u \cap W$. Extend f_{S_u} to a smooth G_K-invariant function f_V on V. Using a $C^\infty(D_K)^{N(K)}$-partition of unity on D_K, there is a function $f_{D_K} \in C^\infty(D_K)^{N(K)}$ and an open neighborhood $W' \subseteq W$ of u in M such that $f_{D_K}|W' = f_M|W'$. Since f_{D_K} is $N(K)$-invariant, it is G_K-invariant.

Because G acts properly on D, there is a slice $\widetilde{S}_u$ at $u \in M \subseteq D_K$ such that $\widetilde{V} = (G \cdot u) \times \widetilde{S}_u$ is a G-invariant open neighborhood of u in D. Since D_K is a smooth submanifold of D, there is an open neighborhood of u in D of the form $\widetilde{W} \times \widetilde{U}$, where $\widetilde{W}$ is an open neighborhood of u in D_K. We may assume that $\widetilde{S}_u = (\widetilde{S}_u \times \widetilde{W}) \times (\widetilde{S}_u \times \widetilde{U})$. So $\widetilde{V} = (G \cdot u) \times (\widetilde{S}_u \times \widetilde{W}) \times (\widetilde{S}_u \times \widetilde{U})$ is a G-invariant open neighborhood of u in D. From the argument of the first paragraph we have found $f_{D_k} \in C^\infty(D_K)^{G_K}$ which extends $f_M \in C^\infty(M)^{G_M}$. Let $\widetilde{f}_{\widetilde{S}_u \times \widetilde{W}} = f_{D_K}|(\widetilde{S}_u \times \widetilde{W})$. Extend $\widetilde{f}_{\widetilde{S}_u \times \widetilde{W}}$ to a smooth function $\widetilde{f}_{\widetilde{S}_u}$ on $\widetilde{S}_u$, which is constant on the fibers of the projection map $(\widetilde{S}_u \times \widetilde{W}) \times (\widetilde{S}_u \times \widetilde{U}) \to (\widetilde{S}_u \times \widetilde{W})$. Extend $\widetilde{f}_{\widetilde{S}_u}$ to a smooth G-invariant function on the G-invariant open neighborhood $\widetilde{V}$. Using a $C^\infty(D)^G$-partition of unity on D, we obtain $\widetilde{f}_D \in C^\infty(D)^G$ and an open neighborhood $\widetilde{W}' \subseteq \widetilde{W}$ of u in D_K such that $\widetilde{f}_D|(\widetilde{W}' \cap W') = f_{D_K}|(\widetilde{W}' \cap W') = f_M|(\widetilde{W}' \cap W')$, which proves the lemma. □

From lemma 3.5.36 and corollary 3.5.35 we obtain

Proposition 3.5.37. *$(C^\infty(M_K), \{\, , \})$ is an almost Poisson algebra.*

Because G_{M_K} is a symmetry of the distributional Hamiltonian system $(M_K, H_{M_K}, \varpi_{M_K}, h|M_K)$ which acts freely and properly on M_K, we may apply regular nonholonomic reduction, see theorem 3.2.15, to obtain a reduced distributional Hamiltonian system $(\overline{M}_K, \overline{H}_{\overline{M}_K}, \overline{\varpi}_{\overline{M}_K}, \overline{h}|\overline{M}_K)$ on the orbit type $\overline{M}_K = \rho(M_K)$ in the G-orbit space $\overline{M}$. Recall that $(\overline{M}_K, C^\infty(\overline{M}_K))$ is a smooth submanifold of the differential space $(\overline{M}, C^\infty(\overline{M}))$. From corollary 3.5.35 and the construction of the reduced almost Poisson algebra $(C^\infty(\overline{M}_K), \{\, , \})$ on $\overline{M}_K$ in §2, it follows that the inclusion map $i_K : \overline{M}_K \to \overline{M}$, which is a smooth map between the differential spaces $(\overline{M}_K, C^\infty(\overline{M}_K))$ and $(\overline{M}, C^\infty(\overline{M}))$, induces a surjective *almost Poisson map* of the almost Poisson algebra $(C^\infty(\overline{M}), \{\, , \})$ onto the almost Poisson algebra $(C^\infty(\overline{M}_K), \{\, , \})$, that is, for every $\overline{f}, \overline{k} \in C^\infty(\overline{M})$ we have $i_K^*\{\overline{f}, \overline{k}\} =$

$\{i_K^* \overline{f}, i_K^* \overline{k}\}$, or equivalently,

$$\{\overline{f}, \overline{k}\}|\overline{M}_K = \{\overline{f}|\overline{M}_K, \overline{k}|\overline{M}_K\}.$$

Moreover, the restriction to $\overline{M}_K$ of the reduced almost Hamiltonian vector field $Y_{\overline{h}}$ on the locally compact subcartesian differential space $(\overline{M}, C^\infty(\overline{M}))$ is the almost Hamiltonian vector field $Y_{\overline{h}|\overline{M}_K}$ on the smooth symplectic manifold $(\overline{M}_K, \overline{\varpi}_{\overline{M}_K})$.

3.6 Conservation laws

In this section we consider a distributional Hamiltonian system (D, H, ϖ, h), which has conservation laws in addition to preservation of energy h. From the *nonholonomic Noether theorem*, see theorem 1.11.47 of chapter 1, a smooth function f on D is a *conservation law* for the distributional Hamiltonian vector field Y_h if and only if $L_{Y_h} f = 0$, that is, f is constant on integral curves of Y_h.

3.6.1 *Momentum map*

Suppose that the distributional Hamiltonian system (D, H, ϖ, h) has a symmetry group G and that the action $\Phi : G \times D \to D$ is proper. Also assume that the distributional Hamiltonian vector field Y_f associated to the conservation law f has a flow $\varphi_t^f : D \to D : u \mapsto \Phi_u(\exp t\,\xi)$ for some $\xi \in \mathfrak{g}$, the Lie algebra of G. In other words, $Y_f = X^\xi$ for some $\xi \in \mathfrak{g}$. Recall that $X^\xi(u) = \frac{\mathrm{d}}{\mathrm{d}t}\big|_{t=0} \Phi_u(\exp t\,\xi)$ for every $u \in D$. Since the maps $\mathfrak{g} \to \mathcal{X}^\infty(D) : \xi \mapsto X^\xi$ and $C^\infty(D) \to \mathcal{X}^\infty(D) : f \mapsto Y_f$ are linear, we obtain that

$$\mathfrak{h} = \{\xi \in \mathfrak{g} \,\big|\, \text{there is } f \in C^\infty(D) \text{ such that } X^\xi \lrcorner\, \varpi = \partial_H f\}$$

is a vector subspace of $\mathfrak{g}$.

Lemma 3.6.1.38. *$\mathfrak{h}$ is an ideal in $\mathfrak{g}$.*

Proof. To prove the lemma it suffices to show that if $X^\xi \lrcorner\, \varpi = \partial_H f$ then

$$X^{\mathrm{Ad}_g \xi} \lrcorner\, \varpi = \partial_H (f \circ \Phi_{g^{-1}}), \tag{21}$$

for every $g \in G$. To prove equation (21) let $Z(u) \in H_u$ where $u \in D$. We now calculate

$$\begin{aligned}
&\langle X^{\mathrm{Ad}_g\xi}(g\cdot u) \,\lrcorner\, \varpi(g\cdot u) \mid T\Phi_g Z(u)\rangle = \varpi(g\cdot u)(T\Phi_g X^{\xi}(u), T\Phi_g Z(u)) \\
&= (\Phi_g^*\varpi)(u)(X^{\xi}(u), Z(u)) \;=\; \varpi(u)(X^{\xi}(u), Z(u)), \\
&\qquad\qquad\qquad \text{since } \varpi \text{ is } G\text{-invariant} \\
&= (\partial_H f(u))Z(u) \;=\; \langle \mathrm{d}f(u) \mid Z(u)\rangle, \\
&\qquad\qquad\qquad \text{by hypothesis and since } Z(u)\in H_u \\
&= \langle \Phi_{g^{-1}}^*(\mathrm{d}f)(g\cdot u) \mid T\Phi_g Z(u)\rangle \;=\; \langle \mathrm{d}(\Phi_{g^{-1}}^* f)(g\cdot u) \mid T\Phi_g Z(u)\rangle \\
&= \langle (\partial_H(f\circ\Phi_{g^{-1}})(g\cdot u)) \mid T\Phi_g Z(u)\rangle,
\end{aligned}$$

since H is G-invariant implies that $T\Phi_g Z(u) \in H_{g\cdot u}$. □

Let $\mathfrak{h}^*$ be the dual space of $\mathfrak{h}$ and let $J : D \to \mathfrak{h}^*$ be a smooth map such that for each $\xi \in \mathfrak{h}$,

$$X^{\xi} \,\lrcorner\, \varpi = \partial_H \langle J \mid \xi\rangle. \tag{22}$$

Here $\langle J \,|\, \xi\rangle\,(u) = \langle J(u)|\xi\rangle$ for every $u \in D$. We shall refer to J as a *momentum map* for the G-action Φ on D. For each $\xi \in \mathfrak{h}$, the function $J^{\xi} = \langle J \mid \xi\rangle$ is the *momentum* associated to ξ. From the nonholonomic Noether theorem, it follows that for every $\xi \in \mathfrak{h}$ the momentum J^{ξ} is a constant of motion for every G-invariant Hamiltonian.

Theorem 3.6.1.39. *For every accessible set L of the distribution H, there exists an affine action $A_L : G \times \mathfrak{h}^* \to \mathfrak{h}^*$ such that*

$$J(g\cdot u) = A_L(g, J(u)) \tag{23}$$

for every $g \in G$ and every $u \in L$.

Proof. Since $\mathfrak{h}$ is an ideal in $\mathfrak{g}$, the adjoint action $\mathrm{Ad} : G \times \mathfrak{g} \to \mathfrak{g}$ preserves $\mathfrak{h}$ and hence induces a G-action on $\mathfrak{h}$ which we also denote by Ad. The *coadjoint action* $\mathrm{Ad}^* : G \times \mathfrak{h}^* \to \mathfrak{h}^*$ on $\mathfrak{h}^*$ is defined by

$$\left\langle \mathrm{Ad}_g^*\alpha \mid \xi\right\rangle = \left\langle \alpha \mid \mathrm{Ad}_{g^{-1}}\xi\right\rangle \tag{24}$$

for every $g \in G$, every $\alpha \in \mathfrak{h}^*$, and every $\xi \in \mathfrak{g}$.

Next we show that, for each $u \in L$ the mapping

$$\sigma_L : G \to \mathfrak{h}^* : g \to \sigma_L(g) = J(\Phi_g(u)) - \mathrm{Ad}_g^* J(u) \tag{25}$$

does not depend on the choice of $u \in L$, that is, $\sigma_L(g)$ is constant on L. Let $\{\xi_i\}$ be a basis of $\mathfrak{h}$. For each $g \in G$, we can write $\mathrm{Ad}_g\xi_i = \sum_j c_{ij}(g)\xi_j$.

Using equation (21) we obtain

$$\begin{aligned}\partial_H \left\langle J\circ\Phi_{g^{-1}} \mid \xi_i\right\rangle (g\cdot u) &= \partial_H(J^{\xi_i}\circ\Phi_{g^{-1}})(g\cdot u) \;=\; T\Phi_g(X^{\xi_i}(u)) \lrcorner\, \varpi \\ &= X^{\mathrm{Ad}_g\xi_i}(g\cdot u)\lrcorner\,\varpi \;=\; \Big(\sum_j c_{ij}(g)X^{\xi_j}(g\cdot u)\Big)\lrcorner\,\varpi \\ &= \Big(\sum_j c_{ij}(g)\partial_H J^{\xi_j}\Big)(g\cdot u) \;=\; \partial_H J^{\sum_j c_{ij}(g)\xi_j}(g\cdot u) \\ &= \partial_H \left\langle J \mid \mathrm{Ad}_g\xi_i\right\rangle (g\cdot u) \;=\; \partial_H\langle \mathrm{Ad}^*_{g^{-1}}J \mid \xi_i\rangle(g\cdot u).\end{aligned}$$

Hence, $\partial_H\{J\circ\Phi_{g^{-1}} - \mathrm{Ad}^*_{g^{-1}}J\}(u) = 0$ for every $u\in L$. Since every pair of points of L can be connected by a piecewise smooth curve whose tangent vector lies in H, it follows that $\sigma_L(g)$ is constant on L. Thus the map σ_L (25) is well defined.

For every $g, g'\in G$ the following calculation shows that

$$\sigma_L(gg') = \sigma_L(g) + \mathrm{Ad}^*_g(\sigma_L(g')).$$

We compute

$$\begin{aligned}\sigma_L(gg') &= J\circ\Phi_{gg'} - \mathrm{Ad}^*_{gg'}J = J\circ\Phi_g\circ\Phi_{g'} - \mathrm{Ad}^*_g\mathrm{Ad}^*_{g'}J \\ &= (J\circ\Phi_g - \mathrm{Ad}^*_gJ)\circ\Phi_{g'} + \mathrm{Ad}^*_g(J\circ\Phi_{g'} - \mathrm{Ad}^*_{g'}J) \\ &= (J\circ\Phi_g - \mathrm{Ad}^*_gJ) + \mathrm{Ad}^*_g(J\circ\Phi_{g'} - \mathrm{Ad}^*_{g'}J) \\ &= \sigma_L(g) + \mathrm{Ad}^*_g\sigma_L(g').\end{aligned}$$

Define the map $A_L : G\times\mathfrak{h}^*\to\mathfrak{h}^*$ by

$$A_L(g,\alpha) = \mathrm{Ad}^*_g\alpha + \sigma_L(g). \tag{26}$$

We now show that A_L is a G-action. If e is the identity element of G, then

$$A_L(e,\alpha) = \mathrm{Ad}^*_e\alpha + \sigma_L(e) = \alpha + J\circ\Phi_e - \mathrm{Ad}^*_eJ = \alpha.$$

Moreover,

$$\begin{aligned}A_L(gg',\alpha) &= \mathrm{Ad}^*_{gg'}\alpha + \sigma_L(gg') \;=\; \mathrm{Ad}^*_g\mathrm{Ad}^*_{g'}\alpha + \sigma_L(g) + \mathrm{Ad}^*_g\sigma_L(g') \\ &= \mathrm{Ad}^*_g(\mathrm{Ad}^*_{g'}\alpha + \sigma_L(g')) + \sigma_L(g) \;=\; A_L(g, A_L(g',\alpha)).\end{aligned}$$

Hence, A_L is a G-action on $\mathfrak{h}^*$.

To finish the proof of the theorem it remains to show that equation (23) is satisfied. We have

$$\begin{aligned}A_L(g, J(u)) &= \mathrm{Ad}^*_gJ(u) + \sigma_L(g) \\ &= \mathrm{Ad}^*_gJ(u) + J\circ\Phi_g(u) - \mathrm{Ad}^*_gJ(u) \;=\; J(g\cdot u).\end{aligned}$$

□

Let L be an accessible set of the distribution H. We assume that $\mathrm{Stab}_L = \{g \in G \big| g \cdot u \in L \text{ for every } u \in L\}$ is a closed subgroup of G and hence is a Lie group. Because G acts properly on D, it follows that Stab_L acts properly on L. Let K be a compact subgroup of G and set $L_K = \{u \in L \big| G_u = K\}$. Then L_K is the intersection of the symmetry type D_K with L. Let $\mathrm{Stab}_{L_K} = \{g \in \mathrm{Stab}_L \big| g \cdot u \in L_K \text{ for every } u \in L_K\}$. Then Stab_{L_K} is a closed subgroup of Stab_L and hence is a Lie group. Moreover, K is a closed normal subgroup of both Stab_{L_K} and Stab_L.

Fact 3.6.1.40. Stab_{L_K} *is a subgroup of the normalizer* $N(K)$ *in* G.

Proof. Suppose that $g \in \mathrm{Stab}_{L_K}$ and $u \in L_K$. Then $g \cdot u \in L_K$, which implies $K = G_{g \cdot u} = gG_u g^{-1} = gKg^{-1}$, since $u \in L_K$ implies $G_u = K$. Therefore $g \in N(K)$. □

Since K acts trivially on D_K, it follows that $G_{D_K} = N(K)/K$ acts freely and properly on D_K, see theorem 2.3.3.7 of chapter 2. Consequently, $G_{L_K} = \mathrm{Stab}_{L_K}/K$ acts freely and properly on L_K.

This result can be refined a bit. Let $H_{L_K} = \{v_u \in H \big| u \in L_K\}$. Then H_{L_K} is a smooth distribution on L_K with symplectic form $\varpi_{L_K} = {i_K}^*\varpi$, where $i_K : L_K \to D$ is the inclusion map. Let M be an accessible set of the distribution H_{L_K} on L_K and set $\mathrm{Stab}_M = \{g \in \mathrm{Stab}_{L_K} \big| g \cdot m \in M \text{ for every } m \in M\}$. Then Stab_M is a closed subgroup of Stab_{L_K} and hence is a Lie group. From $K \subseteq \mathrm{Stab}_M \subseteq \mathrm{Stab}_{L_K}$ and the fact that K is a closed normal subgroup of Stab_M, it follows that $G_M = \mathrm{Stab}_M/K$ is a Lie group. Because K acts trivially on M and G_M is a closed subgroup of G_{L_K}, we obtain that G_M acts freely and properly on M.

For each accessible set L of the distribution H, each compact subgroup K of G, each accessible set M of H_{L_K}, and every $\alpha \in \mathfrak{h}^*$, the set $M \cap J^{-1}(\alpha)$ is preserved by the distributional Hamiltonian vector field Y_{h_M} associated to the G_M-invariant Hamiltonian function $h_M = h|M$ of the G_M-invariant distributional Hamiltonian system (M, H_M, ϖ_M, H_M).

Theorem 3.6.1.41. *For every accessible set* M *of the distribution* H_{L_K} *and every* $\alpha \in \mathfrak{h}^*$, *each connected component of* $J^{-1}(\alpha) \cap M$ *is a smooth submanifold of* M.

The remainder of this section is devoted to the proof of theorem 3.6.1.41.

Above we have shown that the group $G_M = \mathrm{Stab}_M/K$ acts freely and properly on M. Let $\mathfrak{k}$, $\mathfrak{s}_M$, and $\mathfrak{g}_M = \mathfrak{s}_M/\mathfrak{k}$ be the Lie algebras of the Lie groups K, Stab_M, and G_M, respectively. Let

$$\mathfrak{h}_M = (\mathfrak{s}_M \cap \mathfrak{h})/(\mathfrak{k} \cap \mathfrak{h}) \subseteq \mathfrak{g}_M \tag{27}$$

and let $\eta : (\mathfrak{s}_M \cap \mathfrak{h}) \to \mathfrak{h}_M$ be the canonical projection map.

Proposition 3.6.1.42. *The action of G_M on M preserves the symplectic distribution (H_M, ϖ_M) and has a momentum map $J_M : M \to \mathfrak{h}_M^*$. For every $\alpha_M \in \mathfrak{h}_M^*$ every connected component of $J_M^{-1}(\alpha_M)$ is a smooth submanifold of M.*

Proof. Clearly, the action of Stab_M on M preserves the symplectic distribution (H_M, ϖ_M). Since $G_M = \mathrm{Stab}_M/K$ and K acts trivially on M, the G_M-action preserves the symplectic distribution (H_M, ϖ_M).

For each $\xi \in \mathfrak{g}$, we have $X^\xi \lrcorner \varpi = \partial_H J^\xi$. However, $\xi \in \mathfrak{k}$ implies that $X^\xi(u) = 0$ for all $u \in M \subseteq L_K$. Hence $\partial_H J^\xi = 0$ on M. Since M is connected and is contained in an accessible set of H, it follows that the restriction of J^ξ to M is constant. Therefore, for every $\xi \in \mathfrak{k} \cap \mathfrak{h}$ the restriction of J^ξ to M is constant.

Let $\kappa : \mathfrak{k} \cap \mathfrak{h} \to \mathfrak{h}$ be the inclusion map and let $\kappa^* : \mathfrak{h}^* \to (\mathfrak{k} \cap \mathfrak{h})^*$ be its transpose. The restriction of J^ξ to ξ in $\mathfrak{k} \cap \mathfrak{h}$ is given by $\kappa^* \circ J : D \to (\mathfrak{k} \cap \mathfrak{h})^*$. Let $J|M$ be the restriction of J to points in M. From the argument in the paragraph above, we deduce that $\kappa^* \circ J|M$ is constant.

Let $\mu : \mathfrak{k} \cap \mathfrak{h} \to \mathfrak{s}_M \cap \mathfrak{h}$ be the inclusion map. Its transpose $\mu^* : (\mathfrak{s}_M \cap \mathfrak{h})^* \to (\mathfrak{k} \cap \mathfrak{h})^*$ is onto. Hence, there exists $j_M \in (\mathfrak{s}_M \cap \mathfrak{h})^*$ such that $\mu^*(j_M) = \kappa^* \circ J\,|_M$.

Let $\nu : \mathfrak{s}_M \cap \mathfrak{h} \to \mathfrak{h}$ be the inclusion map and let $\nu^* : \mathfrak{h}^* \to (\mathfrak{s}_M \cap \mathfrak{h})^*$ be its transpose. Every element of the kernel of the canonical projection map $\eta : (\mathfrak{s}_M \cap \mathfrak{h}) \to \mathfrak{h}_M$ is $\mathfrak{k} \cap \mathfrak{h}$ is mapped to zero by $(\nu^* \circ J|M - j_M) : M \to (\mathfrak{s}_M \cap \mathfrak{h})^*$. Hence there exists a unique map $J_M : M \to \mathfrak{h}_M^*$ such that

$$\eta^* J_M = \nu^* \circ J|M - j_M. \tag{28}$$

We now show that J_M is a momentum map for the action of G_M on M. For each $\xi \in \mathfrak{s}_M \cap \mathfrak{h}$, we have $\nu(\xi) \in \mathfrak{h}$. Moreover, the G-action on M restricted to the one parameter group $t \to \exp t\nu(\xi)$ is generated by the vector field $X^{\nu(\xi)}$ restricted to M. Here $X^{\nu(\xi)} \lrcorner \varpi = \partial_H \langle J \mid \nu(\xi) \rangle$. Similarly, the G_M-action on M restricted to the one parameter subgroup $t \to \exp t\eta(\xi)$ is generated by the vector field $X_M^{\eta(\xi)}$ on M. Since on M

the G-action of $\exp t\nu(\xi)$ and the G_M-action of $\exp t\eta(\xi)$ coincide, it follows that $X_M^{\eta(\xi)}$ is the restriction of $X^{\nu(\xi)}$ to M. Restricting the equation $X^{\nu(\xi)} \lrcorner\, \varpi = \partial_H \langle J \mid \nu(\xi)\rangle$ to H_M we get

$$\begin{aligned} X_M^{\eta(\xi)} \lrcorner\, \varpi_M &= \partial_{H_M} \langle (J|M) \mid \nu(\xi)\rangle = \partial_{H_M} \langle \nu^* \circ (J|M) \mid \xi\rangle \\ &= \partial_{H_M} \langle (\eta^* J_M + j_M) \mid \xi\rangle \;=\; \partial_{H_M} \langle J_M \mid \eta(\xi)\rangle , \end{aligned}$$

where the last equality above follows because j_M is constant. Hence $J_M : M \to \mathfrak{h}_M^*$ is a momentum map for the action of G_M on M.

Since the action of G_M on M is free, it follows that $X_M^{\xi}(u) \neq 0$ for every $\xi \in \mathfrak{h}_M$ and every $u \in M$. This implies that $\partial_{H_M}\langle J_M \mid \xi\rangle(u) \neq 0$ for every $\xi \in \mathfrak{h}_M$ and every $u \in M$. Hence, the map $\partial_{H_M} J_M : H_M \to \mathfrak{h}_M^*$ is onto. Therefore, for every $\alpha_M \in \mathfrak{h}_M^*$ which lies in the image of J_M, the set $J_M^{-1}(\alpha_M)$ is a local submanifold of M. □

For each $u \in M$, let $H_u = H \cap T_uD$, $H_M = \{H_u \mid u \in M\}$, $H_{M,u} = H_M \cap T_uD$, and

$$H_{M,u}^{\varpi} = \{v \in H_u \mid \varpi(v,w) = 0 \quad \text{for all } w \in H_{M,u}\} . \tag{29}$$

Since M is an accessible set of H_{L_K}, it follows that H_M is the restriction of $H_{L_K} = H \cap TL_K$ to vectors whose base point lies in M. In other words, $H_{M,u} = H_u \cap T_uL_K$.

Because H_{L_K} is a symplectic distribution on L_K, we see that $H_{M,u}^{\varpi} \cap T_uL_K = \{0\}$. Furthermore, $H_{M,u}^{\varpi}$ is K-invariant, which implies that $H_{M,u}^{\varpi} \subseteq T_u^{\perp}L_K$, where $T_u^{\perp}L_K = \{v \in T_uL \big| \int_K T_u\Phi_k v \, \mathrm{d}k = 0\}$. Here $\mathrm{d}k$ is Haar measure on K with $\operatorname{vol} K = 1$. Thus, $H_{M,u}^{\varpi} = H_u \cap T^{\perp}L_K$. From

$$T_uL = T_uL_K \oplus T_u^{\perp}L_K, \tag{30}$$

we obtain the decomposition

$$H_u = H_{M,u} \oplus H_{M,u}^{\varpi}, \tag{31}$$

where $H_{M,u} \subseteq T_uM$ and $H_{M,u}^{\varpi} \subseteq T_u^{\perp}M = \{v \in T_uD \mid \overline{v} = 0\}$. To prove (30) we argue as follows. Each $u \in L_K$ is fixed by the action of $K \subseteq \mathrm{Stab}_{L_K}$. Therefore $T_u\Phi_k : T_uL_K \to T_uL_K$ for every $k \in K$. For $v \in T_uL$ let $\overline{v} = \int_K T_u\Phi_k v \, \mathrm{d}k$. Then $T_u\Phi_k\overline{v} = \overline{v}$ for every $k \in L$. So $\overline{v} \in T_uL_K$. Write $v = \overline{v} + (v - \overline{v})$. Then $\overline{v - \overline{v}} = 0$. So $v - \overline{v} \in T_u^{\perp}L_K$. Therefore (30) holds, because if $w \in T_uL_K \cap T_u^{\perp}L_K$, then $w = \overline{w} = 0$.

Since $\mathfrak{h}$ is an ideal in $\mathfrak{g}$, it is invariant under the adjoint action of $K \subseteq G$ on $\mathfrak{g}$. For each $\xi \in \mathfrak{h}$, its average $\overline{\xi} = \int_K \mathrm{Ad}_k(\xi)\, dk$ is in $\mathfrak{h}$. In addition, the average of $\xi - \overline{\xi}$ is zero. Let

$$\mathfrak{m} = \{\xi \in \mathfrak{h} \mid \mathrm{Ad}_k\xi = \xi \quad \text{for all } k \in K\} \quad \text{and} \quad \mathfrak{n} = \{\zeta \in \mathfrak{h} \mid \overline{\zeta} = 0\}.$$

From $\xi = \overline{\xi} + (\xi - \overline{\xi})$ and the fact that $\mathfrak{m} \cap \mathfrak{n} = \{0\}$, it follows that

$$\mathfrak{h} = \mathfrak{m} \oplus \mathfrak{n}. \tag{32}$$

Lemma 3.6.1.43. *Let $M \subseteq L_K$ be an accessible set of H_{L_K} and let $u \in M$. Then $Y_{J^\xi}(u) \in H_{M,u}$ for every $\xi \in \mathfrak{m}$ and $Y_{J^\zeta}(u) \in H^{\varpi}_{M,u}$ for every $\zeta \in \mathfrak{n}$.*

Proof. To verify the first assertion we argue as follows. Because $u \in M \subseteq L_K$ and $k \in K$, we have $\Phi_k(u) = u$. We now compute

$$\begin{aligned}
\Psi^u_k(Y_{J^\xi}(u)) &= \Psi^u_k(X^\xi(u)) = T\Phi_k(X^\xi(u)) = X^{T_e L_k \xi}(u) \\
&= \frac{d}{dt}\Big|_{t=0} \Phi_{\exp(tT_e L_k \xi)}(u) = \frac{d}{dt}\Big|_{t=0} \Phi_{L_k \exp t\xi}(u) \\
&= \frac{d}{dt}\Big|_{t=0} \Phi_{k\exp(t\xi)k^{-1}}(u) = \frac{d}{dt}\Big|_{t=0} \Phi_{\exp t\mathrm{Ad}_k(\xi)}(u) \\
&= X^{\mathrm{Ad}_k(\xi)}(u) = Y_{J^{\mathrm{Ad}_k(\xi)}}(u).
\end{aligned} \tag{33}$$

If $\xi \in \mathfrak{m}$, then $\mathrm{Ad}_k(\xi) = \xi$ for all $k \in K$. Hence $Y_{J^\xi}(u)$ is K-invariant, which implies that $Y_{J^\xi}(u) \in T_u L_K$. On the other hand, $Y_{J^\xi}(u) \in H_u$ which using $H_{M,u} = H_u \cap T_u L_K$ shows that $Y_{J^\xi}(u) \in H_{M,u}$.

We now verify the second assertion of the lemma. For $\zeta \in \mathfrak{n}$ the average $\overline{\zeta}$ is zero. Since the map $\zeta \mapsto Y_{J^\zeta}(u)$ is linear, equation (33) implies that

$$\begin{aligned}
\overline{Y_{J^\zeta}(u)} &= \int_K \Psi^u_k(Y_{J^\zeta}(u))\, d\mathrm{k} = \int_K \Psi^u_k(Y^{J^{\mathrm{Ad}_k(\zeta)}}(u))\, d\mathrm{k} \\
&= Y^{\overline{\zeta}}(u) = 0.
\end{aligned} \tag{34}$$

Hence, (34) ensures that $Y_{J^\zeta}(u) \in T^{\perp}_u L_K$. But, $Y_{J^\zeta}(u) \in H_u$. So using $H^{\varpi}_{M,u} = H_u \cap T^{\perp} L_K$ we see that $Y_{J^\zeta}(u) \in H^{\varpi}_{M,u}$. □

Proposition 3.6.1.44. *For each $u \in M$, the connected component of $M \cap J^{-1}(J(u))$ and $J_M^{-1}(J_M(u))$ containing u coincide.*

Proof. Equation (28) yields $\eta^* J_M(u) = \nu^*(J(u)) - j_M(u)$. In the proof of proposition 3.6.1.42 we have shown that j_M is locally constant. Since $\eta^* : \mathfrak{h}^*_M \to (\mathfrak{s}_M \cap \mathfrak{h})^*$ is injective and $\nu^* : \mathfrak{h}^* \to (\mathfrak{s}_M \cap \mathfrak{h})^*$ is surjective, it follows that the connected component of $M \cap J^{-1}(J(u))$ containing u is a subset of the connected component of $J_M^{-1}(J_M(u))$ containing u.

By proposition 3.6.1.42, the connected components of $J_M^{-1}(J_M(u))$ are submanifolds of M. Hence, it suffices to show that J is constant on connected components of $J_M^{-1}(J_M(u))$. According to equation (32) we may decompose J into components $J^{\mathfrak{m}} : M \to \mathfrak{m}^*$ and $J^{\mathfrak{n}} : M \to \mathfrak{n}^*$. If $\xi = \xi_{\mathfrak{m}} + \xi_{\mathfrak{n}}$ with $\xi \in \mathfrak{h}$, $\xi_{\mathfrak{m}} \in \mathfrak{m}$, and $\xi_{\mathfrak{n}} \in \mathfrak{n}$, then

$$\langle J \mid \xi \rangle = J^\xi = \langle J^{\mathfrak{m}} \mid \xi_{\mathfrak{m}} \rangle + \langle J^{\mathfrak{n}} \mid \xi_{\mathfrak{n}} \rangle.$$

For $\xi \in \mathfrak{m}$, lemma 3.6.1.43 implies that the vector field Y_{J^ξ} is tangent to M. Hence, the G-action Φ restricted to the one parameter group $t \to \exp t\xi$ preserves M. Consequently, ξ is in the Lie algebra $\mathfrak{s}_M$ of the stability group Stab_M of M. Hence $\xi \in \mathfrak{s}_M \cap \mathfrak{h}$. Using equation (28) we find that

$$\langle J^{\mathfrak{m}} \mid \xi \rangle = J^{\xi} = \langle J_M \mid \eta(\xi) \rangle + \langle j_M \mid \xi \rangle,$$

where $\eta : \mathfrak{s}_M \cap \mathfrak{h} \to \mathfrak{h}_M = (\mathfrak{s}_M \cap \mathfrak{h})/(\mathfrak{k} \cap \mathfrak{h})$ is the canonical projection. The above equation holds for all $\xi \in \mathfrak{m}$, which implies that

$$J^{\mathfrak{m}} = \eta^* J_M + j_M. \tag{35}$$

Since j_M is constant on M, it follows that $J^{\mathfrak{m}}$ is constant on the level sets of J_M.

For $\xi \in \mathfrak{n}$, lemma 3.6.1.43 implies that $Y_{J^\xi}(u) \in H^{\varpi}_{M,u}$ for all $u \in M$. Since $Y_{J^\xi} \mathbin{\lrcorner} \varpi = \mathrm{d}J^{\xi}$, it follows that, for every $w \in H_{M,u}$,

$$\langle \mathrm{d}J^{\xi}(u) \mid w \rangle = \varpi(Y_{J^\xi}(u), w) = 0,$$

where the last equality is a consequence of the definition (29) of $H^{\varpi}_{M,u}$. Because the above equation holds for every $\xi \in \mathfrak{n}$, we deduce that $\mathrm{d}J^{\mathfrak{n}}$ vanishes on TM. Hence, $J^{\mathfrak{n}}$ is constant on connected components of the level sets of $J_M : M \to \mathfrak{h}_M$.

Since $J = J^{\mathfrak{m}} + J^{\mathfrak{n}}$ and both $J^{\mathfrak{m}}$ and $J^{\mathfrak{n}}$ are constant on connected components of level sets of J_M, it follows that J is constant on connected components of level sets of J_M. □

Proof of theorem 3.6.1.41. Proposition 3.6.1.44 ensures that for each $u \in M$ the connected component of $M \cap J^{-1}(J(u))$ and $J_M^{-1}(J_M(u))$ containing u coincide. By proposition 3.6.1.44, the connected components of $J_M^{-1}(J_M(u))$ are manifolds. Hence, the connected components of $M \cap J^{-1}(J(u))$ are also manifolds. This proves theorem 3.6.1.41. □

3.6.2 *Gauge momenta*

A smooth function $f \in C^{\infty}(D)$ is called a *gauge momentum* for the action Φ of G on D if it is G-invariant and its distributional Hamiltonian vector field Y_f is tangent to each G-orbit in D.

Fact 3.6.2.45. *Gauge momenta are constant on integral curves of each distributional Hamiltonian vector field Y_k where $k \in C^{\infty}(D)^G$.*

Proof. Let $f \in C^\infty(D)$ be a gauge momentum function. Then for each $v \in D$ we have $Y_f(u) \in T_u(G \cdot v)$ if and only if there is a $\xi \in \mathfrak{g}$ such that $Y_f(u) = X^\xi(u)$ for every $u \in D$. Therefore for every $u \in D$ we have

$$\begin{aligned}(L_{Y_k} f)(u) &= -(L_{Y_f} k)(u) = -\langle \mathrm{d}k(u) \mid Y_f(u)\rangle = -\langle \mathrm{d}k(u) \mid X^\xi(u)\rangle \\ &= -\frac{\mathrm{d}}{\mathrm{d}t}\Big|_{t=0} k(\Phi_u(\exp t\,\xi)) = -\frac{\mathrm{d}}{\mathrm{d}t}\Big|_{t=0} k(u) = 0.\end{aligned}$$

□

Hence gauge momenta are constant on every accessible set N of the distribution spanned by the distributional Hamiltonian vector fields of G-invariant functions. Since a gauge momentum function is G-invariant, it pushes forward to a smooth function on the orbit space $\overline{D} = D/G$ which is constant on the image of N under the G-orbit map.

From §1 we know that the space $C^\infty(D)^G$ of G-invariant functions on D is an almost Poisson subalgebra of $(C^\infty(D), \{\ \})$. Following the terminology used for Poisson algebgras, we say that $f \in C^\infty(D)^G$ is a *Casimir* of the almost Possion algebra $(C^\infty(D)^G, \{\,,\,\})$ if $\{f, f'\} = 0$ for all $f' \in C^\infty(D)^G$.

Proposition 3.6.2.46. *A function $f \in C^\infty(D)^G$ is a gauge momentum if and only if it is a Casimir in $(C^\infty(D)^G, \{\,,\,\})$.*

Proof. If f is a gauge momentum, then Y_f is tangent to each G-orbit in D. Hence, $\{f, f'\} = L_{Y_f} f' = 0$ for every $f' \in C^\infty(D)^G$. This implies that f is a Casimir in $(C^\infty(D)^G, \{\,,\,\})$. Conversely, if $f \in C^\infty(D)^G$ is a Casimir, then $L_{Y_f} f' = 0$ for all $f' \in C^\infty(D)^G$. This implies that Y_f is tangent to each G-orbit in D. Thus f is a gauge momentum. □

3.7 Lifted actions and the momentum equation

3.7.1 *Lifted actions*

In this subsection we assume that the G action

$$\Phi : G \times D \to D : (g, u) \mapsto T_{\tau_Q(u)} \widetilde{\Phi}_g u$$

is the restriction to D of the *lift* to TQ of an action $\widetilde{\Phi} : G \times Q \to Q$ of G on Q. Here $\tau_Q : TQ \to Q$ is the tangent bundle projection map. In other words, $D \subseteq TQ$ is a G-invariant distribution on Q, and $\Phi_g(u) = T\widetilde{\Phi}_g(u)$ for every $g \in G$ and every $u \in D$. In addition we assume that the action $\widetilde{\Phi}$

is free and proper. The results of this subsection will be used to prove the momentum equation for free and proper actions in subsection two.

The assumption that the G-action $\widetilde{\Phi}$ on Q is free and proper implies that Q is a principal bundle with structure group G. Let $\overline{Q}$ be the space of orbits of the G-action on Q and let $\pi : Q \rightarrow \overline{Q}$ be the G-orbit map, which is the bundle projection of the principal bundle. The bundle $\operatorname{ver} TQ$ of *vertical vectors* of Q consists of vectors tangent to the fibers of $\pi : Q \rightarrow \overline{Q}$. Let k be a G-invariant Riemannian metric on Q and let the *horizontal distribution* $\operatorname{hor} TQ = (\operatorname{ver} TQ)^{\perp}$ be the k-orthogonal complement of $\operatorname{ver} TQ$. We obtain the bundle direct sum decomposition

$$TQ = \operatorname{ver} TQ \oplus_Q \operatorname{hor} TQ. \tag{36}$$

Since k is G-invariant, the distribution $\operatorname{hor} TQ$ is G-invariant. Thus it defines a *connection* on the principal bundle Q. The *connection form* ϑ of this connection is a $\mathfrak{g}$-valued 1-form on Q with kernel $\ker \vartheta = \operatorname{hor} TQ$ such that $\langle \vartheta \mid X_\xi \rangle = \xi$ for every $\xi \in \mathfrak{g}$. Here X_ξ is the vector field on Q whose flow is $\varphi_t(q) = \widetilde{\Phi}_q(\exp t\,\xi)$.

The restriction of the metric k to $\operatorname{hor} TQ$ pushes forward to a metric $\mathrm{k}_{\overline{Q}}$ on $\overline{Q}$. In other words,

$$\mathrm{k}(\operatorname{hor} u_1, \operatorname{hor} u_2) = \mathrm{k}_{\overline{Q}}(T\pi(u_1), T\pi(u_2)).$$

The tangent bundle $T\overline{Q}$ has a 1-form $\overline{\theta}$ such that

$$\langle \overline{\theta} \mid \overline{w} \rangle = \mathrm{k}_{\overline{Q}}(\tau_{T\overline{Q}}(\overline{w}), T\tau_{\overline{Q}}(\overline{w}))$$

for every $\overline{w} \in T\overline{Q}$. Here $\tau_{\overline{Q}} : T\overline{Q} \rightarrow \overline{Q}$ and $\tau_{T\overline{Q}} : T(T\overline{Q}) \rightarrow T\overline{Q}$ are tangent bundle projection maps. The exterior derivative $\mathrm{d}\overline{\theta}$ of $\overline{\theta}$ is the 2-form $\overline{\omega}$, which is a symplectic form on $T\overline{Q}$.

Let D be a $\widetilde{\Phi}$-invariant distribution on Q. The decomposition (36) of TQ gives rise to

$$D = \operatorname{ver} D \oplus_Q \left((\operatorname{ver} D)^{\perp} \cap D \right), \tag{37}$$

where $\operatorname{ver} D = D \cap \operatorname{ver} TQ$. We now make the following

Hypothesis 3.7.1.47. $\operatorname{ver} D$ *is a distribution on* Q.

In other words, $\operatorname{ver} D$ is a subbundle of $\operatorname{ver} TQ$. Let $D_{\overline{Q}} = T\pi(D)$.

In the next few propositions we describe the structure of the orbit space $\overline{D} = D/G$.

For each $q \in Q$, let $\overline{q} = \pi(q)$. We denote the *horizontal lift* from $T_{\overline{q}}\overline{Q}$ to T_qQ by $\mathrm{lift}_q : T_{\overline{q}}\overline{Q} \to T_qQ$. In other words, for each $\overline{u} \in T_{\overline{q}}\overline{Q}$, $\mathrm{lift}_q(\overline{u})$ is the unique vector in $\mathrm{hor}\, T_qQ$ such that $T\pi(\mathrm{lift}_q(\overline{u})) = \overline{u}$.

Proposition 3.7.1.48. *For each $q \in Q$, there exists a unique linear map $\Lambda_q : (D_{\overline{Q}})_{\pi(q)} \to \left((\mathrm{ver}\, D)^{\perp} \cap D\right)_q$ such that*

$$\mathrm{lift}_q(T\pi(u)) + \Lambda_q(T\pi(u)) = u \tag{38}$$

for every $u \in \left((\mathrm{ver}\ D)^{\perp} \cap D\right)_q$. Consequently,

$$(\mathrm{ver}\, D)^{\perp} \cap D = (\mathrm{lift} \oplus \Lambda)(D_{\overline{Q}}). \tag{39}$$

Moreover, the map $\Lambda : Q \to L(D_{\overline{Q}}, (\mathrm{ver}\, D)^{\perp} \cap D) : q \mapsto \Lambda_q$ is smooth, has its image contained in $\mathrm{ver}\, TQ$, and is G-equivariant, that is, $\Phi_g \circ \Lambda_q = \Lambda_{g\cdot q}$ for every $g \in G$ and every $q \in Q$.

Proof. We have $T\pi(\mathrm{ver}\, D) = \{0\}$ and $T\pi\left((\mathrm{ver}\, D)^{\perp} \cap D\right) = D_{\overline{Q}}$. Moreover,

$$\mathrm{ver}((\mathrm{ver}\, D)^{\perp} \cap D) = ((\mathrm{ver}\, D)^{\perp} \cap D) \cap (\mathrm{ver}\, TQ) \subseteq (\mathrm{ver}\, D)^{\perp} \cap (\mathrm{ver}\, D) = \{0\}.$$

Hence the map Λ satisfies (38). Equation (38) ensures that

$$((\mathrm{ver}\, D)^{\perp} \cap D)_q = (\mathrm{ver}\, D)_q \oplus (\mathrm{lift}_q + \Lambda_q)\left((D_{\overline{Q}})_{\pi(q)}\right),$$

which gives (39).

For $u \in T_qQ$, $\mathrm{hor}\, u = \mathrm{lift}_q(T\pi(u))$. If $u \in (\mathrm{ver}\, D)^{\perp} \cap D$, then

$$\mathrm{ver}\, u = u - \mathrm{hor}\, u = u - \mathrm{lift}_q(T\pi(u)) = \Lambda_q(T\pi(u)).$$

Thus the range of Λ is contained in $\mathrm{ver}\, TQ$.

For $g \in G$ and $u \in \left((\mathrm{ver}\, D)^{\perp} \cap D\right)_q \cap T_qQ$, we have

$$\Phi_g(\Lambda_q(T\pi(u))) = \Phi_g(\mathrm{ver}\, u) - \mathrm{ver}\,(\Phi_g(u)),$$

which is an element of $\left((\mathrm{ver}\, D)^{\perp} \cap D\right)_{g\cdot q} \cap \mathrm{ver}\, T_{g\cdot q}Q$. However, the set $\left((\mathrm{ver}\, D)^{\perp} \cap D\right)_{g\cdot q} \cap \mathrm{ver}\, T_{g\cdot q}Q$ has a unique element $\Lambda_{g\cdot q}(T\pi(\Phi_g(u))) = \Lambda_{g\cdot q}(T\pi(u))$. Hence $\Phi_g \circ \Lambda_q = \Lambda_{g\cdot q}$.

Finally, for every vector field $\overline{X}$ on $\overline{Q}$ with values in $D_{\overline{Q}}$, we know that $\Lambda \circ \overline{X} \circ \pi$ is a smooth vector field on Q. This ensures that Λ is smooth. □

Proposition 3.7.1.49. *The bundle $(TQ)/G$ of G-orbits on TQ over the space $\overline{Q}$ of G-orbits on Q is isomorphic to the fiber product $Q[\mathfrak{g}] \times_{\overline{Q}} T\overline{Q}$ of bundles over $\overline{Q}$. Here $Q[\mathfrak{g}]$ is the adjoint bundle of Q over $\overline{Q}$.*

Proof. In the category of vector bundles, fiber products and direct sums are equivalent. Hence, the decomposition of TQ given in (36) is equivalent to $TQ = \mathrm{ver}\, TQ \times_{\overline{Q}} \mathrm{hor}\, TQ$. The lifted action of G on TQ preserves $\mathrm{ver}\, TQ$ and $\mathrm{hor}\, TQ$ separately. Hence

$$(TQ)/G = ((\mathrm{ver}\, TQ)/G) \times_{\overline{Q}} ((\mathrm{hor}\, TQ)/G).$$

Let $\mathrm{Ad} : G \times \mathfrak{g} \to \mathfrak{g}$ be the adjoint action of G on its Lie algebra $\mathfrak{g}$. The mapping $Q \times \mathfrak{g} \to (\mathrm{ver}\, TQ) : (q, \xi) \mapsto X_\xi(q)$ is a bundle isomorphism intertwining the action

$$G \times (Q \times \mathfrak{g}) \to Q \times \mathfrak{g} : (g, (q, \xi)) \mapsto (g \cdot q, \mathrm{Ad}_g \xi),$$

and the restriction to $\mathrm{ver}\, TQ$ of the lifted action of G on TQ. But the G-orbit space $(Q \times \mathfrak{g})/G = Q[\mathfrak{g}]$ is the adjoint bundle of Q. Hence, the bundle $(\mathrm{ver}\, TQ)/G$ is isomorphic to the adjoint bundle $Q[\mathfrak{g}]$.

The restriction to $\mathrm{hor}\, TQ$ of $T\pi : TQ \to T\overline{Q}$ maps $\mathrm{hor}\, TQ$ onto $T\overline{Q}$, and it is constant along G-orbits in $\mathrm{hor}\, TQ$. It induces a vector bundle isomorphism $(\mathrm{hor}\, TQ)/G \to T\overline{Q}$. □

Using proposition 3.7.1.49 we can identify $(TQ)/G$ with $Q[\mathfrak{g}] \times_{\overline{Q}} T\overline{Q}$. Let $\pi_1 : (TQ)/G \to Q[\mathfrak{g}]$ be the projection onto the first factor of $Q[\mathfrak{g}] \times_{\overline{Q}} T\overline{Q}$. By hypothesis 3.7.1.47, $\mathrm{ver}\, D$ is a subbundle of $\mathrm{ver}\, TQ$. Moreover, it is G-invariant. Hence the space $S = \mathrm{ver}\, D/G$ of G-orbits on D is a subbundle of the adjoint bundle $Q[\mathfrak{g}]$.

Proposition 3.7.1.50. *There is a smooth map* $\Sigma : D_{\overline{Q}} \to Q[\mathfrak{g}]$ *such that the space* $\overline{D}$ *of G-orbits on D is diffeomorphic to* $S \times_{\overline{Q}} \mathrm{graph}\, \Sigma$, *which equals*

$$\left\{ (s, \Sigma(\overline{u}), \overline{u}) \in S \times Q[\mathfrak{g}] \times D_{\overline{Q}} \mid \pi_S(s) = \tau_{\overline{Q}}(\overline{u}) \right\}. \tag{40}$$

Proof. Let $\rho : TQ \to (TQ)/G$ be the orbit map for the lifted G action Φ on TQ. Given $\overline{u} \in (D_{\overline{Q}})_{\overline{q}}$, where $\overline{q} \in \overline{Q}$, take any $q \in \pi^{-1}(\overline{q}) \in Q$ and set

$$\Sigma(\overline{u}) = \pi_1(\rho(\Lambda_q(\overline{u}))) \in Q[\mathfrak{g}]. \tag{41}$$

Since $\rho(\Lambda_{g \cdot q}(\overline{u})) = \rho(\Phi_g(\Lambda_q(\overline{u}))) = \rho(\Lambda_q(\overline{u}))$ for every $g \in G$, it follows that $\Sigma(\overline{u})$ does not depend on the choice of $q \in \pi^{-1}(\overline{q})$. Hence the map $\Sigma : D_{\overline{Q}} \to Q[\mathfrak{g}]$ is well defined.

For every vector field $\overline{X}$ on $\overline{Q}$ with values in $D_{\overline{Q}}$, we know that $\Lambda \circ \overline{X} \circ \pi$ is a smooth G-invariant vector field on Q with values in $\mathrm{ver}\, TQ$. Hence it gives rise to a smooth section $\Sigma \circ \overline{X} = \pi_1 \circ \rho \circ \Lambda \circ \overline{X} : \overline{Q} \to Q[\mathfrak{g}]$ of the adjoint bundle $Q[\mathfrak{g}] \to \overline{Q}$. This implies that the map $\Sigma : D_{\overline{Q}} \to Q[\mathfrak{g}]$ is a smooth vector bundle morphism. Equation (38) ensures that the space $\overline{D}$ of G-orbits in D is diffeomorphic to $S \times_{\overline{Q}} \mathrm{graph}\, \Sigma$. □

3.7.2 *Momentum equation*

In this subsection we shall rewrite the reduced dynamics in terms of the momentum equation and the second order differential equation condition, see §8.2 of chapter 1.

From results of §2 it follows that the reduced space $\overline{D}$ is endowed with a symplectic distribution $(H_{\overline{D}}, \varpi_{\overline{D}}, h_{\overline{D}})$ such that the reduced equations of motion are given by the distributional Hamiltonian vector field $Y_{h_{\overline{D}}}$. In other words, the reduced motion $t \mapsto \overline{u}(t) = \rho(u(t))$ is an integral curve of $Y_{h_{\overline{D}}}$. Each curve is uniquely determined by

$$\frac{\mathrm{d}\overline{f}(\overline{u}(t))}{\mathrm{d}t} = (L_{Y_{h_{\overline{D}}}}\overline{f})(\overline{u}(t)) \qquad \text{for every } \overline{f} \in C^\infty(\overline{D}) \tag{42}$$

and the initial condition. The space of smooth functions $C^\infty(\overline{D})$ on the smooth manifold $\overline{D}$ consists of the push forwards $\rho_* f$ of G-invariant functions f on D. In other words, $C^\infty(\overline{D}) = \{\rho_* f \mid f \in C^\infty(D)^G\}$. Hence equation (42), which determines $t \mapsto \overline{u}(t)$, is equivalent to

$$\frac{\mathrm{d}f(u(t))}{\mathrm{d}t} = (L_{Y_h} f)(u(t)) \quad \text{for every } f \in C^\infty(D)^G \tag{43}$$

for $t \mapsto u(t)$ where $\overline{u}(t) = \rho(u(t))$.

Recall that a curve $t \mapsto u(t)$ satisfies the *momentum equation* for a smooth vector field Z on Q if and only if

$$\frac{\mathrm{d}P_Z(u(t))}{\mathrm{d}t} = -(L_{X_{P_Z}} h)(u(t)), \tag{44}$$

where $P_Z(u) = \mathrm{k}(u, Z(\tau_Q(u)))$ and X_{P_Z} is the Hamiltonian vector field corresponding to the function P_Z on the symplectic manifold (TQ, ω). A curve $t \mapsto u(t)$ is said to satisfy the *second order differential equation condition* for $f \subset C^\infty(Q)$ if

$$\frac{\mathrm{d}\tau_Q^* f(u(t))}{\mathrm{d}t} = \langle \mathrm{d}f \mid u(t) \rangle. \tag{45}$$

Theorem 3.7.2.51. *The reduced equations of motion* (42) *are equivalent to the momentum equation* (44) *for G-invariant vector fields on Q with values in D and the second order differential equation condition* (45) *for G-invariant functions $f \in C^\infty(Q)$.*

Proof. It follows from theorem 1.8.2.36 of chapter 1 that the evolution $t \mapsto u(t)$ of the constrained system satisfies the momentum equation for all smooth vector fields on Q with values in D together with the second order differential equation condition for every function $f \in C^\infty(Q)$. Let $t \mapsto u'(t)$ be another curve in D such that $\rho(u'(t)) = \rho(u(t))$ for all t. Then $u'(t) = g(t) \cdot u(t)$ for some smooth curve $t \mapsto g(t) \in G$. For every G-invariant vector field Z on Q, the momentum P_Z is G-invariant. Also $L_{X_{P_Z}} h$ is G-invariant. Hence

$$\begin{aligned}\frac{\mathrm{d}P_Z(u'(t))}{\mathrm{d}t} &= \frac{\mathrm{d}P_Z(g(t)\cdot u(t))}{\mathrm{d}t} = \frac{\mathrm{d}P_Z(u(t))}{\mathrm{d}t} \\ &= -(L_{X_{P_Z}} h)(u(t)) = -(L_{X_{P_Z}} h)(g^{-1}(t)\cdot u'(t)) \\ &= -(L_{X_{P_Z}} h)(u'(t)).\end{aligned}$$

Thus $t \mapsto u'(t)$ satisfies the momentum equation for all smooth vector fields on Q with values in D. Similarly, for every G-invariant function $f \in C^\infty(Q)$,

$$\begin{aligned}\frac{\mathrm{d}\tau_Q^* f(u'(t))}{\mathrm{d}t} &= \frac{\mathrm{d}\tau_Q^* f(g(t)\cdot u(t))}{\mathrm{d}t} = \frac{\mathrm{d}\tau_Q^* f(u(t))}{\mathrm{d}t} \\ &= \langle \mathrm{d}f \mid u(t)\rangle = \langle \mathrm{d}f \mid u'(t)\rangle.\end{aligned}$$

Hence the curve $t \mapsto u'(t)$ satisfies the second order differential equation condition for all G-invariant functions $f \in C^\infty(Q)$.

We now prove the converse. For each $\overline{q} \in \overline{Q}$ there exists an open neighborhood $\overline{U}_1$ of $\overline{q}$ and smooth functions $\{\overline{f}_i\}_{i=1}^m$ on $\overline{Q}$ whose restrictions $\{\overline{f}_i \mid \overline{U}_1\}_{i=1}^m$ form a chart for $\overline{Q}$.

Let $\dim(\mathrm{ver}\, D) = k$. There exists a neighborhood $\overline{U}_2$ of $\overline{q}$ contained in $\overline{U}_1$ such that $\pi^{-1}(\overline{U}_1) = G \times \overline{U}_1$. We can choose G-invariant vector fields $\{Z_i'\}_{i=1}^k$ with values in ver D which are linearly independent in $\pi^{-1}(\overline{U}_2)$. If $\dim \overline{D} = \ell$, there is a neighborhood $\overline{U}$ of $\overline{q}$ contained in $\overline{U}_2$ and ℓ vector fields $\{\overline{Z}_i\}_{i=1}^\ell$ on $\overline{Q}$ with values in $\overline{D}$ which are linearly independent on $\overline{U}$. For each $i = 1, ..., \ell$, let

$$Z_i'' = \Lambda \circ \overline{Z}_i + \mathrm{lift}\, \overline{Z}_i.$$

From equation (39) it follows that the vector fields $\{Z_i''\}_{i=1}^\ell$ have values in $(\mathrm{ver}\, D)^\perp \cap D$. Since Λ and lift are G-invariant mappings, we find that Z_i'' are G-invariant. Thus $\{Z_1', ..., Z_k', Z_1'', ..., Z_\ell''\}$ are G-invariant vector fields on Q with values in D which are linearly independent in $\pi^{-1}(\overline{U})$.

Let $n = \dim D$. Since the metric k is G-invariant, we can use the Gram-Schmidt orthogonalization together with a partition of unity to construct $n = k + \ell$ vector fields $\{Z_i\}_{i=1}^n$ on Q with values in D which are G-invariant and such that at every point $q' \in \pi^{-1}(\overline{U})$, $\{Z_i(q')\}_{i=1}^n$ is an orthonormal basis in $D_{q'}$.

Hence for each $q' \in \pi^{-1}(\overline{U})$, every vector $u \in D_q$ is of the form

$$u = c_1 Z_1(q') + ... + c_n Z_n(q').$$

Since the vector fields $\{Z_i\}_{i=1}^n$ are orthonormal, it follows that

$$c_i = \mathrm{k}(u, Z_i(q')) = P_{Z_i}(u)$$

for every $i = 1, ..., n$. Since the momenta P_Z are G-invariant, they push forward to functions $\rho_* P_Z$ on $(TQ)/G$.

From proposition 3.7.1.50 it follows that $\overline{D}$ is a fiber bundle over $\overline{Q}$. Let $\chi : \overline{D} \to \overline{Q}$ be its projection map. For each $u \in D$, we have $\chi(\rho(u)) = \pi(\tau_Q(u))$. The functions $\{\rho^*\chi^*\overline{f}_1, ..., \rho^*\chi^*\overline{f}_m, P_{Z_1}, ..., P_{Z_n}\}$ are smooth, G-invariant, and independent on $D \cap T(\pi^{-1}(\overline{U}))$. By proposition 3.7.1.48

$$\dim D/G = \dim \overline{Q} + \dim(\mathrm{ver}\ D) + \dim \overline{D} = m + k + \ell = m + n.$$

Hence the $m+n$ functions $\{\chi^*\overline{f}_1, ..., \chi^*\overline{f}_m, \rho_* P_{Z_1}, ..., \rho_* P_{Z_n}\}$ define a chart. This means that, for every function $\overline{f} \in C^\infty(\overline{D})$, there exists a smooth function F on $\mathbb{R}^{m+n}$ such that the restriction of $\overline{f}$ to $\chi^{-1}(\overline{U})$ satisfies

$$\overline{f} \mid \chi^{-1}(\overline{U}) = F(\chi^*\overline{f}_1, ..., \chi^*\overline{f}_m, \rho_* P_{Z_1}, ..., \rho_* P_{Z_n}) \mid \chi^{-1}(\overline{U}).$$

Every G-invariant smooth function f on D is a pull back of a function $\overline{f} \in C^\infty(\overline{D})$ by the orbit map $\rho : D \to \overline{D}$, that is, $f = \rho^*\overline{f}$. Hence the derivative f along $t \mapsto u(t) \in D \cap T(\pi^{-1}(\overline{U}))$ is given by

$$\begin{aligned}\frac{\mathrm{d}f(u(t))}{\mathrm{d}t} &= \frac{\mathrm{d}\overline{f}(\rho(u(t))}{\mathrm{d}t} = \frac{\mathrm{d}F}{\mathrm{d}t}(\chi^*\overline{f}_1, ..., \chi^*\overline{f}_m, \rho_* P_{Z_1}, ..., \rho_* P_{Z_n})(\rho(u(t)))\\ &= \frac{\mathrm{d}F}{\mathrm{d}t}(\rho^*\chi^*\overline{f}_1, ..., \rho^*\chi^*\overline{f}_m, P_{Z_1}, ..., P_{Z_n})(u(t)).\end{aligned}$$

But $\chi \circ \rho = \pi \circ \tau_Q$ implies that $\rho^*\chi^*\overline{f}_i(u) = \tau_Q^*\pi^*\overline{f}_i$ for all $i = 1, ..., m$. Thus $\frac{\mathrm{d}f}{\mathrm{d}t}(u(t))$ is uniquely determined by the momentum equations for the G-invariant vector fields $\{Z_i\}_{i=1}^n$ on Q and the second order differential equation condition for G-invariant functions $\{\pi^*\overline{f}_i\}_{i=1}^m$ on Q. □

From the proof of theorem 3.7.2.51 we see that to determine the reduced equations of motion locally, it suffices to take into account the second order differential equation conditions for G-invariant functions $\{\pi^*\overline{f}_i\}_{i=1}^m$ on Q which push forward to local coordinates on $\overline{Q}$ and the momentum equations for G-invariant vector fields $\{Z_i\}_{i=1}^m$ on Q which form a local moving frame for the distribution D.

Corollary 3.7.2.52. *Let D be a distribution on a Lie group G, k a Riemannian metric on G, and V a smooth function on G which are invariant under the action of G on itself by left multiplication. Then the G-reduced equations of motion for the constrained mechanical system (ℓ, D) with Lagrangian $\ell = \frac{1}{2}k - V$, where k is the kinetic energy k associated to the* *kinetic energy metric* k, V *is the potential energy, and* D *is the constraint distribution, are given by the momentum equations corresponding to left invariant vector fields on* G *spanning* D.

Proof. Since the action of G on itself is transitive, it follows that there are no nontrivial second order differential equation conditions. Hence theorem 3.7.2.51 implies that the G-reduced equations of motion are given by the momentum equations corresponding to invariant vector fields on G spanning D. □

3.8 Notes

When G is compact, proposition 3.1.4.4 follows from proposition A4 of Field [44]. The conclusion of corollary 3.1.4.6 should be compared with theorem 1 of Michel [80], which states that for every smooth action of a compact Lie group G on M, every smooth G-invariant function has m as a critical point if and only if the orbit $G \cdot m$ is an isolated element of the orbit type M_{G_m}/G in the orbit space M/G. We note that Michel's theorem holds for proper actions.

We now give a brief history of the theory of reduction for Hamiltonian and nonholonomic systems. For Hamiltonian systems two cases were considered: the regular case where the symmetry group acts properly and freely on phase space with a momentum map and the singular case where the group acts properly and has a momentum map. The regular case is classical, see [77], [79], [109], and [68]; whereas the singular case is of much more recent origin, see [6], [102], [5], [30], [12], [85], and [87].

In the regular case, the theory of reduction for nonholonomic systems was pioneered by Bates and Śniatycki [13], who showed that the distributional Hamiltonian formulation of the equations of motion is preserved under reduction, see also Cushman and Śniatycki [35]. One difficulty with nonholonomic systems is that symmetries need not give rise to conserved quantities. Thus for a nonholonomic system the reduced space is the space of orbits of the symmetry acting on phase space, whereas for Hamiltonian systems it is the space of orbits of the symmetry on the inverse image of a coadjoint orbit under the momentum map. Bloch, Krishnaprasad, Marsden and Murray [16] have formulated a Lagrangian theory of symmetry reduction, which does not preserve the form of the equations of motion.

For the singular case of reduction for nonholonomic systems the text follows the approach of Śniatycki [104], which is based on Cushman and

Śniatycki [33] for the Hamiltonian case. It uses the concepts of an accessible set of a generalized distribution, the Stefan-Sussmann theorem [111], [112], and differential spaces [100] to reduce the proof to the regular case. Theorem 3.6.1.41 is the nonholonomic analogue of a theorem of Ortega [85] in the Hamiltonian case. The use of the category of subcartesian spaces in the theory of reduction is taken from [106].

In §3 of chapter 3 Neimark and Fufaev [82] call a system satisfying 1–3 of §3 with $G = \mathbb{R}^n$ a nonholonomic Chaplygin system. The generalization to an arbitrary Lie group G and all the results of this section are due to Koiller [59]. The case of a classical mechanical system with $\ell = k - V$, where the kinetic energy k is defined by a Riemannian metric k on Q, and the infinitesimal G-action is k-orthogonal to G, then D is called the *mechanical connection* of the system, see Marsden [75]. The observation that the vector fields associated to the Lagrange derivative of the reduced Lagrangian and that to the reduced Lagrange derivative have the same zeroes and linearization at each zero is due to Korteweg [62], see §13 and §14. Theorem 3.6.1.39 is due to Souriau [109], when L is a symplectic manifold, $H = TL$, and $\mathfrak{h} = \mathfrak{g}$.

Chapter 4

Reconstruction, relative equilibria and relative periodic orbits

4.1 Reconstruction

Suppose that $\Phi : G \times M \to M$ is a smooth proper action of a Lie group G on a smooth manifold M. Let V be a smooth G-invariant vector field on M with flow $\varphi_t : D_t \to M : m \mapsto \varphi_t(m)$, where $\gamma : I_m \to M : t \mapsto \varphi_t(m)$ is an integral curve of V starting at m with maximal domain of definition I_m. Here $D_t = \{m \in M \big| t \in I_m\}$. Since V is G-invariant, $\Phi_g \circ \varphi_t = \varphi_t \circ \Phi_g$ for every $g \in G$. Therefore the flow φ_t induces a reduced flow $\overline{\varphi}_t : \overline{D}_t \to \overline{M} = M/G$ on the space $\overline{M}$ of G-orbits on M. Here $\overline{D}_t = \{\overline{m} \in \overline{M} \big| t \in I_{\overline{m}} = I_m$, where $m \in \overline{m}\}$. The map $\overline{\varphi}_t$ is the flow of a reduced vector field $\overline{V}$ on $\overline{M}$, thought of as the differential space $(\overline{M}, C^\infty(\overline{M}))$, see §4 of chapter 2.

4.1.1 *Reconstruction for proper free actions*

Assume that the G-action Φ is free. Then the reduced space $\overline{M}$ is a smooth manifold and the reduced vector field $\overline{V}$ satisfies $T_m\pi\, V(m) = \overline{V}(\pi(m))$ for every $m \in M$. Here $\pi : M \to \overline{M} : m \mapsto \overline{m}$ is the G-orbit map.

Suppose that $\overline{\gamma} : I_{\overline{m}} \to \overline{M}$ is an integral curve of the reduced vector field $\overline{V}$ on $\overline{M}$ starting at $\overline{m}$. The purpose of reconstruction is to find an integral curve $\gamma : I_m \to M$ of V which starts at $m \in \overline{m}$ such that $\pi \circ \gamma(t) = \overline{\gamma}(t)$ for every $t \in I_m$. To do this we start with an arbitrary smooth curve $\beta : I_m \to M$ such that $\pi(\beta(t)) = \overline{\gamma}(t)$. For each $t \in I_m$ there is a unique element $g(t) \in G$ such that

$$\gamma(t) = g(t) \cdot \beta(t), \tag{1}$$

because the action of G on M is free. Moreover the curve $I_m \to G : t \mapsto g(t)$ is smooth with $g(0) = e$, the identity element of G. Using the inverse

$$G \times \mathfrak{g} \to TG : (g, \xi) \mapsto T_eL_g\xi$$

of the standard left trivialization of TG, where $L_g : G \to G : g' \mapsto gg'$ is left multiplication by g, we transfer the tangent vector $\dot{g}(t) \in T_{g(t)}G$ to an element $\xi(t)$ of the Lie algebra $\mathfrak{g} = T_eG$. Differentiating (1) with respect to t gives

$$\dot{\gamma}(t) = T_{g(t)}\Phi_{\beta(t)}(T_eL_{g(t)}\xi(t)) + T_{\beta(t)}\Phi_{g(t)}\dot{\beta}(t). \tag{2}$$

But $t \mapsto \gamma(t)$ is an integral curve of V. So

$$\dot{\gamma}(t) = V(\gamma(t)) = V(g(t)\cdot\beta(t)) = T_{\beta(t)}\Phi_{g(t)}\,V(\beta(t)), \tag{3}$$

since the vector field V is G-invariant. Moreover,

$$T_{g(t)}\Phi_{\beta(t)}(T_eL_{g(t)}\xi(t)) = X_{\xi(t)}(g(t)\cdot\beta(t)) = T_{\beta(t)}\Phi_{g(t)}X_{\xi(t)}(\beta(t)). \tag{4}$$

Substituting (3) and (4) into (2) and cancelling $T_{\beta(t)}\Phi_{g(t)}$ gives

$$X_{\xi(t)}(\beta(t)) = V(\beta(t)) - \dot{\beta}(t). \tag{5}$$

In order to solve equation (5) for $\xi(t)$ we use a *connection* 1-form Θ *for the* G*-principal bundle* $\pi : M \to \overline{M}$. Specifically, Θ is a smooth $\mathfrak{g}$-valued 1-form on M satisfying

1. for every $(m,\xi) \in M \times \mathfrak{g}$ we have $\Theta(m)(X_\xi(m)) = \xi$;
2. for every $(m,g) \in M \times G$ and every $v_m \in T_mM$, we have

$$\Theta(g\cdot m)(T_m\Phi_g\, v_m) = \mathrm{Ad}_g(\langle\Theta(m)(v_m)\rangle).$$

The linear subspace $\mathrm{hor}_m = \ker\Theta(m)$ in T_mM is complementary to $\mathrm{ver}_m = T_m(G\cdot m) = \ker T_m\pi$ in T_mM. Moreover, $\mathrm{hor} : M \to TM : m \mapsto \mathrm{hor}_m$ and $\mathrm{ver} : M \to TM : m \mapsto \mathrm{ver}_m$ are smooth vector subbundles of TM called the *horizontal* and *vertical distributions*, respectively. From property 1 it follows that hor is invariant under the G action on TM induced from the G-action on M. Assuming that M is paracompact, we can piece together connection 1-forms on local trivializations of the bundle π by means of a partition of unity in $\overline{M}$ to construct a connection 1-form Θ on the G-principal bundle M. Applying $\Theta(\beta(t))$ to both sides of (5) gives

$$\xi(t) = \Theta(\beta(t))(X_{\xi(t)}(\beta(t))) = \Theta(\beta(t))(V(\beta(t)) - \dot{\beta}(t)). \tag{6}$$

Substituting (6) into (5) gives a differential equation satisfied by β, which is what we wanted.

4.1.2 *Reconstruction for nonfree proper actions*

Let H be a compact subgroup of G and let N be a connected component of the isotropy type $M_H = \{m \in M \big| G_m = H\}$. By item 1 of theorem 2.3.3.7, the submanifold N is smooth. It is also invariant under the flow of the vector field V. Because the subgroup $N(H) = \{g \in G \big| gHg^{-1} = H\}$ of G preserves M_H, the free and proper action of $G_{M_H} = N(H)/H$ on M_H is a symmetry of $V|M_H$. More precisely, if $G_N = \{g \in G_{M_H} \big| g \cdot n \in N \text{ for every } n \in N\}$, then G_N acts freely and properly on N and is a symmetry of $V|N$. Thus there is a unique smooth reduced vector field $\overline{V} = \pi_* V$ on $\overline{N} = \pi(N)$, which is contained in the orbit type $\overline{M_{(H)}} = \pi(M_{(H)})$ in the orbit space $\overline{M}$ such that $\overline{V}(\pi(n)) = T_n\pi\, V(n)$ for every $n \in N$. If H is conjugate to H' by an element $g \in G$, then the flow of the vector fields $V|N$ and $V|\Phi_g N$ are conjugate by Φ_g. Therefore the reduced vector field $\overline{V}|\overline{N}$ does not depend on the choice of H in the conjugacy class (H).

The process of *reconstruction* described in §1.1 can be now applied to each connected component of an orbit type in the orbit space.

4.1.3 *Application to nonholonomic systems*

Let us now apply the above theory of reconstruction to a distributional Hamiltonian system (D, H, ϖ, h) which has a symmetry group G that acts properly on D. Let $\gamma : I \to D : t \mapsto u(t)$ be an integral curve of the distributional Hamiltonian vector field Y_h, which starts at $u_0 \in D$.

Let L be an accessible set of the distribution H on D, which passes through u_0. The image of γ lies in L. Thus we can restrict our attention to the distributional Hamiltonian system $(D_L, H_L, \varpi_L, h|L)$. Let $K = G_{u_0}$. Then the stability group $\mathrm{stab}_L = \{g \in G \big| g \cdot u \in L \text{ for every } u \in L\}$, which by hypothesis is a closed subgroup of G, acts properly on L. The group K is a subgroup of stab_L. Let N be a connected component containing u_0 of the submanifold L_K of points of L whose isotropy group in stab_L is K. Then the integral curve γ of $Y_h|N$ is contained in N.

Additional restrictions on the motion γ are provided by the level set of the momentum map $J : D \to \mathfrak{h}^*$. Let $\alpha_N = J(u_0)$. Let stab_N be the stability group of N, which by hypothesis is a closed subgroup of G. The group $G_N = \mathrm{stab}_N/K$ acts freely and properly on N and is a symmetry group of the distributional Hamiltonian system $(N, H_N, \varpi_N, h|N)$. Moreover, G_N has a momentum map $J_N : N \to \mathfrak{h}_N^*$ such that the connected component

M of $N \cap J^{-1}(\alpha_N)$ containing u_0 and the connected component of $J_N^{-1}(\alpha_N)$ coincide. The distributional Hamiltonian system $(M, H_M, \varpi_M, h|M)$ has a symmetry group $G_M = \mathrm{stab}_M/K$, which acts freely and properly on M and has a momentum map $J_M : M \to \mathfrak{h}_M^*$. Let $\beta_M = J_M(u_0)$. The image of the curve γ is contained in a connected component P of $J^{-1}(\beta_M)$. The isotropy group $G_{\beta_M} = \{g \in G_M \,\big|\, \mathrm{Ad}^t_{g^{-1}}\beta_M = \beta_M\}$ acts freely and properly on P with smooth orbit space $\overline{P}$ and orbit map $\rho_P : P \to \overline{P}$, which is a surjective submersion. Thus G_{β_M} is a symmetry of the distributional Hamiltonian system $(P, H_P, \varpi_P, h|P)$, which may be reduced to give a distributional Hamiltonian system $(\overline{P}, \overline{H}_{\overline{P}}, \overline{\varpi}_{\overline{P}}, \overline{h}_{\overline{P}})$, where $\overline{H}_{\overline{P}}$ is a generalized $\overline{\varpi}_{\overline{P}}$-symplectic distribution on $\overline{P}$. Therefore the integral curve γ of $Y_{h|P}$ on P projects under the G_{β_M}-orbit map ρ_P to an integral curve $\overline{\gamma} : I \to \overline{P} : t \mapsto \overline{u(t)}$ of the reduced distributional Hamiltonian vector field $Y_{\overline{h}|\overline{P}}$ on $\overline{P}$, which starts at $\overline{u_0}$. This curve can be reconstructed using the procedure described in §1.1.

4.2 Relative equilibria

In this section we study the *relative equilibria* of a G-invariant vector field V on a smooth manifold M. We assume that the G-action on M is proper.

4.2.1 *Basic properties*

Before we can state what a relative equilibrium is we prove

Lemma 4.2.1.1. *Suppose that G has countably many connected components. The following statements are equivalent.*

1. *The vector $V(m)$ is tangent at m to the G-orbit $G \cdot m$ through m, that is, there is a $\xi \in \mathfrak{g}$ such that $V(m) = X_\xi(m)$.*
2. *The integral curve $\gamma_m : I_m \to M$ of V, which starts at m, is equal to the G-action restricted to a one parameter subgroup $t \mapsto \exp t\,\xi$ of G. In other words, $I_m = \mathbb{R}$ and $\gamma_m(t) = \Phi_m(\exp t\,\xi) = \exp t\,\xi \cdot m$ for every $t \in I_m$.*
3. *$\{\gamma_m(t) \,\big|\, t \in \mathbb{R}\} \subseteq G \cdot m$.*
4. *$\overline{m} = G \cdot m$ is an equilibrium point of the reduced flow on $\overline{M}$, that is, $\overline{\gamma}_{\overline{m}}(t) = \overline{m}$ for every $t \in I_m$, where $m \in \overline{m}$.*

Proof. item 1 $\Longleftrightarrow$ item 2. Suppose that $V(m) = X_\xi(m)$. Let $\varphi_t : D_t \to M$

be the flow of the vector field V. Then

$$X_\xi(\varphi_t(m)) = T_m\varphi_t\, X_\xi(m) = T_m\varphi_t\, V(m) = V(\varphi_t(m)),$$

where the first equality follows because $\Phi_{\exp t\xi} \circ \varphi_t = \varphi_t \circ \Phi_{\exp t\xi}$. Therefore

$$\frac{\mathrm{d}\varphi_t(m)}{\mathrm{d}t} = V(\varphi_t(m)) = X_\xi(\varphi_t(m))$$

and $\varphi_0(m) = m$ implies $\varphi_t(m) = \Phi_{\exp t\xi}(m)$ for every $t \in I_m$. Because $t \mapsto \exp t\xi$ is defined for all $t \in \mathbb{R}$, we deduce that $I_m = \mathbb{R}$. Therefore item 1 $\Rightarrow$ item 2. Differentiating $\gamma_m(t) = \Phi_m(\exp t\xi)$ with respect to t and then setting $t = 0$ gives $V(m) = X_\xi(m)$. Therefore item 2 $\Rightarrow$ item 1.

That item 2 $\Rightarrow$ item 3 is obvious. As is item 3 $\Longleftrightarrow$ item 4.

The implication item 3 $\Rightarrow$ item 1 is obvious if the G-orbit $G \cdot m$ is an embedded submanifold of M. To prove it in general we argue as follows. Let $\mathfrak{q}$ be a complement to $\mathfrak{g}_m = T_eG_m$ in $\mathfrak{g}$. Let Q be a submanifold of G such that $e \in Q$ and $T_eQ = \mathfrak{q}$. Similarly, let E be a complementary subspace in T_mM to $T_m(G\cdot m) = \mathrm{span}\{X_\xi(m) \,\big|\, \xi \in \mathfrak{g}\}$. Let $\mathcal{E}$ be a smooth submanifold of M containing m such that $T_m\mathcal{E} = E$. Then the tangent at (e, m) of the map $\psi : Q \times \mathcal{E} \rightarrow M : (q, x) \mapsto q \cdot x$ is a bijective linear map of $\mathfrak{q} \times E$ onto T_mM. Shrinking Q and $\mathcal{E}$ if necessary, from the inverse function theorem it follows that ψ is a diffeomorphism of $Q \times \mathcal{E}$ onto an open neighborhood U of m in M. Suppose that $\{x_j\}$ is a sequence in $\mathcal{E} \cap (G \cdot m)$ which converges to $x \in \mathcal{E} \cap (G \cdot m)$ both in the topology of $\mathcal{E}$ and in the orbit topology of $G \cdot m$. The latter means that if $x_j = g_j \cdot m$ for a sequence $\{g_j\} \subseteq G$ which converges in G to $g \in G$, then $x = g \cdot m$. Because the tangent to the map $\psi_x : Q \rightarrow M : q \mapsto q \cdot x$ at e is injective, $\mathfrak{q} \cap \mathfrak{g}_x = \{0\}$. Now $x = g \cdot m$ implies that $\mathfrak{g}_x = g\,\mathfrak{g}_m\, g^{-1}$. Therefore $\dim \mathfrak{g}_x = \dim \mathfrak{g}_m = \dim \mathfrak{g} - \dim \mathfrak{q}$. Thus the map $\rho : Q \times G_x \rightarrow G : (q, h) \mapsto q \cdot h$ is a diffeomorphism from an open neighborhood of (e, e) onto an open neighborhood of e in G. Because $g_jg^{-1} \rightarrow e$ in G as $j \rightarrow \infty$, for sufficiently large j we can write $g_jg^{-1} = q_jh_j$ where $q_j \in Q$ and $h_j \in G_x$. Moreover, both of the sequences $\{q_j\}$ and h_j converge to e. So

$$e \cdot x_j = x_j = g_j \cdot m = g_jg^{-1} \cdot x = q_jh_j \cdot x = q_j \cdot x, \tag{7}$$

where the last equality follows because $h_j \in G_x$. Using the injectivity of the mapping ψ from (7) we conclude that $x_j = x$ for sufficiently large j.

Suppose that $\mathcal{E}_0$ is a compact neighborhood of m in $\mathcal{E}$. Let K be a compact subgroup of G. If $\mathcal{E} \cap (K \cdot m)$ is infinite, there is an infinite sequence $\{g_j\}$ of elements of K such that $x_j = g_j \cdot m$ are distinct elements of $\mathcal{E}_0$. Passing to a subsequence we may assume that $g_j \to g$ for some $g \in G$. The sequence $\{x_j\}$ converges to $x \in \mathcal{E}_0$, because $\mathcal{E}_0$ is compact. But this contradicts the conclusion that $x_j = x$ if j is sufficiently large. Because G has a countable number of connected components, it is the union of a countable number of compact subgroups K_n, see theorem 1.9.1 of [42]. Therefore for each positive integer n we know that $\mathcal{E}_0 \cap (K_n \cdot m)$ is a finite set. Thus $\mathcal{E}_0 \cap (G \cdot m)$ is countable.

There is an open interval $I \subseteq I_m$ containing 0 and smooth curves $I \to Q : t \mapsto q(t)$ and $I \to \mathcal{E}_0 : t \mapsto x(t)$ such that $\varphi_t(m) = q(t) \cdot x(t)$ for every $t \in I$ under the assumption that $\gamma_m(t) = \varphi_t(m) \in G \cdot m$ for every $t \in I_m$. Therefore $x(t) \in \mathcal{E}_0 \cap (G \cdot m)$ for every $t \in I$. Because $\mathcal{E}_0 \cap (G \cdot m)$ is countable, this implies that $t \mapsto x(t)$ is constant and therefore $x(t) = x(0) = m$. Hence

$$\varphi_t(m) = q(t) \cdot m \tag{8}$$

for every $t \in I$. Differentiating (8) with respect to t and then setting $t = 0$ gives $V(m) = X_\xi(m)$, where $\xi = q'(0) \in \mathfrak{q} \subseteq \mathfrak{g}$. Therefore item 3 $\Rightarrow$ item 1. □

A point $m \in M$ (and also the integral curve γ_m of the G-invariant vector field V on M which starts at m) is a *relative equilibrium* of V if and only if one (and hence each) of the conditions 1 – 4 of lemma 4.2.1.1 is satisfied. If item 1 is satisfied then we say that $\xi \in \mathfrak{g}$ is a *generator* of the relative equilibrium m of the G-invariant vector field V.

Corollary 4.2.1.2. *An equilibrium point m of V with $V(m) = X_\xi(m)$ is a relative equilibrium if and only if $\xi \in \mathfrak{g}_m$.*

Proof. $\Rightarrow$. Take $\xi = 0$ in item 1.

$\Leftarrow$. If m is an equilibrium point of V, which is a relative equilibrium then $0 = X_\xi(m)$. So $\exp t\xi \cdot m = m$ for every $t \in \mathbb{R}$, that is, $\exp t\xi \in G_m$ for every $t \in \mathbb{R}$. Therefore $\xi \in \mathfrak{g}_m$, which is the Lie algebra of the isotropy group G_m. □

Corollary 4.2.1.3. *If m is a relative equilibrium of V generated by $\xi \in \mathfrak{g}$ and the one parameter group $\mathbb{R} \to G : t \mapsto \exp t\xi$ is periodic with set of periods* $\mathrm{Per}_m = \{t \in \mathbb{R} \,\big|\, \exp t\xi \cdot m = m\}$, *then* Per_m *is equal to* i) $\mathbb{R}$, *or* ii) $\mathbb{Z} \cdot \tau$ *for a unique* $\tau > 0$, *or* iii) $\{0\}$.

Proof. Note that Per_m is a closed subgroup of $\mathbb{R}$. Hence it is one of the three possibilities listed in the statement of the corollary. In case 1 the integral curve γ_m of V starting at m is an equilibrium point of V. In case 2 it is a nonconstant periodic integral curve with primitive period τ. In case 3 the relative equilibrium γ_m is *not* periodic. □

4.2.2 *Quasiperiodic relative equilibria*

A smooth curve $\gamma : \mathbb{R} \to M$ is *quasiperiodic with at most k-frequencies* if and only if there is a smooth map $\Gamma : \mathbb{R}^k/\mathbb{Z}^k \to M$ from the *standard k-torus* $\mathbb{R}^k/\mathbb{Z}^k$ to M and a vector $\nu \in \mathbb{R}^k$, called the *frequency vector*, such that $\gamma(t) = \Gamma(t\,\nu + \mathbb{Z}^k)$ for every $t \in \mathbb{R}$.

Let T be a *torus*, that is, a connected, compact, commutative Lie group. Suppose that $\mathfrak{t}$ is its Lie algebra. Then the exponential map $\exp : \mathfrak{t} \to T$ is a surjective homomorphism of Lie groups from the additive group $(\mathfrak{t}, +)$ onto T. Its kernel $\Lambda = \ker \exp$ is an additive subgroup of the vector space $\mathfrak{t}$. There exist $\{\lambda_j\}_{j=1}^{k} \subseteq \Lambda$, which are linearly independent over $\mathbb{R}$, such that every element λ of Λ can be written uniquely as $\lambda = \sum_{j=1}^{k} z_j\,\lambda_j$ for some $z = (z_1, \ldots, z_k) \in \mathbb{Z}^k$. The mapping $\mathbb{Z}^k \to \Lambda : z \mapsto \sum_{j=1}^{k} z_j\,\lambda_j$ is bijective and is called a $\mathbb{Z}$-*basis* of Λ. Because $\mathbb{Z}^k$ is an integer lattice of $\mathbb{R}^k$, we call Λ an *integer lattice* of $\mathfrak{t}$.

Let $\ell = \dim \mathfrak{t}$. We extend $\{\lambda_j\}_{j=1}^{k}$ to an $\mathbb{R}$-basis $\{\lambda_j\}_{j=1}^{\ell}$ of $\mathfrak{t}$. Consider the mapping $\widetilde{\psi} : \mathbb{R}^\ell \to T : (\theta_1, \ldots, \theta_\ell) \mapsto \exp\big(\sum_{j=1}^{\ell} \theta_j\,\lambda_j\big)$. Then $\widetilde{\psi}$ is a bijective homomorphism of Lie groups from $\big(\mathbb{R}^k/\mathbb{Z}^k\big) \times \mathbb{R}^{\ell-k}$ onto T. Therefore $\widetilde{\psi}^{-1}$ is continuous. Hence $\big(\mathbb{R}^k/\mathbb{Z}^k\big) \times \mathbb{R}^{\ell-k}$ is compact, which implies $\ell = k$. Thus $\widetilde{\psi}$ induces an isomorphism ψ from the Lie group $\mathbb{R}^k/\mathbb{Z}^k$ onto T. Consequently, every $\mathbb{Z}$-basis of Λ is an $\mathbb{R}$-basis of T. Therefore $\{\lambda'_j\}_{i=1}^{k'}$ is a $\mathbb{Z}$-basis of Λ if and only if $k' = \dim \mathfrak{t} = k$ and $\lambda'_i = \sum_{j=1}^{k} A_{ji}\lambda_j$ for every $1 \le i \le k$. The entries (A_{ij}) of the $k \times k$ matrix A are integers. Moreover, A is invertible and its inverse has integer entries. Therefore $\det A = \pm 1$. In other words, $A \in \mathrm{Gl}(k, \mathbb{Z})$. The mapping $\widetilde{\psi}$ above is not unique, the freedom of nonuniqueness is precisely the choice of a $\mathbb{Z}$-basis of Λ.

Lemma 4.2.2.4. *Let G be a Lie group with Lie algebra $\mathfrak{g}$. If $\xi \in \mathfrak{g}$, then $\gamma : \mathbb{R} \to G : t \mapsto \exp t\,\xi$ is either a dense one parameter subgroup of a torus subgroup T of G or the mapping γ is proper.*

Proof. Let T be the closure in G of the set $S = \{\exp t\,\xi \,\big|\, t \in \mathbb{R}\}$. Because S is a commutative, connected subgroup of G so is T. Since T is a closed subgroup of G, it is a Lie group, which is isomorphic to $(\mathbb{R}^k/\mathbb{Z}^k) \times \mathbb{R}^\ell$.

Because the image of γ lies in T, it follows that $\xi = \gamma'(0)$ lies in the Lie algebra $\mathfrak{t}$ of T. Let $(\eta, \zeta) \in \mathbb{R}^k \times \mathbb{R}^\ell$ be the corresponding element in the Lie algebra of $(\mathbb{R}^k/\mathbb{Z}^k) \times \mathbb{R}^\ell$. Then $\exp t\,\xi$ corresponds to $(t\,\eta + \mathbb{Z}^k, t\,\zeta)$ in $(\mathbb{R}^k/\mathbb{Z}^k) \times \mathbb{R}^\ell$. Since S is dense in T, it follows that $U = \{(t\,\eta + \mathbb{Z}^k, t\,\zeta) \,\big|\, t \in \mathbb{R}\}$ is dense in $(\mathbb{R}^k/\mathbb{Z}^k) \times \mathbb{R}^\ell$.

If $\zeta \neq 0$, then the map $\mathbb{R} \rightarrow (\mathbb{R}^k/\mathbb{Z}^k) \times \mathbb{R}^\ell : t \mapsto (t\,\eta + \mathbb{Z}^k, t\,\zeta)$ is proper, because the projection $\mathbb{R}^\ell \rightarrow \mathbb{R} : t\,\zeta \mapsto t$ is proper. Hence the mapping $\mathbb{R} \rightarrow T : t \mapsto \exp t\,\xi$ is proper. Because T is a closed subset of G it follows that γ is proper.

If $\zeta = 0$, then U is contained in the torus subgroup $(\mathbb{R}^k/\mathbb{Z}^k) \times \{0\}$ of $(\mathbb{R}^k/\mathbb{Z}^k) \times \mathbb{R}^\ell$. Because U is dense in $(\mathbb{R}^k/\mathbb{Z}^k) \times \mathbb{R}^\ell$, it follows that $\ell = 0$. Therefore T is a torus subgroup of G. □

An element $\xi \in \mathfrak{g}$ is *elliptic* if and only if the image of $\gamma : \mathbb{R} \rightarrow G : t \mapsto \exp t\,\xi$ is dense in a torus subgroup of G.

Lemma 4.2.2.5. *Let ξ be an elliptic element of the Lie algebra $\mathfrak{g}$ of the Lie group G. Then*

1. *the closure in G of $\{\exp t\,\xi \,\big|\, t \in \mathbb{R}\}$, is a torus subgroup T of G. Also $\xi \in \mathfrak{t}$, the Lie algebra of T.*
2. *For every $\mathbb{Z}$-basis $\{\lambda_j\}_{j=1}^k$ of the integer lattice of $\mathfrak{t}$, the mapping $\widetilde{\psi}$: $\mathbb{R}^k \rightarrow T : (\theta_1, \ldots, \theta_k) \mapsto \exp\big(\sum_{j=1}^k \theta_j\,\lambda_j\big)$ induces an isomorphism ψ : $\mathbb{R}^k/\mathbb{Z}^k \rightarrow T$ of Lie groups from the standard k-torus $\mathbb{R}^k/\mathbb{Z}^k$ onto T.*
3. *Let $\nu = (\nu_1, \ldots, \nu_k) \in \mathbb{R}^k$. Then $U = \{t\,\nu + \mathbb{Z}^k \,\big|\, t \in \mathbb{R}\}$ is dense in $\mathbb{R}^k/\mathbb{Z}^k$ if and only if the real numbers $\{\nu_j\}_{j=1}^k$ are linearly independent over $\mathbb{Q}$.*
4. *Let $\xi = (\xi_1, \ldots, \xi_k)$ with respect to the $\mathbb{R}$-basis $\{\lambda_j\}_{j=1}^k$ of $\mathfrak{t}$. Then the real numbers $\{\xi_j\}_{j=1}^k$ are linearly independent over $\mathbb{Q}$ and $U = \{t\,\xi + \mathbb{Z}^k \,\big|\, t \in \mathbb{R}\}$ is dense in $\mathbb{R}^k/\mathbb{Z}^k$.*

Proof. 1. This is just the definition of elliptic element.
2. This follows from lemma 4.2.2.4.
3. Suppose that $0 = \sum_{j=1}^k q_j\,\nu_j$ for some $q_j \in \mathbb{Q}$, not all zero. Clearing denominators gives $0 = \sum_{j=1}^k z_j\,\nu_j$ for $z_j \in \mathbb{Z}$ not all zero. Therefore we obtain a nonzero linear map $\vartheta : \mathbb{R}^k \rightarrow \mathbb{R} : (\theta_1, \ldots, \theta_k) \mapsto \sum_{j=1}^k z_j\,\theta_j$, which maps $\mathbb{Z}^k$ onto $\mathbb{Z}$. So ϑ induces a nonzero homomorphism $\widetilde{\vartheta} : \mathbb{R}^k/\mathbb{Z}^k \rightarrow \mathbb{R}/\mathbb{Z}$. Clearly, $U \subseteq \ker\widetilde{\vartheta}$. But $\ker\widetilde{\vartheta}$ is a closed subgroup of $\mathbb{R}^k/\mathbb{Z}^k$ of dimension $k-1$. Consequently, U is not dense in $\mathbb{R}^k/\mathbb{Z}^k$. Conversely, if U is not dense in $\mathbb{R}^k/\mathbb{Z}^k$, then item 1 implies that the closure of U in $\mathbb{R}^k/\mathbb{Z}^k$ is a

subtorus S of $\mathbb{R}^k/\mathbb{Z}^k$ of dimension $\ell < k$. Therefore $(\mathbb{R}^k/\mathbb{Z}^k)/S$ is a torus of dimension $k - \ell > 0$. Composing the isomorphism $\lambda : (\mathbb{R}^k/\mathbb{Z}^k)/S \to (\mathbb{R}/\mathbb{Z})^{k-\ell}$ with the projection $(\mathbb{R}/\mathbb{Z})^{k-\ell} \to \mathbb{R}/\mathbb{Z}$ on the first factor, gives a surjective homomorphism μ of $(\mathbb{R}^k/\mathbb{Z}^k)/S$ onto $\mathbb{R}/Z$. Therefore $\vartheta = \mu \circ \lambda : \mathbb{R}^k/\mathbb{Z}^k \to \mathbb{R}/\mathbb{Z}$ is a surjective homomorphism such that $S \subseteq \ker \vartheta$. The tangent map $T\vartheta$ of ϑ is a linear map from $\mathbb{R}^k$ to $\mathbb{R}$ which sends $\mathbb{Z}^k$ to $\mathbb{Z}$. Therefore $T\vartheta = (z_1, \ldots, z_k)$ where $z_j \in \mathbb{Z}$ for $1 \le j \le k$. Because $\nu \in \mathfrak{s}$, the Lie algebra of S, and $\mathfrak{s} \subseteq \ker T\vartheta$, it follows that $0 = \sum_{j=1}^k z_j \nu_j$. This completes the proof of item 3.
4. This follows from item 3, $L = \psi^{-1}(S)$, and the fact that ψ^{-1} is continuous. □

We are now in position to state the following proposition about quasiperiodic relative equilibria.

Proposition 4.2.2.6. *Suppose that m is a relative equilibrium of a G-invariant vector field V on a smooth manifold M generated by $\xi \in \mathfrak{g}$. If ξ is an elliptic element of $\mathfrak{g}$, then the integral curve $\gamma : I_m \to M$ of V which starts at m is quasiperiodic. Moreover, the minimal number of frequencies of γ is equal to the dimension of the closure in M of $\{\gamma(t) \in M \,\big|\, t \in \mathbb{R}\}$.*

Proof. Because m is a relative equilibrium generated by ξ, it follows that $\gamma(t) = \exp t\,\xi \cdot m = \Phi_m(\exp t\,\xi)$. Therefore the closure of $\{\gamma(t) \in M \,\big|\, t \in \mathbb{R}\}$ in M is the image under Φ_m of the closure T in G of the one parameter subgroup generated by ξ. Since ξ is elliptic, T is a torus. If $\{\lambda_j\}_{j=1}^k$, where $k = \dim T$, is a $\mathbb{Z}$-basis of the integer lattice $\Lambda = \ker \exp$ of $\mathfrak{t}$, the Lie algebra of T, then we have a mapping

$$\Gamma : \mathbb{R}^k/\mathbb{Z}^k \to M : (\theta_1, \ldots, \theta_k) \mapsto \exp\big(\sum_{j=1}^k \theta_j \lambda_j\big) \cdot m.$$

With respect to the $\mathbb{R}$-basis $\{\lambda_j\}_{j=1}^k$ of $\mathfrak{t}$ write $\xi = \sum_{j=1}^k \nu_j \lambda_j$. Then the real numbers $\{\nu_j\}_{j=1}^k$ are linearly independent over $\mathbb{Q}$, since ξ is elliptic, see item 3 in lemma 4.2.2.5. From item 4 of lemma 4.2.2.5 it follows that $\{t\,\nu + \mathbb{Z}^k \,\big|\, t \in \mathbb{R}\}$ is dense in $\mathbb{R}^k/\mathbb{Z}^k$. Because $\gamma(t) = \exp t\,\xi \cdot m = \Gamma(t\,\nu + \mathbb{Z}^k)$, we see that $\{\gamma(t) \in M \,\big|\, t \in \mathbb{R}\}$ is dense in $\Gamma(\mathbb{R}^k/\mathbb{Z}^k)$. Since $\Gamma(\mathbb{R}^k/\mathbb{Z}^k)$ is a compact, and therefore closed, subset of M, $\Gamma(\mathbb{R}^k/\mathbb{Z}^k)$ is equal to the closure of $\{\gamma(t) \in M \,\big|\, t \in \mathbb{R}\}$ in M.

Let $\widetilde{\psi} : \mathbb{R}^k \to T$ be the map constructed in item 2 of lemma 4.2.2.5. Then $\Gamma^{-1}(m) = \widetilde{\psi}^{-1}(T \cap G_m)$ is a sublattice Λ_0 of the integer lattice Λ of $\mathfrak{t}$.

Thus there is an induced isomorphism ϑ of Lie groups from $T_0 = \Lambda/\Lambda_0$ onto $T/(T \cap G_m)$. Because $T/(T \cap G_m)$ is a compact, connected and commutative group, it is a torus. In fact, $\widetilde{\Gamma} = \Phi_m \circ \vartheta$ is an injective immersion of T_0 into M. Since T_0 is compact, $\widetilde{\Gamma}$ is an embedding. Let ℓ be the minimal number of frequencies attained by a smooth map $\delta : \mathbb{R}^\ell/\mathbb{Z}^\ell \to M$ with $\gamma(t) = \delta(t\,\omega + \mathbb{Z}^\ell)$, where $\omega \in \mathbb{R}^\ell$. The minimality of ℓ implies that $\{t\,\omega + \mathbb{Z}^\ell \,\big|\, t \in \mathbb{R}\}$ is dense in $\mathbb{R}^\ell/\mathbb{Z}^\ell$, using the proof of item 3 in lemma 4.2.2.5. Therefore the image of δ is equal to the closure of $\{\gamma(t) \in M \,\big|\, t \in \mathbb{R}\}$ in M. So $\delta\big(\mathbb{R}^\ell/\mathbb{Z}^\ell\big) = \widetilde{\Gamma}(T_0)$. By Sard's theorem the set of regular values of δ has full measure. Thus there is a point in $\mathbb{R}^\ell/\mathbb{Z}^\ell$ where the tangent map of δ is surjective. Consequently, $\ell \geq \dim T_0$. Thus $\dim T_0$ is the minimal number of frequencies. □

Corollary 4.2.2.7. *If the G-action is free at m, then $T_0 = \mathbb{R}^k/\mathbb{Z}^k$, $\widetilde{\Gamma} = \Gamma$, and the minimal number of frequencies is k.*

4.2.3 *Runaway relative equilibria*

A continuous curve $\gamma : I \to M$ with I an open interval in $\mathbb{R}$ is a *runaway curve* in M if and only if the map γ is proper, that is, if K is a compact subset of M, then $\gamma^{-1}(K)$ is a compact subset of I. Because γ is continuous, the inverse image of every closed subset of M is a closed subset of I. Since compact subsets of I are closed subsets which are contained in a closed interval $[a, b]$ for some $a, b \in I$, it follows that γ is proper if and only if for every compact subset K of M there are $a, b \in I$ such that $\gamma(t) \in K$ implies $t \in [a, b]$. Equivalently, $t \in I \setminus [a, b]$ implies $\gamma(t) \notin K$. In other words, $t \mapsto \gamma(t)$ runs out of every compact subset of M as t runs to either end of I.

Proposition 4.2.3.8. *Let m be a relative equilibrium of a G-invariant vector field V on a smooth manifold M generated by $\xi \in \mathfrak{g}$. Suppose that the G-action is proper. The following statements are equivalent.*

1. *ξ is not an elliptic element of $\mathfrak{g}$.*
2. *An integral curve $\gamma : I_m \to M$ of the vector field V starting at m is a runaway curve on M.*
3. *γ is not quasiperiodic.*

Proof. item 1 $\Rightarrow$ item 2. Suppose that ξ is not elliptic. Then the one parameter group $\lambda : \mathbb{R} \to G : t \mapsto \exp t\,\xi$ is proper. Because the map Φ_m

is proper by hypothesis, it follows that the map $\Phi_m \circ \lambda : \mathbb{R} \rightarrow M : t \mapsto \exp t\,\xi \cdot m = \gamma(t)$ is proper. Therefore item 1⇒ item 2.

$\neg$ item 3 ⇒ $\neg$ item 2.[1] If γ is quasiperiodic, then its image is contained in $\Gamma(\mathbb{R}^k/\mathbb{Z}^k)$, see the proof of proposition 4.2.2.6. But $\Gamma(\mathbb{R}^k/\mathbb{Z}^k)$ is compact and the map Γ is continuous. Therefore γ is not a runaway curve.

$\neg$ item 1 ⇒ $\neg$ item 3. This follows from item 1 of lemma 4.2.2.5. □

Example 4.2.3.9. Let $G = \mathbb{R} \times \mathbb{C}$ with multiplication

$$(\xi, \zeta) \cdot (\xi', \zeta') = (\xi + \xi', \mathrm{e}^{i\,\xi}\zeta' + \zeta).$$

The only elliptic element of $\mathfrak{g}$ is 0. Let $M = \mathbb{R} \times \mathbb{R} \times \mathbb{C}$ on which G acts by

$$\big((\xi, \zeta), (a, x, z)\big) \mapsto (a, \xi + x, \mathrm{e}^{i\,\xi} z + \zeta).$$

The action of G on M is proper and free. The G-orbit space $\overline{M}$ is $\mathbb{R}$ with orbit map $\pi : M \rightarrow \overline{M} : (a, x, z) \mapsto a$. On M define a vector field V by $V(a, x, z) = (0, a, \mathrm{e}^{i\,x})$. Then V is G-invariant. An integral curve $\gamma : \mathbb{R} \rightarrow M : t \mapsto \big(a(t), x(t), z(t)\big)$ of V starting at $m = \big(a(0), x(0), z(0)\big)$ is given by

$$\begin{aligned} a(t) &= a(0) \\ x(t) &= x(0) + t\,a(0) \\ z(t) &= z(0) + \begin{cases} \frac{1}{i\,a(0)}\,(\mathrm{e}^{ia(0)\,t} - 1)\mathrm{e}^{i\,x(0)}, & \text{if } a(0) \neq 0 \\ t\,\mathrm{e}^{i\,x(0)}, & \text{if } a(0) = 0. \end{cases} \end{aligned}$$

The integral curve $t \rightarrow \gamma(t)$ is a relative equilibrium of V. The $\mathbb{R}$-action on M defined by the flow of V is *not* proper. For if $a(0) \neq 0$ and $t = 2\pi/a(0)$, then $a(t) = a(0)$, $x(t) = x(0) + 2\pi$ and $z(t) = z(0)$. Let $a(0) \rightarrow 0$ and keep $x(0)$ and $z(0)$ bounded. Then $\{(a(0), x(0), z(0))\}$ remains in a compact subset of M, whereas t does not remain in any compact subset of $\mathbb{R}$. □

4.2.4 *Relative equilibria when the action is not free*

The results in §2.2 are not optimal when the action of G at the relative equilibrium $m \in M$ is not free. In order to obtain stronger results we prove

Lemma 4.2.4.10. *If $m \in M$, $t \in I_m$, $g(t) \in G$, and the integral curve $\gamma : I_m \rightarrow M$ of the G-invariant vector field V starting at m satisfies $\gamma(t) = g(t) \cdot m$ for each $t \in I_m$, then $g(t) \in N(G_m)$.*

[1]If p is a proposition, then $\neg p$ is its negation.

Proof. Let $H = G_m$. Then $m \in M_H$. Since the H-isotropy type M_H is invariant under the flow of V, it follows that $g \cdot m$ lies in M_H, where $g = g(t)$. Therefore $H = G_{g \cdot m} = gG_m g^{-1} = gHg^{-1}$. So $g \in N(H)$. □

Let $H = G_m$, $\mathfrak{h} = \mathfrak{g}_m$, and $\mathfrak{n}$ be the Lie algebra of $N(H)$. Suppose that $m \in M$ is a relative equilibrium point of the vector field V generated by $\xi \in \mathfrak{n}$. Setting $g = \exp t\,\xi$, from lemma 4.2.4.10 it follows that $\exp t\,\xi \in N(H)$ for every $t \in \mathbb{R}$. Therefore $\widetilde{\xi} = \xi + \mathfrak{h} \in \mathfrak{n}/\mathfrak{h}$ belongs to the Lie algebra $\mathfrak{g}_{M_H}$ of $G_{M_H} = N(H)/H$. Because G_{M_H} acts freely and properly on the isotropy type M_H, which contains m, we obtain the following proposition whose statement is the same as proposition 4.2.2.6 with G being replaced by G_{M_H}.

Proposition 4.2.4.11. *Suppose that $\widetilde{\xi}$ is an elliptic element of $\mathfrak{g}_{M_H}$, then the corresponding relative equilibrium $\gamma : I_m \to M_H$, which is an integral curve of V starting at m, is quasiperiodic. Moreover, the minimal number of frequencies of γ is equal to the dimension of the closure in M_H of $\{\gamma(t) \in M_H \,\big|\, t \in \mathbb{R}\}$.*

4.2.5 *Other relative equilibria in a G-orbit*

Let m be a relative equilibrium of a G-invariant vector field V on a smooth manifold M, which is generated by $\xi \in \mathfrak{g}$. Let $\gamma_m : I_m \to M$ be an integral curve of V starting at m. For every $t \in I_m$ we have

$$\begin{aligned}\gamma_{g\cdot m}(t) &= g \cdot \gamma_m(t) = (g \exp t\,\xi) \cdot m \\ &= (g \exp t\,\xi\, g^{-1})(g \cdot m) = \exp t\mathrm{Ad}_g\xi \cdot (g \cdot m).\end{aligned}$$

Thus we have proved

Lemma 4.2.5.12. *Let m be a relative equilibrium of V generated by $\xi \in \mathfrak{g}$. Then $g \cdot m$ is a relative equilibrium of V generated by $\mathrm{Ad}_g\xi$.*

4.2.5.1 *When the G-action is free*

Suppose that the G-action Φ on M is *free* at the point $m \in M$. Then the orbit map $\Phi_m : G \to M : g \mapsto g \cdot m$ is a diffeomorphism, using the orbit topology on $G \cdot m$.

Lemma 4.2.5.1.13. *The orbit map Φ_m intertwines the flow φ_t^{ξ} of the vector field X_ξ on G with the flow φ_t of the vector field V on $G \cdot m$.*

Proof. For every $g \in G$ we have

$$\begin{aligned}\varphi_t(\Phi_m(g)) &= \varphi_t(g\cdot m) = g\cdot\varphi_t(m), \quad \text{since } V \text{ is } G\text{-invariant}\\ &= g\cdot(\exp t\,\xi\cdot m), \quad \text{since } \xi \text{ generates the relative equilibrium } m\\ &= (g\exp t\,\xi)\cdot m = \varphi_t^{\xi}(g)\cdot m = \Phi_m(\varphi_t^{\xi}(g)). \qquad (9)\end{aligned}$$

□

Suppose that ξ is an elliptic element of G. Then the closure T in G of $\{\exp t\,\xi \in G \,\big|\, t \in \mathbb{R}\}$ is a torus. Using (9) with $g = e$ we see that the image of T under Φ_m is diffeomorphic to the closure Γ_m in M of $\{\varphi_t(m) \in M \,\big|\, t \in \mathbb{R}\}$. For an arbitrary $g \in G$, equation (9) implies that the image of the coset gT under Φ_m is diffeomorphic to the closure $\Gamma_{g\cdot m}$ in $G\cdot m$ of the integral curve of V starting at $g\cdot m$. Thus we get

Proposition 4.2.5.1.14. *The image of $G/T = \coprod_{g\in G} gT$ under the G-orbit map Φ_m is diffeomorphic to the closure $\Gamma_{G\cdot m}$ in $G\cdot m$ of the set of integral curves of V which start at a point in $G\cdot m$. This latter set is G-invariant.*

Consider the map $\mu : T\times G \to G : (s,g) \mapsto gs$. Since T is a commutative group, μ defines an action of T on G. In more detail, for every $s, s' \in T$ and every $g \in G$ we have

$$\mu(s's, g) = gs's = gss' = \mu(s', gs) = \mu(s', \mu(s,g)).$$

The action μ is free, for if $gs = g$, then $s = e$. Moreover, μ is a proper action since T is compact. Therefore the T-orbit map $\rho : G \to G/T$ realizes G as a T-principal bundle over the T-orbit space G/T, which is a smooth manifold.

Corollary 4.2.5.1.15. *The G-orbit map $\Phi_m : G \to G\cdot m$ induces an isomorphism of the T-principal bundle ρ with the bundle $\Gamma_{G\cdot m} \to G\cdot m$, whose total space $\Gamma_{G\cdot m}$ is the closure in $G\cdot m$ of the set of integral curves of V which start at a point in $G\cdot m$ and whose fiber $\Gamma_{g\cdot m}$ over $g\cdot m$ is the closure in $G\cdot m$ of the integral curve of V starting at $g\cdot m$.*

4.2.5.2 *When the G-action is not free*

Now suppose that the G-action Φ_m is *not* free at the point $m \in M$. Let $H = G_m$ and let M_H be the H-isotropy type. Recall that the Lie group $G_{M_H} = N(H)/H$ acts freely and properly on M_H. Here $N(H)$ is the normalizer of H in G. The G_{M_H}-orbit map

$$\phi_m : G_{M_H} \to G_{M_H}\cdot m = \{\ell H\cdot m \in M_H \,\big|\, \ell \in N(H)\} = N(H)\cdot m \qquad (10)$$

is a diffeomorphism, using the orbit topology on $G{\cdot}m$. Since M_H is invariant under the flow φ_t of V, it follows that $V|M_H$ is a vector field on M_H with flow $\varphi_t|M_H$.

Suppose that m, which lies in M_H, is a relative equilibrium of the $N(H)$-invariant vector field $V|M_H$, which is generated by $\xi \in \mathfrak{n}$, the Lie algebra of $N(H)$. Since $\mathfrak{g}_{M_H}$, the Lie algebra of G_{M_H}, is isomorphic to $\mathfrak{n}/\mathfrak{h}$, where $\mathfrak{h}$ is the Lie algebra of H, we may write $\widetilde{\xi} = \xi + \mathfrak{h}$ for the image of ξ under the projection mapping $\mathfrak{n} \rightarrow \mathfrak{n}/\mathfrak{h}$.

Analogous to lemma 4.2.5.1.13 we have

Lemma 4.2.5.2.16. *The map ϕ_m (10) intertwines the flow $\varphi_t^{\widetilde{\xi}}$ of $X_{\widetilde{\xi}}$ on G_{M_H} with the flow $\varphi_t|M_H$ of $V|M_H$ on M_H.*

Proof. For every $\ell \in N(H)$ we have

$$\begin{aligned}
\varphi_t(\phi_m(\ell H)) &= \varphi_t(\ell H \cdot m) = \varphi_t(\ell \cdot m), \quad \text{since } H = G_m \\
&= \ell \cdot \varphi_t(m), \quad \text{because } V \text{ is } N(H)\text{-invariant} \\
&= \ell \cdot (\exp t\,\xi \cdot m), \quad \text{since } m \text{ is a relative equilibrium} \\
&\qquad\qquad\qquad\qquad \text{of } V|M_H \text{ generated by } \xi \in \mathfrak{n} \\
&= ((\ell \exp t\,\xi)H) \cdot m = \phi_m(\varphi_t^{\widetilde{\xi}}(\ell H)).^2
\end{aligned} \tag{11}$$

□

Suppose that ξ is an elliptic element of $\mathfrak{n}$. Then $\widetilde{\xi}$ is an elliptic element of $\mathfrak{g}_{M_H}$. So the closure in G_{M_H} of $\{\exp t\widetilde{\xi} \in G_{M_H} \,\big|\, t \in \mathbb{R}\}$ is a torus T in G_{M_H}. Using (11) we obtain the following analogue of proposition 4.2.5.1.14.

Proposition 4.2.5.2.17. *The image of $G_{M_H}/T = \coprod_{g \in G_{M_H}} gT$ under the map ϕ_m (10) is the closure $\Gamma_{G_{M_H}\cdot m}$ in $G_{M_H} \cdot m$ of the set of all integral curves of $V|M_H$, which start at a point in $G_{M_H} \cdot m$.*

In addition, we obtain the following analogue of corollary 4.2.5.1.15.

Corollary 4.2.5.2.18. *The G_{M_H} orbit map $\phi_m : G_{M_H} \rightarrow G_{M_H} \cdot m$ induces an isomorphism of the T-principal bundle $\pi : G_{M_H} \rightarrow G_{M_H}/T$ with the bundle $\Gamma_{G_{M_H}\cdot m} \rightarrow G_{M_H} \cdot m$, whose fiber over $\ell H \cdot m$ is the closure $\Gamma_{\ell H \cdot m}$ of the integral curve of $V|M_H$ which starts at $\ell H \cdot m$.*

[2]We justify the last equality in (11) when $\mathfrak{n}$ (and hence $\mathfrak{h}$) is a Lie subalgebra of gl. Then $\exp(\xi + \mathfrak{h}) = (\exp \xi)H$, using the Campbell–Baker–Hausdorff formula and the fact that $\mathfrak{h}$ is an ideal in Lie algebra $\mathfrak{n}$.

Now consider the action of $N(H)$ on $G\times_{N(H)}G_{M_H}$ defined by

$$N(H) \times (G \times G_{M_H}) \to G \times G_{M_H} : (n, (g, \ell H)) \mapsto (gn^{-1}, n\ell H), \quad (12)$$

where $n, \ell \in N(H)$ and $g \in G$. Let us look at the map

$$\widetilde{\vartheta}_m : G \times G_{M_H} \to G \cdot m : (g, \ell H) \mapsto g\ell H \cdot m = g\ell \cdot m.$$

Since $\widetilde{\vartheta}_m$ is invariant under the $N(H)$-action (12), it induces the map

$$\vartheta_m : G\times_{N(H)}G_{M_H} \to G \cdot m : \overline{(g, \ell H)} \mapsto g\ell \cdot m, \quad (13)$$

where $G\times_{N(H)}M_H$ is the orbit space of the $N(H)$-action (12) with orbit map

$$G \times G_{M_H} \to G\times_{N(H)}G_{M_H} : (g, \ell H) \mapsto \overline{(g, \ell H)}.$$

Lemma 4.2.5.2.19. *The map ϑ_m (13) is a diffeomorphism.*

Proof. Clearly ϑ_m is smooth and surjective. Because $\dim G\times_{N(H)}G_{M_H} = \dim G \cdot m$, it follows that ϑ_m is a local diffeomorphism. To see that it is a diffeomorphism, it suffices to check that a fiber of $\widetilde{\vartheta}_m$ is a unique $N(H)$-orbit. Towards this goal suppose that $g\ell \cdot m = g'\ell' \cdot m$ for some $g, g' \in G$ and $\ell, \ell' \in N(H)$. Then $(g^{-1}g') \cdot (\ell' \cdot m) = \ell \cdot m$. But $\ell \cdot m$ and $\ell' \cdot m$ lie in M_H. So $g^{-1}g' = n^{-1} \in N(H)$. Now $(g\ell)^{-1}(g'\ell') \cdot m = m$ implies that $\ell^{-1}(g^{-1}g')\ell' = h \in H$. Therefore $\ell' = (g^{-1}g')^{-1}\ell h$, which implies

$$\ell' H = n\ell h H = n(\ell H). \quad (14)$$

Equation (14) together with $g' = gn^{-1}$ shows that $(g, \ell H)$ and $(g', \ell' H)$ lie in the same $N(H)$-orbit. Hence the fiber of $\widetilde{\vartheta}_m$ is a unique $N(H)$-orbit. □

Recall that $\mathfrak{g}_{M_H} = \mathfrak{n}/\mathfrak{h}$, is the Lie algebra of G_{M_H} and that $\xi \in \mathfrak{n}$. Set $\widetilde{\xi} = \xi + \mathfrak{h}$. The flow $\varphi_t^{\widetilde{\xi}}$ of the vector field $X_{\widetilde{\xi}}$ on G_{M_H} induces a flow $\widetilde{\theta}_t^{\widetilde{\xi}}$ on $G \times G_{M_H}$ defined by $\widetilde{\theta}_t^{\widetilde{\xi}}(g, \ell H) = (g, \ell((\exp t\,\xi H)))$. Since $\widetilde{\theta}_t^{\widetilde{\xi}}$ commutes with the $N(H)$-action (12), it induces a flow $\theta_t^{\widetilde{\xi}}$ on $G\times_{N(H)}G_{M_H}$.

Lemma 4.2.5.2.20. *The diffeomorphism ϑ_m (13) intertwines the flow $\theta_t^{\widetilde{\xi}}$ on $G\times_{N(H)}G_{M_H}$ with the flow φ_t of the vector field V on $G \cdot m$.*

Proof. Let $g \in G$ and $\ell \in N(H)$. Then

$$\begin{aligned}
\vartheta_m(\theta_t^{\widetilde{\xi}}(\overline{(g, \ell H)})) &= \vartheta_m(\overline{(g, \ell \exp t\,\xi H)}) \\
&= g\ell \exp t\,\xi \cdot m \\
&= g\ell \cdot \varphi_t(m) \quad \text{since } \xi \text{ generates the relative equilibrium } m \\
&= \varphi_t(g\ell \cdot m), \quad \text{since } V \text{ is } G\text{-invariant} \\
&= \varphi_t(\vartheta_m(\overline{(g, \ell H)})). \quad (15)
\end{aligned}$$

□

Since ξ is an elliptic element of $\mathfrak{n}$ by hypothesis, it follows that $\widetilde{\xi} = \xi + \mathfrak{h}$ is an elliptic element of $\mathfrak{g}_{M_H} = \mathfrak{n}/\mathfrak{h}$. Therefore the closure T in G_{M_H} of $\{\exp t\widetilde{\xi} \in G_{M_H} \big| t \in \mathbb{R}\}$ is a torus. From (15) it follows that the image under ϑ_m (13) of an orbit of the T-action

$$*: T \times (G\times_{N(H)}G_{M_H}) \to G\times_{N(H)}G_{M_H} : (sH, \overline{(g, \ell H)}) \mapsto \overline{(g, \ell sH)} \quad (16)$$

is diffeomorphic to the closure Γ_m in $G \cdot m$ of an integral curve of V which starts at m. Since $\xi \in \mathfrak{n}$ is elliptic, the closure T of $\{\exp t\xi \in N(H) \big| t \in \mathbb{R}\}$ in $N(H)$ is a torus. Note that every element of the torus in G_{M_H} may be written as sH, where s is an element of the torus in $N(H)$, because $\exp \widetilde{\xi} = (\exp \xi)H$. Define a T-action on $G \cdot m$ by

$$\circ : T \times (G \cdot m) \to G \cdot m : (sH, g \cdot m) \mapsto gs \cdot m. \quad (17)$$

Note that the T-action $\circ$ (17) on $G \cdot m$ is free, for if $sH \circ (g \cdot m) = g \cdot m$, then $gs \cdot m = g \cdot m$, which implies $s \cdot m = m$. So $s \in H$. Therefore $sH = H$, which is the identity element of T. Moreover, T-action $\circ$ is proper, because T is compact. From

$$\begin{aligned} \vartheta_m(sH * \overline{(g, \ell H)}) &= \vartheta_m(sH, \overline{(g, \ell H)}) = \vartheta_m(\overline{(g, \ell sH)}) \\ &= g\ell s \cdot m = s \circ (g\ell \cdot m) = s \circ \vartheta_m(\overline{(g, \ell H)}) \end{aligned} \quad (18)$$

it follows that Γ_m is an orbit of the T-action (17) on $G \cdot m$. Thus we obtain

Proposition 4.2.5.2.21. *The image under the map ϑ_m (13) of all the orbits of the T-action (16) is diffeomorphic to the closure $\Gamma_{G\cdot m}$ in $G \cdot m$ of the union of all integral curves of V, which start at some point in $G \cdot m$. Moreover, the set $\Gamma_{G\cdot m}$ is the union of all the orbits of the T-action (17), which pass through some point in $G \cdot m$.*

Showing that the T-action (16) is free and proper is a special case of the following situation. Let L be a closed subgroup of a Lie group G and let K be another Lie group. Suppose that G and K act by $\cdot$ and $*$, respectively, on a smooth manifold Q and that the actions of L and K commute. Let L act on $G \times Q$ by

$$\cdot : L \times (G \times Q) \to G \times Q : (\ell, (g, q)) \mapsto (g\ell^{-1}, \ell \cdot q)$$

with orbit space $G\times_L Q$ and let K act on $G \times Q$ by

$$* : K \times (G \times Q) \to G \times Q : (k, (g, q)) \mapsto (g, k * q).$$

Then the K and L-actions on $G \times Q$ commute, since the L and K-actions on Q commute. Therefore there is an induced K-action $*$ on the L-orbit space $G\times_L Q$.

Lemma 4.2.5.2.22. *If the K-action on Q is free and proper, then so is the induced K-action on $G\times_L Q$.*

Proof. Suppose that $(g,q) = k*(g\ell^{-1}, \ell\cdot q)$ for some $\ell \in L$. Then $(g,q) = (g\ell^{-1}, \ell\cdot(k*q))$. But this implies $\ell = e$ and $q = k*q$. But K acts freely on Q. Therefore $k = e$. Consequently, the induced K-action $*$ on $G\times_L Q$ is free.

To show that this action is proper we argue as follows. Let $\rho : G\times Q \rightarrow G\times_L Q$ be the orbit map of the L-action $\circ$. Suppose that $\{x_j\}$ is a sequence in $G\times_L Q$ which converges to $x \in G\times_L Q$. Then there is a sequence $\{(g_j, q_j)\}$ in $G\times Q$ with $\rho(g_j, q_j) = x_j$, which converges to $(g,q) \in G\times Q$ with $\rho(g,q) = x$. Suppose that there is a sequence $\{k_j\}$ in K such that $k_j * x_j \rightarrow x'$. So there is a sequence $\{\ell_j\}$ in L such that $k_j * \ell_j \cdot (g_j, q_j) \rightarrow (g', q')$, where $\rho(g', q') = x'$. Then $g_j\ell_j^{-1} \rightarrow g'$ and $\ell_j \cdot k_j * q_j \rightarrow q'$. Because $g_j \rightarrow g$, we see that $\ell_j = (g')^{-1}g_j \rightarrow (g')^{-1}g = \ell$, which lies in L since L is a closed subgroup of G. Since the action of K on Q is proper, there is a subsequence $\{k_{j_n}\}$ of $\{k_j\}$ which converges in K to k. Therefore $k_{j_n} * (\ell_{j_n} \cdot q_{j_n}) \rightarrow k*(\ell\cdot q) = q'$. Consequently, the induced K-action on $G\times_L Q$ is proper. □

Applying lemma 4.2.5.2.22 with $L = N(H)$, $K = T$, and $Q = G_{M_H}$, we conclude that the T-action (16) on $G\times_{N(H)} G_{M_H}$ is free and proper. In addition, the T-action (17) is free and proper. Therefore we obtain the following corollary of proposition 4.2.5.2.21.

Corollary 4.2.5.2.23. *The diffeomorphism ϑ_m (13) induces an isomorphism of the T-principal bundle $\widetilde{\rho} : G\times_{N(H)} G_{M_H} \rightarrow (G\times_{N(H)} G_{M_H})/T$ with the T-principal bundle $\rho : G\cdot m \rightarrow (G\cdot m)/T$, whose fiber over $\rho(g\cdot m)$ is $\Gamma_{g\cdot m}$, the closure in $G\cdot m$ of the integral curve of V which starts at $g\cdot m$.*

4.2.6 *Smooth families of quasiperiodic relative equilibria*

Let F be the image under the G orbit map π of the set of relative equilibrium points of a G-invariant vector field V on a smooth manifold M. In this subsection we find a smooth manifold in F, whose preimage under π is a smooth family of quasiperiodic integral curves of V.

To see some of the difficulties involved in finding smooth families of quasiperiodic orbits, consider the following situation. Let T be a torus in G with Lie algebra $\mathfrak{t}$. For each $\xi \in \mathfrak{t}$ let T_ξ be the closure in T of $\{\exp t\xi \,\big|\, t \in \mathbb{R}\}$. Then T_ξ is the smallest closed subgroup of T such that ξ belongs to its Lie algebra $\mathfrak{t}_\xi$ of T_ξ. Suppose that $\dim T > 1$. Then for every $1 \le k \le \dim T$ the set $\{\xi \in \mathfrak{t} \,\big|\, \dim T_\xi = k\}$ is dense in $\mathfrak{t}$. This shows that the mapping $\xi \mapsto T_\xi$ is highly discontinuous – even the mapping $\xi \mapsto \dim T_\xi$

is everywhere discontinuous. So if G contains tori T of dimension greater than 1, then one needs a very detailed description of the dependence of T_ξ on the parameter ξ to show that the closure of a quasiperiodic orbit of the G-invariant vector field V corresponding to a relative equilibrium generated by ξ depends smoothly on ξ.

If one drops the requirement that such quasiperiodic orbits depend on a minimal number of frequencies, then, under mild hypotheses, we shall find smooth families of tori on which the motion is quasiperiodic.

4.2.6.1 *Elliptic, regular, and stably elliptic elements of* $\mathfrak{g}$

We start by defining certain subspaces of the Lie algebra $\mathfrak{g}$ of the Lie group G, which will play an important role later on.

Let G be a Lie group with Lie algebra $\mathfrak{g}$. For any $\xi \in \mathfrak{g}$, the *centralizer* $\mathfrak{g}_\xi$ *of* ξ *in* $\mathfrak{g}$ is $\{\eta \in \mathfrak{g} \,\big|\, [\xi, \eta] = 0\}$. Note that $\mathfrak{g}_\xi$ is the Lie algebra of the *centralizer* G_ξ of ξ *in* G, namely, $\{g \in G \,\big|\, \mathrm{Ad}_g \xi = \xi\}$. An element $\xi \in \mathfrak{g}$ is *regular* if and only if there is an open neighborhood U of ξ in $\mathfrak{g}$ such that $\dim \mathfrak{g}_\xi \le \dim \mathfrak{g}_\eta$ for every $\eta \in U$. Because $\dim \mathfrak{g}_\eta \le \dim \mathfrak{g}_\xi$ for every $\eta \in \mathfrak{g}$ near ξ, it follows that ξ is a regular element of $\mathfrak{g}$ if and only if $\dim \mathfrak{g}_\eta$ is constant for every $\eta \in \mathfrak{g}$ near ξ. Recall that $\xi \in \mathfrak{g}$ is *elliptic* if and only if $\{\exp t\xi \in G \,\big|\, t \in \mathbb{R}\}$ is dense in a torus subgroup of G. An element ξ in $\mathfrak{g}$ is *stably elliptic* if there is an open neighborhood U of ξ in $\mathfrak{g}$ such that every $\eta \in U$ is an elliptic element of $\mathfrak{g}$.

Lemma 4.2.6.1.24. *Let G be a Lie group with Lie algebra $\mathfrak{g}$ and let $\xi \in \mathfrak{g}$. The following statements are equivalent.*

1. *ξ is a regular, stably elliptic element of $\mathfrak{g}$.*
2. *The centralizer $\mathfrak{g}_\xi$ of ξ in $\mathfrak{g}$ is equal to the Lie algebra of a torus subgroup of G.*

Proof. item 1 $\Rightarrow$ item 2. Let $\mathrm{ad}_\xi : \mathfrak{g} \to \mathfrak{g} : \eta \mapsto [\xi, \eta]$. Then $\exp t\,\mathrm{ad}_\xi = \mathrm{Ad}_{\exp t\xi}$ for all $t \in \mathbb{R}$, see theorem 1.5.2a) of [42] . If ξ is an elliptic element of $\mathfrak{g}$, then the closure in G of $\{\exp t\xi \in G \,\big|\, t \in \mathbb{R}\}$ is a torus subgroup of G. Because the adjoint representation $\mathrm{Ad} : G \to \mathrm{Gl}(\mathfrak{g}, \mathbb{R}) : g \mapsto \mathrm{Ad}_g$ is a continuous homomorphism, it follows that $\{\exp t\,\mathrm{ad}_\xi \in \mathrm{Gl}(\mathfrak{g}, \mathbb{R}) \,\big|\, t \in \mathbb{R}\}$ is contained in a compact subgroup K of $\mathrm{Gl}(\mathfrak{g}, \mathbb{R})$. Averaging an arbitrary inner product on $\mathfrak{g}$ over K, we obtain a K-invariant inner product β on $\mathfrak{g}$, see corollary 4.2.2 in [42]. Consequently, $t \mapsto \exp t\xi$ is a one parameter group of β-orthogonal transformations of $(\mathfrak{g}, \beta)$. Therefore ad_ξ is a β-antisymmetric linear transformation. This implies that $\mathrm{im}\,\mathrm{ad}_\xi$ is the β-orthogonal complement of $\ker \mathrm{ad}_\xi$ in $\mathfrak{g}$. Therefore

$$\mathfrak{g} = \mathrm{im}\,\mathrm{ad}_\xi \oplus \ker \mathrm{ad}_\xi \tag{19}$$

and $\mathrm{ad}_\xi|\mathrm{im}\,\mathrm{ad}_\xi$ is a bijective linear map of $\mathrm{im}\,\mathrm{ad}_\xi$ into itself.

Consider the mapping $\psi : G \times \mathfrak{g}_\xi \to \mathfrak{g} : (g, \zeta) \mapsto \mathrm{Ad}_g \zeta$. Then its tangent at (e, ξ) is $T_{(e,\xi)}\psi : \mathfrak{g} \times \mathfrak{g}_\xi \to \mathfrak{g} : (\rho, \eta) \mapsto [\rho, \xi] + \eta$. Because of (19) we see that $T_{(e,\xi)}\psi$ is surjective. Hence, by the implicit function theorem there is an open neighborhood U of ξ in $\mathfrak{g}$ and real analytic mappings $\varphi : U \subseteq \mathfrak{g} \to G$ and $\chi : U \subseteq \mathfrak{g} \to \mathfrak{g}_\xi$ such that $\varphi(\xi) = e$, $\chi(\xi) = \xi$, and

$$\eta = \mathrm{Ad}_{\varphi(\eta)}\chi(\eta) \tag{20}$$

for every $\eta \in U$. So every element η of $\mathfrak{g}$ near ξ is conjugate by an element $\varphi(\eta)^{-1}$ in the *identity component* G^0 of G to an element $\chi(\eta)$ in $\mathfrak{g}_\xi$.

Let $\zeta \in \mathfrak{g}_\xi$. Then the linear mappings ad_ξ and ad_ζ commute. Therefore ad_ζ leaves $\mathfrak{g}_\xi = \ker \mathrm{ad}_\xi$ and $\mathrm{im}\,\mathrm{ad}_\xi$ invariant. If ζ is sufficiently close to ξ in $\mathfrak{g}_\xi$, then $\mathrm{ad}_\zeta|\mathrm{im}\,\mathrm{ad}_\xi$ is bijective. Consequently, $\mathfrak{g}_\zeta \subseteq \mathfrak{g}_\xi$. From the hypothesis that ξ is a regular element of $\mathfrak{g}$, it follows that $\mathfrak{g}_\zeta = \mathfrak{g}_\xi$ for all $\zeta \in \mathfrak{g}_\xi$ sufficiently close to ξ.

Let $\eta \in \mathfrak{g}_\xi$ be sufficiently close to 0. Then $\zeta = \xi + \eta$ is sufficiently close to ξ in $\mathfrak{g}_\xi$ so that $\mathfrak{g}_\zeta = \mathfrak{g}_\xi$. Thus for $\rho \in \mathfrak{g}_\xi \subseteq \mathfrak{g}_\zeta$ we have

$$0 = [\zeta, \rho] = [\xi + \eta, \rho] = [\eta, \rho],$$

since $\rho \in \mathfrak{g}_\xi$. Because the map $\mathfrak{g}_\xi \to \mathfrak{g} : \eta \mapsto [\eta, \rho]$ is linear, it follows that $\mathfrak{g}_\xi$ is commutative.

Since $T = \exp \mathfrak{g}_\xi$ is the identity component of a closed subgroup of G, it is also a closed subgroup of G. If for $\zeta \in \mathfrak{g}_\xi$, the closure in G of $\{\exp t\zeta \,\big|\, t \in \mathbb{R}\}$ is a torus subgroup of G, then it is a torus subgroup of T. If ξ is a regular stably elliptic element of $\mathfrak{g}$, then it is a stably elliptic element of $\mathfrak{t}$. Because T is a connected commutative Lie group, it is isomorphic to $(\mathbb{R}/\mathbb{Z})^k \times \mathbb{R}^\ell$. But the Lie algebra of $(\mathbb{R}/\mathbb{Z})^k \times \mathbb{R}^\ell$ contains stably elliptic elements only if $\ell = 0$. This proves item 1 $\Rightarrow$ item 2.

item 2 $\Rightarrow$ item 1. Suppose that item 2 holds. Then for every $\eta \in \mathfrak{g}_\xi$ we have $\exp t\eta \in T$ for every $t \in \mathbb{R}$. Hence the closure in T of $\{\exp t\,\eta \,\big|\, t \in \mathbb{R}\}$ is a connected commutative subgroup of T and hence is a torus subgroup of T. Consequently, η is an elliptic element of $\mathfrak{g}$. Because $\xi \in \mathfrak{g}_\xi$, it follows that ξ is an elliptic element of $\mathfrak{g}$. Because every element ζ of $\mathfrak{g}$ near ξ is conjugate to an element of $\mathfrak{g}_\xi$, we conclude that ζ is elliptic. So ξ is stably elliptic, which implies that $\mathfrak{g}_\xi \subseteq \mathfrak{g}_\eta$ for every $\eta \in \mathfrak{g}_\xi$. Since T is commutative, $\mathfrak{g}_\xi$ is commutative. Therefore ξ is a regular element of $\mathfrak{g}$. This proves item 2 $\Rightarrow$ item 1. □

Let $\mathfrak{g}^{\mathrm{se}}$, $\mathfrak{g}^{\mathrm{reg}}$, and $\mathfrak{g}^{\mathrm{rse}}$ be the set of stably elliptic, regular, and regular stably elliptic elements of $\mathfrak{g}$, respectively.

Lemma 4.2.6.1.25. $\mathfrak{g}^{\mathrm{rse}}$ *is an open dense subset of* $\mathfrak{g}^{\mathrm{se}}$. *Moreover, for every* $\xi \in \mathfrak{g}^{\mathrm{rse}}$ *there is an open neighborhood* U *of* ξ *in* $\mathfrak{g}$ *and a real analytic mapping* $\theta : U \subseteq \mathfrak{g} \to G$ *with* $\theta(\xi) = e$ *such that* $\mathfrak{g}_\zeta = \mathfrak{g}_\xi$, *where* $\zeta = \mathrm{Ad}_{\theta(\eta)}\eta$ *and* $\eta \in U$.

Proof. By definition the sets $\mathfrak{g}^{\mathrm{se}}$ and $\mathfrak{g}^{\mathrm{reg}}$ are open subsets of $\mathfrak{g}$. If $\xi \in \mathfrak{g}^{\mathrm{se}}$ but is not an element of $\mathfrak{g}^{\mathrm{reg}}$, then there is an $\eta \in \mathfrak{g}$ arbitrarily close to ξ such that $\dim \mathfrak{g}_\eta < \dim \mathfrak{g}_\xi$. Because $\dim \mathfrak{g}$ is finite, we can repeat this argument only finitely many times, after which we obtain an element $\eta \in \mathfrak{g}$, which is arbitrarily close to ξ, such that locally $\dim \mathfrak{g}_\eta$ is minimal. Therefore $\eta \in \mathfrak{g}^{\mathrm{reg}}$. So $\mathfrak{g}^{\mathrm{rse}}$ is an open dense subset of $\mathfrak{g}^{\mathrm{se}}$.

If we write $\theta(\eta) = \varphi(\eta)^{-1}$, then from $\eta = \mathrm{Ad}_{\varphi(\eta)}\chi(\eta)$ (20) it follows that $\zeta = \mathrm{Ad}_{\theta(\eta)}\eta = \chi(\eta) \in \mathfrak{g}_\xi$. □

Let C be a connected component of $\mathfrak{g}^{\mathrm{rse}}$.

Corollary 4.2.6.1.26. *For any* $\xi, \eta \in C$ *there is an element* g *in the* identity component G^0 *of* G *such that* $\mathfrak{g}_\eta = \mathrm{Ad}_g \mathfrak{g}_\xi$. *In particular, for* $\xi \in C$, *the dimension of* $\mathfrak{g}_\xi$ *is constant, say* k *and the map* $\xi \mapsto \mathfrak{g}_\xi$ *from* C *into the Grassmannian manifold of* k*-dimensional subspaces of* $\mathfrak{g}$ *is real analytic.*

Proof. We say that two elements ξ and η of $\mathfrak{g}$ are related if $\mathfrak{g}_\xi$ is conjugate to $\mathfrak{g}_\eta$ by $g \in G^0$. Being related is an equivalence relation on G. By lemma 4.2.6.1.25 nearby elements of C are equivalent. Since C is connected, it follows that $\mathfrak{g}_\eta$ is conjugate to $\mathfrak{g}_\xi$ for every $\xi, \eta \in C$ by an element of G. Therefore $\dim \mathfrak{g}_\xi$ is constant for every $\xi \in C$. For η near ξ in C, we know that $\mathfrak{g}_\eta$ is conjugate to $\mathfrak{g}_\xi$ by $g \in G^0$, which depends real analytically on η. Therefore the mapping $\eta \mapsto \mathfrak{g}_\eta$ is real analytic in an open neighborhood of ξ in C. In particular, the map $\xi \mapsto \mathfrak{g}_\xi$ is real analytic on C. □

Example 4.2.6.1.27.

1. If G^0 is a compact subgroup of G, then $\mathfrak{g}^{\mathrm{se}} = \mathfrak{g}$. Moreover, $\mathfrak{g}^{\mathrm{rse}} = \mathfrak{g}^{\mathrm{reg}}$. If $\xi \in \mathfrak{g}^{\mathrm{reg}}$, then $\mathfrak{g}_\xi$ is a maximal abelian subspace of $\mathfrak{g}$ and $\exp \mathfrak{g}_\xi$ is a maximal torus in G^0. All maximal tori in G^0 are conjugate by an element of G^0. In particular, they have the same dimension, called the rank of G^0, see theorem 3.7.1 of [42]. Conversely, if every element of $\mathfrak{g}$ is elliptic, then G^0 is compact.

2. A proper Euclidean motion of $\mathbb{R}^2$ is an affine mapping $(A, a) : \mathbb{R}^2 \to \mathbb{R}^2 : x \mapsto Ax + a$, where $(A, a) \in \mathrm{SO}(2) \times \mathbb{R}^2$. The set of all proper Euclidean motions forms the *Euclidean group* $\mathrm{E}(2)$ whose multiplication is given by composition of affine maps. The Lie algebra of $\mathrm{E}(2)$ is $\mathrm{e}(2)$, which is the set of all infinitesimal Euclidean motions, namely, the set of all affine maps $(\Xi, \xi) : \mathbb{R}^2 \to \mathbb{R}^2 : x \mapsto \Xi x + \xi$, where $(\Xi, \xi) \in \mathrm{so}(2) \times \mathbb{R}^2$. If $\Xi \neq 0$, then Ξ is a bijective linear map and (Ξ, ξ) is an infinitesimal rotation about $p = -\Xi^{-1}\xi$. Let $\mathrm{E}(2)$ act on $\mathrm{e}(2)$ by conjugation. Then every element of the isotropy group $\mathrm{E}(2)_{(\Xi,\xi)}$ leaves p fixed. Therefore $\mathrm{E}(2)_{(\Xi,\xi)}$ is the *group of rotations about* p, which is a circle. Consequently, $(\Xi, \xi) \in \mathrm{e}(2)^{\mathrm{rse}}$ if and only if $\Xi \neq 0$. This shows that $\mathrm{e}(2)^{\mathrm{rse}}$ is an open dense subset of $\mathrm{e}(2)$.

3. Let $\mathrm{Sl}(2, \mathbb{R})$ be the set of 2×2 real matrices with determinant 1. Its Lie algebra $\mathrm{sl}(2, \mathbb{R})$ is the set of 2×2 real matrices $\xi = \begin{pmatrix} a & b \\ c & -a \end{pmatrix}$. Then $\mathrm{sl}(2, \mathbb{R})^{\mathrm{rse}} = \{\xi \in \mathrm{sl}(2, \mathbb{R}) \,\big|\, a^2 + bc < 0\}$ is the disjoint union of two open convex cones in $\mathbb{R}^3$. If $a^2 + bc = 0$, then ξ is elliptic if and only if $\xi = 0$. □

4.2.6.2 *When the G-action is free and proper*

Suppose that Φ is free and proper action of a Lie group G on a smooth manifold M with orbit map $\pi : M \to \overline{M} = M/G : m \mapsto \overline{m}$. Let V be a G-invariant smooth vector field on M with flow φ_t. Since Φ is free and proper, the G-orbit space $\overline{M}$ is a smooth manifold and the G-invariant vector field V induces a smooth vector field $\overline{V}$ on $\overline{M}$ with flow $\overline{\varphi}_t$.

Let

$$F = \{\overline{m} \in \overline{M} \,\big|\, \overline{V}(\overline{m}) = 0\} = \{\overline{m} \in \overline{M} \,\big|\, \overline{\varphi}_t(\overline{m}) = \overline{m}\}.$$

Then F is a smooth submanifold of $\overline{M}$. Therefore $\pi^{-1}(F)$ is a G-invariant, locally closed smooth submanifold of M. So $\pi|\pi^{-1}(F) : \pi^{-1}(F) \to F$ is a principal G-bundle. Because $\pi^{-1}(F)$ consists of all of the relative equilibria of V, it is invariant under the flow of V. Additionally, $V|\pi^{-1}(F)$ is complete.

Lemma 4.2.6.2.28. *Let $m \in \pi^{-1}(F)$ and let $\xi(m)$ be the unique element of $\mathfrak{g}$ which generates the relative equilibrium of $V|\pi^{-1}(F)$ through m. Then the mapping $\pi^{-1}(F) \to \mathfrak{g} : m \mapsto \xi(m)$ is smooth.*

Proof. Write $N = \pi^{-1}(F)$. Let $D : N \to TN$ be the smooth distribution D on N defined by $m \mapsto D_m = \mathrm{span}\{X_\xi(m) \in T_mN \,\big|\, \xi \in \mathfrak{g}\} = T_e\Phi_m\mathfrak{g}$. Because V is a G-invariant vector field, $V|N$ is a smooth map from N into

TN, which has values in D. The map $\alpha : N \times \mathfrak{g} \to D : (m, \xi) \mapsto X_\xi(m)$ is smooth. α is surjective by definition. It is injective, for if $X_\xi(m) = X_\eta(n)$, then applying the tangent bundle projection $\tau : TN \to N$ to both sides of the preceding equation gives $m = n$, since X_ξ and X_η are smooth vector fields on N. But then $X_\xi(m) = X_\eta(m)$, which implies $X_{\xi-\eta}(m) = 0$. So for all $t \in \mathbb{R}$ we have $\exp t(\xi - \eta) \cdot m = m$. However, by hypothesis the G action on M is free. Therefore $\xi = \eta$. Additionally, α has a bijective tangent map. Consequently, the map α is a diffeomorphism. Let $\pi_2 : N \times \mathfrak{g} \to N : (m, \xi) \mapsto \xi$. Then the mapping $\pi_2 \circ \alpha^{-1} \circ V|N$, which sends m to $\xi(m)$, is smooth. □

Let

$$F^{\mathrm{rse}} = \{f \in F \big| \, \xi(m) \in \mathfrak{g}^{\mathrm{rse}} \text{ for some } m \in \pi^{-1}(\{f\})\}.$$

Because $\mathfrak{g}^{\mathrm{rse}}$ is invariant under conjugation by elements of G and $\xi(g \cdot m) = \mathrm{Ad}_g \xi(m)$, it follows that $\xi(m) \in \mathfrak{g}^{\mathrm{rse}}$ for every $m \in \pi^{-1}(\{f\})$, where $f \in F$. Since the mapping $\pi^{-1}(F) \to \mathfrak{g} : m \mapsto \xi(m)$ is continuous, we deduce that $\xi^{-1}(\mathfrak{g}^{\mathrm{rse}})$ is an G-invariant open subset of M. Therefore, $F^{\mathrm{rse}} = \pi(\xi^{-1}(\mathfrak{g}^{\mathrm{rse}}))$ is an open subset of F.

Proposition 4.2.6.2.29. *Let $m_0 \in \pi^{-1}(F)$ such that $f_0 = \pi(m_0) \in F^{\mathrm{rse}}$. Set $\mathfrak{t} = \mathfrak{g}_{\xi(m_0)}$ and let $T = \exp \mathfrak{t}$. Then there is an open neighborhood U of f_0 in F^{rse} and a smooth section $\sigma : U \subseteq F^{\mathrm{rse}} \to \pi^{-1}(U)$ of the bundle $\pi|\pi^{-1}(U) : \pi^{-1}(U) \to U$ such that $\xi(\sigma(f)) \in \mathfrak{t}$ for every $f \in U$.*

Proof. Because $\pi|\pi^{-1}(F) : \pi^{-1}(F) \to F$ is a smooth fibration, there is an open neighborhood U_1 of f_0 in F^{rse} and a smooth section $\sigma_1 : U_1 \to \pi^{-1}(U_1)$ of $\pi|\pi^{-1}(U) : \pi^{-1}(U_1) \to U_1$. From lemma 4.2.6.1.25 there is an open neighborhood $\widetilde{V}$ of $\xi(m_0)$ in $\mathfrak{g}$ and a real analytic mapping $\theta : \widetilde{V} \subseteq \mathfrak{g} \to G$ such that $\mathrm{Ad}_{\theta(\xi')}\xi' \in \mathfrak{t}$ for every $\xi' \in \widetilde{V}$. Take $m = \sigma_1(f)$, $\xi' = \xi(\sigma_1(f))$, and $g = \theta(\xi')$. Set $\sigma(f) = \theta(\xi') \cdot \xi'$ and let $U = \sigma_1^{-1}(\xi^{-1}(\widetilde{V}))$. Then U is an open neighborhood of f_0 in F^{rse} such that the map $\sigma : U \to \pi^{-1}(U) : f \to \sigma(f)$ is defined and is smooth. Because

$$\pi(g \cdot m) = \pi(m) = \pi(\sigma_1(f)) = f,$$

we see that σ is a smooth section of the bundle $\pi|\pi^{-1}(U)$. For every $f \in U$ we have $\sigma(f) = \theta(\xi') \cdot \xi'$, which implies $\xi(\sigma(f)) = \mathrm{Ad}_{\theta(\xi')}(\xi') \in \mathfrak{t}$. □

We use the notation of proposition 4.2.6.2.29. We have a free and proper action of the torus T on $U \times G$ defined by

$$* : T \times (U \times G) \to U \times G : (s, (f, g)) \mapsto (f, gs) \tag{21}$$

and a free and proper action of T on $\pi^{-1}(U)$ defined by

$$\bullet : T \times \pi^{-1}(U) \to \pi^{-1}(U) : (s, g \cdot \sigma(f)) \mapsto gs \cdot \sigma f. \tag{22}$$

The mapping

$$\lambda : U \times G \to \pi^{-1}(U) : (f, g) \mapsto g \cdot \sigma(f) \tag{23}$$

is a diffeomorphism, which intertwines the T-actions $*$ and $\bullet$ because

$$\lambda(s * (f, g)) = \lambda(f, gs) = gs \cdot \sigma(f) = s\bullet(g \cdot \sigma(f)) = s\bullet\lambda(f, g).$$

Lemma 4.2.6.2.30. *The mapping* λ (23) *intertwines the* $\mathbb{R}$*-action*

$$\circ : \mathbb{R} \times (U \times G) \to U \times G : (t, (f, g)) \mapsto \big(f, g \exp t\, \xi(\sigma(f))\big) \tag{24}$$

with the flow $\varphi_t|\pi^{-1}(U)$ *of the* G*-invariant vector field* $V|\pi^{-1}(U)$.

Proof. For $(t, (f, g)) \in \mathbb{R} \times (U \times G)$ we have

$$\begin{aligned} \lambda(t \circ (f, g)) &= \lambda(f, g \exp t\, \xi(\sigma(f))) = g \exp t\, \xi(\sigma(f)) \cdot \sigma(f) \\ &= g \cdot \varphi_t(\sigma(f)), \quad \text{since } \sigma(U) \subseteq \pi^{-1}(U) \\ &= \varphi_t(g \cdot \sigma(f)), \quad \text{since } V|\pi^{-1}(U) \text{ is } G\text{-invariant} \\ &= \varphi_t(\lambda(f, g)). \end{aligned}$$

□

Thus there is a smooth mapping $U \subseteq F^{\mathrm{rse}} \to \mathfrak{t} : f \mapsto \xi(\sigma(f))$ such that $V(\sigma(f)) = \xi(\sigma(f))$ for every $f \in U$.

Let $\mathcal{C}$ be the connected component of F^{rse}, which contains f_0 and let $\mathcal{N}$ be the connected component of $\{m \in \pi^{-1}(\mathcal{C}) \big| \, \xi(m) \in \mathfrak{t}\}$ which contains m_0. Then $\mathcal{N}$ is invariant under the restriction $\widetilde{\Phi}$ of the G-action $\Phi : G \times M \to M$ to $N(T) \times \mathcal{N}$, because for $(n, m) \in N(T) \times \mathcal{N}$ we have

$$\xi(n \cdot m) = \mathrm{Ad}_n \xi(m) \in \mathrm{Ad}_n \mathfrak{t} = \mathfrak{t}. \tag{25}$$

The last equality in (25) holds because $n \in N(T)$. Since $\mathcal{N}$ is a locally closed subset of M, the proper free G-action Φ induces a proper free $N(T)$-action $\widetilde{\Phi}$ on $\mathcal{N}$ with orbit map $\widetilde{\pi} : \mathcal{N} \to \mathcal{C}$.

Lemma 4.2.6.2.31. $\widetilde{\pi} : \mathcal{N} \to \mathcal{C}$ *is a principal* $N(T)$*-bundle.*

Proof. We describe the bundle $\widetilde{\pi}$ using local trivializations and transition maps on their overlaps as follows. Let $\widetilde{m} \in \mathcal{N}$ and $\widetilde{f} = \widetilde{\pi}(\widetilde{m}) \in \mathcal{C}$. Let $\mathcal{C}_\sigma$ be the open subset of $\mathcal{C}$ containing $\widetilde{f}$ and let $\sigma : \mathcal{C}_\sigma \to \pi^{-1}(\mathcal{C}_\sigma)$ be the smooth section of $\pi|\pi^{-1}(\mathcal{C}_\sigma)$ constructed in proposition 4.2.6.2.29. Then for every $f \in \mathcal{C}_\sigma$, we have $\xi(\sigma(f)) \in \mathfrak{g}_{\xi(\widetilde{m})} = \mathfrak{t}$, since $\sigma(f) \in \mathcal{N}$. Consequently, $\sigma(\mathcal{C}_\sigma) \subseteq \mathcal{N}$. Because there is a unique G-orbit through $\sigma(f)$, we have $\widetilde{\pi}(\sigma(f)) = \pi(\sigma(f)) = f$, where the last equality holds because σ is a section of $\pi|\pi^{-1}(\mathcal{C}_\sigma)$. Therefore $\sigma(\mathcal{C}_\sigma) \subseteq \widetilde{\pi}^{-1}(\mathcal{C}_\sigma)$ and σ is a smooth section of $\widetilde{\pi}|\widetilde{\pi}^{-1}(\mathcal{C}_\sigma)$.

Consider the map

$$\lambda_\sigma : \mathcal{C}_\sigma \times N(T) \to \widetilde{\pi}^{-1}(\mathcal{C}_\sigma) : (f,n) \mapsto n \cdot \sigma(f). \tag{26}$$

Then λ_σ is a diffeomorphism, which is a local trivialization of $\widetilde{\pi}|\widetilde{\pi}^{-1}(\mathcal{C}_\sigma)$ because

$$\widetilde{\pi}\big(\lambda_\sigma(f,n)\big) = \widetilde{\pi}(n \cdot \sigma(f)) = \widetilde{\pi}(\sigma(f)) = f$$

is the projection map $\mathcal{C}_\sigma \times N(T) \to \mathcal{C}_\sigma$.

Since $\{\mathcal{C}_\sigma\}$ forms an open covering of $\mathcal{C}$, to finish the description of the bundle $\widetilde{\pi}$ it remains to find the transition map between the overlap of two local trivializations λ_σ and λ_τ. This we do as follows. First note that $\xi(\mathcal{N})$ is a connected component of $\mathfrak{g}^{\mathrm{rse}}$. Using lemma 4.2.6.1.25 we see that for $f \in \mathcal{C}_\sigma \cap \mathcal{C}_\tau$ there is a $g = g_{\tau\sigma}(f) \in G^0$, which depends smoothly on f with $g_{\tau\sigma}(\widetilde{f}) = e$, such that $\xi(\tau(f)) = \mathrm{Ad}_g \xi(\sigma(f))$. Therefore

$$\mathfrak{t} = \mathfrak{g}_{\xi(\tau(f))} = \mathrm{Ad}_g\, \mathfrak{g}_{\xi(\sigma(f))} = \mathrm{Ad}_g\, \mathfrak{t}, \tag{27}$$

since $\sigma(f), \tau(f) \in \mathcal{N}$ implies that $\xi(\sigma(f))$ and $\xi(\tau(f))$ both lie in $\mathfrak{t} \subseteq \mathfrak{g}^{\mathrm{rse}}$. From (27) and $\exp(\mathrm{Ad}_g \xi) = \mathrm{Ad}_g(\exp \xi)$, we get $g\, T\, g^{-1} = T$, which implies $g \in N(T)$. Because σ and τ are sections of $\widetilde{\pi}|(\widetilde{\pi}^{-1}(\mathcal{C}_\sigma \cap \mathcal{C}_\tau))$, for $f \in \mathcal{C}_\sigma \cap \mathcal{C}_\tau$ we have $\widetilde{\pi}(\sigma(f)) = \widetilde{\pi}(\tau(f)) = f$. Therefore $\sigma(f)$ and $\tau(f)$ lie in the same $N(T)$-orbit. So there is an $n = n(f) \in N(T)$, depending smoothly on f, such that $\tau(f) = n \cdot \sigma(f)$. But then $\xi(\tau(f)) = \mathrm{Ad}_n(\xi(\sigma(f)))$. Using the fact that $\xi(\tau(f)) = \mathrm{Ad}_g(\xi(\sigma(f)))$, we get $\mathrm{Ad}_{g^{-1}n}(\xi(\sigma(f))) = \xi(\sigma(f))$, which implies

$$X_{\xi(\sigma(f))}(g^{-1} n \cdot \sigma(f)) = X_{\mathrm{Ad}_{g^{-1}n}(\xi(\sigma(f)))}(\sigma(f)) = X_{\xi(\sigma(f))}(\sigma(f)). \tag{28}$$

Applying the tangent bundle projection map $T\mathcal{N} \to \mathcal{N}$ to both sides of (28) gives $g^{-1} n \cdot \sigma(f) = \sigma(f)$. But the $N(T)$-action on $\mathcal{N}$ is free, so $g^{-1} n = e$, that is, $n(f) = g_{\tau\sigma}(f)$ for every $f \in \mathcal{C}_\sigma \cap \mathcal{C}_\tau$. The transition map between the overlap of $\mathcal{C}_\tau \times N(T)$ and $\mathcal{C}_\sigma \times N(T)$ is

$$\varphi_{\tau\sigma} = \lambda_\sigma^{-1} \circ \lambda_\tau : (\mathcal{C}_\tau \cap \mathcal{C}_\sigma) \times N(T) \to (\mathcal{C}_\tau \cap \mathcal{C}_\sigma) \times N(T) : (f,n) \mapsto (f, n \cdot g_{\tau\sigma}(f)),$$

because

$$\lambda_\tau(f,n) = n \cdot \tau(f) = n g_{\tau\sigma}(f) \cdot \sigma(f) = \lambda_\sigma(f, n g_{\tau\sigma}(f)).$$

Note that the map $g_{\tau\sigma} : \mathcal{C}_\tau \cap \mathcal{C}_\sigma \to N(T) : f \mapsto g_{\tau\sigma}(f)$ is smooth. □

Corollary 4.2.6.2.32. *Corresponding to the principal $N(T)$-bundle $\widetilde{\pi} : \mathcal{N} \to \mathcal{C}$ is the G-bundle $\widehat{\pi} : G \cdot \mathcal{N} \to \mathcal{C}$, where $\widehat{\pi} = \pi|(G \cdot \mathcal{N})$.*

Proof. Take the local trivializations to be

$$\mu_\sigma : \mathcal{C}_\sigma \times G \to \widehat{\pi}^{-1}(\mathcal{C}_\sigma) : (f, g) \mapsto g \cdot \sigma(f)$$

and the transition map between the overlap of $\mathcal{C}_\sigma \times G$ and $\mathcal{C}_\tau \times G$ to be

$$\phi_{\tau\sigma} : (\mathcal{C}_\sigma \cap \mathcal{C}_\tau) \times G \to (\mathcal{C}_\sigma \cap \mathcal{C}_\tau) \times G :(f, g) \mapsto (f, gg_{\tau\sigma}(f)).$$

□

We now construct a T-bundle associated to the G-bundle $\widehat{\pi}$ constructed in corollary 4.2.6.2.32. On $G \cdot \mathcal{N}$ consider the T-action $\bullet$ defined by

$$\bullet : T \times (G \cdot \mathcal{N}) \to G \cdot \mathcal{N} : (s, g \cdot m) \mapsto gs \cdot m. \tag{29}$$

Let $\overline{\pi} : G \cdot \mathcal{N} \to \mathcal{C}$ be its orbit map. The set $\{f\} \times gT = \{(f, gs) \in \mathcal{C}_\sigma \times G \big| \, s \in T\}$ is an orbit of the T-action (29) on $G \cdot \mathcal{N}$ and thus is a fiber of the bundle $\overline{\pi}$. The map

$$\mu_\sigma : \mathcal{C}_\sigma \times gT \to \overline{\pi}^{-1}(\mathcal{C}_\sigma) : (f, gT) \mapsto gT \cdot \sigma(f)$$

is a diffeomorphism and is in fact a local trivialization of the bundle $\overline{\pi}$ because

$$\overline{\pi} \circ \mu_\sigma(f, gT) = \overline{\pi}(gT \cdot \sigma(f)) = \overline{\pi}(\sigma(f)) = \widehat{\pi}(\sigma(f)) = f.$$

The second to last equality above follows because the T-orbit of $\sigma(f) \in \mathcal{N}$ is contained in the unique G-orbit through $\sigma(f)$. The last equality follows because σ is a smooth section of $\widehat{\pi}|\widehat{\pi}^{-1}(\mathcal{C})$. To complete the description of the T-bundle $\overline{\pi}$, we note that the transition map between the overlap of $\mathcal{C}_\sigma \times gT$ and $\mathcal{C}_\tau \times gg_{\tau\sigma}T$ is given by

$$\psi_{\tau\sigma} = \mu_\sigma^{-1} \circ \mu_\tau : (\mathcal{C}_\sigma \cap \mathcal{C}_\tau) \times gT \to (\mathcal{C}_\sigma \cap \mathcal{C}_\tau) \times gg_{\tau\sigma}T : (f, gT) \mapsto (f, gg_{\tau\sigma}(f)T),$$

since

$$\begin{aligned} \mu_\tau(f, gT) &= gT \cdot \tau(f) = gTg_{\tau\sigma}(f) \cdot \sigma(f) \\ &= gg_{\tau\sigma}(f)T \cdot \sigma(f), \quad \text{because } g_{\tau\sigma}(f) \in N(T) \\ &= \mu_\sigma(f, gg_{\tau\sigma}(f)T). \end{aligned}$$

The T-bundle $\overline{\pi}$ need *not* be a principal bundle. We now develop a criterion to say when it is. Let $Z(T) = \{g \in G \big| \, gsg^{-1} = s \text{ for every } s \in T\}$ be the centralizer of T in G. Clearly $Z(T)$ is a closed subgroup of G. From

$sTs^{-1} = T$ for every $s \in Z(T)$ it follows that $Z(T)$ is a subgroup of $N(T)$. In addition, $Z(T)$ is a normal subgroup of $N(T)$.[3]

On $G \cdot \mathcal{N}$ define an action of $Z(T)$ by

$$\circ : Z(T) \times (G \cdot \mathcal{N}) \to G \cdot \mathcal{N} : (z, g \cdot m) \mapsto gz^{-1} \cdot m. \tag{30}$$

This action is free and proper. Let $\widetilde{\mathcal{C}}$ be the orbit space $(G \cdot \mathcal{N})/Z(T)$ of the action $\circ$ with orbit map $\pi^{\circ} : G \cdot \mathcal{N} \to \widetilde{\mathcal{C}}$. The free and proper action of $N(T)$ on $G \cdot \mathcal{N}$ given by

$$\bullet : N(T) \times (G \cdot \mathcal{N}) \to G \cdot \mathcal{N} : (n, g \cdot m) \mapsto ng \cdot m \tag{31}$$

commutes with the $Z(T)$-action $\circ$ and therefore induces an $N(T)$-action on $\widetilde{\mathcal{C}}$, whose orbit map is $\breve{\pi} : \widetilde{\mathcal{C}} \to \mathcal{C}$. Let $W(T) = N(T)/Z(T)$. Observe that $\breve{\pi}$ is a $W(T)$-bundle over $\mathcal{C}$, because

$$\widetilde{\mathcal{C}}/W(T) = \big((G \cdot \mathcal{N})/Z(T)\big)/\big(N(T)/Z(T)\big) = (G \cdot \mathcal{N})/N(T) = \mathcal{C}.$$

We have

Lemma 4.2.6.2.33. *$W(T)$ is a discrete group.*

Proof. The Lie algebra of $N(T)$ is $\{\eta \in \mathfrak{g} \big| \, [\xi, \eta] \in \mathfrak{t}$ for every $\xi \in \mathfrak{t}\, \}$. If ξ is a regular element of $\mathfrak{t}$, then $\mathfrak{g} = \mathrm{im}\, \mathrm{ad}_{\xi} \oplus \mathfrak{t}$ and $\mathrm{ad}_{\xi} | \mathrm{im}\, \mathrm{ad}_{\xi}$ is a bijective linear map of $\mathrm{im}\, \mathrm{ad}_{\xi}$ onto itself. Therefore $\mathrm{ad}_{\xi}\eta \in \mathfrak{t}$ if and only if $\eta \in \mathfrak{t}$. Thus the Lie algebra of $N(T)$ is $\mathfrak{t}$. This implies that the identity component of $N(T)$ is T. Therefore $N(T)/T$ is a discrete group and consequently $Z(T)/T$ is also. Hence $\big(N(T)/T\big)/\big(Z(T)/T\big) = N(T)/Z(T) = W(T)$ is a discrete group. □

$W(T)$ is the *Weyl group* of T. Because $N(T)$ acts on T by conjugation, there is an induced effective action of $W(T)$ on T. Therefore $W(T)$ may be viewed as the automorphism group $\mathrm{Aut}(T)$ of T.

Because $W(T)$ is discrete, we obtain

Corollary 4.2.6.2.34. *$\breve{\pi} : \widetilde{\mathcal{C}} \to \mathcal{C}$ is a covering map with fiber $W(T)$.*

---

[3] To see this we argue as follows. For every $z \in Z(T)$, every $n \in N(T)$, and every $t \in T$ we have

$$\begin{aligned}(nzn^{-1})t(nzn^{-1})^{-1} &= nz(n^{-1}tn)z^{-1}n^{-1} = n(zt'z^{-1})n^{-1}, \\ &\qquad \text{where } t' = n^{-1}tn \in T \text{ since } n \in N(T) \\ &= nt'n^{-1}, \ \text{since } z \in Z(T) \\ &= t.\end{aligned}$$

Therefore $nzn^{-1} \in Z(T)$, which implies $nZ(T)n^{-1} \subseteq Z(T)$ for every $n \in N(T)$. But then $Z(T) \subseteq n^{-1}Z(T)n = (n^{-1})Z(T)(n^{-1})^{-1} \subseteq Z(T)$. So $(n^{-1})Z(T)(n^{-1})^{-1} = Z(T)$ for every $n^{-1} \in N(T)$, which proves the assertion.

Suppose that $\gamma : [0,1] \rightarrow \mathcal{C}$ is a closed curve which begins and ends at $f \in \mathcal{C}$. Because $W(T)$ is discrete, γ lifts to a *unique* curve $\widetilde{\gamma} : [0,1] \rightarrow \widetilde{\mathcal{C}}$, where $\widetilde{\gamma}(0) = \widetilde{f}_1$ and $\widetilde{\gamma}(1) = \widetilde{f}_2$ with $\widetilde{f}_1, \widetilde{f}_2 \in \breve{\pi}^{-1}(\{f\})$. Since $\breve{\pi}^{-1}(\{f\}) = W(T)$, there is a unique element $w(\gamma) \in W(T)$ such that $\widetilde{f}_2 = w(\gamma)f_1$. If γ_1 and γ_2 are two homotopic closed curves in $\mathcal{C}$ which begin and end at f, then their lifts $\widetilde{\gamma}_1$ and $\widetilde{\gamma}_2$ are homotopic closed curves in $\widetilde{\mathcal{C}}$ which begin at $\widetilde{f}_1$ and end at $\widetilde{f}_2$. Therefore, $w(\gamma_1) = w(\gamma_2)$. So $w(\gamma)$ depends only on the homotopy class $[\gamma]$ of γ based at f. If $\gamma_2\,\gamma_1$ is the concatenation of γ_1 with γ_2, formed by joining the end point of γ_1 to the beginning point of γ_2, then $w(\gamma_2\,\gamma_1) = w(\gamma_2)\,w(\gamma_1)$. So we have a homomorphism $\mu_{\breve{\pi}} : \pi_1(\mathcal{C}, f) \rightarrow W(T) : [\gamma] \mapsto w([\gamma])$, called the *monodromy homomorphism* of the $W(T)$-bundle $\breve{\pi}$.

Corollary 4.2.6.2.35. *The bundle $\breve{\pi} : \widetilde{\mathcal{C}} \rightarrow \mathcal{C}$ is a principal $W(T)$-bundle if and only if its monodromy homomorphism $\mu_{\breve{\pi}}$ is trivial.*

Proof. $\Rightarrow$. If $\mu_{\breve{\pi}}$ is trivial, then $\widetilde{C}$ is a product bundle $\mathcal{C} \times W(T)$, which is clearly a principal $W(T)$-bundle.

$\Leftarrow$. Suppose that $\breve{\pi}$ is a principal $W(T)$-bundle. Then the monodromy homomorphism $\mu_{\breve{\pi}} : \pi_1(\mathcal{C}, f) \rightarrow W(T) : [\gamma] \rightarrow w([\gamma])$ is defined and is continuous. Since $W(T)$ is discrete, it follows from continuity that $\mu_{\breve{\pi}}$ is constant. Because $\mu_{\breve{\pi}}$ is a homomorphism, this constant is the identity element e of $W(T)$. In other words, $\mu_{\breve{\pi}}(\pi_1(\mathcal{C}, f)) = e$, that is, $\mu_{\breve{\pi}}$ is trivial. □

We return to looking for a criterion to decide when the T-bundle $\overline{\pi} : G \cdot \mathcal{N} \rightarrow \mathcal{C}$ is a principal bundle. Since the $N(T)$ and $Z(T)$ actions (31) and (30), respectively, commute with the T-action (29), the T-bundle $\overline{\pi}$ can be factored into the composition of the $Z(T)$-bundle $\pi^\circ : G \cdot \mathcal{N} \rightarrow \widetilde{\mathcal{C}}$ followed by the $W(T)$-bundle $\breve{\pi} : \widetilde{\mathcal{C}} \rightarrow \mathcal{C}$. In other words, $\overline{\pi} = \breve{\pi} \circ \pi^\circ$. Because π° is a principal bundle by construction, $\overline{\pi}$ is a principal T-bundle if and only if $\breve{\pi}$ is a principal $W(T)$-bundle. But this is the case if and only if the monodromy homomorphism $\mu_{\breve{\pi}}$ is trivial.

We now recall how to define the monodromy homomorphism of the T-bundle $\overline{\pi}$. Let γ be a closed curve in $\mathcal{C}$ starting and ending at f. Transporting a fiber T of the bundle $\overline{\pi}$ along $\mathcal{C}$ gives an automorphism $\mu_{\overline{\pi}}([\gamma])$ of the fiber $\overline{\pi}^{-1}(\{f\}) = T$, which depends only on the homotopy class $[\gamma]$ of γ. The map $\mu_{\overline{\pi}} : \pi_1(\mathcal{C}, f) \rightarrow \mathrm{Aut}(T) : [\gamma] \mapsto \mu_{\overline{\pi}}([\gamma])$ is a continuous homomorphism, called the monodromy homomorphism of $\overline{\pi}$.

Next we find a relation between the monodromy homomorphism $\mu_{\overline{\pi}}$ of the bundle $\overline{\pi}$ and the monodromy homomorphism $\mu_{\breve{\pi}}$ of the bundle $\breve{\pi}$. Because $\pi^\circ(\overline{\pi}^{-1}(\gamma)) = \breve{\pi}^{-1}(\gamma)$, the T-automorphism $\mu_{\overline{\pi}}([\gamma])$ induces the $W(T)$-automorphism $\mu_{\breve{\pi}}([\gamma])$. Since $\mathrm{Aut}(T) = W(T)$, every T-automorphism $\mu_{\overline{\pi}}([\gamma])$ is of the form $\mathrm{Ad}_{\widetilde{w}([\gamma])}$ for a unique $\widetilde{w}([\gamma]) \in W(T)$. Because this latter T-automorphism induces the $W(T)$-automorphism given by multiplication by $\widetilde{w}([\gamma])$, it follows that $\mu_{\overline{\pi}}([\gamma]) = \mathrm{Ad}_{\mu_{\breve{\pi}}([\gamma])}$. Therefore $\mu_{\overline{\pi}}$ is trivial if and only if $\mu_{\breve{\pi}}$ is trivial. So we have proved

Proposition 4.2.6.2.36. $\overline{\pi} : G\cdot\mathcal{N} \to \mathcal{C}$ *is a principal T-bundle if and only if its monodromy homomorphism $\mu_{\overline{\pi}}$ is trivial.*

Example 4.2.6.2.37.

1. Let G be a compact noncommutative Lie group with Lie algebra $\mathfrak{g}$. Then $\mathfrak{g}^{\mathrm{se}} = \mathfrak{g}$ and $\mathfrak{g}^{rse} = \mathfrak{g}^{\mathrm{reg}}$. For $\xi \in \mathfrak{g}^{\mathrm{reg}}$ the tori $T = \exp \mathfrak{g}_\xi$ are the maximal tori of G, which are conjugate to one another, see theorem 3.7.1 in [42]. The Weyl group $W(T)$ is a finite group, which is nontrivial.

2. If $G = \mathrm{E}(2)$, the group of proper Euclidean motions of the plane, see example 2 in §2.6.1.4, then a maximal torus T is a circle group and maximal tori and conjugate to one another. Thus the Weyl group is trivial. Therefore the torus fibration $\overline{\pi}^{-1}(\mathfrak{g}^{\mathrm{rse}})$ is a principal torus bundle. □

4.2.6.3 *When the G-action is proper but not free*

Suppose that the G-action Φ on M is proper, but not necessarily free. In this subsection we give results which are analogous to those found in §2.6.2. Most of the proofs are omitted, because they are similiar to the ones given in §2.6.2.

Let $m \in M$ and let $H = G_m$. The H-isotropy type M_H is a smooth submanifold of M, on which the Lie group $G_{M_H} = N(H)/H$ acts freely and properly. Let π_{M_H} be the G_{M_H}-orbit map. The image $\overline{M_H}$ of M_H under π_{M_H} is a smooth submanifold of the differential space $(\overline{M}, C^\infty(\overline{M}))$, where $\overline{M}$ is the space of G-orbits on M.

Let F_{M_H} be the smooth submanifold of $\overline{M_H}$ consisting of all fixed points of the reduced flow of the G_{M_H}-invariant vector field $V|M_H$. Here V is a G-invariant vector field on M. Then $\pi^{-1}_{M_H}(F_{M_H})$ is the set of all relative equilibria of $V|M_H$ on M_H and thus is invariant under G_{M_H}-action. Let $\mathfrak{g}_{M_H}$ be the Lie algebra of G_{M_H}. For each $m \in \pi^{-1}_{M_H}(F_{M_H})$ let $\widetilde{\xi}(m) \in \mathfrak{g}_{M_H}$

be the generator of the relative equilibrium m of the vector field $V|M_H$. Then

Lemma 4.2.6.3.38. *The map* $\pi_{M_H}^{-1}(F_{M_H}) \to \mathfrak{g}_{M_H} : m \mapsto \widetilde{\xi}(m)$ *is smooth.*

Proof. The proof is analogous to that of lemma 4.2.6.2.28. □

Let $F_{M_H}^{\mathrm{rse}}$ be $\{f \in F_{M_H} \,\big|\, \widetilde{\xi}(m) \in \mathfrak{g}_{M_H}^{\mathrm{rse}}$ for some $m \in \pi_{M_H}^{-1}(f)\ \}$. Then $F_{M_H}^{\mathrm{rse}}$ is an open subset of F_{M_H}. Analogous to proposition 4.2.6.2.29 we have

Proposition 4.2.6.3.39. *Let* $m_0 \in \pi_{M_H}^{-1}(F_{M_H})$ *such that* $f_0 = \pi_{M_H}(m_0) \in F_{M_H}^{\mathrm{rse}}$. *Let* $\mathfrak{t}$ *be the centralizer of* $\widetilde{\xi}(m_0)$ *in* $\mathfrak{g}_{M_H}$ *and let* $T = \exp \mathfrak{t}$. *Then there is an open neighborhood* U *of* f_0 *in* $F_{M_H}^{\mathrm{rse}}$ *and a smooth section* $\sigma : U \subseteq F_{M_H}^{\mathrm{rse}} \to \pi_{M_H}^{-1}(U)$ *of the bundle* $\pi_{M_H}|\pi_{M_H}^{-1}(U)$ *such that* $\widetilde{\xi}(\sigma(f)) \in \mathfrak{t}$ *for every* $f \in U$.

Proof. See the proof of proposition 4.2.6.2.29. □

Let

$$\lambda : U \times G_{M_H} \to \pi_{M_H}^{-1}(U) : (f, g) \mapsto g \cdot \sigma(f).$$

Then λ intertwines the $\mathbb{R}$-action

$$\mathbb{R} \times (U \times G_{M_H}) \to U \times G_{M_H} : (t, (f, g)) \mapsto \big(f, g \exp t\, \widetilde{\xi}(\sigma(f))\big)$$

with the flow $\varphi_t|\pi_{M_H}^{-1}(U)$ of the G_{M_H}-invariant vector field $V|\pi_{M_H}^{-1}(U)$.

Let $\mathcal{C}^{M_H}$ be the connected component of $F_{M_H}^{\mathrm{rse}}$ which contains f_0 and let $\mathcal{N}_{M_H}$ be the connected component of $\{m \in \pi_{M_H}^{-1}(\mathcal{C}^{M_H}) \big| \widetilde{\xi}(m) \in \mathfrak{t}\}$, which contains m_0. Since $\mathcal{N}_{M_H}$ is a locally closed subset of M_H, the proper free action of G_{M_H} on M_H induces a proper free action on $\mathcal{N}_{M_H}$ of the normalizer $N(T)$ of T in G_{M_H}. Let $\widetilde{\pi}_{M_H} : \mathcal{N}_{M_H} \to \mathcal{C}^{M_H}$ be the $N(T)$-orbit map. Analogous to lemma 4.2.6.2.31 we have

Lemma 4.2.6.3.40. $\widetilde{\pi}_{M_H} : \mathcal{N}_{M_H} \to \mathcal{C}^{M_H}$ *is a principal* $N(T)$*-bundle.*

Corresponding to the $N(T)$-bundle $\widetilde{\pi}_{M_H}$ we have a G_{M_H}-bundle $\widehat{\pi}_{M_H} : G \cdot \mathcal{N}_{M_H} \to \mathcal{C}^{M_H}$ with projection map $\widehat{\pi}_{M_H} = \pi_{M_H}|(G \cdot \mathcal{N}_{M_H})$ and local trivializations

$$\mu_\sigma : \mathcal{C}_\sigma^{M_H} \times G_{M_H} \to \pi_{M_H}^{-1}(\mathcal{C}_\sigma^{M_H}) : (f, g) \mapsto g \cdot \sigma(f).$$

Here $\mathcal{C}_\sigma^{M_H}$ is the open set U containing $\widetilde{f} = \pi_{M_H}(\widetilde{m})$ with $\widetilde{m} \in \mathcal{N}_{M_H}$ and $\sigma : \mathcal{C}_\sigma^{M_H} \to \widetilde{\pi}_{M_H}^{-1}(\mathcal{C}_\sigma^{M_H})$ the smooth section constructed in proposition

4.2.6.3.39. Moreover, $\widetilde{\xi}(\sigma(f)) \in \mathfrak{t}$ for every $f \in \mathcal{C}_\sigma^{M_H}$. The transition map between the overlap of $\mathcal{C}_\sigma^{M_H} \times G_{M_H}$ and $\mathcal{C}_\tau^{M_H} \times G_{M_H}$ is

$$\phi_{\tau\sigma} : (\mathcal{C}_\sigma^{M_H} \cap \mathcal{C}_\tau^{M_H}) \times G_{M_H} \to (\mathcal{C}_\sigma^{M_H} \cap \mathcal{C}_\tau^{M_H}) \times G_{M_H} : (f, g) \mapsto (f, g g_{\tau\sigma}(f)),$$

where

$$g_{\tau\sigma} : (\mathcal{C}_\sigma^{M_H} \cap \mathcal{C}_\tau^{M_H}) \to N(T) : f \to g_{\tau\sigma}(f)$$

is the smooth map such that the transition map between the overlap of the local trivializations $\lambda_\sigma : \mathcal{C}_\sigma^{M_H} \times N(T) \to \widetilde{\pi}_{M_H}^{-1}(\mathcal{C}_\sigma^{M_H})$ and $\lambda_\tau : \mathcal{C}_\tau^{M_H} \times N(T) \to \widetilde{\pi}_{M_H}^{-1}(\mathcal{C}_\tau^{M_H})$ of the principal $N(T)$-bundle $\widetilde{\pi}_{M_H} : G \cdot \mathcal{N}_{M_H} \to \mathcal{C}^{M_H}$ is given by

$$\varphi_{\tau\sigma} : (\mathcal{C}_\sigma^{M_H} \cap \mathcal{C}_\tau^{M_H}) \times N(T) \to (\mathcal{C}_\sigma^{M_H} \cap \mathcal{C}_\tau^{M_H}) \times N(T) : (f, n) \mapsto (f, n g_{\tau\sigma}(f)).$$

As in §2.6.2 we can construct a T-bundle on $G_{M_H} \cdot \mathcal{N}_{M_H}$ associated to the G_{M_H}-bundle $\widehat{\pi}_{M_H}$ using the T-action

$$\bullet : T \times (G_{M_H} \cdot \mathcal{N}_{M_H}) \to G_{M_H} \cdot \mathcal{N}_{M_H} : (sH, g \cdot m) \mapsto gs \cdot m.$$

Let $\overline{\pi}_{M_H}$ be the T-orbit map of the action $\bullet$. As before the T-bundle $\overline{\pi}_{M_H} : G_{M_H} \cdot \mathcal{N}_{M_H} \to \mathcal{C}^{M_H}$ need not be a principal bundle. Indeed, an argument shows that $\overline{\pi}_{M_H}$ is a principal T-bundle if and only if its monodromy homomorphism $\mu_{\overline{\pi}_{M_H}} : \pi_1(\mathcal{C}^{M_H}, f) \to \mathrm{Aut}(T)$ is trivial.

4.3 Relative periodic orbits

In this section we study the relative periodic orbits of a G-invariant vector field V on a smooth manifold. We assume that the G-action on M is proper and that $\pi : M \to \overline{M}$ is its G-orbit map.

4.3.1 *Basic properties*

Before we define what a relative periodic orbit of V is we prove

Lemma 4.3.1.41. *Let $\tau \in I_m \setminus \{0\}$, where I_m is the maximal open interval in $\mathbb{R}$ where the integral curve $\gamma_m : I_m \to M$ of V starting at m is defined. The following statements are equivalent.*
1. *There is a $\tau \neq 0$ and an element $s \in G$ such that*

$$\varphi_\tau(m) = s \cdot m. \tag{32}$$

Here $\varphi : D \subseteq \mathbb{R} \times M \to M : (t, m) \mapsto \gamma_m(t)$ is the flow of V.

2. $I_m = \mathbb{R}$, $\tau > 0$, and there is an element $s \in G$ [4] *such that*

$$\varphi_t(m) = s^n \cdot \varphi_{t-n\,\tau}(m) = \varphi_{t-n\,\tau}(s^n \cdot m) \tag{33}$$

for every $t \in [n\tau, (n+1)\tau]$ and every $n \in \mathbb{Z}$.[5]
3. $\overline{m} \in \overline{M}$ is a periodic point of period τ of the reduced flow $\overline{\varphi}_t$ of V on $\overline{M}$, that is, $\overline{\varphi}_\tau(\overline{m}) = \overline{m}$.

Proof. We only need to prove item 1 $\Rightarrow$ item 2 as the other implications are straightforward. Suppose that item 1 holds and that $\tau > 0$. Observe that $t \in [n\tau, (n+1)\tau]$ if and only if $t - n\,\tau \in [0, \tau]$. The right hand side of (33) defines a piece of an integral curve γ of V on $[n\tau, (n+1)\tau]$ such that $\gamma(n\tau) = s^n \cdot m$ and $\gamma((n+1)\tau) = s^n \cdot \varphi_\tau(m) = s^n \cdot s \cdot m = s^{(n+1)} \cdot m$. As n ranges over $\mathbb{Z}$, these pieces fit together to form an integral curve γ of V defined on $\mathbb{R}$ such that $\gamma(0) = m$. When $\tau < 0$ a similar argument can be used. □

An integral curve γ of the vector field V, which starts at $m \in M$, is a *relative periodic orbit* of V of *relative period* τ if one of the statemtents in lemma 4.3.1.41 holds. If item 1 holds, then we call $s \in G$ the *shift element* of the relative periodic orbit. In addition, we say that m is a *relative periodic point* of *relative period* τ.

4.3.2 *Quasiperiodic relative periodic orbits*

Let G be a Lie group with $s \in G$. Let $\langle s \rangle = \{s^n \in G \,\big|\, n \in \mathbb{Z}\}$. Then $\langle s \rangle$ is the smallest subgroup of G containing s. Let S be the closure in G of $\langle s \rangle$. We say that S is the *closed subgroup of* G *generated by* s.

Lemma 4.3.2.42. *Either the mapping $\rho : \mathbb{Z} \to S \subseteq G : n \mapsto s^n$ is proper or S is a compact subgroup of G.*

Proof. Suppose that ρ is a proper map. Then for every compact subset K of G we have $\rho^{-1}(K)$ is a compact, and thus finite, subset F of $\mathbb{Z}$. Let n be an integer not in F. Then $s^n \notin K$. Therefore S is not contained in any compact subset of G. Consequently, S is not a compact subgroup of G. This shows that if S is a compact subgroup of G, then ρ is not a proper mapping. □

[4] The element s can be chosen to be equal to the element s in 1 and is unique up to multiplication on the right by an element of G_m.
[5] If $\tau < 0$ then $\varphi_t(m) = s^{-n} \cdot \varphi_{t+n\,\tau}(m) = \varphi_{t+n\,\tau}(s^{-n} \cdot m)$.

Proposition 4.3.2.43. *Suppose that S is a compact subgroup of G. Then*

1. *$T = S^0$ is a torus subgroup of G.*
2. *There is a unique minimal positive integer n_0 such that $s^{n_0} \in T$. The map $\psi : \mathbb{Z} \to S/T : n \mapsto s^n T$ induces an isomorphism $\overline{\psi} : \mathbb{Z}/n_0\,\mathbb{Z} \to S/T$. The dimension of T is 0 if and only if $s^{n_0} = e$. When this is the case $S = \mathbb{Z}/n_0\,\mathbb{Z}$.*
3. *Let $\mathfrak{t}$ be the Lie algebra of T. Then $s^{n_0} = \exp \xi$ for some $\xi \in \mathfrak{t}$. The element ξ is determined modulo the integer lattice $\Lambda = \ker \exp$ in $\mathfrak{t}$. If $\{\lambda_j\}_{j=1}^k$ is a $\mathbb{R}$-basis of Λ and $\xi = \sum_{j=1}^k \xi_j\,\lambda_j$, then the elements of $\{\xi_j\}_{j=1}^k$ are linearly independent real numbers over $\mathbb{Q}$.*

Proof. 1. As S is a closed subgroup of G, it is a Lie group. Moreover, S is commutative, since $\langle s \rangle$ is. The identity component S^0 of S is connected. Therefore S^0 is a connected commutative Lie subgroup of G and consequently is a torus T.

2. Suppose that the homomorphism ψ is injective and that $\{n_i\}$ is a sequence of integers which converges to $\pm\infty$ such that s^{n_i} converges to some element $g \in G$. Since S is closed, $g \in S$. Because gT is an open neighborhood of g in S, we have $s^{n_i} \in gT$ for all i sufficiently large. Therefore $g \in s^{n_i}T$ for i sufficiently large. Since $s^n T$ are connected components of S, we obtain $\psi(n_i) = \psi(n_j)$ for all sufficiently large i and j. But this contradicts the hypothesis that ψ is injective. Therefore the map $\rho : \mathbb{Z} \to G : n \mapsto s^n$ is proper if ψ is injective. But this contradicts the hypothesis that S is a compact subgroup of G, using lemma 4.3.2.42. Therefore the map ψ is not injective.

So there is a unique integer n_0 such that $\ker \psi = n_0\,\mathbb{Z}$. In other words, n_0 is the unique positive integer such that $t = s^{n_0} \in T$. Because $\langle s \rangle$ is dense in S, each connected component of S contains an s^n for some $n \in \mathbb{Z}$. Therefore the homomorphism $\overline{\psi}$ is injective and induces an isomorphism $\overline{\psi} : Z/n_0\,\mathbb{Z} \to S/T$. We have $s^n \in T$ if and only if $n \in n_0\,\mathbb{Z}$. Because $\langle s \rangle$ is dense in S, the group $\langle t \rangle$ is dense in T. Since T is a connected commutative Lie group, T is isomorphic to $(\mathbb{R}^k/\mathbb{Z}^k) \times \mathbb{R}^\ell$ for some $(k, \ell) \in \mathbb{Z}_{\ge 0} \times \mathbb{Z}_{\ge 0}$. Let $(a + \mathbb{Z}^k, b)$ be the element of $(\mathbb{R}^k/\mathbb{Z}^k) \times \mathbb{R}^\ell$ which corresponds to t under the given isomorphism. Then for every $n \in \mathbb{Z}$, the element $(n\,a + \mathbb{Z}^k, n\,b)$ corresponds to t^n. If $b \ne 0$ then $\{(n\,a + \mathbb{Z}^k, n\,b)\}_{n\in\mathbb{Z}}$ is not dense in $(\mathbb{R}^k/\mathbb{Z}^k) \times \mathbb{R}^\ell$. But this contradicts the density of $\langle t \rangle$ in T. Therefore $b = 0$. Thus T is a torus subgroup of G. The other assertions in statement 2 are straightforward to prove.

3. Because $\exp : \mathfrak{t} \to T$ is a surjective homomorphism, there is a unique $\xi \in \mathfrak{t}$ modulo $\Lambda = \ker \exp$ such that $t = s^{n_0} = \exp \xi$. Since exp induces an isomorphism from $\mathfrak{t}/\Lambda$ onto T, it follows that $\{n\,\xi + \Lambda\}_{n \in \mathbb{Z}}$ is dense in $\mathfrak{t}/\Lambda$. However, the map

$$\mathbb{R}^k/\mathbb{Z}^k \to \Lambda : (\theta_1, \ldots, \theta_k) \mapsto \sum_{j=1}^{k} \theta_j\,\lambda_j$$

is an isomorphism. Arguing as in the proof of item 4 of lemma 4.2.2.5, we conclude that $\{\xi_j\}_{j=1}^{k}$ are linearly independent over $\mathbb{Q}$. □

An element s of G is *elliptic* if the closed subgroup of G generated by s is compact.

Proposition 4.3.2.44. *Suppose that $\gamma : \mathbb{R} \to M : t \mapsto \varphi_t(m)$ is a relative periodic orbit of V, which starts at m, has relative period $\tau > 0$, and has a shift element s, which is elliptic. Then*

1. γ is quasiperiodic with at least $k+1$ frequencies, where $k = \dim T$. In particular, we have a smooth map

$$\begin{aligned} &\Gamma : \mathbb{R}^{k+1}/\mathbb{Z}^{k+1} \to M : \qquad (34)\\ &\theta = (\theta_1, \ldots, \theta_{k+1}) \mapsto \exp\big(\textstyle\sum_{j=1}^{k} \theta_j\,\lambda_j - \theta_{k+1}\,\xi\big) \cdot \varphi_{\theta_{k+1} n_0 \tau}(m). \end{aligned}$$

with frequency vector $\nu = (\nu_1, \ldots \nu_{k+1})$, where $\nu_j = \xi_j/(n_0\tau)$ for $1 \le j \le k$ and $\nu_{k+1} = 1/n_0\tau$.
2. The closure of $\gamma(\mathbb{R})$ in M is the image of Γ. As $\Gamma^{-1}(\{m\})$ is a closed subgroup of $\mathbb{R}^{k+1}/\mathbb{Z}^{k+1}$, the group T_0 given by the quotient of $\mathbb{R}^{k+1}/\mathbb{Z}^{k+1}$ by $\Gamma^{-1}(\{m\})$ is a torus. The map Γ (34) induces a smooth embedding $\overline{\Gamma} : T_0 \to M$, whose image is the closure in M of $\gamma(\mathbb{R})$. The minimal number of frequencies of γ is $\dim T_0$.

Proof. 1. For $1 \le j \le k+1$ let $\theta'_j = \theta_j + n_j$ with $n_j \in \mathbb{Z}$. Then

$$\begin{aligned} \Gamma(\theta') &= \exp\big(\sum_{j=1}^{k} \theta_j\,\lambda_j - \theta_{k+1}\,\xi\big) \exp(-n_{k+1}\,\xi) \cdot \varphi_{\theta_{k+1} n_0 \tau}(\varphi_{n_{k+1} n_0 \tau}(m)) \\ &= \exp\big(\sum_{j=1}^{k} \theta_j\,\lambda_j - \theta_{k+1}\,\xi\big) \varphi_{\theta_{k+1} n_0 \tau}(s^{-n_{k+1} n_0} \cdot \varphi_{n_{k+1} n_0 \tau}(m)), \\ &\qquad \text{since } s^{n_0} = \exp \xi \\ &= \exp\big(\sum_{j=1}^{k} \theta_j\,\lambda_j - \theta_{k+1}\,\xi\big) \cdot \varphi_{\theta_{k+1} n_0 \tau}(m) = \Gamma(\theta). \end{aligned}$$

The third equality above follows using (33) with $t = n\tau$ and $n = n_{k+1}n_0$. Therefore Γ is a well defined smooth map from $\mathbb{R}^{k+1}/\mathbb{Z}^{k+1}$ into M.

2. If $\theta_j = t\xi_j/n_0\tau$ for $1 \le j \le k$ and $\theta_{k+1} = t/n_0\tau$, then

$$\sum_{j=1}^{k} \theta_j\, \lambda_j - \theta_{k+1}\, \xi = \frac{t}{n_0\tau}\big(\sum_{j=1}^{k} \xi_j\, \lambda_j - \xi\big) = 0. \tag{35}$$

In view of the definition of the map Γ (34), equation (35) implies that $\gamma(t) = \Gamma(t\,\nu + \mathbb{Z}^{k+1})$ for all $t \in \mathbb{R}$, where ν has components $\nu_j = \xi_j/n_0\tau$ for $1 \le j \le k$ and $\nu_{k+1} = 1/n_0\tau$. Because $\{1, \{\xi_j\}_{j=1}^{k}\}$ are linearly independent over $\mathbb{Q}$, it follows that the components of ν are linearly independent over $\mathbb{Q}$. Therefore, using item 4 of lemma 4.2.2.5, we see that $\{t\,\nu + \mathbb{Z}^{k+1} \in \mathbb{R}^{k+1}/\mathbb{Z}^{k+1} \big| \, t \in \mathbb{R}\}$ is dense in $\mathbb{R}^{k+1}/\mathbb{Z}^{k+1}$. Because the map Γ is continuous, it follows that $\Gamma\big(\mathbb{R}^{k+1}/\mathbb{Z}^{k+1}\big)$ is equal to the closure of $\gamma(\mathbb{R})$ in M.

Since $\gamma(0) = m$, the map

$$\Xi : (T \times \mathbb{R}) \times M \to M : ((g,t),m) \mapsto g \cdot \varphi_t(m) = \varphi_t(g \cdot m)$$

defines a local smooth action of $T \times \mathbb{R}$ on M. Therefore its orbit map $\Xi_m : T \times \mathbb{R} \to M : (g,t) \mapsto \Xi((g,t),m)$ induces a smooth immersion of $(T \times \mathbb{R})/(T \times \mathbb{R})_m$ into M. Consequently, Γ induces a smooth embedding of $T_0 = (\mathbb{R}^{k+1}/\mathbb{Z}^{k+1})/\Gamma^{-1}(m)$ into M. An argument similar to the one used in the proof of proposition 4.2.2.6 shows that the minimal number of frequencies is $\dim T_0$. □

Corollary 4.3.2.45. *If $\tau = \inf\{t \in \mathbb{R}_{>0} \big| \, \gamma(t) \in G \cdot m\}$ and the G-action at m is free, then the map Γ (34) is a smooth embedding and the minimal number of frequencies is $k+1$.*

Proof. Because the G-action is free at m, so is the T-action. Then $\Gamma(\theta) = m$ implies $\varphi_{\theta_{k+1}n_0\tau}(m) \in T \cdot m \subseteq G \cdot m$. Since the T-action is free at m, there is an $n \in \mathbb{Z}$ such that $\theta_{k+1}n_0\tau = n\tau$, which implies that $\varphi_{\theta_{k+1}n_0\tau}(m) = s^n \cdot m \in T \cdot m$. Again because of the freeness of the T-action at m, there is a $q \in \mathbb{Z}$ such that $n = qn_0$. Consequently, $\theta_{k+1} = q \in \mathbb{Z}$. However, $\exp\xi = s^{n_0}$ and therefore $\exp(-\theta_{k+1}\xi) = s^{-qn_0}$. From $\varphi_{\theta_{k+1}n_0\tau}(m) = s^n \cdot m = s^{qn_0}$ and the definition of the map Γ (34), we get $\exp\big(\sum_{j=1}^{k} \theta_j\, \lambda_j\big) \cdot m = m$. Using the freeness of the T-action at m once more, we deduce that $\theta_j \in \mathbb{Z}$ for every $1 \le j \le k$. Therefore $\Gamma^{-1}(\{m\}) = \{e\}$, so $T_0 = \mathbb{R}^{k+1}/\mathbb{Z}^{k+1}$. □

4.3.3 *Runaway relative period orbits*

Proposition 4.3.3.46. *Let m be a relative periodic point of relative period $\tau > 0$ with shift element $s \in G$. Then the following statements are equivalent.*

1. *The element s is not elliptic.*
2. *The integral curve $\gamma : \mathbb{R} \to M$ of the G-invariant vector field V, which starts at m, is a runaway curve in M.*
3. *The integral curve γ of V starting at m is not quasiperiodic.*

Proof. item $\Rightarrow$ item 2. If item 1 holds, then from lemma 4.3.2.42 it follows that the mapping $\psi : \mathbb{Z} \to M : n \mapsto s^n \cdot m$ is proper. Let K be a compact subset of M. Then $K' = \{\varphi_{-s}(x) \in M \big| \, x \in K, \; s \in [0, \tau]\}$ is equal to the image of the compact set $[-\tau, 0] \times K$ under the continuous map $\varphi|([-\tau, 0] \times K')$. Therefore K' is a compact subset of M. Because the mapping ψ is proper, there is a positive integer N such that $|n| \leq N$ whenever $s^n \cdot m \in K'$. Consequently, if $t \in [n\tau, (n+1)\tau]$ and $\varphi_t(m) \in K$, then $s^n \cdot m = \varphi_{-(t-n\,\tau)}(\varphi_t(m)) \in K'$. Thus $|n| \leq N$, which implies $|t| = |(t - n\,\tau) + n\,\tau| \leq \tau + N\,\tau$. Consequently, the integral curve $\gamma : \mathbb{R} \to M$ is a proper mapping. A similar argument works when $\tau < 0$.

$\neg$ item 3 $\Rightarrow$ $\neg$ item 2. If γ is quasiperiodic, then its image is contained in the image of a torus under a continuous map, and thus is contained in a compact subset of M. Thus γ is not a runaway curve.

item 3 $\Rightarrow$ item 1. This implication follows from item 1 of proposition 4.3.2.43. □

Without additional assumptions on the G-action, not much can be deduced that is in the spirit of proposition 4.3.3.46. For instance, let V be a complete vector field on M with flow φ_t and let $\tau > 0$. Then $\Psi : \mathbb{Z} \times M \to M : (n, m) \mapsto \varphi_{n\tau}(m)$ defines an action of $G = (\mathbb{Z}, +)$, which is discrete. Clearly, every integral curve of V is a $\mathbb{Z}$-relative periodic orbit with relative period τ. But it is certainly not the case that for every dynamical system every integral curve is either quasiperiodic or runs away.

4.3.4 *When the G-action is not free*

The results of proposition 4.3.3.46 are not optimal when the proper G-action is not free at m. Let $H = G_m$.

From (33) and lemma 4.2.5.12 with $t = \tau$ and $g = s$ it follows that $s \in N(H)$. Because the Lie group $G_{M_H} = N(H)/H$ acts freely on the H-isotropy type M_H containing m, we obtain the following proposition from proposition 4.3.3.46.

Proposition 4.3.4.47. *Suppose that m is a relative periodic point with shift element s. Assume that sH is an elliptic element of G_{H_M}. Then the integral curve $\gamma : \mathbb{R} \to M_H$ of the G_{M_H}-invariant vector field $V|M_H$, which starts at m is quasiperiodic. If $\tau = \inf\{t \in \mathbb{R}_{>0} \big| \, \varphi_t(m) \in G \cdot m\}$, then the map Γ (34) is a smooth embedding of $\mathbb{R}^{k+1}/\mathbb{Z}^{k+1}$ into M. Here $k = \dim T$, where T is the torus in G_{M_H} formed by taking the identity component of the closure of $\langle sH \rangle$ in G_{M_H}. The minimal number of frequencies of γ is $k+1$.*

Corollary 4.3.4.48. *If G_{M_H} is compact, then every relative periodic orbit of $V|M_H$ is quasiperiodic.*

4.3.5 *Other relative periodic orbits in the $(G \times \mathbb{R})$-orbit*

If m is a relative periodic point of the vector field V on M with relative period τ and shift element s, then for $g \in G$ we have

$$\varphi_\tau(g \cdot m) = g \cdot \varphi_\tau(m) = g \cdot s \cdot m = (gsg^{-1}) \cdot g \cdot m. \tag{36}$$

This shows that $g \cdot m$ is a relative periodic point with relative period τ and with shift element $\mathrm{Ad}_g s$.

An integral curve of V starting at $g \cdot m$ is periodic, quasiperiodic, or runaway, if and only if the integral curve of V starting at m has these properties, respectively. In addition, the closures in M of V-orbits starting at a point in $G \cdot m$ define a smooth fibration of the closure of the V flow out of $G \cdot m$, that is, the closure in M of $\{\varphi_t(G \cdot m) \in M \big| \, t \in \mathbb{R}\}$.

Let us give some more detail when we are in the situation where $\tau = \inf\{t \in \mathbb{R}_{>0} \big| \, \varphi_t(m) \in G \cdot m\}$ and the G-action on M is free at m. In other words, $G_m = \{e\}$. This is more complicated than the case of relative equilibria because the closure of a V-orbit is not generated by a torus subgroup of G, since a V-orbit is not contained in a G-orbit.

For $g, g' \in G$ and $t, t' \in \mathbb{R}$ we have $g \cdot \varphi_t(m) = g' \cdot \varphi_{t'}(m)$ if and only if $\varphi_{t'-t}(m) = (g')^{-1} g \cdot m \in G \cdot m$ if and only if there is an $n \in \mathbb{Z}$ such that $t' - t = n\tau$ and $(g')^{-1} g = s^n$. Define a $\mathbb{Z}$-action on $G \times \mathbb{R}$ by

$$\mathbb{Z} \times (G \times \mathbb{R}) \to G \times \mathbb{R} : (n, (g, t)) \mapsto (gs^{-n}, t + n\tau) \tag{37}$$

Then the mapping $G \times \mathbb{R} \rightarrow M : (g,t) \mapsto g \cdot \varphi_t(m)$ induces a smooth embedding $\Psi : (G \times \mathbb{R})/\mathbb{Z} \rightarrow (G \times \mathbb{R}) \cdot m$. The space $(G \times \mathbb{R}) \cdot m$ is the V flow out of the G-orbit $G \cdot m$. The $\mathbb{Z}$-orbit map $\rho : G \times \mathbb{R} \rightarrow (G \times \mathbb{R})/\mathbb{Z}$ intertwines the G-action on $G \times \mathbb{R}$, given by multiplication on the left in the first factor, with the induced smooth G-action on $(G \times \mathbb{R})/\mathbb{Z}$. Also it intertwines the $\mathbb{R}$-action on $G \times \mathbb{R}$, given by addition on the second factor, with a smooth $\mathbb{R}$-action on $(G \times \mathbb{R})/\mathbb{Z}$. The map Ψ intertwines the G and $\mathbb{R}$-actions on $(G \times \mathbb{R})/\mathbb{Z}$ with the G-action and the flow of V on $(G \times \mathbb{R}) \cdot m$, respectively. Note that the projection map $G \times \mathbb{R}$ onto the first factor induces the bundle projection map from $(G \times \mathbb{R})/\mathbb{Z}$ onto the circle $\mathbb{R}/\tau\mathbb{Z}$. Using the map Ψ, the preceding bundle can be identified with the image of the relative periodic orbits of V starting at a point in $G \cdot m$ under the G-orbit map π.

Consider the $\mathbb{R}^{k+1}$-action on $G \times \mathbb{R}$ defined by

$$\begin{aligned} \bullet : \mathbb{R}^{k+1} \times (G \times \mathbb{R}) \rightarrow G \times \mathbb{R} : \\ (\theta, (g,t)) \mapsto \big(g \exp(\textstyle\sum_{j=1}^{k} \theta_j \, \lambda_j - \theta_{k+1} \, \xi), t + \theta_{k+1} n_0 \tau\big). \end{aligned} \qquad (38)$$

If $\theta \in \mathbb{Z}^{k+1}$ then using $\exp \xi = s^{n_0}$ we get $\theta \bullet (g,t) = (g \, s^{-\theta_{k+1} n_0}, t + \theta_{k+1} n_0 \tau)$, which is equal to $\theta_{k+1} n_0 \circ (g,t)$, where $\circ$ is the $\mathbb{Z}$ action defined by

$$\circ : \mathbb{Z} \times (G \times \mathbb{R}) \rightarrow G \times \mathbb{R} : (n, (g,t)) \mapsto (g, t+n).$$

Therefore we have an induced action of $\mathbb{R}^{k+1}/\mathbb{Z}^{k+1}$ on $(G \times \mathbb{R})/\mathbb{Z}$. Moreover, this $\mathbb{R}^{k+1}/\mathbb{Z}^{k+1}$-action commutes with the G-action on $(G \times \mathbb{R})/\mathbb{Z}$ induced from the G action on $G \times \mathbb{R}$ given by mulitiplication of the first factor on the left.

If $\theta \bullet (g,t) = (t + n\tau, g \, s^{-n})$ for some $n \in Z$, then

$$s^n = \exp(-\sum_{j=1}^{k} \theta_j \, \lambda_j + \theta_{k+1} \, \xi) \in T, \qquad (39)$$

which implies that $n = q n_0$ for some $q \in \mathbb{Z}$. But then $n\tau = \theta_{k+1} n_0 \tau$, implies that $\theta_{k+1} = q$. Then (39) together with $\exp \xi = s^{n_0}$ implies that $\sum_{j=1}^{k} \theta_j \, \lambda_j \in \Lambda$, the integer lattice in $\mathfrak{t}$. Therefore $\theta_j \in \mathbb{Z}$ for every $1 \leq j \leq k$. This shows that the $\mathbb{R}^{k+1}/\mathbb{Z}^{k+1}$-action on $(G \times \mathbb{R})/\mathbb{Z}$ is free. It is proper because $\mathbb{R}^{k+1}/\mathbb{Z}^{k+1}$ is compact.

Let $\{\xi_j\}_{j=1}^{k}$ be the components of $\xi \in \mathfrak{t}$ with respect to the $\mathbb{Z}$-basis $\{\lambda_j\}_{j=1}^{k}$ of the integer lattice Λ of $\mathfrak{t}$. Then

$$\sum_{j=1}^{k} \theta_j \, \lambda_j - \theta_{k+1} \, \xi = \sum_{j=1}^{k} (\theta_j - \theta_{k+1} \, \xi_j) \lambda_j .$$

Consequently, the vector field V on $(G \times \mathbb{R}) \cdot m$ is equal to the element $\dot{\theta}$ in the Lie algebra $\mathbb{R}^{k+1}$ of $\mathbb{R}^{k+1}/\mathbb{Z}^{k+1}$, which is defined by

$$\dot{\theta} = \begin{cases} \dot{\theta}_{k+1}\, \xi_j \text{ for } 1 \le j \le k \\ 1/(n_0 \tau) \text{ for } j = k+1. \end{cases} \tag{40}$$

Because $\{1, \{\xi_j\}_{j=1}^{k}\}$ are linearly independent over $\mathbb{Q}$, it follows from item 3 of lemma 4.2.2.5 that the orbits of (40) are dense in $\mathbb{R}^{k+1}/\mathbb{Z}^{k+1}$. Therefore the orbits of the $\mathbb{R}^{k+1}/\mathbb{Z}^{k+1}$-action on $(\mathbb{R} \times G) \cdot m$ are equal to the closure of the set of integral curves of V in $(\mathbb{R} \times G) \cdot m$.

Now assume that the G-action is not free at m. Suppose that sH is an ellipitic element of $G_{M_H} = N(H)/H$. Then the V-flow out of $G \cdot m$ is a fiber bundle over $G/N(H)$ with typical fiber equal to the V-flow out of $N(H) \cdot m$. Because the action of G_{M_H} is free at m, we can use the construction of the preceding paragraphs to show that the closure of the V-orbits, starting on $N(H) \cdot m$, define a G_{M_H}-invariant smooth principal $\mathbb{R}^{k+1}/\mathbb{Z}^{k+1}$-bundle of the V flow out of $N(H) \cdot m$. Using lemma 4.2.5.2.22 with $L = N(H)$, $K = \mathbb{R}^{k+1}/\mathbb{Z}^{k+1}$ and $Q = (G_{M_H} \times \mathbb{R})/\mathbb{Z}$, we obtain

Proposition 4.3.5.49. *Suppose that m is a relative periodic point with shift element $s \in G$. Suppose that sH is an elliptic element of G_{M_H}. Then the closure of the V-orbits in the V flow out $(G \times \mathbb{R}) \cdot m$ of $G \cdot m$ define a smooth G-invariant principal $\mathbb{R}^{k+1}/\mathbb{Z}^{k+1}$ fibration of $(G \times \mathbb{R}) \cdot m$. The vector field V on $(G \times \mathbb{R}) \cdot m$ is equal to the element $\dot{\theta}$ (40) in the Lie algebra $\mathbb{R}^{k+1}$ of $\mathbb{R}^{k+1}/\mathbb{Z}^{k+1}$.*

4.3.6 *Smooth families of quasiperiodic relative periodic orbits*

In this section we consider a smooth family of periodic orbits of the reduced flow $\overline{\varphi}_t$ on the orbit space $\overline{M}$ corresponding to a G-invariant vector field V on M with flow φ_t.

4.3.6.1 *Elliptic, regular, and stably elliptic elements of G*

In this subsubsection we prove group theoretic analogues of the lemmas 4.2.6.1.24 – 4.2.6.1.26.

Let G be a Lie group with Lie algebra $\mathfrak{g}$. For $s \in G$, the *centralizer* $\mathfrak{g}_s$ of s in $\mathfrak{g}$ is $\{\eta \in \mathfrak{g} \,\big|\, \mathrm{Ad}_s \eta = \eta\}$. The *centralizer* G_s of s in G is $\{g \in G \,\big|\, g\, s = s\, g\}$. The group G_s is a closed subgroup of G and hence is a Lie group with

Lie algebra $\mathfrak{g}_s$. The element s is an *elliptic* element of G if and only if the closure of $\{s^n \in G \,\big|\, n \in \mathbb{Z}\}$ is contained in a compact subgroup of G. The element s is a *regular* element of G if there is an open neighborhood U of s in G such that $\dim \mathfrak{g}_s \le \dim \mathfrak{g}_{s'}$ for every $s' \in U$. Because $\dim \mathfrak{g}_{s'} \le \dim \mathfrak{g}_s$ for every s' in G near s, we see that s is regular if and only if $\dim \mathfrak{g}_{s'}$ is constant for all s' near s. The element s is *stably elliptic* if there is an open neighborhood U of s in G such that every $s' \in U$ is an elliptic element of G.

Lemma 4.3.6.1.50. *Let G be a Lie group with Lie algebra $\mathfrak{g}$. The following statements are equivalent.*

1. *s is a regular and stably elliptic element of G.*

2. *The centralizer $\mathfrak{g}_s$ of s in $\mathfrak{g}$ is the Lie algebra of a torus subgroup T of G. Moreover, there is a nonzero integer n such that $s^n \in T$.*

Proof. The proof of this lemma is analogous to the proof of lemma 4.2.6.1.24. We give the details.

property 1$\Rightarrow$ property 2. Suppose that s is an elliptic element of G, then the closure of $S = \{s^n \in G \,\big|\, n \in \mathbb{Z}\}$ in G is a compact subgroup of $\mathrm{Gl}(\mathfrak{g}, \mathbb{R})$. Because the map $\mathrm{Ad} : G \to \mathrm{Gl}(\mathfrak{g}, \mathbb{R})$ is a continuous homomorphism of Lie groups, the group $K = \mathrm{Ad}(S)$ is a compact subgroup of $\mathrm{Gl}(\mathfrak{g}, \mathbb{R})$. Averaging an arbitrary inner product on $\mathfrak{g}$ over K gives a K-invariant inner product β on $\mathfrak{g}$. Hence Ad_s is a β-orthogonal linear transformation of $(\mathfrak{g}, \beta)$ onto itself. Therefore, $\mathrm{im}\,(\mathrm{I} - \mathrm{Ad}_s)$ is the β-orthogonal complement of $\ker(\mathrm{I} - \mathrm{Ad}_s) = \mathfrak{g}_s$. So

$$\mathfrak{g} = \mathrm{im}\,(\mathrm{I} - \mathrm{Ad}_s) \oplus \mathfrak{g}_s \tag{41}$$

and $\mathrm{I} - \mathrm{Ad}_s$ is a bijective linear map of $\mathrm{im}\,(\mathrm{I} - \mathrm{Ad}_s)$ onto itself.

The tangent of the mapping

$$\Psi : G \times \mathfrak{g}_s \to \mathfrak{g} : (g, \xi) \mapsto (g \exp \xi \, g^{-1}) s^{-1}$$

at $(e, 0)$ is

$$T_{(e,0)}\Psi : \mathfrak{g} \times \mathfrak{g}_s \to \mathfrak{g} : (\eta, \xi) \mapsto (Id - \mathrm{Ad}_s)\eta + \xi.$$

In view of (41) it follows that $T_{(e,0)}\Psi$ is surjective. Thus by the implicit function theorem there is an open neighborhood U of e in G and a real analytic mappings $\varphi : U \subseteq G \to G$ and $\chi : U \subseteq G \to \mathfrak{g}_s$ such that $\varphi(e) = e$, $\chi(e) = 0$, and

$$u = \varphi(u)(\exp \chi(u) s)\varphi(u)^{-1}\, s^{-1} \tag{42}$$

for every $u \in U$. Writing $s' = us$, equation (42) implies that every element s' of G near s is conjugate to an element of $G_s^0\, s$ by means of an element of G which depends real analytically on s'. Note that $G_s^0\, s$ ($= s\, G_s^0$) is the connected component of G_s which contains s.

Let $s' \in G_s$. From the fact that $\mathrm{Ad} : G \to \mathrm{Gl}(\mathfrak{g}, \mathbb{R})$ is a group homomorphism, it follows that the linear mappings $\mathrm{Ad}_{s'}$ and Ad_s commute. Consequently, $\mathrm{Ad}_{s'}$ leaves $\mathfrak{g}_s$ and $\mathrm{im}\,(\mathrm{I} - \mathrm{Ad}_s)$ invariant. If s' is sufficiently close to s, then the restriction of $\mathrm{I} - \mathrm{Ad}_{s'}$ to $\mathrm{im}\,(\mathrm{I} - \mathrm{Ad}_s)$ is bijective. Therefore $\mathfrak{g}_{s'} = \ker(\mathrm{I} - \mathrm{Ad}_{s'}) \subseteq \mathfrak{g}_s$.

If s is a regular element of G, then for $s' \in G_s$ sufficiently close to s we have $\mathfrak{g}_{s'} = \mathfrak{g}_s$. So $\mathrm{Ad}_{(s' s^{-1})} = \mathrm{Ad}_{s'} \circ \mathrm{Ad}_{s^{-1}} = id_{\mathfrak{g}_s}$. In other words, $\mathrm{Ad}_{(s' s^{-1})}\eta = \eta$ for every $\eta \in \mathfrak{g}_s$. Differentiating the last equality in the direction of $\xi \in \mathfrak{g}_s$ gives $[\xi, \eta] = 0$ for every $\xi, \eta \in \mathfrak{g}_s$. Therefore $\mathfrak{g}_s$ is commutative, which implies that G_s^0 is commutative. Hence $G_s^0 = \exp \mathfrak{g}_s$, see theorem 1.12.1 of [42].

Because the identity component G_s^0 of G_s is an open and closed subgroup of the closed subgroup G_s of G, the subgroup $T = \exp \mathfrak{g}_s = G_s^0$ is a closed subgroup of G. Since $T \subseteq G_s$, we see that $H = \bigcup_{n \in \mathbb{Z}} s^n\, T$ is a commutative subgroup of G, which is the smallest subgroup of G that contains both s and T. In addition, the map

$$\lambda : \mathbb{Z} \to H/T : n \mapsto s^n\, T$$

is a homomorphism of groups. Recall that S is the closed subgroup of G generated by s. Because s is elliptic, there is a smallest positive integer n_0 such that $s^{n_0} \in S^0$. Since $S^0 \subseteq G_s$ and S^0 is connected, it follows that $S^0 \subseteq G_s^0 = T$. Therefore $n_0 \in \ker \lambda$, which implies that $\ker \lambda$ is a nontrivial subgroup of $\mathbb{Z}$. So there is a unique positive integer n_1 such that $\ker \lambda = n_1\, \mathbb{Z}$. Consequently, the group H/T is isomorphic to $\mathbb{Z}/(n_1\, \mathbb{Z})$. Note that n_1 is the smallest positive integer such that $s^{n_1} \in T$ and that n_0 is a positive multiple of n_1. Because $H = \bigcup_{0 \le n < n_1} s^n\, T$ and T is closed in G, it follows that H is a closed subgroup of G and hence is a Lie group with connected components $s^n\, T$ for $0 \le n < n_1$.

Since H is a closed subgroup of G, an element $h \in H$ is an elliptic element of G if and only if it is an elliptic element of H. Let $h = sk$, where $k \in T$. For any $n \in \mathbb{Z}$, we have $n = qn_1 + r$, where $q \in \mathbb{Z}$ and $0 \le r < n$. Because s commutes with k we get

$$h^n = s^n k^n = (sk)^r (s^{n_1} k^{n_1})^q,$$

where $s^{n_1} \in T$. Therefore h is an elliptic element of H if and only if $h^{n_1} = s^{n_1}\, k^{n_1}$ is an elliptic element of T. But the mapping $G \to G : k \mapsto$

k^{n_1} is open. So elements of the form h^{n_1} for all h near s in G form an open neighborhood of s^{n_1} in T. If s is a stably elliptic element of G, and hence one of H, then s^{n_1} is a stably elliptic element of T. Because T is a connected, commutative Lie group, it is isomorphic to $(\mathbb{R}/\mathbb{Z})^k \times \mathbb{R}^\ell$ for some $k, \ell \in \mathbb{Z}_{\geq 0}$. But $(\mathbb{R}/\mathbb{Z})^k \times \mathbb{R}^\ell$ contains stably elliptic elements only if $\ell = 0$. Therefore T is a torus. Thus property 1$\Rightarrow$ property 2.

Suppose that property 2 holds. Because $\mathfrak{t} = \mathfrak{g}_s$, for every $\xi \in \mathfrak{t}$ we have $s(\exp \xi)s^{-1} = \exp \mathrm{Ad}_s \xi = \exp \xi$. Thus s commutes with every element of $T = \exp \mathfrak{t}$. It follows that $H = \bigcup_{n \in \mathbb{Z}} s^n T$ is a commutative group. H is compact because there is a minimal positive integer n_1 such that $s^{n_1} \in T$, which implies that H/T is isomorphic to $\mathbb{Z}/n_1\mathbb{Z}$. Therefore every element of H is elliptic. In particular, s is elliptic. Because every element of G near s is conjugate to an element $s' \in sT \subseteq H$, we deduce that s' is elliptic. Therefore s is a stably elliptic element of G.

Since every element $s' \in H$ near s commutes with s, we get $\mathrm{Ad}_{s'}$ commutes with Ad_s. Therfore $\mathrm{Ad}_{s'}$ maps $\mathrm{im}(\mathrm{I} - \mathrm{Ad}_s)$ into itself. Since $\mathrm{I} - \mathrm{Ad}_s$ is a bijective linear map of $\mathrm{im}(\mathrm{I} - \mathrm{Ad}_s)$ into itself, it follows that $\mathrm{I} - \mathrm{Ad}_{s'}$ is a bijective linear map of $\mathrm{im}(\mathrm{I} - \mathrm{Ad}_s)$ into itself. Therefore $\dim \mathfrak{g}_{s'} \leq \dim \mathfrak{g}_s$, if s' is sufficiently close to s. Because every $s'' \in G$ near s is conjugate to an $s' \in H$ near s, we have $\dim \mathfrak{g}_{s''} = \dim \mathfrak{g}_{s'} \leq \dim \mathfrak{g}_s$. Therefore s is a regular element of G. So property 2 $\Rightarrow$ property 1. □

Let G^{se}, G^{reg}, and G^{rse} be the set of stably elliptic, regular, and regular stably elliptic elements of G, respectively. By definition G^{se} and G^{reg} are open subsets of G. Therefore $G^{\mathrm{rse}} = G^{\mathrm{se}} \cap G^{\mathrm{reg}}$ is open. To show that G^{rse} is dense in G^{se} we argue as follows. Suppose that $s \in G^{\mathrm{se}}$ but $s \notin G^{\mathrm{reg}}$. Then there is an $s' \in G^{\mathrm{se}}$ arbitrarily close to s such that $\dim \mathfrak{g}_{s'} < \dim \mathfrak{g}_s$. As $\dim \mathfrak{g}_s$ is finite, we can repeat this argument only finitely many times. We eventually obtain an $s' \in G^{\mathrm{se}}$ arbitrarily close to s such that $\dim \mathfrak{g}_{s'}$ is minimal. Then $s' \in G^{\mathrm{reg}}$.

Lemma 4.3.6.1.51. *For every $s \in G^{\mathrm{rse}}$ there is an open neighborhood U of s in G and a real analytic mapping $\theta : U \to G$ such that for every $s' \in U$ with $s'' = \mathrm{Ad}_{\theta(s')} s'$ we have $G^0_{s''} = G^0_s$ and $s'' G^0_{s''} = s G^0_s$.*

Proof. Write $s' = us$. Then by the implicit function theorem, see the proof of property 2 in lemma 4.3.6.1.50, there is an open neighborhood $\widetilde{U}$ of e in G and real analytic mappings $\varphi : \widetilde{U} \subseteq G \to G$ and $\chi : \widetilde{U} \subseteq G \to \mathfrak{g}$ such that $\varphi(e) = e$, $\chi(e) = 0$, and

$$u = \varphi(u)\big((\exp \chi(u))s\big)\varphi(u)^{-1}s^{-1}. \tag{43}$$

Let $U = \widetilde{U}s$, $\theta(s') = \varphi(u)^{-1}$, and $s'' = \mathrm{Ad}_{\theta(s')}s'$. Then equation (43) is equivalent to $s'' = (\exp \chi(u))s = s(\exp \chi(u))$. Therefore $s'' \in sG_s^0$. If s'' is sufficiently close to s in G, then $\mathfrak{g}_s = \mathfrak{g}_{s''} = \mathrm{Ad}_{\theta(s')}\mathfrak{g}_{s'}$, $G_s^0 = G_{s''}^0$, and $sG_s^0 = s''G_s^0 = s''G_{s''}^0$. The last equality follows because s'' to the connected component sG^0 of G_s which contains s. □

Let C be a connected component of G^{rse}.

Proposition 4.3.6.1.52. *Let $s, s' \in C$. Then*

1. *There is a $g \in G^0$ such that $\mathfrak{g}_{s'} = \mathrm{Ad}_g \mathfrak{g}_s$.*
2. $\dim \mathfrak{g}_s$ *is constant, say k, for every $s \in C$. The assignment $s \mapsto \mathfrak{g}_s$ defines a real analytic mapping from C into the Grassman manifold of k-dimensional vector subspaces of $\mathfrak{g}$.*
3. *The positvie integer n_1 such that $s^{n_1} \in G_s^0 = \exp \mathfrak{g}_s$ is the same for all $s \in C$.*

Proof. 1. Two elements $s, s' \in C$ are said to be related if there is a $g \in G^0$ such that $\mathfrak{g}_{s'} = \mathrm{Ad}_g \mathfrak{g}_s$. Being related in an equivalence relation on C. By lemma 4.3.6.1.51, nearby elements of C are related. Since C is connected, we obtain $\mathfrak{g}_{s'}$ is conjugate to $\mathfrak{g}_s$ by an element of G^0 for every $s, s' \in C$.

2. For s' near s, we know that $\mathfrak{g}_{s'}$ is conjugate to $\mathfrak{g}_s$ by $g = g(s') \in G^0$, which depends real analytically on s, the mapping $s' \mapsto \mathfrak{g}_{s'}$ is real analytic in an open neighborhood of s. Therefore the map $s \mapsto \mathfrak{g}_s$ is real analytic on C.

3. The constancy of n_1 is proved in a way similar to that given in part 1 above. □

Example 4.3.6.1.53.
1. If G is a compact Lie group, then every element of G is elliptic. Therefore $G^{\mathrm{se}} = G$ and $G^{\mathrm{rse}} = G^{\mathrm{reg}}$. If G is connected and $s \in G^{\mathrm{reg}}$, then $s \in G_s^0$, see proposition 3.1.3 of [42]. Therefore the smallest nonnegative integer n such that $s^n \in G_s^0$ is equal to 1. Every $s \in G$ belongs to a maximal torus T in G_s. Therefore $T \subseteq G_s^0$ and $T = G_s^0$ if $s \in G^{\mathrm{reg}}$. If G is compact and connected, then all tori in G^{rse} are maximal tori of dimension equal to the rank of G.

2. Let $s = (A, a) \in G = \mathrm{E}(2)$, the group of proper *Euclidean motions* of $\mathbb{R}^2$. Then s is not elliptic if $A = \mathrm{I}$ and $a \neq 0$. If $A \neq \mathrm{I}$, then $\mathrm{I} - A$ is invertible and s has a unique fixed point $p = (\mathrm{I} - A)^{-1}a$. In this case, the

isotropy group of s is equal to the *group of all rotations about* p, which is a circle subgroup of E(2). Therefore $G^{\mathrm{rse}} = \{(A, a) \in \mathrm{E}(2) \big| A \neq \mathrm{I}\}$ is an open dense subset of $G^{\mathrm{se}} = G$. Therefore every element of G^{rse} lies in a torus.

3. Let $G = \mathrm{Sl}(2, \mathbb{R})$. Then $G^{\mathrm{rse}} = \{A \in \mathrm{Sl}(2, \mathbb{R}) \big| |\mathrm{tr}\, A| < 2\}$. Since the subgroup $\mathrm{SO}(2, \mathbb{R})$ meets each conjugacy class of elements of G exactly once, G^{rse} is connected and all the tori given by $\exp \mathfrak{g}_s$, where $s \in G^{\mathrm{rse}}$, are conjugate to one another. □

4.3.6.2 *When the G-action is free*

Suppose that the G-action Φ on M is free and proper with orbit map $\pi : M \to \overline{M} = M/G : m \mapsto \overline{m}$. Let V be a smooth G-invariant vector field on M with flow φ_t. Since Φ is free and proper, the G-orbit space $\overline{M}$ is a smooth manifold and the vector field V induces a smooth vector field $\overline{V}$ on $\overline{\varphi}_t$.

Let P be the set of points $\overline{m} \in \overline{M}$ such that $\mathbb{R} \to \overline{M} : t \mapsto \overline{\varphi}_t(\overline{m})$ is a nonconstant periodic integral curve of $\overline{V}$, which starts at $\overline{m}$ and has minimal positive period $\tau(\overline{m})$. Moreover, suppose that the map $\tau : P \to \mathbb{R}_{>0}$ is continuous.[6] Then P is a smooth submanifold of $\overline{M}$. So $\pi^{-1}(P)$ is a locally closed smooth submanifold of M, which is G-invariant. Therefore $\pi|\pi^{-1}(P) : \pi^{-1}(P) \to P$ is a principal G-bundle. Because $\pi^{-1}(P)$ is the *set* of all *relative periodic orbits* of V, it is invariant under the flow φ_t of V. In addition, the vector field $V|\pi^{-1}(P)$ is complete. By definition P is invariant under the reduced flow $\overline{\varphi}_t$ of the reduced vector field $\overline{V}$ and $\overline{V}$ does not have any zeroes on P.

Lemma 4.3.6.2.54. *The integral curves of $\overline{V}|P$ define the fibers of a smooth principal $\mathbb{R}/\mathbb{Z}$-bundle $\nu : P \to \overline{P}$. Here $\overline{P}$ is the orbit space of the $\mathbb{R}/\mathbb{Z}$-action generated by the flow of $V|P$. Moreover, there is a smooth function $\mu : \overline{P} \to \mathbb{R}_{>0}$ such that $\tau = \mu \circ \nu$.*

Proof. Let $p \in P$. Since $\overline{V(p)} \neq 0$, there is a smooth codimension 1 submanifold S of P containing p such that $T_pP = T_pS \oplus \mathrm{span}\{\overline{V}(p)\}$. By the implicit function theorem there is an open neighborhood W of p in S and an $\varepsilon > 0$ such that

$$\overline{\varphi} : (-\varepsilon, \varepsilon) \times W \to U \subseteq P : (t, s) \mapsto \overline{\varphi}_t(s)$$

[6] Note that if $\overline{m} \in P$, then $\overline{\varphi}_u(\overline{m}) \in P$, because $t \mapsto \overline{\varphi}_t(\overline{\varphi}_u(\overline{m}))$ is a periodic integral curve of $\overline{V}$ of minimal positive period $\tau(\overline{m})$.

is a diffeomorphism onto U, which is an open neighborhood of p in P. For any $u \in U$ we write $\overline{\varphi}^{-1}(u) = (t(u), s(u))$, where $t : U \to (-\varepsilon, \varepsilon) : u \mapsto t(u)$ and $s : U \to W : u \mapsto s(u)$ are smooth maps.

Because $\overline{\varphi}_{\tau(p)}(p) = p$ for every $p \in U$, the set $W_1 = \overline{\varphi}^{-1}_{\tau(p)}(U)$ is an open neighborhood of p in W. For $\widetilde{s} \in W_1$, we have $\widetilde{u} = \overline{\varphi}_{\tau(p)}(\widetilde{s}) \in U$ and $s(\widetilde{u}) = \overline{\varphi}_{\mu(\widetilde{s})}(\widetilde{s}) \in W$, where $\mu(\widetilde{s}) = \tau(p) - t(\widetilde{u})$ is a smooth function of $\widetilde{s}$ with $\mu(p) = \tau(p)$. From $\overline{\varphi}_{\mu(\widetilde{s})}(\widetilde{s}) \in W$ and $\overline{\varphi}_{\tau(\widetilde{s})}(\widetilde{s}) = \widetilde{s}$, it follows that $\overline{\varphi}_{\mu(s)-\tau(s)}(s) \in W$ for every $s \in W_1$. Therefore $\mu(s) = \tau(s)$ if $|\mu(s) - \tau(s)| < \varepsilon$. Because $\mu - \tau$ is a continuous function on W_1, which vanishes at p, we conclude that τ is equal to the smooth function μ on some open neighborhood W_2 of p in W_1.

The mapping $\Psi : \mathbb{R}/\mathbb{Z} \times W_2 \to P : (t, s) \mapsto \overline{\varphi}_{\tau(s)\,t}(s)$ is well defined, smooth, injective, and has bijective tangent map at every point in its domain. Therefore by the inverse function theorem, Ψ is a diffeomorphism onto an open neighborhood U of p in P. Moreover, for each $s \in W_2$, we see that Ψ maps the circle $(\mathbb{R}/\mathbb{Z}) \times \{s\}$ onto the $\overline{V}|P$ orbit through $\Psi(s, 0)$ in U. Because $\tau(\Psi(t, s)) = \tau(s)$ for all $(t, s) \in \mathbb{R}/\mathbb{Z} \times W_2$, we find that τ is a smooth function on U, which is constant on $\overline{V}|P$-orbits in U. Since every $p \in P$ has an open neighborhood U, which is invariant under the flow $\overline{\varphi}_t$ and on which τ is a smooth function, which is constant on $\overline{V}|P$-orbits, it follows that $\tau : P \to \mathbb{R}_{>0}$ is a smooth function, which is constant on $\overline{V}|P$-orbits. The mapping

$$\widetilde{\Psi} : \mathbb{R}/\mathbb{Z} \times P \to P : (t, p) \mapsto \overline{\varphi}_{\tau(p)\,t}(\tau(p))$$

defines a proper free $\mathbb{R}/\mathbb{Z}$-action on P, whose orbits coincide with the orbits of $\overline{V}|P$. □

Example 4.3.6.2.55.
Let $R = \begin{pmatrix} \cos 2\pi\,\frac{p}{q} & -\sin 2\pi\,\frac{p}{q} \\ \sin 2\pi\,\frac{p}{q} & \cos 2\pi\,\frac{p}{q} \end{pmatrix}$, where $q \in \mathbb{Z}_{\ge 2}$, $p \in \mathbb{Z}_{\ge 0}$, $0 < p < q$, and p and q have no common divisor greater than 1. On $\mathbb{R}^2 \times \mathbb{R}$ define a $\mathbb{Z}$-action by

$$\phi : \mathbb{Z} \times (\mathbb{R}^2 \times \mathbb{R}) \to \mathbb{R}^2 \times \mathbb{R} : (n, (x, t)) \mapsto (R^n x, t + n).$$

Let P be the $\mathbb{Z}$-orbit space $(\mathbb{R}^2 \times \mathbb{R})/\mathbb{Z}$ with orbit map $\pi : \mathbb{R}^2 \times \mathbb{R} \to P$. Then π intertwines the vector field $V = \frac{\partial}{\partial t}$ with a real analytic vector field $\overline{V}$ on P, whose integral curve through $\pi(x(0), t(0))$ is $t \mapsto \pi(x(0), t(0) + t)$. When $x(0) \neq 0$, then all the integral curves of $\overline{V}$ on P are periodic with minimal positive period 1; whereas when $x(0) = 0$, the integral curve of $\overline{V}$

has minimal positive period q. Thus the minimal positive period function $\tau : P \to \mathbb{R}_{>0}$ of $\overline{V}$ is discontinuous. To avoid this *subperiodic behavior*, we have made the hypothesis that the minimal positive period function is continuous on the set of relative periodic orbits. □

Lemma 4.3.6.2.56. *Let $s(m)$ be the shift element corresponding to the relative periodic orbit $t \mapsto \varphi_t(m)$ of relative period $\tau(m)$ through $m \in \pi^{-1}(P)$. Then the mapping $\pi^{-1}(P) \to G : m \mapsto s(m)$ is smooth.*

Proof. Write $N = \pi^{-1}(P)$. The mapping

$$\Psi : G \times N \to N \times N : (g, m) \mapsto (m, g \cdot m)$$

is an injective immersion, because the G-action is free and proper. Therefore $Q = \Psi(G \times N)$ is a smooth closed submanifold of $N \times N$, on which Ψ has a smooth inverse $\Psi^{-1} : Q \to G \times N$. The map

$$\psi : N \to N \times N : m \mapsto (m, \varphi_{\tau(m)}(m))$$

is smooth and its image is contained in Q. Therefore $\mu = \pi_1 \circ \Psi^{-1} \circ \psi$ is a smooth map from N to G, where $\pi_1 : G \times N \to N$ is the projection map on the first factor. Moreover, μ is the assignment $m \mapsto s(m)$ as desired. □

Let $P^{\mathrm{rse}} = \{p \in P \big| \, s(m) \in G^{\mathrm{rse}} \text{ for some } m \in \pi^{-1}(\{p\})\}$. Since G^{rse} is invariant under conjugation by element of G and $s(g \cdot m) = \mathrm{Ad}_g s(m)$, it follows that $p \in P^{\mathrm{rse}}$ if and only if $s(m) \in G^{\mathrm{rse}}$ for every $m \in \pi^{-1}(\{p\})$. So P^{rse} is an open subset of P. Let $\overline{P}^{\mathrm{rse}} = \nu(P^{\mathrm{rse}})$, where $\nu : P \to \overline{P}$ is the orbit map of the smooth fibration of P into periodic orbits corresponding to relative periodic orbits of V, see lemma 4.3.6.2.54. Then $\overline{P}^{\mathrm{rse}}$ is an open subset of $\overline{P}$. Because P^{rse} is invariant under the reduced flow of $\overline{V}|P$, it follows that $P^{\mathrm{rse}} = \nu^{-1}(\overline{P}^{\mathrm{rse}})$.

Proposition 4.3.6.2.57. *Let $\chi = \nu \circ \pi : M \to \overline{P}^{\mathrm{rse}}$. For $m_0 \in M$ let $\overline{p}_0 = \chi(m_0) \in \overline{P}^{\mathrm{rse}}$. Let $s_0 = s(\overline{p}_0)$ and set $T = G^0_{s_0}$. Then there is an open neighborhood U of $\overline{p}_0 \in \overline{P}^{\mathrm{rse}}$ and a smooth section $\sigma : U \subseteq \overline{P}^{\mathrm{rse}} \to \chi^{-1}(U) \subseteq M$ of the fibration $\chi|\chi^{-1}(U)$ such that for every $\overline{p} \in U$ the identity component of the centralizer in G of $s(\sigma(\overline{p}))$ is a torus T and $s(\sigma(\overline{p}))T = s_0 T$.*

Proof. Because $\chi|\chi^{-1}(\overline{P})$ is a smooth fibration, there is an open neighborhood U_1 of $\overline{p}_0$ in $\overline{P}^{\mathrm{rse}}$ and a smooth section $\sigma_1 : U_1 \to \chi^{-1}(U)$ of $\chi|\chi^{-1}(U)$. From lemma 4.3.6.1.50 there is an open neighborhood $\widetilde{V}$ of s_0 in G and a real analytic mapping $\theta : \widetilde{V} \to G$ such that if $s' \in \widetilde{V}$ and $s'' = \theta(s')s'\theta(s')^{-1}$, then $G^0_{s''} = G^0_{s_0} = T$ and $s''T = s_0T$. If we write $m = \sigma_1(\overline{p})$, $u = s(\sigma_1(\overline{p}))$, $g = \theta(u)$, and set $\sigma(\overline{p}) = \theta(s(\sigma_1(\overline{p}))) \cdot \sigma_1(\overline{p})$, then $G^0_{s(\sigma(\overline{p}))} = T$ and $s(\sigma(\overline{p}))T = s_0T$ for every $\overline{p} \in U = \{\overline{p}' \in U_1 \big| \, s(\sigma_1(\overline{p}')) \in \widetilde{V}\}$, because $s(g \cdot m) = \mathrm{Ad}_g s(m)$. Now U is an open neighborhood of $\overline{p}_0$ in $\overline{P}^{\mathrm{rse}}$ and $\sigma : U \rightarrow \chi^{-1}(U)$ is a smooth section of $\chi|\chi^{-1}(U)$. □

The map

$$G \times \mathbb{R} \times U \rightarrow \chi^{-1}(U) : (g, t, \overline{p}) \mapsto g \cdot \varphi_t(\sigma(\overline{p})) \tag{44}$$

is a surjective local diffeomorphism onto $\chi^{-1}(U) = \pi^{-1}(\nu^{-1}(U))$. If

$$g \cdot \varphi_t(\sigma(\overline{p})) = g' \cdot \varphi_t(\sigma(\overline{p}')), \tag{45}$$

then applying the bundle projection map $\chi|\chi^{-1}(\overline{P}^{\mathrm{rse}})$ to both sides of (45) gives $\overline{p} = \overline{p}'$. Then (45) is equivalent to $(g')^{-1}g \cdot \sigma(\overline{p}) = \varphi_{t'-t}(\sigma(\overline{p}))$, which in turn is equivalent to $t' - t = n\tau(\pi(\sigma(\overline{p})))$ and $(g')^{-1}g = s(\sigma(\overline{p}))^n$ for some $n \in \mathbb{Z}$. Using lemma 4.3.6.2.54 and the fact that σ is a section of the bundle $\chi|\chi^{-1}(U)$, we get

$$\tau(\pi(\sigma(\overline{p}))) = \mu(\nu(\pi(\sigma(\overline{p})))) = \mu(\chi(\sigma(\overline{p}))) = \mu(\overline{p}).$$

Therefore the map (44) induces a diffeomorphism ψ from $(G \times \mathbb{R} \times U)/\mathbb{Z}$ into $\chi^{-1}(U)$, where $(G \times \mathbb{R} \times U)/\mathbb{Z}$ is the orbit space of the $\mathbb{Z}$-action

$$\begin{aligned} \circ : \mathbb{Z} \times (G \times \mathbb{R} \times U) \rightarrow G \times \mathbb{R} \times U : \\ (n, (g, t, \overline{p})) \mapsto \big(g\, s(\sigma(\overline{p}))^{-n}, t + n\,\mu(\overline{p}), \overline{p}\big). \end{aligned} \tag{46}$$

Using the notation of proposition 4.3.6.2.57, let $\{\lambda_j\}_{j=1}^k$ be a $\mathbb{Z}$-basis for the integer lattice Λ of $\mathfrak{t}$, the Lie algebra of T. In view of lemma 4.3.6.1.50 we have $s(\sigma(\overline{p}))^{n_1} \in T$ for every $\overline{p} \in U$. Because $\exp : \mathfrak{t} \rightarrow T$ is a local diffeomorphism, shrinking U is necessary, we obtain a smooth map $\xi : U \rightarrow \mathfrak{t}$ by requiring $s(\sigma(\overline{p})) = \exp \xi(\overline{p})$.

Define an $\mathbb{R}^{k+1}$-action on $G \times \mathbb{R} \times U$ by

$$\begin{aligned} \bullet : \mathbb{R}^{k+1} \times (G \times \mathbb{R} \times U) \rightarrow G \times \mathbb{R} \times U : (\theta, (g, t, \overline{p})) \mapsto \\ \big(g \cdot \exp(\textstyle\sum_{j=1}^k \theta_j\, \lambda_j - \theta_{k+1}\, \xi(\overline{p})), t + \theta_{k+1} n_1\, \mu(\overline{p}), \overline{p}\big). \end{aligned} \tag{47}$$

If $\theta \in \mathbb{Z}^{k+1}$, then from $s(\sigma(\overline{p}))^{n_1} = \exp \xi(\overline{p})$ and (47) it follows that

$$\theta \bullet (g, t, \overline{p}) = (g \cdot s(\sigma(\overline{p}))^{-\theta_{k+1} + n_1}, t + \theta_{k+1} n_1\, \mu(\overline{p}), \overline{p}) = \theta_{k+1} n_1 \circ (g, t, \overline{p}),$$

where the last equality follows because $\theta_{k+1} n_1 \in \mathbb{Z}$.

Consequently, we have an induced $\mathbb{R}^{k+1}/\mathbb{Z}^{k+1}$-action on $(G\times\mathbb{R}\times U)/\mathbb{Z}$, which commutes with the G-action on $(G\times\mathbb{R}\times U)/\mathbb{Z}$ induced by the G-action

$$G\times(G\times\mathbb{R}\times U)\to G\times\mathbb{R}\times U : (g',(g,t,\overline{p}))\mapsto(g'g,t,\overline{p}). \qquad (48)$$

If $\theta\bullet(g,t,\overline{p}) = \big(g\cdot s(\sigma(\overline{p}))^{-n}, t+n\,\mu(\overline{p}),\overline{p}\big)$ for some $n\in\mathbb{Z}$, then

$$s(\sigma(\overline{p}))^{-n} = \exp\big(-\sum_{j=1}^{k}\theta_j\,\lambda_j + \theta_{k+1}\,\xi(\overline{p})\big)\in T, \qquad (49)$$

which implies $n = qn_1$ for some $q\in\mathbb{Z}$. But then $q\,\mu(\overline{p}) = \theta_{k+1}n_1\,\mu(\overline{p})$, which implies $\theta_{k+1} = q$. Now (49) together with $\exp\xi(\overline{p}) = s(\sigma(\overline{p}))^{n_1}$ implies that $\sum_{j=1}^{k}\theta_j\,\lambda_j\in\Lambda$. Therefore $\theta_j\in\mathbb{Z}$ for every $1\le j\le k$. Consequently, the induced $\mathbb{R}^{k+1}/\mathbb{Z}^{k+1}$-action on $(G\times\mathbb{R}\times U)/\mathbb{Z}$ is free. This action is proper, because $\mathbb{R}^{k+1}/\mathbb{Z}^{k+1}$ is compact. The diffeomorphism $\psi:(G\times\mathbb{R}\times U)/\mathbb{Z}\to\chi^{-1}(U)$ intertwines the $\mathbb{R}^{k+1}/\mathbb{Z}^{k+1}$-action on $(G\times\mathbb{R}\times U)/\mathbb{Z}$ with a uniquely defined $\mathbb{R}^{k+1}/\mathbb{Z}^{k+1}$-action on $\chi^{-1}(U)$, which commutes with the G-action on $\chi^{(}U)$.

Let $\{\xi_j(\overline{p})\}_{j=1}^{k}$ be the components of $\xi(\overline{p})$ with respect to the $\mathbb{Z}$-basis $\{\lambda_j\}_{j=1}^{k}$ of the integer lattice Λ of $\mathfrak{t}$. Then from

$$\sum_{j=1}^{k}\theta_j\,\lambda_j - \theta_{k+1}\,\xi(\overline{p}) = \sum_{j=1}^{k}\big(\theta_j - \theta_{k+1}\,\xi_j(\overline{p})\big)\lambda_j$$

it follows that on $\chi^{-1}(U)$ the vector field V on $(G\times\mathbb{R})\cdot m$ is equal to the element $\dot{\theta}$ of the Lie algebra $\mathbb{R}^{k+1}$ of $\mathbb{R}^{k+1}/\mathbb{Z}^{k+1}$, where

$$\dot{\theta} = \begin{cases} \theta_{k+1}\,\xi_j(\overline{p}) \text{ when } 1\le j\le k \\ 1/(n_1\mu(\overline{p})) \text{ when } j=k+1. \end{cases} \qquad (50)$$

We summarize the above discussion in the following lengthy proposition.

Proposition 4.3.6.2.58. *Let $\mathcal{C}$ be a connected component of $\overline{P}^{\mathrm{rse}}$, where $\nu: P^{\mathrm{rse}}\to\overline{P}^{\mathrm{rse}}$ is the principal $\mathbb{R}/\mathbb{Z}$-fibration of the set of relative periodic orbits in $\overline{P}^{\mathrm{rse}}\subseteq\overline{M}$. Let $\chi:\pi(P^{\mathrm{rse}})\to\overline{P}^{\mathrm{rse}}$ be $(\nu\circ\pi)|\pi^{-1}(P^{\mathrm{rse}})$. For each $m\in\chi^{-1}(\mathcal{C})$ let $T(m)=G^0_{s(m)}$, where $s(m)$ is the shift element corresponding to the integral curve $t\mapsto\varphi_t(m)$ of the G-invariant vector field V starting at m. The disjoint union over all $m\in\chi^{-1}(\mathcal{C})$ of the V-flow out of $T(m)\cdot m$, that is, $\varphi_t(T(m)\cdot m)$, define a G invariant smooth fibration of $\chi^{-1}(\mathcal{C})$, whose fiber is diffeomorphic to a torus of dimension* $\dim T+1$.

Let $m_0\in M$ such that $\overline{p}_0=\chi(m_0)\in\mathcal{C}$. Let $\mathfrak{t}$ be the Lie algebra of $T = T(m_0)$. Then there is an open neighborhood U of $\overline{p}_0$ in $\mathcal{C}$, a smooth section

$\sigma : U \to \chi^{-1}(U)$, and a smooth map $\xi : U \to \mathfrak{t}$ such that $G^0_{s(\sigma(\overline{p}))} = T$ and $s(\sigma(\overline{p}))^n = \exp \xi(\overline{p})$ for every $\overline{p} \in U$. For any $m \in \chi^{-1}(\mathcal{C})$ we let n_1 be the smallest positive integer such that $s(m)^{n_1} \in T$.

Let $\{\lambda_j\}_{j=1}^k$ be a $\mathbb{Z}$-basis of the integer lattice Λ of $\mathfrak{t}$ and let $\{\xi_j(\overline{p})\}_{j=1}^k$ be the components of $\xi(\overline{p})$ with respect to this basis. Then the $\mathbb{R}^{k+1}$-action (47) on $G \times \mathbb{R} \times U$ induces a $\mathbb{R}^{k+1}/\mathbb{Z}^{k+1}$-action on $G \times \mathbb{R} \times U$, which is intertwined by the map

$$G \times \mathbb{R} \times G \times U \to \chi^{-1}(U) : (g, t, \overline{p}) \mapsto g \cdot \varphi_t(\sigma(\overline{p}))$$

with a G-invariant proper free action of $\mathbb{R}^{k+1}/\mathbb{Z}^{k+1}$ on $\chi^{-1}(U)$, whose orbits are the fibers of the toral fibration given in the first paragraph above. In addition, the vector field V on $\chi^{-1}(U)$ is $\dot{\theta}$, where $\dot{\theta}$ lies in the Lie algebra $\mathbb{R}^{k+1}$ of $\mathbb{R}^{k+1}/\mathbb{Z}^{k+1}$ and is given by (50).

As in proposition 4.2.6.2.36, the Weyl group $W(T)$ of T measures the obstruction to $\chi^{-1}(\mathcal{C})$ being a G-invariant principal $\mathbb{R}^{k+1}/\mathbb{Z}^{k+1}$-bundle.

4.3.6.3 *When the G-action is not free*

Suppose that the G-action on M is proper, but is not necessarily free. Let $m \in M$ and let $H = G_m$. Let M_H be the smooth submanifold of M of points of isotropy H and let $\overline{M_H}$ be its image in the G-orbit space $\overline{M}$ under the G-orbit map $\pi : M \to \overline{M} : m \mapsto \overline{m}$. Let $P_{\overline{M_H}}$ be the smooth submanifold of points $\overline{m} \in \overline{M_H}$ which are starting points of a nonconstant periodic orbits of minimal positive period $\tau(\overline{m})$ of the reduced flow of the $G_{M_H} = N(H)/H$-invariant vector field $V|M_H$. We also assume that $\tau : P_{\overline{M_H}} \to \mathbb{R}_{>0}$ is continuous, and thus is smooth. Let $\nu_{M_H} : P_{\overline{M_H}} \to \overline{P}_{\overline{M_H}}$ be the principal $\mathbb{R}/\mathbb{Z}$-bundle whose fiber is a periodic orbit of the reduced flow of $V|M_H$. Analogous to the definition of $\overline{P}^{\mathrm{rse}}_{\overline{M_H}}$ in §3.6.2, we may define $P^{\mathrm{rse}}_{\overline{M_H}}$. Let $\mathcal{C}^{M_H}$ be a connected component of $\overline{P}^{\mathrm{rse}}_{\overline{M_H}}$ and let $\chi_{M_H} : \pi(\mathcal{C}^{M_H}) \to \overline{P}^{\mathrm{rse}}_{\overline{M_H}}$ be $\nu_{M_H} \circ \pi_{M_H}$, where $\pi_{M_H} : M_H \to \overline{M_H}$ is the G_{M_H}-orbit map. Using the fact that G_{M_H} acts freely and properly on M_H and the results of §3.6.2 with G replaced by G_{M_H} and M replaced by M_H, we obtain the following analogue of proposition 4.3.6.2.58.

Proposition 4.3.6.3.59. *For each $m \in \chi^{-1}_{M_H}(\mathcal{C}^{\mathcal{M}_{\mathcal{H}}})$ let $T(m) = (G_{M_H})^0_{s(m)}$, where $s(m)$ is the shift element corresponding to the integral curve $t \mapsto \varphi_t(m)$ of the G_{M_H}-invariant vector field $V|M_H$ starting at m. Then the disjoint union over all $m \chi^{-1}_{M_H}(\mathcal{C}^{M_H})$ of the V-flow out* *of* $T(m) \cdot m$, *that is,* $\varphi_t(T(m) \cdot m)$, *define a* G_{M_H} *invariant smooth fibration of* $\chi_{M_H}^{-1}(\mathcal{C}^{M_H})$, *whose fiber is diffeomorphic to a torus of dimension* $\dim T(m) + 1$.

Let $m_0 \in M_H$ *such that* $\overline{p}_0 = \chi_{M_H}(m_0) \in \mathcal{C}_{M_H}$. *Let* $\mathfrak{t}$ *be the Lie algebra of* $T = T(m_0)$. *Then there is an open neighborhood* U *of* $\overline{p}_0$ *in* $\mathcal{C}^{M_H}$, *a smooth section* $\sigma : U \to \chi_{M_H}^{-1}(U)$, *and a smooth map* $\xi : U \to \mathfrak{t}$ *such that* $(G_{M_H})^0_{s(\sigma(\overline{p}))} = T$ *and* $s(\sigma(\overline{p}))^n = \exp \widetilde{\xi}(\overline{p})$ *for every* $\overline{p} \in U$. *For any* $m \in \chi^{-1}(\mathcal{C}^{M_H})$ *we let* n_1 *be the smallest positive integer such that* $s(m)^{n_1} \in T(m)$ *for every* $m \in \chi_{M_H}^{-1}(\mathcal{C}^{M_H})$.

Let $\{\lambda_j\}_{j=1}^k$ *be a* $\mathbb{Z}$*-basis of the integer lattice* Λ *of* $\mathfrak{t}$ *and let* $\{\widetilde{\xi}_j(\overline{p})\}_{j=1}^k$ *be the components of* $\widetilde{\xi}(\overline{p})$ *with respect to this basis. Then the* $\mathbb{R}^{k+1}$*-action* (47) *on* $G_{M_H} \times \mathbb{R} \times U$ *induces a* $\mathbb{R}^{k+1}/\mathbb{Z}^{k+1}$*-action on* $G_{M_H} \times \mathbb{R} \times U$, *which is intertwined by the map*

$$G_{M_H} \times \mathbb{R} \times U \to \chi_{M_H}^{-1}(U) : (g, t, \overline{p}) \mapsto g \cdot \varphi_t(\sigma(\overline{p}))$$

with a G_{M_H}*-invariant proper free action of* $\mathbb{R}^{k+1}/\mathbb{Z}^{k+1}$ *on* $\chi_{M_H}^{-1}(U)$, *whose orbits are the fibers of the toral fibration given in the first paragraph above. In addition, the vector field* V *on* $G \times \mathbb{R} \times U$ *is* $\dot{\theta}$, *where* $\dot{\theta}$ *lies in the Lie algebra* $\mathbb{R}^{k+1}$ *of* $\mathbb{R}^{k+1}/\mathbb{Z}^{k+1}$ *and is given by* (50) *with* ξ *replaced by* $\widetilde{\xi}$.

As in proposition 4.2.6.2.36, *the Weyl group* $W(T)$ *of* T *measures the obstruction to* $\chi_{M_H}^{-1}(\mathcal{C}^{M_H})$ *being a* G_{M_H}*-invariant principal* $\mathbb{R}^{k+1}/\mathbb{Z}^{k+1}$*-bundle.*

4.4 Notes

The reconstruction procedure given in the text follows Marsden [75], see §7.2. Most of the results on relative equilibria and relative periodic orbits can be found in the literature, especially Field [44] and Krupa [64]. Our discussion is based on that of Duistermaat [41]. The denseness of a straight line on an affine torus with frequency vector having components which are linearly independent over the rational numbers is due to Kronecker [63]. When the Lie group G is compact, the statement of proposition 4.2.2.6 about the minimal number of frequencies of a quasiperiodic relative equilibrium generated by an elliptic element of the Lie algebra of G is due to proposition B1 of Field [44]. A proof of Sard's theorem may be found in Sard [97]. Proposition 4.2.4.11 is a consequence of proposition B1 of Field [44]. When the Lie group G is compact, corollary 4.3.4.48 follows from proposition B2 of Field [44]. For a free action of a connected compact Lie group proposition 4.3.6.2.58 was proved by Hermans [54], who applied it to the nonholonomically constrained system of a ball rolling in an axially symmetric bowl.

Chapter 5

Carathéodory's sleigh

In this chapter we will discuss the classical nonholonomically constrained system known as Carathéodory's sleigh. In order to illustrate the theory given in chapters 1, 2, and 4, we will derive the equations of motion in five different ways, construct the reduced system in three different ways, and carry out reconstruction explicitly.

5.1 Basic set up

Carathéodory's sleigh is a planar rigid body (where the plane contains the vertical axis) with a sharp strongly convex chisel edge in a vertical plane, which makes contact with a horizontal plane at its lowest point P. The sleigh is kept on the horizontal plane by a constant vertical gravitational force. It moves so that the instantaneous direction is in the vertical plane of the chisel edge, which does not slip sideways. In addition, we assume that the center of mass lies in the plane of the sleigh and is kept at a fixed height above the horizontal plane. This latter condition can be arranged by attaching to the sleigh two massless outriggers at an angle out of the plane of the sleigh, each having a single point of contact with the horizontal plane, which glide without friction. Since no forces act on the point of contact of either outrigger, the outriggers have no effect on the motion of the sleigh.

5.1.1 *Configuration space*

Suppose that in the *reference position* in $\mathbb{R}^3$ the sleigh is in the vertical e_1–e_3 plane with its center of mass at the origin O and that its point of contact P with the horizontal e_1–e_2 plane has coordinates $(-\ell, 0, 0)$.

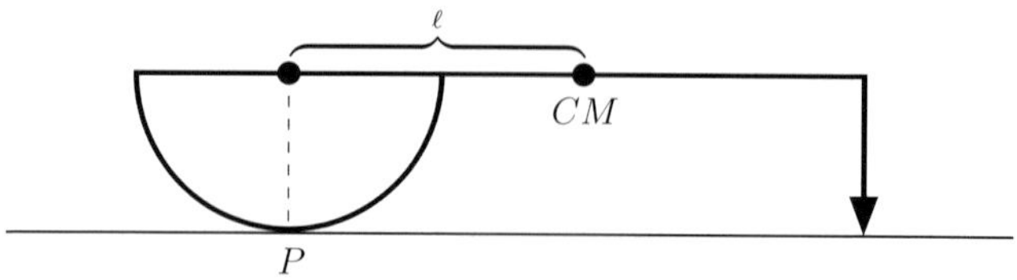

Fig. 5.1 Carathéodory's sleigh (without the outriggers). ℓ is the distance from the center of mass CM along the skate to a point directly over the point of contact P of the chisel edge with the horizontal plane.

The position of the moving sleigh is obtained by applying a proper Euclidean motion $(A, a) : \mathbb{R}^3 \to \mathbb{R}^3 : \xi \mapsto A\xi + a$ to the reference sleigh. Here A is a rotation of $\mathbb{R}^3$ about the vertical e_3-axis and $a = (x, 0) \in \mathbb{R}^2 \times \{0\} \subseteq \mathbb{R}^3$. Corresponding to each A is a counter clockwise rotation $R_\theta = \left(\begin{smallmatrix} \cos\theta & -\sin\theta \\ \sin\theta & \cos\theta \end{smallmatrix}\right)$ of the horizontal plane $\mathbb{R}^2 \times \{0\} = \mathbb{R}^2$ through an angle $\theta \in \mathbb{R}/2\pi\mathbb{Z} = S^1$. The map

$$Q = S^1 \times \mathbb{R}^2 \to \mathrm{E}(2) : q = (\theta, x) \mapsto (R_\theta, x)$$

is a diffeomorphism, which identifies the *configuration space* Q of the sleigh with the *Euclidean group* $\mathrm{E}(2)$.

5.1.2 *Kinetic energy*

Let μ be the mass distribution of the sleigh in its reference position. Suppose that μ is a finite positive measure whose support $\mathcal{S}$ is the sleigh, which is a compact subset of $\mathbb{R}^3$. Then $m = \int_{\mathcal{S}} \mu(\mathrm{d}\xi)$ is the *mass* of the sleigh. In addition, since the center of mass of the reference sleigh is the origin, we have $\int_{\mathcal{S}} \xi\, \mu(\mathrm{d}\xi) = 0$, Now $A\xi + a$ is a point on the moving sleigh which has *velocity* $\dot{A}\xi + \dot{a}$ and *kinetic energy*

$$T = \tfrac{1}{2} \int_{\mathcal{S}} \langle \dot{A}\xi + \dot{a}, \dot{A}\xi + \dot{a} \rangle\, \mu(\mathrm{d}\xi).$$

Here $\langle\ ,\ \rangle$ is the Euclidean inner product on $\mathbb{R}^3$. From $\int_{\mathcal{S}} \xi\, \mu(\mathrm{d}\xi) = 0$ it follows that

$$\int_{\mathcal{S}} \dot{A}\xi\, \mu(\mathrm{d}\xi) = \dot{A}\Big(\int_{\mathcal{S}} \xi\, \mu(\mathrm{d}\xi)\Big) = 0.$$

Therefore $T = T_{\mathrm{rot}} + T_{\mathrm{trans}}$, where

$$T_{\mathrm{trans}} = \tfrac{1}{2} \int_{\mathcal{S}} \langle \dot{a}, \dot{a} \rangle\, \mu(\mathrm{d}\xi) = \tfrac{1}{2}\, (\dot{x}_1^2 + \dot{x}_2^2) \int_{\mathcal{S}} \mu(\mathrm{d}\xi) = \tfrac{1}{2}\, m(\dot{x}_1^2 + \dot{x}_2^2)$$

is the *translational kinetic energy* and

$$T_{\mathrm{rot}} = \tfrac{1}{2} \int_{\mathcal{S}} \langle \dot{A}\xi, \dot{A}\xi \rangle \, \mu(\mathrm{d}\xi)$$

is the *rotational kinetic energy* of the sleigh both about its center of mass. Since A is a rotation

$$T_{\mathrm{rot}} = \tfrac{1}{2} \int_{\mathcal{S}} \langle A^{-1}\dot{A}\xi, A^{-1}\dot{A}\xi \rangle \, \mu(\mathrm{d}\xi).$$

From $A = \left(\begin{smallmatrix} R_\theta & 0 \\ 0 & 1 \end{smallmatrix}\right)$ we get

$$A^{-1}\dot{A}\xi = \dot{\theta} \begin{pmatrix} 0 & -1 & 0 \\ 1 & 0 & 0 \\ 0 & 0 & 0 \end{pmatrix} \xi = \dot{\theta}(e_3 \times \xi).$$

Here $\times$ is the vector product on $\mathbb{R}^3$. Therefore $T_{\mathrm{rot}} = \frac{1}{2} I \dot{\theta}^2$, where $I = \int_{\mathcal{S}} \langle e_3 \times \xi, e_3 \times \xi \rangle \, \mu(\mathrm{d}\xi)$ is the *moment of inertia* of the reference sleigh about the vertical e_3-axis. Consequently, the *kinetic energy* of the moving sleigh is

$$T = \tfrac{1}{2} I\dot{\theta}^2 + \tfrac{1}{2} m(\dot{x}_1^2 + \dot{x}_2^2). \tag{1}$$

5.1.3 *Nonholonomic constraint*

Because the sleigh is rigid and its chisel edge has the same instantaneous direction as the vertical plane of the sleigh, the point of contact of the edge follows a pursuit curve relative to the center of mass of the sleigh, that is,

$$\ell\dot{\theta} = -\dot{x}_1 \sin\theta + \dot{x}_2 \cos\theta. \tag{2}$$

Another way to formulate (2) is to say that the velocity of a motion $\mathbb{R} \rightarrow Q : t \mapsto q(t) = (\theta(t), x_1(t), x_2(t))$ of the sleigh must lie in the kernel of the *constraint 1-form*

$$\varphi(q) = \varphi(\theta, x_1, x_2) = \ell \, \mathrm{d}\theta + \sin\theta \, \mathrm{d}x_1 - \cos\theta \, \mathrm{d}x_2 \tag{3}$$

on configuration space Q. Physically, φ is the infinitesimal work done on the edge of the sleigh to keep it from slipping sideways.

The *constraint distribution* D on configuration space Q is the kernel $\ker\varphi$ of the constraint 1-form φ (3). Evaluating $\mathrm{d}\varphi = \cos\theta \, \mathrm{d}\theta \wedge \mathrm{d}x_1 + \sin\theta \, \mathrm{d}\theta \wedge \mathrm{d}x_2$ on the vectors

$$v_1 = \frac{\partial}{\partial\theta} - \ell\sin\theta \frac{\partial}{\partial x_1} + \ell\cos\theta \frac{\partial}{\partial x_2} \quad \text{and} \quad v_2 = \cos\theta \frac{\partial}{\partial x_1} + \sin\theta \frac{\partial}{\partial x_2}$$

gives $\mathrm{d}\varphi(v_1, v_2) = 1$. Therefore φ is a *contact 1-form* on Q. Hence $\ker\varphi$ is a *contact structure* on Q. Consequently the accessible set of the constraint distribution D on Q is all of Q.[1]

[1] According to corollary 1.10.46 of chapter 1, this implies that the accessible set of the distribution H on D is all of D.

5.2 Equations of motion

We derive the equations of motion of Carathéodory's sleigh in five different ways: first, using the Lagrange-d'Alembert equations of §3.2 of chapter 1; second, using nonholonomic Dirac brackets of §7.2 of chapter 1; third, using the Lagrange-d'Alembert equations in a trivialization of §4.1 of chapter 1; fourth, using almost Poisson brackets of §7.2 of chapter 1 computed in a trivialization; and finally using the distributional Hamiltonian formulation of §6 in chapter 1.

5.2.1 *Lagrange-d'Alembert equations*

From the Lagrange-d'Alembert principle of §3.2 of chapter 1 it follows that the equation of motion of the sleigh is

$$\delta L = \lambda \varphi. \tag{4}$$

Here δL is the *Lagrange derivative* of the *Lagrangian*

$$L : TQ \to \mathbb{R} : (q, \dot{q}) = (\theta, x_1, x_2, \dot{\theta}, \dot{x}_1, \dot{x}_2) \mapsto T = \tfrac{1}{2}\, I\dot{\theta}^2 + \tfrac{1}{2}\, m(\dot{x}_1^2 + \dot{x}_2^2) \tag{5}$$

and φ is the constraint 1-form (3). The unknown multiplier λ in (4) is determined by the *nonholonomic constraint*

$$\ell\dot{\theta} + \dot{x}_1 \sin\theta - \dot{x}_2 \cos\theta = 0. \tag{6}$$

Writing (4) out we obtain

$$\begin{aligned}\frac{\mathrm{d}}{\mathrm{d}t}\left(\frac{\partial L}{\partial \dot{\theta}}\right) \mathrm{d}\theta + \frac{\mathrm{d}}{\mathrm{d}t}\left(\frac{\partial L}{\partial \dot{x}_1}\right) \mathrm{d}x_1 + \frac{\mathrm{d}}{\mathrm{d}t}\left(\frac{\partial L}{\partial \dot{x}_2}\right) \mathrm{d}x_2 \\ = \lambda\ell \;\mathrm{d}\theta + \lambda \sin\theta \;\mathrm{d}x_1 - \lambda\cos\theta \;\mathrm{d}x_2.\end{aligned}$$

Using the definition of L (5), the above equation is equivalent to the system

$$\begin{aligned} I\ddot{\theta} &= \lambda\ell \\ m\ddot{x}_1 &= \lambda \sin\theta \\ m\ddot{x}_2 &= -\lambda\cos\theta. \end{aligned} \tag{7}$$

To determine the multiplier λ, we multiply the first equation in (7) by ℓ and then differentiate (6). We get

$$\ell^2\lambda = -\ell I\ddot{\theta} = I\dot{\theta}\dot{x}_1 \cos\theta - I\ddot{x}_1 \sin\theta - I\dot{\theta}\dot{x}_2 \sin\theta + I\ddot{x}_2 \cos\theta.$$

Using the second and third equations in (7) we obtain

$$\lambda = -\frac{mI}{I + m\ell^2}\,\dot{\theta}(\dot{x}_1 \cos\theta + \dot{x}_2 \sin\theta). \tag{8}$$

Thus the *equations of motion* of Carathédory's sleigh are

$$\begin{aligned}\ddot{\theta} &= -\frac{m\ell}{I+m\ell^2}\,\dot{\theta}(\dot{x}_1\cos\theta + \dot{x}_2\sin\theta)\\ \ddot{x}_1 &= -\frac{I}{I+m\ell^2}\,\dot{\theta}\sin\theta(\dot{x}_1\cos\theta + \dot{x}_2\sin\theta)\\ \ddot{x}_2 &= \frac{I}{I+m\ell^2}\,\dot{\theta}\cos\theta(\dot{x}_1\cos\theta + \dot{x}_2\sin\theta).\end{aligned} \tag{9}$$

5.2.2 *Nonholonomic Dirac brackets*

Next we derive the equations of motion of the sleigh using nonholonomic Dirac brackets from §7.2 of chapter 1.

The *Hamiltonian* of the *unconstrained* sleigh on T^*Q with coordinates $(q,p) = (\theta, x_1, x_2, p_\theta, p_{x_1}, p_{x_2})$ and bundle projection map $\tau : T^*Q \to Q : (q,p) \mapsto q$ is

$$h = \tfrac{1}{2m}(p_{x_1}^2 + p_{x_2}^2) + \tfrac{1}{2I}p_\theta^2. \tag{10}$$

This Hamiltonian is obtained from the Lagrangian L (5) using the *Legendre transformation*

$$(\dot{\theta}, \dot{x}_1, \dot{x}_2) \mapsto (p_\theta, p_{x_1}, p_{x_2}) = \left(\frac{\partial L}{\partial\dot{\theta}}, \frac{\partial L}{\partial\dot{x}_1}, \frac{\partial L}{\partial\dot{x}_2}\right) = (I\dot{\theta}, m\dot{x}_1, m\dot{x}_2).$$

The *canonical symplectic form* $\omega = \mathrm{d}\theta\wedge\mathrm{d}p_\theta + \mathrm{d}x_1\wedge\mathrm{d}p_{x_1} + \mathrm{d}x_2\wedge\mathrm{d}p_{x_2}$ on T^*Q induces the bundle isomorphism $\omega^\sharp : T^*Q \to TQ$, where

$$\omega^\sharp(\mathrm{d}\theta) = -\tfrac{\partial}{\partial p_\theta} \qquad \omega^\sharp(\mathrm{d}x_1) = -\tfrac{\partial}{\partial p_{x_1}} \qquad \omega^\sharp(\mathrm{d}x_2) = -\tfrac{\partial}{\partial p_{x_2}}$$

$$\omega^\sharp(\mathrm{d}p_\theta) = \tfrac{\partial}{\partial\theta} \qquad \omega^\sharp(\mathrm{d}p_{x_1}) = \tfrac{\partial}{\partial x_1} \qquad \omega^\sharp(\mathrm{d}p_{x_2}) = \tfrac{\partial}{\partial x_2}.$$

The nonholonomic constraint (6) gives rise to a distribution D on Q defined by the kernel of the constraint 1-form φ (3). As a submanifold of T^*Q the pushforward $(\omega^\sharp)^{-1}(D)$ of the *constraint distribution* D is the 0-level set of the *constraint function*

$$C = \frac{m\ell}{I}\,p_\theta + p_{x_1}\sin\theta - p_{x_2}\cos\theta. \tag{11}$$

To find the equations of motion of the sleigh we use equation (88) of §7.2 of chapter 1 for nonholonomic Dirac brackets. In our case this formula reads

$$\{f|D, g|D\} = \{f,g\}|D + \frac{1}{\{C,\varphi\}}\left(-\{f,C\}\{g,\varphi\} + \{f,\varphi\}\{g,C\}\right)|D \tag{12}$$

for every $f, g \in C^\infty(T^*Q)$, where

$$\{C, \varphi\}(d) = \omega(d)\big(\omega^\sharp(d)(\mathrm{d}C(d)), \omega^\sharp(d)(\varphi(d))\big)$$

by definition. But

$$\omega^\sharp(d)(\mathrm{d}C(d)) = \frac{m\ell}{I}\frac{\partial}{\partial\theta} - \big(p_{x_1}\cos\theta + p_{x_2}\sin\theta\big)\frac{\partial}{\partial p_\theta} + \sin\theta\frac{\partial}{\partial x_1} - \cos\theta\frac{\partial}{\partial x_2}$$

and

$$\omega^\sharp(d)(\varphi(d)) = -\ell\frac{\partial}{\partial p_\theta} - \sin\theta\frac{\partial}{\partial p_{x_1}} + \cos\theta\frac{\partial}{\partial p_{x_2}}.$$

So $\{C, \varphi\}(d) = -\frac{I+m\ell^2}{I}$. Therefore (12) becomes

$$\{f|D, g|D\} = \{f, g\}|D - \frac{I}{I+m\ell^2}\left(-\{f, C\}\{g, \varphi\} + \{f, \varphi\}\{g, C\}\right)|D. \quad (13)$$

This allows us to compute the almost Poisson bracket of two smooth functions on the constraint distribution D.

A straightforward calculation using (13) gives the following nonholonomic almost Poisson brackets

$$\begin{aligned}
\{p_{x_1}, p_\theta\} &= -\tfrac{I}{I+m\ell^2}\sin\theta(p_{x_1}\cos\theta + p_{x_2}\sin\theta)\\
\{p_{x_2}, p_\theta\} &= \tfrac{I}{I+m\ell^2}\cos\theta(p_{x_1}\cos\theta + p_{x_2}\sin\theta)
\end{aligned}$$

$$\begin{array}{lll}
\{x_1, p_{x_1}\} = 1 - \tfrac{I}{I+m\ell^2}\sin^2\theta & \{x_2, p_{x_1}\} = \tfrac{I}{I+m\ell^2}\sin\theta\cos\theta & \\
\{x_1, p_{x_2}\} = \tfrac{I}{I+m\ell^2}\sin\theta\cos\theta & \{x_2, p_{x_2}\} = 1 - \tfrac{I}{I+m\ell^2}\cos^2\theta & \\
\{x_1, p_\theta\} = -\tfrac{\ell I}{I+m\ell^2}\sin\theta & \{x_2, p_\theta\} = \tfrac{\ell I}{I+m\ell^2}\cos\theta & \\
\{\theta, p_{x_1}\} = -\tfrac{m\ell}{I+m\ell^2}\cos\theta & \{\theta, p_{x_2}\} = \tfrac{m\ell}{I+m\ell^2}\cos\theta & \{\theta, p_\theta\} = \tfrac{I}{I+m\ell^2},
\end{array}$$

where all the coordinate functions $(\theta, x_1, x_2, p_\theta, p_{x_1}, p_{x_2})$ on T^*Q are considered to be functions on D. All other nonholonomic Dirac brackets between the restriction of the coordinate functions to D vanish identically. Thus we have determined the *structure matrix of the almost Poisson structure* $\{\ ,\ \}$ on $C^\infty(D)$.

Using the Hamiltonian h (10) for Carathéodory's sleigh and the above bracket relations, we find that its motions are the integral curves of the nonholonomic Hamiltonian vector field Y_h on D, which satisfy

$$\begin{aligned}
\dot{\theta} &= \{\theta, h\} = \frac{1}{I} p_\theta \\
\dot{x_1} &= \{x_1, h\} = \frac{1}{m} p_{x_1} \\
\dot{x_2} &= \{x_2, h\} = \frac{1}{m} p_{x_2} \\
\dot{p}_\theta &= -\frac{I}{m(I+m\ell^2)}\big(-p_{x_1}\sin\theta + p_{x_2}\cos\theta\big)\big(p_{x_1}\cos\theta + p_{x_2}\sin\theta\big) \\
\dot{p}_{x_1} &= -\frac{I}{m\ell(I+m\ell^2)}\sin\theta\big(-p_{x_1}\sin\theta + p_{x_2}\cos\theta\big)\big(p_{x_1}\cos\theta + p_{x_2}\sin\theta\big) \\
\dot{p}_{x_2} &= \frac{I}{m\ell(I+m\ell^2)}\cos\theta\big(-p_{x_1}\sin\theta + p_{x_2}\cos\theta\big)\big(p_{x_1}\cos\theta + p_{x_2}\sin\theta\big).
\end{aligned} \tag{14}$$

Because $p_\theta = \frac{I}{m\ell}\big(-p_{x_1}\sin\theta + p_{x_2}\cos\theta\big)$, the fourth equation in (14) is automatically satisfied. Note that the first order equations of motion (14) give rise to the second order equations (9) obtained from the Lagrange-d'Alembert principle.

5.2.3 *Lagrange-d'Alembert in a trivialization*

In this subsection we calculate the equations of motion of Carathéodory's sleigh using a *trivialization of the constraint distribution.*

Let

$$\begin{aligned}
\lambda_D : Q \times \mathbb{R}^2 &\to D \subseteq TQ : (q,c) = ((\theta,x),c) \mapsto c^Q(\theta,x) \\
&= c_2\tfrac{\partial}{\partial\theta} + (c_1\cos\theta - \ell c_2\sin\theta)\tfrac{\partial}{\partial x_1} + (c_1\sin\theta + \ell c_2\cos\theta)\tfrac{\partial}{\partial x_2}.
\end{aligned} \tag{15}$$

Since $c^Q(\theta,x) \,\lrcorner\, \varphi = 0$, the image of λ_D is contained in $\ker\varphi = D$. But holding q fixed, the map λ_D becomes $(\lambda_D)_q : \mathbb{R}^2 \to D_q$, which is a linear isomorphism. So λ_D is the inverse of a trivialization of the constraint distribution D. If we interpret tangent vectors to Q as velocity vectors, then (15) amounts to

$$\begin{aligned}
\dot{\theta} &= c_2 \\
\dot{x}_1 &= c_1\cos\theta - \ell c_2\sin\theta \\
\dot{x}_2 &= c_1\sin\theta + \ell c_2\cos\theta.
\end{aligned} \tag{16}$$

Solving the last two equations in (16) for c_1 and ℓc_2 gives

$$\begin{aligned}
c_1 &= \dot{x}_1\cos\theta + \dot{x}_2\sin\theta \\
\ell c_2 &= -\dot{x}_1\sin\theta + \dot{x}_2\cos\theta = \ell\dot{\theta}.
\end{aligned} \tag{17}$$

The last equality in the second equation of (17) follows from the defining equation (6) of D. Therefore the *Lagrangian* L (5) *in the trivialization* λ_D^{-1} is the smooth function $\mathcal{L}: Q \times \mathbb{R}^2 \to \mathbb{R}$, where

$$\mathcal{L}(q,c) = L \circ \lambda_D(q,c) = \tfrac{1}{2}\, mc_1^2 + (I + m\ell^2)c_2^2. \tag{18}$$

Given a smooth curve $\gamma : \mathbb{R} \to Q$ let $c : \mathbb{R} \to \mathbb{R}^2 : t \mapsto c(t)$ be the smooth curve defined by the condition $\lambda_D(\gamma(t), c(t)) = \dot{\gamma}(t)$. From proposition 1.4.6 of chapter 1, it follows that the Lagrange derivative of $\mathcal{L}$ along the curve $\mathbb{R} \to Q \times \mathbb{R}^2 : t \mapsto (\gamma(t), c(t))$ is

$$\begin{aligned}\langle \delta\mathcal{L}(\gamma(t), c(t)) | a\rangle &= \langle \frac{\mathrm{d}}{\mathrm{d}t}\frac{\partial \mathcal{L}}{\partial c}(\gamma(t), c)\Big|_{c=c(t)} |a\rangle - \langle \frac{\partial \mathcal{L}}{\partial q}(q, c(t))\Big|_{q=\gamma(t)} |a^Q_{\gamma(t)}\rangle \\ &\quad - \langle \frac{(\mathcal{L}\circ\lambda_D^{-1})}{\partial \dot{q}}(\gamma(t), \dot{q})\Big|_{\dot{q}=\dot{\gamma}(t)} | \big[c^Q(t), a^Q\big]_{\gamma(t)}\rangle, \end{aligned} \tag{19}$$

for every $a \in \mathbb{R}^2$. But

$$\frac{\mathrm{d}}{\mathrm{d}t}\left(\frac{\partial \mathcal{L}}{\partial c}\right) = \left(\frac{\mathrm{d}}{\mathrm{d}t}\frac{\partial \mathcal{L}}{\partial c_1}, \frac{\mathrm{d}}{\mathrm{d}t}\frac{\partial \mathcal{L}}{\partial c_2}\right) = \big(m\dot{c}_1, (I+m\ell^2)\dot{c}_2\big), \qquad \frac{\partial \mathcal{L}}{\partial q} = 0,$$

$$\frac{\partial(\mathcal{L}\circ\lambda_D^{-1})}{\partial \dot{q}} = \left(\frac{\partial L}{\partial \dot{x}_1}, \frac{\partial L}{\partial \dot{x}_2}\right) = (m\dot{x}_1, m\dot{x}_2),$$

and

$$[a^Q, b^Q] = (a_1b_2 - a_2b_1)\big(\sin\theta \frac{\partial}{\partial x_1} - \cos\theta \frac{\partial}{\partial x_2}\big) \tag{20}$$

for any $a, b \in \mathbb{R}^2$, see (25) below. Therefore for every $a = (a_1, a_2) \in \mathbb{R}^2$ we have

$$\begin{aligned}\langle \delta\mathcal{L}|a\rangle &= \langle \frac{\mathrm{d}}{\mathrm{d}t}\left(\frac{\partial \mathcal{L}}{\partial c}\right)|a\rangle - \langle \frac{\partial L}{\partial \dot{x}} | [c^Q, a^Q]\rangle \\ &= \langle (m\dot{c}_1, (I+m\ell^2)\dot{c}_2) | (a_1, a_2)\rangle - \langle (m\dot{x}_1, m\dot{x}_2) | (c_1a_2 - c_2a_1)(\sin\theta, -\cos\theta)\rangle \\ &= \langle \big(m\dot{c}_1 - m\ell c_2^2, (I+m\ell^2)\dot{c}_2 + m\ell c_1c_2\big) | (a_1, a_2)\rangle, \end{aligned} \tag{21}$$

using (16). The *Lagrange-d'Alembert principle in a trivialization* states that $\delta\mathcal{L} = 0$. Consequently, the equations of motion of Carathéodory's sleigh in the trivialization λ_D^{-1} are

$$\begin{aligned}\dot{c}_1 &= \ell c_2^2 \\ \dot{c}_2 &= -\tfrac{m\ell}{I+m\ell^2}\, c_1c_2. \end{aligned} \tag{22}$$

An explanation why $\dot{c}$ only depends on c is given in the text after equation (73) below.

5.2.4 *Almost Poisson bracket form*

In this subsection we find the equation of motion of Carathéodory's sleigh using almost Poisson brackets on $C^\infty(D)$, which we compute using the trivialization λ_D^{-1}, see (15).

5.2.4.1 *H and ϖ in a trivialization*

First we determine the *symplectic distribution* (H, ϖ) for the sleigh *in the trivialization* λ_D^{-1}, see §6.2 of chapter 1.

In the trivialization λ_D^{-1} the distribution H is given by

$$H_u = T_{(q,c)}\lambda_D(D_q \times \mathbb{R}^2), \tag{23}$$

where $u = \lambda_D(q,c) \in D$, see (39) of §6.2 of chapter 1.

We now find an expression for the symplectic form ϖ in the trivialization λ_D^{-1}. Let $X_i(q) = \lambda_D(q, e_i)$, be a vector field on Q with values in D. For $a \in \mathbb{R}^2$, $a^Q(q) = \sum_{i=1}^2 a_i\, X_i(q)$ is a vector field on Q with values in D. From (23) it follows that every vector field on D with values in H can be written as $a^Q_{\rightarrow}(u) + b^Q_{\uparrow}(u)$, where $a^Q_{\rightarrow}(u) = T_{(q,c)}\lambda_D(a^Q(q), 0) = a^Q(q)$ and $b^Q_{\uparrow}(u) = T_{(q,c)}\lambda_D(0,b)$ for some $a, b \in \mathbb{R}^2$, see (45) of §6.2 of chapter 1. We compute $a^Q_{\rightarrow}$ and $b^Q_{\uparrow}$ as follows. Using

$$u = \lambda_D(\theta, x, c) = (\dot\theta, \dot x_1, \dot x_2) = (c_2, c_1\cos\theta - \ell c_2 \sin\theta, c_1 \sin\theta + \ell c_2 \cos\theta)$$

it follows from the definition (15) of the map λ_D that its tangent is given by

$$\begin{aligned} T_{(\theta,x,c)}\lambda_D(\dot\theta, \dot x_1, \dot x_2, \dot c) &= (\dot\theta, \dot x_1, \dot x_2, \ddot\theta, \ddot x_1, \ddot x_2) = \big(c_2, c_1\cos\theta - \ell c_2\sin\theta, \\ &c_1\sin\theta + \ell c_2\cos\theta, \dot c_2, \dot c_1\cos\theta - c_1\dot\theta\sin\theta - \ell\dot c_2\sin\theta - \ell c_2\dot\theta\cos\theta, \\ &\quad \dot c_1\sin\theta + c_1\dot\theta\cos\theta + \ell\dot c_2\cos\theta - \ell c_2\dot\theta\sin\theta\big). \end{aligned} \tag{24}$$

Therefore

$$\begin{aligned} a^Q_{\rightarrow}(u) &= T_{(\theta,x,c)}\lambda_D(a^Q(\theta,x),0) = a^Q(\theta,x) \\ &= a_1\big(\cos\theta\frac{\partial}{\partial x_1} + \sin\theta\frac{\partial}{\partial x_2}\big) + a_2\big(\frac{\partial}{\partial\theta} - \ell\sin\theta\frac{\partial}{\partial x_1} + \ell\cos\theta\frac{\partial}{\partial x_2}\big) \end{aligned} \tag{25}$$

and

$$\begin{aligned} b^Q_{\uparrow}(u) &= T_{(\theta,x,c)}\lambda_D(0,b) \\ &= \dot b_1\big(\cos\theta\frac{\partial}{\partial \dot x_1} + \sin\theta\frac{\partial}{\partial \dot x_2}\big) + \dot b_2\big(\frac{\partial}{\partial\dot\theta} - \ell\sin\theta\frac{\partial}{\partial \dot x_1} + \ell\cos\theta\frac{\partial}{\partial \dot x_2}\big). \end{aligned}$$

Since the *Lagrangian* $\mathcal{L}$ (18) *in the trivialization* λ_D^{-1} is the *kinetic energy*

$$k(q)(c^Q(q), c^Q(q)) = \tfrac{1}{2}\, mc_1^2 + \tfrac{1}{2}\,(I + m\ell^2)c_2^2$$

of the sleigh, the kinetic energy metric k of the moving frames a^Q, b^Q on Q in the trivialization λ_D^{-1}, which are associated to $a, b \in \mathbb{R}^2$, is

$$\mathrm{k}(q)(a^Q(q), b^Q(q)) = m\, a_1 b_1 + (I + m\ell^2) a_2 b_2. \tag{26}$$

Note that k is *constant* on Q. From equation (52) of §6.2 of chapter 1, we obtain

$$\begin{aligned}\varpi(u)\big(a^Q_{\rightarrow}(u) + b^Q_{\uparrow}(u), \widetilde{a}^Q_{\rightarrow}(u) + \widetilde{b}^Q_{\uparrow}(u)\big) &= \mathrm{k}(q)(a^Q(q), \widetilde{b}^Q(q)) \\ -\mathrm{k}(q)(\widetilde{a}^Q(q), b^Q(q)) &- \sum_{i=1}^{2} c_i\, \mathrm{d}\alpha_i(q)(a^Q(q), \widetilde{a}^Q(q)),\end{aligned} \tag{27}$$

where $\alpha_i(q) = \mathrm{k}^\flat(q)(X_i(q))$. Using equation (51) of §6.2 of chapter 1, we obtain

$$\begin{aligned}&\sum_{i=1}^{2} c_i\, \mathrm{d}\alpha_i(q)(a^Q(q), \widetilde{a}^Q(q)) \\ &= \sum_{n,m=1}^{2} a_n \widetilde{a}_m \big(L_{X_n}(\mathrm{k}(c^Q, X_m)) - L_{X_m}(\mathrm{k}(c^Q, X_n)) - \mathrm{k}(c^Q, [X_n, X_m])\big) \\ &= -\mathrm{k}(c^Q, [a^Q, \widetilde{a}^Q]), \text{ since } \mathrm{k}(c^Q, a^Q) \text{ and } \mathrm{k}(c^Q, \widetilde{a}^Q) \text{ are} \\ &\qquad\qquad \text{constant functions on } Q \\ &= -(a_1\widetilde{a}_2 - a_2\widetilde{a}_1)(\lambda_D^{-1})^*\mathrm{k}(v_1, v_2), \text{ where } v_1 = c_2 \tfrac{\partial}{\partial\theta} + (c_1\cos\theta - \ell c_2 \sin\theta)\tfrac{\partial}{\partial x_1} \\ &\qquad + (c_1 \sin\theta + \ell c_2\cos\theta)\tfrac{\partial}{\partial x_2} \text{ and, } v_2 = \sin\theta \tfrac{\partial}{\partial x_1} - \cos\theta\tfrac{\partial}{\partial x_2}, \\ &\qquad \text{using (16) and (20)} \\ &= -m(a_1\widetilde{a}_2 - a_2\widetilde{a}_1)\big(\sin\theta(c_1\cos\theta - \ell c_2\sin\theta) - \cos\theta(c_1\sin\theta + \ell c_2\cos\theta)\big) \\ &= m\ell c_2\,(a_1\widetilde{a}_2 - a_2\widetilde{a}_1).\end{aligned} \tag{28}$$

Therefore from (26), (27), and (28) we obtain

$$\begin{aligned}&\varpi(u)(a^Q_{\rightarrow}(u) + b^Q_{\uparrow}(u), \widetilde{a}^Q_{\rightarrow}(u) + \widetilde{b}^Q_{\uparrow}(u)) \\ &= m(a_1\widetilde{b}_1 - b_1\widetilde{a}_1) + (I + m\ell^2)(a_2\widetilde{b}_2 - b_2\widetilde{a}_2) - m\ell c_2(a_1\widetilde{a}_2 - a_2\widetilde{a}_1).\end{aligned} \tag{29}$$

5.2.4.2 *Almost Poisson bracket of* c_1 *and* c_2

In this subsection we compute the *almost Poisson bracket* of the smooth functions c_1 and c_2 (17) using the trivialization λ_D^{-1} (15).

First we calculate the distributional Hamiltonian vector field $Y_{c_1} = a^Q_{\rightarrow} + b^Q_{\uparrow}$ associated to the function c_1. For every $\widetilde{a}^Q_{\rightarrow} + \widetilde{b}^Q_{\uparrow}$ with $\widetilde{a}, \widetilde{b} \in \mathbb{R}^2$ we have

$$\langle \mathrm{d}c_1 | \widetilde{a}^Q_{\rightarrow} + \widetilde{b}^Q_{\uparrow} \rangle = \varpi(Y_{c_1}, \widetilde{a}^Q_{\rightarrow} + \widetilde{b}^Q_{\uparrow}) = \varpi(a^Q_{\rightarrow} + b^Q_{\uparrow}, \widetilde{a}^Q_{\rightarrow} + \widetilde{b}^Q_{\uparrow})$$

by definition of Y_{c_1}. In other words,

$$\widetilde{b}_1 = m(a_1 \widetilde{b}_1 - b_1 \widetilde{a}_1) + (I + m\ell^2)(a_2 \widetilde{b}_2 - b_2 \widetilde{a}_2) - m\ell c_2 (a_1 \widetilde{a}_2 - a_2 \widetilde{a}_1) \quad (30)$$

for every $\widetilde{a}, \widetilde{b} \in \mathbb{R}^2$. Successively setting one of the variables $\widetilde{a}_1$, $\widetilde{a}_2$, $\widetilde{b}_1$, and $\widetilde{b}_2$ equal to 1 and the others equal to 0 we get $(a_1, a_2, b_1, b_2) = (\frac{1}{m}, 0, 0, -\frac{\ell c_2}{I + m\ell^2})$. Therefore

$$\begin{aligned} Y_{c_1} &= \frac{1}{m}(1,0)^Q_{\rightarrow} - \frac{\ell c_2}{I + m\ell^2}(0,1)^Q_{\uparrow} && (31) \\ &= \tfrac{1}{m}\Big(\cos\theta \tfrac{\partial}{\partial x_1} + \sin\theta \tfrac{\partial}{\partial x_2}\Big) - \tfrac{\ell c_2}{I + m\ell^2}\Big(\tfrac{\partial}{\partial \dot{\theta}} - \ell \sin\theta \tfrac{\partial}{\partial \dot{x}_1} + \ell\cos\theta \tfrac{\partial}{\partial \dot{x}_2}\Big). && (32) \end{aligned}$$

Consequently,

$$\{c_1, c_2\} = -L_{Y_{c_1}} c_2 = -L_{Y_{c_1}} \dot{\theta} = \frac{\ell}{I + m\ell^2} c_2. \quad (33)$$

5.2.4.3 *Equations of motion*

The *equations of motion* of Carathéodory's sleigh *in almost Poisson bracket form* using the trivialization λ_D^{-1} of the constraint distribution D are

$$\begin{aligned} \dot{c}_1 &= \{c_1, \mathcal{L}\} = (I + m\ell^2) c_2 \{c_1, c_2\} = \ell c_2^2 \\ \dot{c}_2 &= \{c_2, \mathcal{L}\} = -m c_1 \{c_1, c_2\} = -\tfrac{m\ell}{I + m\ell^2} c_1 c_2. \end{aligned} \quad (34)$$

Here $\mathcal{L} = \frac{1}{2} m c_1^2 + \frac{1}{2}(I + m\ell^2) c_2^2$, the Lagrangian in the trivialization λ_D^{-1}, is thought of as a function on the constraint distribution D.

5.2.5 *Distributional Hamiltonian system*

In this subsection we derive the equations of motion of Carathéodory's sleigh from its distributional Hamiltonian formulation as a nonholonomically constrained system on the Lie group E(2) with left invariant constraint 1-form and left invariant kinetic energy.

5.2.5.1 *Lie group model*

The *configuration space* of Carathéodory's sleigh is the Lie group E(2) of proper Euclidean motions on $\mathbb{R}^2$ with Euclidean inner product $\langle\ ,\ \rangle$. The kinetic energy of the unconstrained sleigh $\mathcal{S}$ is

$$T = T_{\mathrm{rot}} + T_{\mathrm{trans}}, \quad (35)$$

where the *rotational kinetic energy* about the center of mass of the sleigh is

$$
\begin{aligned}
T_{\mathrm{rot}} &= \tfrac{1}{2}\int_{\mathcal{S}} \langle \dot{A}x, \dot{A}x\rangle \,\mathrm{d}m = \tfrac{1}{2}\int_{\mathcal{S}} \langle A^{-1}\dot{A}x, A^{-1}\dot{A}x\rangle \,\mathrm{d}m \\
&= \tfrac{1}{2}\int_{\mathcal{S}} \langle \Omega x, \Omega x\rangle \,\mathrm{d}m, \quad \text{where } \Omega = A^{-1}\dot{A} \in \mathrm{so}(2) \\
&= \tfrac{1}{2}\Big(\int_{\mathcal{S}} \langle x, x\rangle \,\mathrm{d}m\Big) i(\Omega)^2 = \tfrac{1}{2} I\,(\!(\, i(\Omega), i(\Omega)\,)\!) = \tfrac{1}{2}\langle\langle \mathcal{I}(\Omega), \Omega\rangle\rangle. \qquad (36)
\end{aligned}
$$

Here $I = \int_{\mathcal{S}} \langle x, x\rangle \,\mathrm{d}m$ is the moment of inertia of the sleigh about the vertical axis about its center of mass. $\Omega = i(\Omega)E$, where $\{E = \left(\begin{smallmatrix} 0 & -1 \\ 1 & 0 \end{smallmatrix}\right)\}$ is a basis of so(2). The mapping $i : \mathrm{so}(2) \rightarrow \mathbb{R} : X = xE \mapsto x$ is an isometry from $(\mathrm{so}(2), \langle\langle\ ,\ \rangle\rangle)$, where $\langle\langle X, Y\rangle\rangle = -\frac{1}{2}\,\mathrm{tr}XY$, to $(\mathbb{R}, (\!(\ ,\ )\!)\,)$, where $(\!(x, y)\!) = xy$. The *translational kinetic energy* of the center of mass of the sleigh is

$$
\begin{aligned}
T_{\mathrm{trans}} &= \tfrac{1}{2}\int_{\mathcal{S}} \langle \dot{a}, \dot{a}\rangle \,\mathrm{d}m = \tfrac{1}{2}\int_{\mathcal{S}} \langle A^{-1}\dot{a}, A^{-1}\dot{a}\rangle \,\mathrm{d}m \\
&= \tfrac{1}{2}\Big(\int_{\mathcal{S}} \mathrm{d}m\Big)\langle b, b\rangle, \quad \text{where } b = A^{-1}\dot{a} \\
&= \tfrac{1}{2} m\,\langle b, b\rangle. \qquad (37)
\end{aligned}
$$

Thus the kinetic energy of the sleigh is a homogeneous quadratic function on e(2) with coordinates (Ω, b) given by

$$T = \tfrac{1}{2}\langle\langle \mathcal{I}(\Omega), \Omega\rangle\rangle + \tfrac{1}{2}\langle b, b\rangle. \qquad (38)$$

Here $\mathcal{I} : \mathrm{so}(2) \rightarrow \mathrm{so}(2)$ is the *generalized moment of inertia* map $i \circ I \circ i^{-1}$. Hence the kinetic energy T determines an inner product on e(2) given by

$$\mathrm{k} : \mathrm{e}(2) \times \mathrm{e}(2) \rightarrow \mathbb{R} : \big((\Omega, b), (\Omega', b')\big) \mapsto \langle\langle \mathcal{I}(\Omega), \Omega'\rangle\rangle + m\langle b, b'\rangle, \qquad (39)$$

which extends to a left invariant Riemannian metrix k on E(2) called the *kinetic energy metric*.

We now impose the nonholonomic constraint that the velocity $(\dot{A}, \dot{a})$ of the sleigh lies in the kernel of the *constraint* 1*-form* ϕ on E(2) defined by

$$\phi(A, a) = \ell\langle\langle \mathrm{d}A, AE\rangle\rangle - \langle \mathrm{d}a, Ae_2\rangle. \qquad (40)$$

Here $\{e_1, e_2\}$ is the standard basis of $\mathbb{R}^2$.

Lemma 5.2.5.1.1. *The* 1*-form* ϕ (40) *is the same as the* 1*-form* φ (3).

Proof. To see this let $A = \left(\begin{smallmatrix} \cos\theta & -\sin\theta \\ \sin\theta & \cos\theta \end{smallmatrix}\right) \in \mathrm{SO}(2)$, where $\theta \in \mathbb{R}/2\pi\mathbb{Z}$, and let $a = \left(\begin{smallmatrix} x \\ y \end{smallmatrix}\right) \in \mathbb{R}^2$. Then $A^{-1}\,\mathrm{d}A = E\,\mathrm{d}\theta$ and

$$\langle A^{-1}\,\mathrm{d}a, e_2\rangle = \langle \begin{pmatrix} x \\ y \end{pmatrix}, Ae_2\rangle = \langle \begin{pmatrix} \mathrm{d}x \\ \mathrm{d}y \end{pmatrix}, \begin{pmatrix} -\sin\theta \\ \cos\theta \end{pmatrix}\rangle.$$

So

$$\phi = \ell\langle\langle E, E\rangle\rangle\,\mathrm{d}\theta - \langle \begin{pmatrix} \mathrm{d}x \\ \mathrm{d}y \end{pmatrix}, \begin{pmatrix} -\sin\theta \\ \cos\theta \end{pmatrix}\rangle = \ell\,\mathrm{d}\theta + \sin\theta\,\mathrm{d}x - \cos\theta\,\mathrm{d}y = \varphi.$$

□

Let

$$\pi : \mathrm{E}(2) \times \mathrm{e}(2) \to \mathrm{E}(2) : (A, a, \Omega, b) \mapsto (A, a). \tag{41}$$

As a submanifold of $\mathrm{E}(2) \times \mathrm{e}(2)$ with coordinates (A, a, Ω, b) the *constraint manifold* $D = \ker \pi^*\phi$ is defined by

$$0 = \ell\langle\langle A^{-1}\dot{A}, E\rangle\rangle - \langle A^{-1}\dot{a}, e_2\rangle = \ell\langle\langle \Omega, E\rangle\rangle - \langle b, e_2\rangle = i(\Omega)\ell - b_2,$$

that is,

$$D = \{(A, a, \ell^{-1}b_2E, b) \in \mathrm{E}(2) \times \mathrm{e}(2) \,\big|\, (A, a) \in \mathrm{E}(2),\ b \in \mathbb{R}^2\}. \tag{42}$$

5.2.5.2 *The distribution H and its symplectic form ϖ_H*

In this subsection we determine the *distribution* $H = TD \cap \ker \pi^*\phi$ on the constraint manifold D. Here π is the mapping (41), which is the pull back of the bundle projection $T\mathrm{E}(2) \to E(2) : (A, a, \dot{A}, \dot{a}) \mapsto (A, a)$ by the inverse of the left trivialization

$$\lambda : T\mathrm{E}(2) \to \mathrm{E}(2) \times \mathrm{e}(2) : (A, a, \dot{A}, \dot{a}) \mapsto (A, a, \Omega = A^{-1}\dot{A}, b = A^{-1}\dot{a}). \tag{43}$$

Until further mention we assume that $\ell \neq 0$.

Let $u = (A, a, \ell^{-1}b_2E, b) \in D$ and let (Y, y), $(Z, z) \in \mathrm{e}(2)$. The vector $v_u = (AY, Ay, Z, z) \in T_u(\mathrm{E}(2) \times \mathrm{e}(2))$ lies in T_uD if and only if

$$0 = \langle\langle \dot{\Omega}, E\rangle\rangle - \langle \dot{b}, e_2\rangle = \ell\langle\langle Z, E\rangle\rangle - \langle z, e_2\rangle = \ell i(Z) - z_2,$$

that is, $Z = \ell^{-1}z_2E$. The vector v_u lies in $\ker \pi^*\phi$ if and only if

$$0 = \ell\langle\langle Y, E\rangle\rangle - \langle y, e_2\rangle = \ell i(Y) - y_2,$$

that is, $Y = \ell^{-1}y_2E$. Therefore

$$H_u = \{(A(\ell^{-1}y_2E), Ay, \ell^{-1}z_2E, z) \in T_u(\mathrm{E}(2) \times \mathrm{e}(2)) \,\big|\, y, z \in \mathbb{R}^2\}, \tag{44}$$

which is 4-dimensional. Thus H is a distribution on D.

Next we find the *symplectic form* ϖ_H *on* H. The pull back of the canonical 1-form $\theta_{\mathrm{E}(2)}$ on $T^*\mathrm{E}(2)$ by the kinetic energy metric k is the 1-form θ on $T\mathrm{E}(2)$. Pulling θ back by λ^{-1} (43) gives the 1-form $\widetilde{\theta}$ on $\mathrm{E}(2)\times \mathrm{e}(2)$ defined by

$$\widetilde{\theta}(A,a,\Omega,b)\big(AY,Ay,Z,z\big) = \mathrm{k}\big((\Omega,b),(Y,y)\big).$$

Thus $\widetilde{\omega} = -\,\mathrm{d}\widetilde{\theta}$ is a symplectic form on $\mathrm{E}(2)\times\mathrm{e}(2)$. Explicitly,

$$\begin{aligned}
&\widetilde{\omega}(A,a,\Omega,b)\big((AY,Ay,Z,z),(AY',Ay',Z',z')\big)\\
&\quad = \mathrm{k}((Z,z),(Y',y')) - \mathrm{k}((Z',z'),(Y,y)) + \mathrm{k}((\Omega,b),[(Y,y),(Y',y')]),\\
&\qquad\qquad\qquad \text{see appendix A of [30]}\\
&\quad = ((\,I(i(Z)),i(Y')\,)) + m\langle z,y'\rangle - ((\,I(i(Z')),i(Y)\,)) - m\langle(z',y\rangle\\
&\qquad\qquad + m\langle b, Yy' - Y'y\rangle,
\end{aligned} \tag{45}$$

since the Lie bracket in $\mathrm{e}(2)$ is $[(Y,y),(Y',y')] = ([Y,Y'],Yy'-Y'y) = (0,Yy'-Y'y)$. Here we have used the fact that the Lie bracket $[Y,Y']$ in $\mathrm{so}(2)$ is 0. Restricting $\widetilde{\omega}(u)$ to $H_u\times H_u$ gives the symplectic form $\varpi_H(u)$ on H_u for every $u\in D$. In other words, for $u=(A,a,\ell^{-1}b_2E,b)\in D$ and $v_u = (A(\ell^{-1}y_2E),Ay,\ell^{-1}z_2E,z)$ and $v'_u = (A(\ell^{-1}y'_2E),Ay',\ell^{-1}z'_2E,z')\in H_u$ we have

$$\begin{aligned}
\varpi_H(u)(v_u,v'_u) &= ((\,I(i(\ell^{-1}z'_2E)),i(\ell^{-1}y_2E)\,)) - ((\,I(i(\ell^{-1}z_2E)),i(\ell^{-1}y'_2E)\,))\\
&\qquad + m(z'_1y_1 - z_1y'_1) - m(z_2y'_2 - y_2z'_2) + m\langle b,\ell^{-1}y_2Ey' - \ell^{-1}y'_2Ey\rangle\\
&= (I\ell^{-2}+m)(y_2z'_2 - z_2y'_2) + m(y_1z'_1 - z_1y'_1)\\
&\qquad\qquad - m\ell^{-1}b_2(y_1y'_2 - y_2y'_1).
\end{aligned} \tag{46}$$

5.2.5.3 *Equations of motion*

For the distributional Hamiltonian system (D,H,ϖ_H,T) corresponding to Carathéodory's sleigh the *distributional Hamiltonian vector field* Y_T on D, which governs the motion of the sleigh, satisfies

$$(Y_T \,\lrcorner\, \varpi_H)(u) = \mathrm{d}T(u)|H_u \tag{47}$$

for every $u\in D$. Because $Y_T(u)\in H_u$ we may write

$$Y_T(u) = (\ell^{-1}z_2AE, Az, \ell^{-1}w_2E, w) \tag{48}$$

for some $z,w\in\mathbb{R}^2$. At the point $u=(A,a,\ell^{-1}b_2E,b)$ in the constraint distribution D, the kinetic energy of the sleigh is

$$\begin{aligned}
T = T_u(\ell^{-1}b_2E,b) &= \tfrac{1}{2}\,\langle\langle \mathcal{I}(\ell^{-1}b_2E),\ell^{-1}b_2E\rangle\rangle + \tfrac{1}{2}\,m\langle b,b\rangle\\
&= \tfrac{1}{2}\,mb_1^2 + \tfrac{1}{2}\,(m+I\ell^{-2})b_2^2.
\end{aligned} \tag{49}$$

Therefore

$$\mathrm{d}T = mb_1\,\mathrm{d}b_1 + (m + I\ell^{-2})b_2\,\mathrm{d}b_2. \tag{50}$$

Evaluating both sides of (47) at $(\ell^{-1}z_2'AE, Az', \ell^{-1}w_2'E, w') \in H_u$ and using (46) gives

$$\begin{aligned} mb_1w_1' + (m + I\ell^{-2})b_2w_2' = (m + I\ell^{-2})(z_2w_2' - w_2z_2') + m(z_1w_1' - w_1z_1') \\ - m\ell^{-1}b_2(z_1z_2' - z_2z_1'), \end{aligned} \tag{51}$$

for every $z', w' \in \mathbb{R}^2$. Successively setting all but one of the variables z_1', z_2', w_1', w_2' in (51) equal to 0 and the remaining variable equal 1 gives $(z_1, z_2, w_1, w_2) = (b_1, b_2, \ell^{-1}b_2^2, -\frac{m\ell^{-1}}{m+I\ell^{-2}}\, b_1b_2)$. Therefore

$$Y_T(A, a, \ell^{-1}b_2E, b) = \Big(\ell^{-1}b_2AE, Ab, -\frac{m}{I+m\ell^2}b_1b_2E, \begin{pmatrix} \ell^{-1}b_2^2 \\ -\frac{m\ell}{I+m\ell^2}b_1b_2 \end{pmatrix}\Big).$$

Thus the integral curves of Y_T satisfy

$$\begin{aligned} \dot{A} &= A(\ell^{-1}b_2E) \\ \dot{a} &= Ab \\ \begin{pmatrix} \dot{b}_1 \\ \dot{b}_2 \end{pmatrix} &= \begin{pmatrix} \ell^{-1}b_2^2 \\ -\frac{m\ell}{I+m\ell^2}b_1b_2 \end{pmatrix}. \end{aligned} \tag{52}$$

We have omitted the equation $\dot{\Omega} = \ell^{-1}\dot{b}_2E = -\frac{m}{I+m\ell^2}b_1b_2E$ from (52) as it follows from the equation for $\dot{b}_2$ in (52). From (17) we find that

$$\begin{pmatrix} c_1 \\ \ell c_2 \end{pmatrix} = \begin{pmatrix} \cos\theta & \sin\theta \\ -\sin\theta & \cos\theta \end{pmatrix} \begin{pmatrix} \dot{x}_1 \\ \dot{x}_2 \end{pmatrix} = A^{-1}\dot{a} = b. \tag{53}$$

Thus the third equation in (52) becomes

$$\begin{aligned} \dot{c}_1 &= \ell c_2^2 \\ \dot{c}_2 &= -\frac{m\ell}{I+m\ell^2}\, c_1c_2, \end{aligned} \tag{54}$$

which is just equation (22). Note that equation (54) holds when $\ell = 0$. This is the reason for introducing the variables c_1 and c_2 in (53).

5.3 Reduction of the E(2) symmetry

In this section we show that Carathéodory's sleigh has an E(2) symmetry. After reducing this symmetry, we calculate the equations of motion of the sleigh after reducing this symmetry in three ways. First, we use the momentum equations of §8 of chapter 1; second, we use almost Poisson brackets of §7 of chapter 1, and third, we use the distributional Hamiltonian approach of §3 of chapter 3.

5.3.1 *The* E(2) *symmetry*

Physically it is obvious that the sleigh is invariant under horizontal translations and rotations about a vertical axis through its center of mass. Mathematically, there is an action of E(2) on configuration space $Q = S^1 \times \mathbb{R}^2$ given by

$$\phi : \mathrm{E}(2) \times Q \to Q : \big((R_\vartheta, y), (\theta, x)\big) \mapsto (\theta + \vartheta, R_\vartheta x + y), \tag{55}$$

where $R_\vartheta = \left(\begin{smallmatrix} \cos\vartheta & -\sin\vartheta \\ \sin\vartheta & \cos\vartheta \end{smallmatrix}\right) \in \mathrm{SO}(2,\mathbb{R})$ and $\vartheta \in S^1 = \mathbb{R}/2\pi\,\mathbb{Z}$. This action lifts to an action of E(2) on T^*Q defined by

$$\begin{aligned} \widetilde{\phi} : \mathrm{E}(2) \times T^*Q \to T^*Q : \big((R_\vartheta, y), (\theta, x, p_x, p_\theta)\big) &\mapsto \\ (R_\vartheta, y) \cdot (\theta, x, p_x, p_\theta) = (\theta + \vartheta, R_\vartheta x + y, R_\vartheta p_x, p_\theta). \end{aligned} \tag{56}$$

Since the constraint distribution $D \subseteq T^*Q$, which is defined by $p_\theta = \frac{I}{m\ell}(-p_{x_1}\sin\theta + p_{x_2}\cos\theta)$, is invariant under $\widetilde{\phi}$, we have an induced E(2)-action

$$\Phi : \mathrm{E}(2) \times D \to D : \big((R_\vartheta, y), (\theta, x, p_x)\big) \mapsto (\theta + \vartheta, R_\vartheta x + y, R_\vartheta p_x), \tag{57}$$

using (θ, x, p_x) as coordinates on D.

Proposition 5.3.1.2. *The* E(2) *action* Φ (57) *is a symmetry of Carathéodory's sleigh.*

Proof. Clearly the Hamiltonian h (10) is invariant under Φ (57), as is the constraint 1-form φ (3). Since the constraint distribution $D = \ker\varphi$, it is also invariant. It remains to show that the equations of motion (14) are invariant under Φ. This follows from lemma 3.2.7 of chapter 3 directly. We deduce this from (14) as follows. First note that the functions

$$f_1(\theta, x, p_x) = -p_{x_1}\sin\theta + p_{x_2}\cos\theta \ \text{ and } \ f_2(\theta, x, p_x) = p_{x_1}\cos\theta + p_{x_2}\sin\theta$$

are invariant under Φ. Now

$$\begin{aligned} &Y_h((R_\vartheta, y) \cdot (\theta, x, p_x)) = Y_h(\theta + \vartheta, R_\vartheta x + y, R_\vartheta p_x) \\ &\quad = \Big(\frac{1}{m\ell} f_1(\theta + \vartheta, R_\vartheta x + y, R_\vartheta p_x), \frac{1}{m} R_\vartheta p_x, \\ &\qquad\qquad \frac{I}{m\ell(I + m\ell^2)} f_1 \cdot f_2(\theta + \vartheta, R_\vartheta x + y, R_\vartheta p_x) \begin{pmatrix} -\sin(\theta+\vartheta) \\ \cos(\theta+\vartheta) \end{pmatrix}\Big) \\ &\quad = \Big(\frac{1}{m\ell} f_1(\theta, x, p_x), R_\vartheta\big(\frac{1}{m} p_x\big), R_\vartheta\Big(\frac{I}{m\ell(I + m\ell^2)} f_1 \cdot f_2(\theta, x, p_x) \begin{pmatrix} -\sin\theta \\ \cos\theta \end{pmatrix}\Big)\Big) \\ &\quad = \Big(\frac{\mathrm{d}\theta}{\mathrm{d}t}, R_\vartheta\big(\frac{\mathrm{d}x}{\mathrm{d}t}\big), R_\vartheta\big(\frac{\mathrm{d}p_x}{\mathrm{d}t}\big)\Big) = \frac{\mathrm{d}}{\mathrm{d}t}\big((R_\vartheta, y) \cdot (\theta, x, p_x)\big). \end{aligned}$$

□

5.3.2 *The momentum equation*

In this subsection we use the momentum equation of §8 of chapter 1 to reduce the E(2) symmetry from the equations of motion of the sleigh.

Before we can apply the momentum equation, we need to verify

Lemma 5.3.2.3. *The* E(2)*-action* (57) *on the constraint distribution D is free and proper.*

Proof. The action is free, for if $(\theta+\vartheta, R_\vartheta x + y, R_\vartheta p_x) = (\theta, x, p_x)$, then $\theta = 0 \bmod 2\pi$, which implies $x = R_\vartheta x + y = x + y$, that is, $y = 0$.

The action Φ (57) is proper for if the sequence $\{(\theta_n, x_n, (p_x)_n, \theta_n + \vartheta_n, R_{\vartheta_n} x_n + y_n, R_{\vartheta_n}(p_x)_n)\}_{n=1}^{\infty}$ converges to $(\theta', x', p'_x, \varphi'', x'', p''_x)$, then the sequence $\{\vartheta_n\}$ converges to $\varphi'' - \theta'$, the sequence $\{y_n\}$ converges to $y' = x'' - R_{\vartheta'} x'$, and the sequence $\{R_{\vartheta_n}(p_x)_n\}$ converges to $R_{\vartheta'} p'_x$. In other words,

$$(\varphi'', x'', p''_x) = (\theta' + \vartheta', R_{\vartheta'} x' + y', R_{\vartheta'} p'_x) = \Phi_{(R_{\vartheta'}, y')}(\theta', x', p'_x).$$

□

From lemma 5.3.2.3 it follows that the *reduced space* $\overline{D} = D/\mathrm{E}(2)$ of orbits of the action Φ is a smooth manifold.

Lemma 5.3.2.4. *The reduced space $\overline{D}$ is diffeomorphic to $\mathbb{R}^2$.*

Proof. Using the coordinates $(\theta, x, p_x) = (\theta, x_1, x_2, p_{x_1}, p_{x_2})$ for D, consider the map

$$\pi : D \to \mathbb{R}^2 : (\theta, x, p_x) \mapsto R_{-\theta} p_x = \begin{pmatrix} p_{x_1}\cos\theta + p_{x_2}\sin\theta \\ -p_{x_1}\sin\theta + p_{x_2}\cos\theta \end{pmatrix}.$$

Thus the derivative of π at $(\theta, x, p_x) \in D$ is the 5×2 matrix

$$\begin{pmatrix} -p_{x_1}\cos\theta + p_{x_2}\cos\theta & 0 & 0 & \cos\theta & \sin\theta \\ -p_{x_1}\cos\theta - p_{x_2}\sin\theta & 0 & 0 & -\sin\theta & \cos\theta \end{pmatrix},$$

which has rank 2, because its $1, 2; 4, 5$ minor is 1. Consequently, the tangent of π is surjective at every point of D. In addition, π is constant on E(2)-orbits, because

$$\begin{aligned} \pi((R_\vartheta, y)\cdot(\theta, x, p_x)) &= \pi(\theta+\vartheta, R_\vartheta x + y, R_\vartheta p_x) \\ &= R_{-(\theta+\vartheta)}(R_\vartheta p_x) = R_{-\theta} p_x = \pi(\theta, x, p_x). \end{aligned}$$

To finish the proof it suffices to show that a fiber of π is a single E(2)-orbit on D. For then π induces a mapping $\widetilde{\pi} : D/\mathrm{E}(2) \to \mathbb{R}^2$, which is injective and smooth. Because π is surjective, $\widetilde{\pi}$ is a diffeomorphism. Now suppose that $\pi(\theta, x, p_x) = \pi(\theta', x', p'_x)$. Then $R_{-\theta'} p'_x = R_{-\theta} p_x$. So $p'_x = R_{\theta-\theta'} p_x$. From

$$\begin{aligned}(R_{\theta'-\theta}, x' - R_{\theta'-\theta} x) \cdot (\theta, x, p_x) \\ = (\theta + \theta' - \theta, R_{\theta'-\theta} x + x' - R_{\theta'-\theta} x, R_{\theta'-\theta} p_x) = (\theta', x', p'_x)\end{aligned}$$

it follows that (θ, x, p_x) and (θ', x', p'_x) lie on the same E(2) orbit on D. Thus a fiber of π is a unique E(2)-orbit. □

We now use the momentum equation of §7.2 of chapter 3 to compute the E(2)-reduced equation of motion of Carathéodory's sleigh.

First we find a basis of E(2)-invariant vector fields on the constraint distribution D. Fix a point $q = (\theta, x)$ in the configuration space $Q = S^1 \times \mathbb{R}^2$. Differentiating the orbit map

$$\phi_q : \mathrm{E}(2) \to Q : (R_\vartheta, y) \mapsto \phi_{(R_\vartheta, y)}(q)$$

associated to the E(2)-action ϕ (55) at the identity element $e = (I, 0)$ of E(2) gives

$$T_e\phi_q : \mathrm{e}(2) \to T_qQ : \left(\dot{\vartheta}\left(\begin{smallmatrix} 0 & -1 \\ 1 & 0 \end{smallmatrix}\right), \dot{y}\right) \mapsto \dot{\vartheta}\left(\tfrac{\partial}{\partial \theta} - x_2 \tfrac{\partial}{\partial x_1} + x_1 \tfrac{\partial}{\partial x_2}\right) + \dot{y}_1 \tfrac{\partial}{\partial x_1} + \dot{y}_2 \tfrac{\partial}{\partial x_2}.$$

Thus the vector fields

$$Z_\theta = \frac{\partial}{\partial \theta} - x_2 \frac{\partial}{\partial x_1} + x_1 \frac{\partial}{\partial x_2}, \quad Z_{x_1} = \frac{\partial}{\partial x_1}, \quad \text{and } Z_{x_2} = \frac{\partial}{\partial x_2}$$

generate the module of E(2)-invariant vector fields on Q over the ring of smooth E(2)-invariant functions on Q. But none of the vector fields Z_θ, Z_{x_1}, or Z_{x_2} leave the constraint distribution $D = \ker \varphi$ (3) invariant. However, a short calculation shows that the vector fields

$$Z_1 = \cos\theta\, Z_{x_1} + \sin\theta\, Z_{x_2} = \cos\theta \frac{\partial}{\partial x_1} + \sin\theta \frac{\partial}{\partial x_2}$$

and

$$\begin{aligned}Z_2 &= Z_\theta + (x_2 - \ell \sin\theta) Z_{x_1} - (x_1 - \ell\cos\theta) Z_{x_2} \\ &= \frac{\partial}{\partial \theta} - \ell \sin\theta \frac{\partial}{\partial x_1} + \ell\cos\theta \frac{\partial}{\partial x_2}\end{aligned}$$

lie in $\ker \varphi$, are linearly independent on D, and are E(2)-invariant.

Recall that the canonical 1-form on T^*Q is $\theta_Q = p_\theta\, \mathrm{d}\theta + p_{x_1}\, \mathrm{d}x_1 + p_{x_2}\, \mathrm{d}x_2$. Therefore the momenta corresponding to the infinitesimal symmetries Z_1 and Z_2 are

$$P_{Z_1} = Z_1 \lrcorner\, \theta_Q = p_{x_1}\cos\theta + p_{x_2}\sin\theta \tag{58a}$$

and

$$P_{Z_2} = Z_2 \lrcorner\, \theta_Q = p_\theta - \ell p_{x_1} \sin\theta + \ell p_{x_2} \cos\theta \tag{58b}$$
$$= \frac{I + m\ell^2}{m\ell}(-p_{x_1} \sin\theta + p_{x_2} \cos\theta),$$

since $p_\theta = \frac{I}{m\ell}(-p_{x_1}\sin\theta + p_{x_2}\cos\theta)$. On $(T^*Q, \omega_Q = \mathrm{d}\theta \wedge \mathrm{d}p_\theta + \mathrm{d}x_1 \wedge \mathrm{d}p_{x_1} + \mathrm{d}x_2 \wedge \mathrm{d}p_{x_2})$ the Hamiltonian vector fields corresponding to the momenta P_{Z_1} and P_{Z_2} are

$$X_{P_{Z_1}} = \cos\theta \frac{\partial}{\partial x_1} + \sin\theta \frac{\partial}{\partial x_2} - (p_{x_1}\sin\theta + p_{x_2}\cos\theta)\frac{\partial}{\partial p_\theta}$$

and

$$X_{P_{Z_2}} = \frac{I + m\ell^2}{m\ell}\Big(-\sin\theta \frac{\partial}{\partial x_1} + \cos\theta \frac{\partial}{\partial x_2} + (p_{x_1}\cos\theta + p_{x_2}\sin\theta)\frac{\partial}{\partial p_\theta}\Big),$$

respectively. Therefore the momentum equations on $\overline{D} = \mathbb{R}^2$ are

$$\dot{P}_{Z_1} = -L_{X_{P_{Z_1}}} h = (-p_{x_1}\sin\theta + p_{x_2}\cos\theta)\frac{1}{I}p_\theta$$
$$= \frac{1}{m\ell}(-p_{x_1}\sin\theta + p_{x_2}\cos\theta)^2,$$
$$\text{since } p_\theta = \tfrac{I}{m\ell}(-p_{x_1}\sin\theta + p_{x_2}\cos\theta)$$
$$= \frac{m\ell}{(I + m\ell^2)^2}P_{Z_2}^2, \quad \text{using the definition of } P_{Z_2}$$

and

$$\dot{P}_{Z_2} = -L_{X_{P_{Z_2}}} h$$
$$= -\frac{I + m\ell^2}{m\ell}(p_{x_1}\cos\theta + p_{x_2}\sin\theta)\frac{1}{m\ell}(-p_{x_1}\sin\theta + p_{x_2}\cos\theta)$$
$$= -\frac{1}{m\ell}P_{Z_1}P_{Z_2}, \quad \text{using the definition of } P_{Z_1} \text{ and } P_{Z_2}.$$

Lemma 5.3.2.5. *Rescaling the momentum equations gives the* E(2)*-reduced equations of motion of Carathéodory's sleigh.*

Proof. Let $c_1 = \alpha \dot{P}_{Z_1}$ and $c_2 = \dot{P}_{Z_2}$ for some $\alpha, \beta \in \mathbb{R}$. Then

$$\begin{aligned} \dot{c}_1 &= \alpha \dot{P}_{Z_1} = \tfrac{\alpha m\ell}{(I+m\ell^2)^2}P_{Z_2}^2 = \tfrac{\alpha m\ell}{\beta^2(I+m\ell^2)^2}\,c_2^2 \\ \dot{c}_2 &= \beta\dot{P}_{Z_2} = -\tfrac{\beta}{m\ell}P_{Z_1}P_{Z_2} = -\tfrac{1}{\alpha m\ell}c_1c_2. \end{aligned} \tag{59}$$

But (22) reads

$$\begin{aligned} \dot{c}_1 &= \ell c_2^2 \\ \dot{c}_2 &= -\tfrac{m\ell}{I+m\ell^2}c_1c_2. \end{aligned} \tag{60}$$

Comparing (59) with (60) we obtain $\frac{\alpha m\ell}{\beta^2(I+m\ell^2)^2} = \ell$ and $\frac{1}{\alpha m\ell} = \frac{m\ell}{I+m\ell^2}$. Therefore $\alpha = \frac{I+m\ell^2}{(m\ell)^2}$ and $\beta = \frac{1}{\ell\sqrt{m(I+m\ell^2}}$ is the desired rescaling. □

5.3.3 E(2)-*reduced equations of motion*

To remove the E(2) symmetry (55) of Carathéodory's sleigh we use the Lie group model of §2.5 and follow the pattern of reduction of symmetries for distributional Hamiltonian systems described in §3 of chapter 3. This involves constructing certain distributions V, $V \cap H$, and U on the constraint manifold D, whose definitions we recall below. Until further mention in this subsection we assume that $\ell \neq 0$.

The E(2)-action on E(2) of left multiplication lifts to an action on $T\mathrm{E}(2)$, which using the trivialization λ^{-1} (43) is given by

$$\begin{aligned}\psi : \mathrm{E}(2) \times (\mathrm{E}(2) \times \mathrm{e}(2)) &\rightarrow \mathrm{E}(2) \times \mathrm{e}(2) : \\ \big((A', a'), (A, a, \Omega, b)\big) &\mapsto (A'A, A'a + a', \Omega, b).\end{aligned}$$

This action induces an E(2)-action on the constraint distribution D (42) given by

$$\begin{aligned}&\Psi : \mathrm{E}(2) \times D \rightarrow D : \\ &\quad \big((A', a'), u\big) = (A, a, \ell^{-1} b_2 E, b) \mapsto (A'A, A'a + a', \ell^{-1} b_2 E, b).\end{aligned} \tag{61}$$

The reduced space $\overline{D} = D/\mathrm{E}(2)$ of E(2) orbits on D is clearly $\mathbb{R}^2$ and the associated orbit map $\rho : D \subseteq \mathrm{E}(2) \times D \rightarrow \overline{D} : u \mapsto b$ is a surjective submersion. In other words, the maps ρ and

$$T_u\rho : T_u D \rightarrow T_{\rho(u)}\overline{D} : (AY, Ay, \ell^{-1} z_2 E, z) \mapsto z, \tag{62}$$

for every $u \in D$ are both surjective. Fix $u \in D$. Differentiating the E(2)-orbit map

$$\Psi_u : \mathrm{E}(2) \rightarrow D : (A', a') \mapsto (A'A, A'a + a', \ell^{-1} b_2 E, b)$$

at the identity element $e = (I, 0)$ of E(2) gives

$$T_u\Psi_u : \mathrm{e}(2) = T_e\mathrm{E}(2) \rightarrow T_u D : (A', a') \mapsto (\dot{A}'A, \dot{A}'a + \dot{a}, 0, 0).$$

Let $\mathcal{O}_u$ be the E(2) orbit of the action Ψ through $u \in D$. Then $T_u\mathcal{O}_u = T_e\Psi_u\,\mathrm{e}(2)$. Thus the distribution $V : D \rightarrow TD : u \mapsto V_u$ where

$$V_u = T_u\mathcal{O}_u = \{(\lambda\, AE, \lambda\, Ea + \dot{a}', 0, 0) \in T_u D \,\big|\, (\lambda E, a) \in \mathrm{e}(2),\ \lambda \in \mathbb{R}\}.$$

We now find the distribution $V \cap H$, where $u \mapsto H_u$ is the distribution H on D with H_u determined by (44). The vector $w_u = (AY, Ay, Z, z)$ lies in V_u if and only if

$$(\lambda AE, \lambda Ea + \dot{a}', 0, 0) = (AY, Ay, Z, z) \tag{63}$$

for some (Y, y), $(Z, z) \in \mathrm{e}(2)$. From (63) and the fact that SO(2) is abelian it follows that $(Z, z) = (0, 0)$, $Y = \lambda E$, and $y = A^{-1}(\lambda Ea + \dot{a}')$. For

$(A(\lambda E), \lambda E a + \dot{a}', 0, 0)$ to lie in H_u we must have $Y = \ell^{-1} y_2 E$, see (44). In other words,

$$\lambda = \ell^{-1} y_2 = \ell^{-1} \langle y, e_2 \rangle = \ell^{-1} \big(\lambda \langle E A^{-1} a, e_2 \rangle + \langle A^{-1} \dot{a}', e_2 \rangle \big)$$
$$= \ell^{-1} \big(\lambda \langle A^{-1} a, E e_2 \rangle + \langle A^{-1} \dot{a}', e_2 \rangle \big) = \ell^{-1} \big(-\lambda \langle A^{-1} a, e_1 \rangle + \langle A^{-1} \dot{a}', e_2 \rangle \big).$$

Therefore $\langle \dot{a}', A e_2 \rangle = \lambda(\ell - \langle a, A e_1 \rangle)$, which determines a 2 dimensional subspace $V_u \cap H_u$ of H_u, which is spanned by the linearly independent vectors $(AE, Aw, 0, 0)$ and $(0, Ae_1, 0, 0)$. Here

$$w = A^{-1} E a + (\ell - \langle a, A e_1 \rangle) e_2$$
$$= \begin{pmatrix} \langle A^{-1} E a, e_1 \rangle \\ \langle A^{-1} E a, e_2 \rangle \end{pmatrix} + \begin{pmatrix} 0 \\ \ell - \langle a, A e_1 \rangle \end{pmatrix} = \begin{pmatrix} \langle a, A e_2 \rangle \\ \ell \end{pmatrix}. \tag{64}$$

Thus $V \cap H$ is a distribution on D. We now prove

Lemma 5.3.3.6. *Let $u = (A, a, (\ell^{-1} b_2 E, b)) \in D_u$ with $b_2 \neq 0$. Then $V_u \cap H_u$ is a symplectic subspace of $(H_u, \varpi_H(u))$. If $b_2 = 0$, then $V_u \cap H_u$ is a Lagrangian subspace of $(H_u, \varpi_H(u))$.*

Proof. Let $u = (A, a, \ell^{-1} b_2 E, b) \in D$. Let $v_u = \lambda(AE, Aw, 0, 0) + \mu(0, e_1, 0, 0)$ and $v'_u = \lambda'(AE, Aw, 0, 0) + \mu'(0, e_1, 0, 0)$ be linearly independent vectors in $V_u \cap H_u$, that is, $\lambda\mu' - \lambda'\mu \neq 0$. Then using (45), we get

$$\varpi_H(u)(v_u, v'_u) = m \langle b, \lambda E(\lambda' w + \mu' e_1) - \lambda' E(\lambda w + \mu e_1) \rangle$$
$$= m b_2 (\lambda\mu' - \lambda'\mu). \tag{65}$$

The conclusions of the lemma follow immediately. □

By definition the distribution U on D is given by $u \mapsto U_u = H_u \cap (V_u \cap H_u)^{\varpi_H(u)}$. The following calculation determines U_u. The vector $v'_u = (A(\ell^{-1} y'_2 E, Ay', \ell^{-1} z'_2 E, z'))$, where $y', z' \in \mathbb{R}^2$ lies in H_u. It lies in U_u if and only if for every $\lambda, \mu \in \mathbb{R}$

$$0 = \varpi_H(u)\big((A(\lambda E), A(\lambda w + \mu e_1), 0, 0), (A(\ell^{-1} y'_e E), Ay', \ell^{-1} z'_2 E, z')\big)$$
$$= \mathrm{k}\big((\ell^{-1} z'_2 E, z'), (\lambda E, \lambda w + \mu e_1)\big) + \mathrm{k}\big((\ell^{-1} b_2 E, b), [(\lambda E, \lambda w + \mu e_1), (\ell^{-1} y'_2 E, y')]\big)$$
$$= \langle\langle \mathcal{I}(\ell^{-1} z'_2 E), \lambda E \rangle\rangle + m \langle z', \lambda w + \mu e_1 \rangle + m \langle b, \lambda E y' - \ell^{-1} y'_2 E(\lambda w + \mu e_1) \rangle$$
$$= I \lambda \ell^{-1} z'_2 + m z'_1 (\lambda w_1 + \mu e_1) + m z'_2 \lambda w_2 + \langle \begin{pmatrix} b_1 \\ b_2 \end{pmatrix}, \begin{pmatrix} -\lambda y'_2 + \ell^{-1} y'_2 w_2 \\ \lambda y'_1 - \lambda \ell^{-1} y'_2 w_1 - \mu \ell^{-1} y'_2 \end{pmatrix} \rangle$$
$$= \lambda \ell^{-1} I z'_2 + m z'_1 (\lambda w_1 + \mu) + m \lambda \ell z'_2 + m b_2 (\lambda y'_1 - \ell^{-1} y'_2 \mu - \lambda \ell^{-1} y'_2 w_1), \tag{66}$$

since $w_2 = \ell$ by (64). Successively setting one of the variables λ or μ in (66) equal to 1 and the other equal to 0 gives $(z'_1, z'_2) = (\ell^{-1} b_2 y'_2, -\frac{m\ell}{I + m\ell^2} b_2 y'_1)$. Therefore

$$U_u = \big\{ \big(A(\ell^{-1} y'_2 E), Ay', -\tfrac{m}{I + m\ell^2} b_2 y'_1 E, \begin{pmatrix} \ell^{-1} b_2 y'_2 \\ -\frac{m\ell}{I + m\ell^2} b_2 y'_1 \end{pmatrix}\big) \in H_u \big| \, y' \in \mathbb{R}^2 \big\}, \tag{67}$$

which shows that U is a distribution on D.

Next we calculate the reduced subspace $\overline{H}_b$. By definition

$$\overline{H}_{b=\rho(u)} = T_u\,\rho(U_u) = \big\{ b_2 \begin{pmatrix} \ell^{-1} y_2' \\ -\frac{m\ell}{I+m\ell^2} y_1' \end{pmatrix} \in \mathbb{R}^2 \,\big|\, y' \in \mathbb{R}^2 \big\}.$$

The last equality follows from (62) and (67). When $b_2 = 0$, it follows that $\overline{H}_b = \{0\}$. This occurs when $u = (A, a, 0, \binom{b_1}{0}) \in D$, for then $U_u = \{\big(A(\ell^{-1}y_2'E), Ay', 0, 0\big) \in H_u \big| y' \in \mathbb{R}^2\}$ and therefore $U_u \subseteq \ker T_u\rho$. Because $\dim \overline{H}_b$ changes, the mapping $\overline{H}$ given by $b \mapsto \overline{H}_b$ does *not* define a distribution on the E(2)-reduced phase space $\overline{D} = \mathbb{R}^2$, but does define a *reduced generalized distribution* on $\overline{D}$.

Finally, we compute the *reduced symplectic form* $\overline{\varpi}_{\overline{H}}$ on the reduced generalized distribution $\overline{H}$. Fix $u = (A, a, \ell^{-1}b_2E, b) \in D$. The 2-form $\overline{\varpi}_{\overline{H}}$ is defined by

$$\overline{\varpi}_{\overline{H}}(\rho(u))(T_u\,\rho(v_u), T_u\,\rho(v_u')) = \varpi_H(u)(v_u, v_u'), \tag{68}$$

where $v_u, v_u' \in U_u$. Suppose that $b_2 \neq 0$. Let

$$v_u = \big(A(\ell^{-1}y_2E), Ay, -\frac{m}{I+m\ell^2}(b_2y_1E), \begin{pmatrix} \ell^{-1}b_2y_2 \\ -\frac{m\ell}{I+m\ell^2}b_2y_1 \end{pmatrix}\big)$$

and

$$v_u = \big(A(\ell^{-1}y_2'E), Ay, -\frac{m}{I+m\ell^2}(b_2y_1'E), \begin{pmatrix} \ell^{-1}b_2y_2' \\ -\frac{m\ell}{I+m\ell^2}b_2y_1' \end{pmatrix}\big)$$

be vectors in U_u. Then

$$z = T_u\,\rho(v_u) = \begin{pmatrix} \ell^{-1}b_2y_2 \\ -\frac{m\ell}{I+m\ell^2}b_2y_1 \end{pmatrix} \text{ and } z' = T_u\,\rho(v_u') = \begin{pmatrix} \ell^{-1}b_2y_2' \\ -\frac{m\ell}{I+m\ell^2}b_2y_1' \end{pmatrix}.$$

Using (68) and (46) we get

$$\begin{aligned}
\overline{\varpi}_{\overline{H}}(b) &= \varpi_H(u)(v_u, v_u') \\
&= (\ell^{-2}I + m)(y_2z_2' - z_2y_2') + m(y_1z_1' - z_1y_1') - m\ell^{-1}b_2(y_1y_2' - y_2y_1') \\
&= (\ell^{-2}I + m)b_2^{-1}\ell(z_1z_2' - z_2z_1') + \frac{I+m\ell^2}{\ell}b_2^{-1}(z_1z_2' - z_2z_1') \\
&\qquad - \frac{I+m\ell^2}{\ell}b_2^{-1}(z_1z_2' - z_2z_1') \\
&= \frac{I+m\ell^2}{\ell}b_2^{-1}(z_1z_2' - z_2z_1').
\end{aligned} \tag{69}$$

Hence $\overline{\varpi}_{\overline{H}}$ is a symplectic form on $\mathbb{R}^2$ with coordinates (b_1, b_2) when $b_2 \neq 0$. If $b_2 = 0$, then $\overline{H}_{(b_1,0)} = \{0\}$. Thus the statement that $\overline{\varpi}_{\overline{H}}(b_1, 0)$ is nondegenerate is vacuously satisfied.

We are now in position to derive the E(2)-*reduced equations of motion* for Carathéodory's sleigh. Since the kinetic energy T (49) on D (42) is invariant under the E(2)-action Ψ (61), it induces a function

$$\overline{T}(b) = \tfrac{1}{2}\, m b_1^2 + \tfrac{1}{2}\,(m + I\ell^{-2}) b_2^2 \tag{70}$$

on the reduced phase space $\overline{D} = \mathbb{R}^2$. Because the vector field Y_T (52) on D is invariant under the E(2)-action Ψ, it induces a vector field

$$Y_{\overline{T}}(b = \rho(u)) = T_u\rho\, Y_T(u) = \begin{pmatrix} \ell^{-1} b_2^2 \\ -\frac{m\ell}{I+m\ell^2} b_1 b_2 \end{pmatrix}. \tag{71}$$

In other words, on $\mathbb{R}^2$ with coordinates (b_1, b_2) the integral curves of the E(2)-*reduced vector field* $Y_{\overline{T}}$ satisfy

$$\begin{aligned} \dot{b}_1 &= \ell^{-1} b_2^2 \\ \dot{b}_2 &= -\tfrac{m\ell}{I+m\ell^2}\, b_1 b_2 \end{aligned} \tag{72}$$

In terms of the variables $c = (c_1, c_2) = (b_1, \ell^{-1} b_2)$ equation (72) may be written as

$$\begin{aligned} \dot{c}_1 &= \ell c_2^2 \\ \dot{c}_2 &= -\tfrac{m\ell}{I+m\ell^2}\, c_1 c_2. \end{aligned} \tag{73}$$

Note that (73) holds when $\ell = 0$. This explains why the equations of motion of Carathéodory's sleigh (22) in the trivialization λ_D^{-1} depend only on c, namely, they are the E(2)-reduced equations of motion.

The distributional form of the E(2)-reduced Hamiltonian vector field on the E(2)-reduced phase space $\mathbb{R}^2$ with coordinates (b_1, b_2) corresponding to the E(2)-reduced Hamiltonian function $\overline{T}$ (70) is

$$\mathrm{d}\overline{T}(b) = Y_{\overline{T}}(b) \,\lrcorner\, \overline{\varpi}_{\overline{H}}(b), \tag{74}$$

that is, for every vector field Z on $\mathbb{R}^2$

$$\mathrm{d}\overline{T}\, Z(b) = \overline{\varpi}_{\overline{H}}(b)(Y_{\overline{T}}(b), Z(b)). \tag{75}$$

When $b_2 \neq 0$, using (69) we see that equation (75) is equivalent to

$$m b_1 Z^1 + (m + I\ell^{-2}) m b_2 Z^2 = \frac{I + m\ell^2}{\ell} b_2^{-1} (Y_{\overline{T}}^1 Z^2 - Y_{\overline{T}}^2 Z^1), \tag{76}$$

for every $Z = (Z^1, Z^2) \in \mathbb{R}^2$. Successively setting one of the variables Z^1, Z^2 equal to 1 and the other equal to 0 gives $(Y_{\overline{T}}^1(b), Y_{\overline{T}}^2(b)) = (\ell^{-1} b_2^2, -\frac{m\ell}{I+m\ell^2} b_1 b_2)$. When $b_2 = 0$ we obtain $Y_{\overline{T}}(b_1, 0) = (0, 0)$, because $Y_{\overline{T}}(b_1, 0) \in \overline{H}_{(b_1,0)} = \{0\}$. Therefore

$$Y_{\overline{T}}(b) = \ell^{-1} b_2^2 \frac{\partial}{\partial b_1} - \frac{m\ell}{I + m\ell^2} b_1 b_2 \frac{\partial}{\partial b_2}, \tag{77}$$

or in terms of the variables c_1 and c_2

$$Y_{\overline{T}}(c) = \ell c_2^2 \frac{\partial}{\partial c_1} - \frac{m\ell}{I + m\ell^2} c_1 c_2 \frac{\partial}{\partial c_2}, \tag{78}$$

as in (22) and (73).

5.3.3.1 *The* E(2)*-reduced equations using almost Poisson brackets*

On $\mathbb{R}^2$ with coordinates (c_1, c_2) we have an almost Poisson bracket $\{\ ,\ \}$ whose structure matrix is

$$\begin{pmatrix} \{c_1,c_1\} & \{c_1,c_2\} \\ \{c_2,c_1\} & \{c_2,c_2\} \end{pmatrix} = \begin{pmatrix} 0 & \frac{\ell c_2}{I+m\ell^2} \\ -\frac{\ell c_2}{I+m\ell^2} & 0 \end{pmatrix},$$

see (33). Since the E(2) reduced kinetic energy is $\overline{T} = \frac{1}{2}\, mc_1^2 + \frac{1}{2}\,(I+m\ell^2)c_2^2$, the E(2)-reduced equations of motion of Carathéodory's sleigh are

$$\begin{aligned} \dot{c}_1 &= \{c_1, \overline{T}\} = \ell\, c_2^2 \\ \dot{c}_2 &= \{c_2, \overline{T}\} = -\tfrac{m\ell}{I+m\ell^2}\, c_1 c_2. \end{aligned} \tag{79}$$

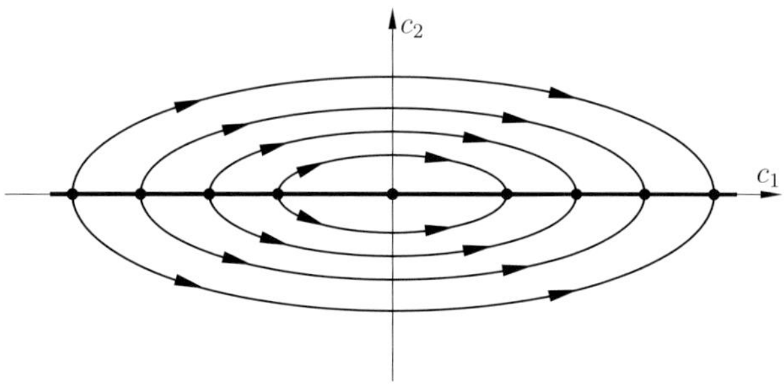

Fig. 5.2 The integral curves of the E(2)-reduced vector field $Y_{\overline{T}}$ on $\mathbb{R}^2$.

5.4 Motion on the E(2) reduced phase space

Here we discuss the qualitative properties of the E(2)-reduced motion of Carathéodory's sleigh, which is governed by the vector field

$$Y_{\overline{T}}(c_1, c_2) = \ell c_2^2 \,\frac{\partial}{\partial c_1} - \frac{m\ell}{I + m\ell^2} c_1 c_2 \,\frac{\partial}{\partial c_2}. \tag{80}$$

Theorem 5.4.7. *The* E(2)*-reduced vector field* $Y_{\overline{T}}$ (80) *on* $\mathbb{R}^2$ *has the following properties.*

A. *When* $\ell = 0$*, every point of the reduced phase space* $\mathbb{R}^2$ *is an equilibrium point of* $Y_{\overline{T}}$.

B. *When* $\ell \neq 0$,

1. *the* E(2)*-reduced kinetic energy* $\overline{T} = \frac{1}{2}\, mc_1^2 + \frac{1}{2}\,(I + m\ell^2)c_2^2$ *is an integral of* $Y_{\overline{T}}$*;*

2. *the line* $\{c_2 = 0\}$ *consists of equilibrium points of* $Y_{\overline{T}}$*;*

3. *every integral curve of* $Y_{\overline{T}}$ *is asymptotic to an equilibrium point on* $\{c_2 = 0\}$ *as* $t \rightarrow \pm\infty$*;*

4. *on* $\mathbb{R}^2 \setminus \{c_2 = 0\}$ *with symplectic form* $\mathrm{d}c_1 \wedge \mathrm{d}c_2$*, the* E(2) *reduced vector field is Hamiltonian, after rescaling time.*

Proof. A. This assertion is clear.

B. We now verify the statements.

1. The calculation

$$L_{Y_{\overline{T}}}\overline{T} = mc_1\dot{c}_1 + (I + m\ell^2)c_2\dot{c}_2 = m\ell c_1 c_2^2 - m\ell c_1 c_2^2 = 0$$

proves the assertion.

2. This follows from the definition (80) of $Y_{\overline{T}}$.

3. The c_2-axis is transverse to every positive valued level set of $\overline{T}$. Since $Y_{\overline{T}}(c_1, -c_2) = Y_{\overline{T}}(c_1, c_2)$, it suffices to look at integral curves $t \mapsto c(t) = (c_1(t), c_2(t))$ of $Y_{\overline{T}}$ of energy $\frac{1}{2}\, mh^2$, $h > 0$, which cross the positive c_2-axis at time t_0, that is, $(c_1(t_0), c_2(t_0)) = (0, \frac{hk}{\sqrt{\ell}})$, where $k^2 = \frac{m\ell}{I+m\ell^2}$. We find $c(t)$ as follows. Solving

$$\tfrac{1}{2}\, mc_1^2 + \tfrac{1}{2}\,(I + m\ell^2)c_2^2 = \tfrac{1}{2}\, mh^2 \tag{81}$$

for c_2 gives $\ell c_2^2 = k^2(h^2 - c_1^2)$. Separating variables in $\frac{\mathrm{d}c_1}{\mathrm{d}t} = \ell c_2^2 = k^2(h^2 - c_1^2)$ and integrating yields

$$c_1(t) = h \tanh(hk^2(t - t_0)) \tag{82}$$

and therefore

$$c_2(t) = \frac{hk}{\sqrt{\ell}}\operatorname{sech}(hk^2(t - t_0)), \tag{83}$$

using (81). From (82) and (83) it follows that $\lim_{t\rightarrow\pm\infty} c(t) = (\pm h, 0)$.

4. Define a new time scale s by $\frac{\mathrm{d}s}{\mathrm{d}t} = \ell c_2$. Using the time scale s the integral curves of $Y_{\overline{T}}$ on $\mathbb{R}^2 \setminus \{c_2 = 0\}$ satisfy

$$\begin{aligned}
\frac{\mathrm{d}c_1}{\mathrm{d}s} &= \frac{\mathrm{d}c_1}{\mathrm{d}t}\frac{\mathrm{d}t}{\mathrm{d}s} = \ell c_2^2\,\frac{1}{\ell c_2} = c_2 \\
\frac{\mathrm{d}c_2}{\mathrm{d}s} &= \frac{\mathrm{d}c_2}{\mathrm{d}t}\frac{\mathrm{d}t}{\mathrm{d}s} = -\frac{m\ell}{I + m\ell^2}c_1c_2\frac{1}{\ell c_2} = -\lambda^2 c_1,
\end{aligned}$$

which is Hamiltonian on $(\mathbb{R}^2 \setminus \{c_2 = 0\}, \mathrm{d}c_1 \wedge \mathrm{d}c_2)$ with Hamiltonian $H(c_1, c_2) = \frac{1}{2}\lambda^2 c_1^2 + \frac{1}{2}c_2^2$. The Hamiltonian vector field X_H on $\mathbb{R}^2 \setminus \{c_2 = 0\}$ is *incomplete*. To see this consider an integral curve $s \mapsto \gamma(s)$, which starts at $(0, \lambda h)$. Then $\gamma(s) = (c_1(s), c_2(s)) = (h \sin \lambda s, \lambda h \cos \lambda s)$. When $s = \pi/2\lambda$, we have $c_2(\pi/2\lambda) = 0$. Thus γ runs off $\mathbb{R}^2 \setminus \{c_2 = 0\}$ in *finite* time. □

5.5 Reconstruction

In this section we reconstruct the motion of Carathéodory's sleigh from the E(2)-reduced motion, which we have described in §4.

5.5.1 *Relative equilibria*

First we assume that $\ell = 0$. Then every point (c_1, c_2) of the reduced space $\mathbb{R}^2$ is an equilbrium point of the E(2)-reduced vector field $Y_{\overline{T}}(c)$ (78).

Therefore the reconstructed motion of the sleigh is given by the action Φ (57) on the constraint distribution D restricted to a one parameter subgroup of E(2). Every one parameter subgroup of E(2) is either a translation of $\mathbb{R}^2$ with constant velocity or a rotation about some fixed point p in $\mathbb{R}^2$ with constant rotational velocity $\dot{\theta} = c_2 \neq 0$.

Suppose that $\dot{\theta} = c_2 = 0$. Then θ is constant and equation (17) reads

$$\frac{\mathrm{d}}{\mathrm{d}t}\begin{pmatrix} x_1 \\ x_2 \end{pmatrix} = \begin{pmatrix} \cos\theta & -\sin\theta \\ \sin\theta & \cos\theta \end{pmatrix}\begin{pmatrix} c_1 \\ 0 \end{pmatrix} = c_1 \begin{pmatrix} \cos\theta \\ \sin\theta \end{pmatrix}.$$

Therefore

$$\begin{pmatrix} x_1(t) \\ x_2(t) \end{pmatrix} = tc_1 \begin{pmatrix} \cos\theta \\ \sin\theta \end{pmatrix} + \begin{pmatrix} x_1(0) \\ x_2(0) \end{pmatrix}.$$

We determine the point p as follows. Let ξ be a fixed point of the reference sleigh. The horizontal projection of the velocity $v = \dot{A}\xi + \dot{a}$ of the moving sleigh is

$$\begin{aligned} v_{\mathrm{hor}} &= \frac{\mathrm{d}}{\mathrm{d}t}\begin{pmatrix} \cos\theta & -\sin\theta \\ \sin\theta & \cos\theta \end{pmatrix}\begin{pmatrix} \xi_1 \\ \xi_2 \end{pmatrix} + \frac{\mathrm{d}}{\mathrm{d}t}\begin{pmatrix} x_1 \\ x_2 \end{pmatrix} = \dot{\theta}\begin{pmatrix} -\sin\theta & -\cos\theta \\ \cos\theta & -\sin\theta \end{pmatrix}\begin{pmatrix} \xi_1 \\ \xi_2 \end{pmatrix} + \begin{pmatrix} \dot{x}_1 \\ \dot{x}_2 \end{pmatrix} \\ &= \dot{\theta}\begin{pmatrix} -\sin\theta & -\cos\theta \\ \cos\theta & -\sin\theta \end{pmatrix}\begin{pmatrix} \xi_1 \\ \xi_2 \end{pmatrix} + \begin{pmatrix} \cos\theta & -\sin\theta \\ \sin\theta & \cos\theta \end{pmatrix}\begin{pmatrix} c_1 \\ \ell c_2 \end{pmatrix}, \end{aligned} \tag{84}$$

using (17). If $\dot{\theta} \neq 0$, then $v = (v_{\mathrm{hor}}, 0) = 0$ if and only if

$$p = \begin{pmatrix} \xi_1 \\ \xi_2 \end{pmatrix} = \dot{\theta}^{-1} \begin{pmatrix} \sin\theta & -\cos\theta \\ \cos\theta & \sin\theta \end{pmatrix} \begin{pmatrix} \cos\theta & -\sin\theta \\ \sin\theta & \cos\theta \end{pmatrix} \begin{pmatrix} c_1 \\ \ell c_2 \end{pmatrix}$$
$$= \dot{\theta}^{-1} \begin{pmatrix} 0 & -1 \\ 1 & 0 \end{pmatrix} \begin{pmatrix} c_1 \\ \ell c_2 \end{pmatrix} = \begin{pmatrix} -\ell \\ c_2^{-1} c_1 \end{pmatrix}. \tag{85}$$

This means that the sleigh either 1) rigidly translates with constant velocity along a straight line through the horizontal projection of its center of mass or 2) rotates rigidly with constant angular velocity about the point p, which lies on line through the point of contact P of the sleigh and perpendicular to the chisel edge. Case 1) occurs when $\dot{\theta} = c_2 = 0$ and case 2) when $\dot{\theta} = c_2 \neq 0$.

Now assume that $\ell \neq 0$ and suppose that (c_1, c_2) lies on the c_1-axis, which consists of equilibrium points of the E(2)-reduced vector field. From $\dot{\theta} = c_2 = 0$ it follows that θ is constant. Then $\dot{c}_1 = c_2 = 0$, which implies that c_1 is constant. Hence

$$\frac{\mathrm{d}}{\mathrm{d}t} \begin{pmatrix} x_1 \\ x_2 \end{pmatrix} = \begin{pmatrix} \cos\theta & -\sin\theta \\ \sin\theta & \cos\theta \end{pmatrix} \begin{pmatrix} c_1 \\ 0 \end{pmatrix} = c_1 \begin{pmatrix} \cos\theta \\ \sin\theta \end{pmatrix}.$$

Therefore the reconstructed motion of the center of mass of the sleigh corresponding to the relative equilibrium $(c_1, 0)$ is rectilinear with constant velocity if $c_1 \neq 0$ or is stationary if $c_1 = 0$.

5.5.2 *General motions*

Now suppose that $c_2 \neq 0$ in addition to $\ell \neq 0$. From (82) and (83) we find that $t \mapsto (c_1(t), c_2(t))$, where

$$\begin{aligned} c_1(t) &= \ell\kappa c_2(0) \tanh rt \\ c_2(t) &= c_2(0)\,\mathrm{sech}\, rt, \end{aligned} \tag{86}$$

is an integral curve of the E(2)-reduced vector field with starting point $(0, c_2(0))$. Here $\kappa^2 = 1 + \frac{I}{m\ell^2}$, $r = \pm\frac{c_2(0)}{\kappa}$, and $\pm$ is the sign of $c_2(0)$.[2] Integrating $\dot{\theta} = c_2$, using the second equation in (86), gives

$$\begin{aligned} \theta(t) - \theta(0) &= \pm\kappa r \int_0^t \frac{1}{\mathrm{sech}\, rt}\,\mathrm{d}t = \pm 2\kappa r \int_0^t \frac{1}{\mathrm{e}^{rs} + \mathrm{e}^{-rs}}\,\mathrm{d}s \\ &= \pm 2\kappa r \int_1^{\mathrm{e}^{rt}} \frac{1}{\eta + \eta^{-1}} \frac{1}{r\eta}\,\mathrm{d}\eta, \quad \text{where } \eta = \mathrm{e}^{rt} \\ &= \pm 2\kappa \int_1^{\mathrm{e}^{rt}} \frac{1}{1+\eta^2}\,\mathrm{d}\eta = \pm 2\kappa \left(\tan^{-1}(\mathrm{e}^{rt}) - \pi/4\right). \end{aligned} \tag{87}$$

[2] Compared to (82) and (83) we have $r = hk^2$, $\kappa^2 = \frac{1}{\ell k^2}$, and $c_2(0) = \frac{hk}{\sqrt{\ell}}$.

From (87) we obtain

$$\theta(+\infty) - \theta(-\infty) = \pm\kappa[(\pi/2 - \pi/4) - (0 - \pi/4)] = \pm\kappa\pi. \tag{88}$$

Substituting (86) and (87) into

$$\dot{x} = \frac{\mathrm{d}}{\mathrm{d}t}\begin{pmatrix} x_1 \\ x_2 \end{pmatrix} = \begin{pmatrix} \cos\theta & -\sin\theta \\ \sin\theta & \cos\theta \end{pmatrix}\begin{pmatrix} c_1 \\ \ell c_2 \end{pmatrix} \tag{89}$$

we get an explicit formula for $\dot{x}$ as a function of t. Integrating gives x as a function of t. For $t \to \pm\infty$, we see that $\dot{x}(t)$ converges exponentially fast to

$$\dot{x}(\pm\infty) = \pm\ell|\dot{\theta}(0)|(\cos\theta(\pm\infty), \sin\theta(\pm\infty)).$$

Thus asymptotically the center of mass moves along a straight line with constant velocity. These *asymptotic motions* are *relative equilibria.*

We retain the hypothesis $c_2 \neq 0$. For a given kinetic energy, all the E(2)-reduced motions can be mapped onto each other by a time translation together with a time reversal $(c_1, c_2, t) \mapsto (c_1, -c_2, -t)$ if the signs of c_2 differ. This implies that the reconstructed motions are mapped to each other either by a time translation (and possibly a time reversal) and an application of the E(2) symmetry group on the constraint distribution. Because the E(2)-reduced solutions of different kinetic energy are mapped onto each other by a constant time rescaling $t \mapsto \tau t$, we obtain

Proposition 5.5.2.8. *For given values of ℓ, m, and I, every trajectory of the center of mass and the point of contact of the sleigh is congruent to another such trajectory, except for the relative equilibria corresponding to $(c_1, 0)$. When $c_1 \neq 0$, the motion is asymptotic to trajectories with $c_2 \neq 0$; while when $c_1 = 0$ the sleigh is stationary.*

5.5.3 *Motion of a material point on the sleigh*

In this subsection we describe the motion of a material point on the sleigh. Throughout we assume that $\ell \neq 0$ and $c_2 \neq 0$.

Let ξ be a material point on the reference sleigh. The curve

$$\gamma : \mathbb{R} \to \mathbb{R}^3 : t \mapsto \gamma(t) = A(t)\xi + a(t) \tag{90}$$

gives the motion of ξ on the sleigh with respect to a spatially fixed coordinate system. Since the center of mass of the moving sleigh has a constant height above the horizontal plane, and the matrix $A(t) = \begin{pmatrix} R_{\theta(t)} & 0 \\ 0 & 1 \end{pmatrix}$ is a rotation around the vertical axis, $(A(t)\xi + a(t))_3$ is constant. Therefore we need only consider the curve

$$\gamma_{\rm hor} : \mathbb{R} \to \mathbb{R}^2 : t \mapsto \gamma(t)_{\rm hor},$$

where $\gamma(t)_{\rm hor}$ is the horizontal projection of $\gamma(t)$.

Recall that in §5.1 we have shown that the horizontal component of the velocity of the moving material point ξ is

$$\begin{aligned} v_{\rm hor} = \dot{\gamma}_{\rm hor} &= R_{\theta(t)}\xi_{\rm hor} + \dot{x}(t) \\ &= -\dot{\theta}\begin{pmatrix} \sin\theta & \cos\theta \\ -\cos\theta & \sin\theta \end{pmatrix}\begin{pmatrix} \xi_1 \\ \xi_2 \end{pmatrix} + \begin{pmatrix} \cos\theta & -\sin\theta \\ \sin\theta & \cos\theta \end{pmatrix}\begin{pmatrix} c_1 \\ \ell c_2 \end{pmatrix}. \end{aligned} \tag{91}$$

Hence the velocity of ξ is 0 if and only if $v_{\rm hor} = 0$, that is, if and only if $\xi_1 = -\ell$ and $c_2\xi_2 - c_1 = 0$ with $c_2 = \dot{\theta} \neq 0$.

Now every solution of the E(2)-reduced equations of motion

$$\begin{aligned} \dot{c}_1 &= \ell c_2^2 \\ \dot{c}_2 &= -\tfrac{m\ell}{I+m\ell^2}\, c_1 c_2 \end{aligned} \tag{92}$$

intersects the line $c_2\xi_2 - c_1 = 0$ at a unique time t_0. Therefore the curve $\gamma_{\rm hor}$ is a smooth immersion of $\mathbb{R} \setminus \{t_0\}$ into $\mathbb{R}^2$ with a singularity at $t = t_0$.

Let $\left\{\mu = \begin{pmatrix} \cos\theta(t) \\ \sin\theta(t) \end{pmatrix}, \ \nu = \begin{pmatrix} -\sin\theta(t) \\ \cos\theta(t) \end{pmatrix}\right\}$ be a time dependent basis of $\mathbb{R}^2$. Then

$$\dot{\mu} = \dot{\theta}\,\nu = c_2(t)\,\nu \ \text{ and } \ \dot{\nu} = -\dot{\theta}\,\mu = -c_2(t)\,\mu. \tag{93}$$

To determine the nature of the singularity of $\gamma_{\rm hor}$ at $t = t_0$ we take the first and second derivatives of

$$\langle \dot{\gamma}_{\rm hor}, \mu \rangle = -c_2(t)\xi_2 + c_1(t) \ \text{ and } \ \langle \dot{\gamma}_{\rm hor}, \nu \rangle = 0 \tag{94}$$

and then evaluate at $t = t_0$, when $\gamma_{\rm hor}$ crosses the line $c_2\xi_2 - c_1 = 0$. First we verify (94). From (91) and the definition of μ we get

$$\begin{aligned} \langle \dot{\gamma}_{\rm hor}, \mu \rangle &= \langle -\dot{\theta}\begin{pmatrix} \sin\theta & \cos\theta \\ -\cos\theta & \sin\theta \end{pmatrix}\begin{pmatrix} \xi_1 \\ \xi_2 \end{pmatrix} + \begin{pmatrix} \cos\theta & -\sin\theta \\ \sin\theta & \cos\theta \end{pmatrix}\begin{pmatrix} c_1 \\ \ell c_2 \end{pmatrix}, \begin{pmatrix} \cos\theta \\ \sin\theta \end{pmatrix}\rangle \\ &= \langle -\dot{\theta}\begin{pmatrix} \cos\theta & \sin\theta \\ -\sin\theta & \cos\theta \end{pmatrix}\begin{pmatrix} \sin\theta & \cos\theta \\ -\cos\theta & \sin\theta \end{pmatrix}\begin{pmatrix} \xi_1 \\ \xi_2 \end{pmatrix} + \begin{pmatrix} c_1 \\ \ell c_2 \end{pmatrix}, \begin{pmatrix} \cos\theta & \sin\theta \\ -\sin\theta & \cos\theta \end{pmatrix}\begin{pmatrix} \cos\theta \\ \sin\theta \end{pmatrix}\rangle \\ &= \langle \dot{\theta}\begin{pmatrix} 0 & -1 \\ 1 & 0 \end{pmatrix}\begin{pmatrix} \xi_1 \\ \xi_2 \end{pmatrix} + \begin{pmatrix} c_1 \\ \ell c_2 \end{pmatrix}, \begin{pmatrix} 1 \\ 0 \end{pmatrix}\rangle = -\dot{\theta}\xi_2 + c_1(t) = -c_2(t)\xi_2 + c_1(t) \end{aligned}$$

and similarly

$$\langle \dot{\gamma}_{\rm hor}, \nu \rangle = \dot{\theta}\xi_1 + \ell c_2(t) = c_2(t)(\xi_1 + \ell) = 0.$$

Thus (94) holds. Now

$$\frac{\mathrm{d}}{\mathrm{d}t}\langle\dot\gamma_{\mathrm{hor}},\mu\rangle = -\dot c_2\xi_2 + \dot c_1 = \frac{m\ell}{I+m\ell^2}\,\xi_2 c_1(t)c_2(t) + \ell c_2^2(t),$$

using (92), which evaluated at $t = t_0$ gives

$$\begin{aligned}\frac{\mathrm{d}}{\mathrm{d}t}\langle\dot\gamma_{\mathrm{hor}},\mu\rangle(t_0) &= \frac{m\ell}{I+m\ell^2}\,c_1(t_0)c_2^{-1}(t_0)c_1(t_0)c_2(t_0) + \ell c_2^2(t_0)\\ &\qquad \text{since } c_2(t_0)\xi_2 - c_1(t_0) = 0\\ &= \frac{m\ell}{I+m\ell^2}\,c_1^2(t) + \ell\, c_2^2(t_0) > 0.\end{aligned}$$

But

$$\begin{aligned}\frac{\mathrm{d}}{\mathrm{d}t}\langle\dot\gamma_{\mathrm{hor}},\mu\rangle &= \langle\ddot\gamma_{\mathrm{hor}},\mu\rangle + \langle\dot\gamma_{\mathrm{hor}},\dot\mu\rangle = \langle\ddot\gamma_{\mathrm{hor}},\mu\rangle + \langle\dot\gamma_{\mathrm{hor}},c_2(t)\,\nu\rangle\\ &= \langle\ddot\gamma_{\mathrm{hor}},\mu\rangle, \quad \text{using the second equation in (94).}\end{aligned}$$

Therefore

$$\langle\ddot\gamma_{\mathrm{hor}}(t_0),\mu(t_0)\rangle > 0. \tag{95}$$

Differentiating $\langle\dot\gamma_{\mathrm{hor}},\nu\rangle = 0$, we get

$$\begin{aligned}0 &= \frac{\mathrm{d}}{\mathrm{d}t}\langle\dot\gamma_{\mathrm{hor}},\nu\rangle = \langle\ddot\gamma_{\mathrm{hor}},\nu\rangle + \langle\dot\gamma_{\mathrm{hor}},\dot\nu\rangle = \langle\ddot\gamma_{\mathrm{hor}},\nu\rangle - c_2(t)\,\langle\dot\gamma_{\mathrm{hor}},\mu\rangle\\ &= \langle\ddot\gamma_{\mathrm{hor}},\nu\rangle - c_2(t)\big(c_1(t) - \xi_2 c_2(t)\big),\end{aligned} \tag{96}$$

which evaluated at $t = t_0$ gives

$$\langle\ddot\gamma_{\mathrm{hor}}(t_0),\nu(t_0)\rangle = 0. \tag{97}$$

In other words, $\ddot\gamma_{\mathrm{hor}}(t_0)$ is a multiple of $\mu(t_0)$. Differentiating (96) once again gives

$$\begin{aligned}0 &= \langle\dddot\gamma_{\mathrm{hor}},\nu\rangle + \langle\ddot\gamma_{\mathrm{hor}},\dot\nu\rangle - \dot c_2\,\langle\dot\gamma_{\mathrm{hor}},\mu\rangle - c_2(t)\langle\ddot\gamma_{\mathrm{hor}},\mu\rangle - c_2(t)\langle\dot\gamma_{\mathrm{hor}},\dot\mu\rangle\\ &= \langle\dddot\gamma_{\mathrm{hor}},\nu\rangle - 2c_2(t)\langle\ddot\gamma_{\mathrm{hor}},\mu\rangle + \frac{m\ell}{I+m\ell^2}c_1(t)c_2(t)\,\langle\dot\gamma_{\mathrm{hor}},\mu\rangle - c_2^2(t)\,\langle\dot\gamma_{\mathrm{hor}},\nu\rangle,\end{aligned}$$

which, because $\langle\dot\gamma_{\mathrm{hor}}(t_0),\nu(t_0)\rangle = 0 = \langle\dot\gamma_{\mathrm{hor}}(t_0),\mu(t_0)\rangle$, becomes

$$\langle\dddot\gamma_{\mathrm{hor}}(t_0),\nu(t_0)\rangle = 2c_2(t_0)\,\langle\ddot\gamma_{\mathrm{hor}}(t_0),\mu(t_0)\rangle \neq 0,$$

using (95). Therefore $\dddot\gamma_{\mathrm{hor}}(t_0)$ and $\ddot\gamma_{\mathrm{hor}}(t_0)$ are linearly independent. This together with $\dot\gamma_{\mathrm{hor}}(t_0) = 0$ proves

Proposition 5.5.3.9. *The planar curve* $\gamma_{\mathrm{hor}} : \mathbb{R} \to \mathbb{R}^2 : t \mapsto \gamma_{\mathrm{hor}}(t)$ *has an ordinary cusp singularity at* $t = t_0$ *when the integral curve of the* $\mathrm{E}(2)$*-reduced vector field* (92) *crosses the line* $c_2\xi_2 - c_1 = 0$.

Proposition 5.5.3.10. *The curve $\mathbb{R} \to \mathbb{R}^2 : t \mapsto P(t)$ traced by the point of contact of the sleigh with the horizontal plane has self intersections if $\theta(+\infty) - \theta(-\infty)$ is larger than 2π, that is, if $\kappa^2 = 1 + \frac{I}{m\ell^2} > 2$.*

Proof. This follows because asymptotically the center of mass of the sleigh moves along straight lines, which subtend an angle $\theta(+\infty) - \theta(-\infty)$. When this angle is larger than 2π the pursuit curve traced by the point of contact, which tracks the center of mass, must have self intersections. □

When $\kappa = 1$, then $I = 0$, that is, the mass of the sleigh is concentrated along a vertical straight line through its center of mass. Differentiating

$$\frac{\mathrm{d}}{\mathrm{d}t}\begin{pmatrix} x_1 \\ x_2 \end{pmatrix} = \begin{pmatrix} \cos\theta & -\sin\theta \\ \sin\theta & \cos\theta \end{pmatrix}\begin{pmatrix} c_1 \\ \ell c_2 \end{pmatrix}$$

and using (92) gives

$$\begin{aligned} \ddot{x}_1 &= -\dot{\theta} c_2 \sin\theta + \dot{c}_1 \cos\theta - \dot{\theta}\ell c_2 \cos\theta - \ell \dot{c}_2 \sin\theta \\ &= -c_1 c_2 \sin\theta + \ell c_2^2 \cos\theta - \ell c_2^2 \cos\theta - \kappa c_1 c_2 \sin\theta = 0 \end{aligned}$$

and

$$\begin{aligned} \ddot{x}_2 &= \dot{\theta} c_1 \cos\theta + \dot{c}_1 \sin\theta - \dot{\theta}\ell c_2 \sin\theta + \ell \dot{c}_2 \cos\theta \\ &= c_1 c_2 \cos\theta + \ell c_2^2 \sin\theta - \ell c_2^2 \sin\theta + \kappa c_1 c_2 \cos\theta = 0. \end{aligned}$$

In other words, the center of mass of the sleigh moves in a straight line with constant velocity. Because the point of contact follows a pursuit curve relative to the center of mass, we deduce that the point of contact curve is a tratrix.

5.6 Notes

In §6 of [24] Chaplygin treated the skate for the first time. Chaplygin's skate is a massless wheel whose horizontal axle is rigidly attached to a rigid body like a pizza knife. It is assumed that the wheel has a sharp edge which prevents it from slipping sideways in a direction perpendicular to the plane of the edge. In [21] Carathéodory published a detailed investigation of a special case of Chaplygin's skate, called Carathéodory's sleigh. Here he assumed that the center of mass of the sleigh lies in the plane of the chisel edge. He did not mention Chaplygin.

We now show that the dynamics of Chaplygin's skate and Carathéodory's sleigh are the same. Assume that in the reference position the center of mass of the skate lies over the origin of $\mathbb{R}^2$, that is, $x = (0, 0)$.

Moreover, assume that in the reference position the sharp edge of the wheel of the skate is parallel to the e_1–e_3 plane. Write $-(\ell, d)$ for the coordinates of the point of contact Q of the sharp edge of the skate in the reference position with the horizontal plane. Because

$$Q = -R_\theta \begin{pmatrix} \ell \\ d \end{pmatrix} + x = \begin{pmatrix} -\ell\cos\theta + d\sin\theta + x_1 \\ -\ell\sin\theta - d\cos\theta + x_2 \end{pmatrix},$$

the nonholonomic constraint for the skate is:

$$\dot{Q} = \begin{pmatrix} (\ell\sin\theta + d\cos\theta)\dot{\theta} + \dot{x}_1 \\ (-\ell\cos\theta + d\sin\theta)\dot{\theta} + \dot{x}_2 \end{pmatrix}$$

is a multiple of $u_\theta = \binom{\cos\theta}{\sin\theta}$. This condition is equivalent to

$$0 = \det(\dot{Q}, u_\theta) = \ell\dot{\theta} + \dot{x}_1\sin\theta - \dot{x}_2\cos\theta, \qquad (98)$$

which is the same as the nonholonomic constraint (2) for Carathéodory's sleigh. Because the equations of motion of the skate are given in terms of its kinetic energy and the nonholonomic constraint (98), the equations of motion of the skate depend only on the parmeters ℓ, m, and I. Thus the dynamics of the skate and the sleigh are the same, where the point of contact of the chisel edge of the sleigh lies on a line perpendicular to the sharp edge of the skate and through the center of rotation p of the skate, see (85), when $\dot{\theta} \neq 0$.

Chaplygin's skate is discussed in chapter II §4 of Neimark and Fufaev [82] who in a footnote on page 71 noted that "Chaplygin stated and solved in quadratures the problem [Chaplygin's skate] in this section in 1911". Several classic sources treat Chaplygin's skate without mentioning Chaplygin, for instance, pages 465–470 of Hamel [51] and page 334 of Rosenberg [95]. For a modern treatment see Bates [10] and page 126 in Koiller [59].

Our treatment of removing the E(2)-symmetry of Carathéodory's sleigh, which is a special case of regular nonholonomic reduction treated in chapter 2, is new, see also [34].

The relative equilibria with $\ell = 0$ in §5.1 are described in Chaplygin [24]. Equation (88) was obtained by Carathéodory [21]. Integrating (89) completes the integration by quadratures of Chaplygin's skate, and thereby Carathéodory's sleigh, as observed by Chaplygin [24]. Proposition 5.5.2.8 is due to Carathéodory [21]. In [21] Carathéodory illustrated the conclusion of proposition 5.5.3.10 by giving pictures of the curve traced out by the point of contact of the sleigh when $\kappa = 1,\ 1,5,\ 2$, and 3. The fact that when $\kappa = 1$ the point of contact of the sleigh traces out a tratrix was known to Carathéodory, see [21].

Chaplygin's skate is often called a dynamic hatchet planimeter. See Foote [45] for an explanation how one can use a hatchet planimeter to measure the area enclosed by a simple planar curve.

Chapter 6

Smooth strongly convex rolling rigid body

In this chapter we treat the example of a smooth strongly convex rigid body rolling without slipping on a horizontal plane under the influence of a constant vertical gravitational force. We will use traditional notation, which sometimes clashes with that of the preceding chapters.

6.1 Basic set up

Let $\mathcal{B}$ be a *reference body* in a fixed reference frame in $\mathbb{R}^3$. Assume that its mass distribution is given by a nonnegative finite measure dm such that $\mathrm{m} = \int_{\mathcal{B}} dm > 0$ is the *mass* of $\mathcal{B}$. Suppose that the *center of mass* of $\mathcal{B}$ is the origin in $\mathbb{R}^3$, that is,

$$\int_{\mathcal{B}} x\, dm = 0. \tag{1}$$

The position $x(t)$ of the point x on the *moving body* at time t is given by applying a Euclidean motion

$$x \mapsto A(t)x + a(t) = x(t)$$

to $x \in \mathcal{B}$. Here $A(t) \in \mathrm{SO}(3)$ is a rotation and $a(t) \in \mathbb{R}^3$ is a translation vector. In other words, the *configuration space* of the moving body is identified with the 3-dimensional group $\mathrm{E}(3)$ of *Euclidean motions*

$$(A, a) : \mathbb{R}^3 \rightarrow \mathbb{R}^3 : x \mapsto Ax + a. \tag{2}$$

In particular, as a set $\mathrm{E}(3)$ equals $\mathrm{SO}(3) \times \mathbb{R}^3$, but as a group $\mathrm{E}(3)$ has multiplication defined by

$$(A, a) \cdot (A', a') = (AA', Aa' + a), \quad (A, a), (A', a') \in \mathrm{SO}(3) \times \mathbb{R}^3. \tag{3}$$

E(3) is a Lie group with Lie algebra $\mathrm{e}(3) = \{(X, b) \in \mathrm{so}(3) \times \mathbb{R}^3\}$, whose Lie bracket is

$$[(X, b), (X', b')] = ([X, X'], Xb' - X'b).^1 \tag{4}$$

The bracket $[\,,\,]$ on the right hand side of (4) is the Lie bracket in so(3), that is, $[X, X'] = XX' - X'X$.

Denoting differentiation with respect to time by $\dot{}$, we see that $\dot{x} = \dot{A}x + \dot{a}$ is the velocity of a point on the moving body with respect to the fixed reference frame. Thus the *kinetic energy* of the moving body is

$$T = \tfrac{1}{2} \int_{\mathcal{B}} \langle \dot{A}x + \dot{a}, \dot{A}x + \dot{a} \rangle \, dm. \tag{5}$$

Here $\langle\,,\,\rangle$ is the standard Euclidean inner product on $\mathbb{R}^3$. Because the center of mass of the reference body is the origin, we find that $T = T_{\mathrm{rot}} + T_{\mathrm{trans}}$, where the *rotational kinetic energy* is

$$T_{\mathrm{rot}} = \tfrac{1}{2} \int_{\mathcal{B}} \langle \dot{A}x, \dot{A}x \rangle \, dm \; = \; \tfrac{1}{2} \int_{\mathcal{B}} \langle A^{-1}\dot{A}x, A^{-1}\dot{A}x \rangle \, dm \tag{6}$$

and the *translational kinetic energy* is

$$T_{\mathrm{trans}} = \tfrac{1}{2} \int_{\mathcal{B}} \langle \dot{a}, \dot{a} \rangle \, dm \; = \; \tfrac{1}{2} \int_{\mathcal{B}} \langle A^{-1}\dot{a}, A^{-1}\dot{a} \rangle \, dm \; = \; \tfrac{1}{2} \, \mathrm{m} \, \langle b, b \rangle. \tag{7}$$

Here $b = A^{-1}\dot{a} \in \mathbb{R}^3$ and m is the mass of the reference body.

We identify the 3×3 skew symmetric matrix $\omega^{\times} = A^{-1}\dot{A}$ with the vector $\omega \in \mathbb{R}^3$ using the mapping

$$i : \mathrm{so}(3) \rightarrow \mathbb{R}^3 : \omega^{\times} = \begin{pmatrix} 0 & -\omega_3 & \omega_2 \\ \omega_3 & 0 & -\omega_1 \\ -\omega_2 & \omega_1 & 0 \end{pmatrix} \mapsto \omega = \begin{pmatrix} \omega_1 \\ \omega_2 \\ \omega_3 \end{pmatrix}, \tag{8}$$

whose inverse is

$$^{\times} : \mathbb{R}^3 \rightarrow \mathrm{so}(3) : \omega \mapsto \omega^{\times}. \tag{9}$$

[1] Equation (4) is obtained by first differentiating

$$(A, a) \cdot (\exp tX', tb') \cdot (A, a)^{-1} = (A \exp tX' A^{-1}, -(A \exp tX' A^{-1})a + tAb' + a)$$

with respect to t and then evaluating at $t = 0$. This gives

$$\mathrm{Ad}_{(A,a)}(X', b') = (\mathrm{Ad}_A X', -(\mathrm{Ad}_A X')\, a + Ab').$$

Replacing (A, a) in the above equation by $(\exp tX, tb)$ and then differentiating the result with respect to t and finally evaluating at $t = 0$ gives

$$[(X, b), (X', b')] = \mathrm{ad}_{(X,b)}(X', b') \; = \; ([X, X'], Xb' - X'b).$$

Fact 6.1.1. *The maps i and $^\times$ have the following properties.*

1. $i : (\mathrm{so}(3), \langle\langle\, , \,\rangle\rangle) \to (\mathbb{R}^3, \langle\, , \,\rangle)$ *is an isometry.* $\langle\langle\, , \,\rangle\rangle$ *is the Killing metric on* $\mathrm{so}(3)$ *defined by* $\langle\langle X, Y\rangle\rangle = -\frac{1}{2}\,\mathrm{tr}(XY)$, *for every* $X, Y \in \mathrm{so}(3)$. *Equivalently,* $\langle x, y\rangle = \langle\langle x^\times, y^\times\rangle\rangle$.
2. *For every* $(\omega^\times, x) \in \mathrm{so}(3) \times \mathbb{R}^3$, *we have* $\omega^\times x = \omega \times x$. *Here* $\times$ *is the vector product on* $\mathbb{R}^3$.
3. *For every* $(A, \omega^\times) \in \mathrm{SO}(3) \times \mathrm{so}(3)$, *we have* $\mathrm{Ad}_A \omega^\times = A\omega^\times A^{-1} = (A\omega)^\times$.
4. *For every* $x, y \in \mathbb{R}^3$, *we have* $(x \times y)^\times = [x^\times, y^\times]$.

Using the *Grassmann identity*

$$\xi \times (\eta \times \zeta) = \langle \xi, \zeta\rangle\, \eta - \langle \xi, \eta\rangle\, \zeta \tag{10}$$

for vectors ξ, η, $\zeta \in \mathbb{R}^3$, we see that equation (6) can be written as

$$T_{\mathrm{rot}} = \tfrac{1}{2} \int_{\mathcal{B}} \langle \omega \times x, \omega \times x\rangle\, dm = \tfrac{1}{2} \int_{\mathcal{B}} \big(\langle \omega, \omega\rangle \langle x, x\rangle - \langle \omega, x\rangle^2\big)\, dm, \tag{11}$$

because

$$\begin{aligned}\langle \omega \times x, \omega \times x\rangle &= \langle \omega, x \times (\omega \times x)\rangle \\ &= \langle \omega, \langle x, x\rangle \omega - \langle x, \omega\rangle x\rangle = \langle \omega, \omega\rangle\, \langle x, x\rangle - \langle \omega, x\rangle^2 .\end{aligned}$$

Let $M = \int_{\mathcal{B}} x \otimes x\, dm$ be the tensor of second moments of the mass distribution dm. As a 3×3 semidefinite symmetric matrix, M has entries $M_{ij} = \int_{\mathcal{B}} x_i x_j\, dm$. Let $I = \mathrm{tr}(M)\,\mathrm{id} - M$. Then

$$\begin{aligned}\langle I(\omega), \omega\rangle &= \mathrm{tr}(M)\langle \omega, \omega\rangle - \langle M\omega, \omega\rangle \\ &= \int_{\mathcal{B}} \big(\langle x, x\rangle \langle \omega, \omega\rangle - \langle (x \otimes x)\omega, \omega\rangle\big)\, dm \\ &= \int_{\mathcal{B}} \big(\langle \omega, \omega\rangle\, \langle x, x\rangle - \langle \omega, x\rangle^2\big)\, dm = 2\, T_{\mathrm{rot}}.\end{aligned} \tag{12}$$

I is the *moment of inertia tensor* of $\mathcal{B}$. In a suitable orthonormal basis, which defines the *principal axis frame*, $M = \mathrm{diag}(M_1, M_2, M_3)$, where $M_j \ge 0$ for $j = 1, 2, 3$. Hence $I = \mathrm{diag}(I_1, I_2, I_3)$, where $I_1 = M_2 + M_3$, $I_2 = M_1 + M_3$, and $I_3 = M_1 + M_2$. The I_j, $j = 1, 2, 3$ are the *principal moments of inertia*. They satisfy the inequalities

$$0 \le I_1 \le I_2 + I_3, \quad 0 \le I_2 \le I_1 + I_3, \quad \text{and} \quad 0 \le I_3 \le I_1 + I_2. \tag{13}$$

To avoid I being singular we will assume that $I_j > 0$ for $j = 1, 2, 3$. This means that the mass of $\mathcal{B}$ is not concentrated along a straight line through its center of mass.

The potential energy of the moving body in a constant vertical gravitation field of strength g is

$$V(A, a) = \int_{\mathcal{B}} \mathrm{g}\, \langle Ax + a, e_3\rangle\, dm = \mathrm{mg}\, \langle a, e_3\rangle, \tag{14}$$

where e_3 is the third standard basis vector of $\mathbb{R}^3$.

6.2 Unconstrained motion

To describe the unconstrained rolling rigid body, we specify its Lagrangian ℓ. As usual ℓ is a function on the velocity space $T\mathrm{E}(3)$ of the Euclidean group $\mathrm{E}(3)$ and, as expected, it is the difference of the kinetic and potential energies. Letting (A, a) be coordinates on $\mathrm{E}(3)$ and $\big((A, a), (\dot{A}, \dot{a})\big)$ coordinates on $T\mathrm{E}(3)$, the *Lagrangian* is

$$\ell((A, a), (\dot{A}, \dot{a})) = \tfrac{1}{2}\,\langle\langle \mathcal{I}(\omega^\times), \omega^\times\rangle\rangle - \mathrm{mg}\,\langle a, e_3\rangle, \tag{15}$$

where $\omega^\times = A^{-1}\dot{A}$. Here $\mathcal{I} = i^{-1}\circ I\circ i : \mathrm{so}(3) \to \mathrm{so}(3)$ is the *generalized moment of inertia tensor* on $\mathrm{so}(3)$. The *trivialization* of $T\mathrm{E}(3)$ we use is the mapping

$$\begin{aligned} &T\mathrm{E}(3) \to \mathrm{E}(3) \times (\mathbb{R}^3 \times \mathbb{R}^3) : \\ &\quad \big((A, a), (\dot{A}, \dot{a})\big) \mapsto \big((A, a), (\omega, b)\big), \end{aligned}$$

where $\omega^\times = A^{-1}\dot{A}$ and $b = A^{-1}\dot{a}$. The inverse of this trivialization is the mapping

$$\begin{aligned} &\lambda : \mathrm{E}(3) \times \mathbb{R}^3 \times \mathbb{R}^3 \to T\mathrm{E}(3) : \\ &\quad \big((A, a), (\omega, b)\big) \mapsto \big((A, a), (\dot{A} = A\omega^\times, \dot{a} = Ab)\big). \end{aligned} \tag{16}$$

Pulling back ℓ by the map λ we obtain the Lagrangian in the trivialization λ^{-1}

$$\mathcal{L} = \ell \circ \lambda = T - V : \mathrm{E}(3) \times (\mathbb{R}^3 \times \mathbb{R}^3) \to \mathbb{R}, \tag{17}$$

where

$$T = T((A, a), (\omega, b)) = \tfrac{1}{2}\,\langle I(\omega), \omega\rangle + \tfrac{1}{2}\,\mathrm{m}\,\langle b, b\rangle \tag{18}$$

is the *kinetic energy* and

$$V = V((A, a), (\omega, b)) = \mathrm{mg}\,\langle a, e_3\rangle \tag{19}$$

is the *potential energy*.

Lemma 6.2.2. *At $(\omega, b) \in T_{(A,a)}\mathrm{E}(3)$ the Lagrange derivative $\delta\mathcal{L}$ of the Lagrangian $\mathcal{L}$ (17) is the 1-form*

$$\begin{aligned} &\delta\mathcal{L}\big((A, a), (\omega, b)\big) : \mathbb{R}^3 \times \mathbb{R}^3 \to \mathbb{R} : \\ &(\eta, \beta) \mapsto \langle I\dot{\omega} - I\omega \times \omega \,|\, \eta\rangle + \langle \mathrm{m}\,\dot{b} - \mathrm{m}\,b \times \omega + \mathrm{mg}\,u \,|\, \beta\rangle. \end{aligned} \tag{20}$$

Proof. We use (17) of proposition 1.4.6 in chapter 1, namely,

$$\langle \delta\mathcal{L}(\gamma(t),c(t))\,|\,e\rangle = \langle \frac{\mathrm{d}}{\mathrm{d}t}\Big(\frac{\partial \mathcal{L}}{\partial c}\Big)(\gamma(t),c)\Big|_{c=c(t)}\,|\,e\rangle - \langle \frac{\partial \mathcal{L}}{\partial q}(q,c(t))\Big|_{q=\gamma(t)}\,|\,e^Q_{\gamma(t)}\rangle$$
$$- \langle \frac{\partial \mathcal{L}}{\partial c}(\gamma(t),c)\Big|_{c=c(t)}\,|\,\lambda^{-1}_{\gamma(t)}\big(\big[c(t)^Q,e^Q\big]_{\gamma(t)}\big)\rangle, \qquad (21)$$

where $q \in Q = \mathrm{E}(3)$, $\gamma(t) = (A(t),a(t)) \in \mathrm{E}(3)$, $c(t) = (\omega(t),b(t))$ and $e = (\eta,\beta) \in \mathbb{R}^3 \times \mathbb{R}^3$. The first term in (21) is

$$\langle \frac{\mathrm{d}}{\mathrm{d}t}\Big(\frac{\partial \mathcal{L}}{\partial c}\Big)(\gamma(t),c)\Big|_{c=c(t)}\,|\,e\rangle = \frac{\mathrm{d}}{\mathrm{d}t}\big(\langle I\omega(t),\eta\rangle + \mathrm{m}\,\langle b(t),\beta\rangle\big), \qquad (22)$$

which corresponds to the linear form $(I\dot{\omega},\mathrm{m}\,\dot{b})$. The vector field on $\mathrm{E}(3)$, which corresponds to $(\eta,\beta) \in \mathbb{R}^3 \times \mathbb{R}^3$ under the inverse of the trivialization λ (16), is

$$e^Q : \mathrm{E}(3) \rightarrow \mathbb{R}^3 \times \mathbb{R}^3 : (A,a) \mapsto (\eta,\beta). \qquad (23)$$

Therefore the second term in (21) is

$$-\langle \frac{\partial \mathcal{L}}{\partial q}(q,c(t))\Big|_{q=\gamma(t)}\,|\,e^Q_{\gamma(t)}\rangle = \frac{\mathrm{d}}{\mathrm{d}t}\Big|_{t=0} V(\gamma(t)) = \mathrm{mg}\,\langle \dot{a},e_3\rangle$$
$$= \mathrm{mg}\,\langle A\beta,e_3\rangle = \mathrm{mg}\,\langle u\,|\,\beta\rangle, \qquad (24)$$

which corresponds to the linear form $(0,\mathrm{mg}\,u)$. The vector field on $\mathrm{E}(3)$ corresponding to $(\omega(t),b(t))$ is

$$c(t)^Q : \mathrm{E}(3) \rightarrow \mathbb{R}^3 \times \mathbb{R}^3 : (A,a) \mapsto (\omega(t),\beta(t)). \qquad (25)$$

The Lie bracket of (25) with (23) is the vector field

$$\mathrm{E}(3) \rightarrow \mathbb{R}^3 \times \mathbb{R}^3 : (A,a) \mapsto \big(\omega(t)\times\eta, \omega(t)^{\times}\beta - \eta^{\times}\times b(t)\big),$$

see (4). Therefore the third term in (21) is

$$-\langle \frac{\partial \mathcal{L}}{\partial c}(\gamma(t),c)\Big|_{c=c(t)}\,|\,\lambda^{-1}_{\gamma(t)}\big(\big[c(t)^Q,e^Q\big]_{\gamma(t)}\rangle$$
$$= -\langle (I\omega,\mathrm{m}\,b)\,|\,[(\omega,b),(\eta,\beta)]\rangle = -\langle (I\omega,\mathrm{m}\,b)\,|\,(\omega\times\eta,\omega\times\beta-\eta\times b)\rangle$$
$$= -\langle (I\omega\times\omega,\mathrm{m}\,b\times\omega)\,|\,(\eta,\beta)\rangle. \qquad (26)$$

Adding (22), (24), and (26) together gives (20). □

6.3 Constraint distribution

In this section we describe the constraint distribution of a rigid body which rolls without slipping on a horizontal plane. We also find its accessible sets.

Let $\mathcal{S}$ be the boundary of the reference rigid body $\mathcal{B}$. By hypothesis $\mathcal{S}$ is a smooth compact strongly convex[2] surface in $\mathbb{R}^3$. Let $n(s)$ be the *inward* unit normal to $\mathcal{S}$ at the point s. The map $n : \mathcal{S} \to S^2 : s \mapsto n(s)$ is the *Gauss map* of the surface $\mathcal{S}$. Because $\mathcal{S}$ is smooth, compact, and strongly convex, the Gauss map is a diffeomorphism.

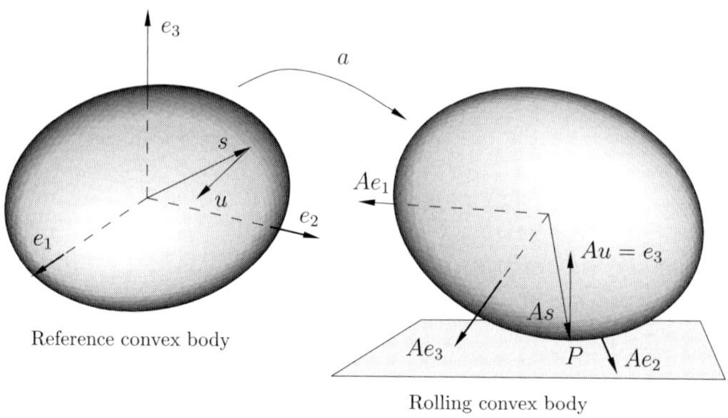

Fig. 6.1 The reference and moving strongly convex rigid body.

The condition that the body lies on a horizontal plane means that if s is the point on the reference surface such that the inward normal to the surface at the position As is pointing vertically upward, that is, $A\,n(s) = e_3$, then the vector $P = A\,s + a$ lies in the horizontal plane. Because $A\,n(s) = e_3$ is equivalent to $n(s) = A^{-1}\,e_3 = u$, we see that $s : S^2 \to \mathcal{S} : u \mapsto s(u)$ is the *inverse* of the *Gauss map*. Therefore the condition that the body lies on a horizontal plane can be written as

$$f(A,\, a) = \langle A\,s(u) + a,\, e_3 \rangle = 0. \tag{27}$$

This constraint on the position coordinates of the moving body is a *holonomic constraint*. The zero set N of the function $f : \mathrm{E}(3) \to \mathbb{R}$, called the *position space* of the rolling body, is a smooth submanifold of $\mathrm{E}(3)$ which is diffeomorphic to $\mathrm{SO}(3) \times \mathbb{R}^2$.

[2]That is, the tangent space at a point of $\mathcal{S}$ meets $\mathcal{S}$ only at that point.

In order that the moving body rolls without slipping, the *nonholonomic constraint*

$$\dot{P} = \dot{A}s(u) + \dot{a} = 0 \tag{28}$$

must hold. In words, equation (28) states that the velocity (with respect to the fixed reference frame) of the point of contact of the moving body with the horizontal plane vanishes. Equation (28) determines the *constraint distribution* D on $\mathrm{E}(3)$ defined by

$$D_{(A,a)} = \Big\{(\dot{A},\dot{a}) \in T_{(A,a)}\mathrm{E}(3) \big| \dot{A}s(u) + \dot{a} = 0\Big\}. \tag{29}$$

Using the inverse of the trivialization (16), the constraint distribution becomes

$$\Big\{(A, a, \omega, b) \in \mathrm{E}(3) \times \mathbb{R}^6 \big| \, \omega \times s(u) + b = 0\Big\}, \tag{30}$$

where $u = A^{-1}e_3$. In what follows when we refer only to the manifold structure of D we shall use the term *constraint manifold*.

We now give an alternative description of D. Let $(\dot{A},\dot{a}) \in T_{(A,a)}\mathrm{E}(3)$. Then for some $(Y, y) \in \mathrm{e}(3)$ we have $(\dot{A},\dot{a}) = (AY, Ay)$. Define an $\mathbb{R}^3$-valued 1-form φ on $\mathrm{E}(3)$ by

$$\begin{aligned} \varphi(A,a)(AY,Ay) &= (\mathrm{d}A \otimes s(u) + \mathrm{d}a)(AY, Ay) \\ &= (A^{-1}\,\mathrm{d}A \otimes s(u))Y + (A^{-1}\,\mathrm{d}a)y \\ &= Y(s(u)) + y. \end{aligned} \tag{31}$$

Then the constraint distribution D is $\ker\varphi$. To see this apply A^{-1} to the equation $\dot{A}s(u) + \dot{a} = 0$ defining the rolling constraint. Note φ is invariant under the action of $\mathrm{SO}(3)$ on $\mathrm{E}(3)$ given by $(B, (A, a)) \to (BA, a)$.

Having identified the constraint distribution, we find its accessible sets.

Lemma 6.3.3. *We have $D \subseteq \ker \mathrm{d}f$. At every point of $\mathrm{E}(3)$ the kernel of $\mathrm{d}f$ is spanned by D and the Lie brackets of smooth vector fields which take values in D.*

Proof. Differentiating f (27) along a curve in D gives

$$\dot{f} = \langle \dot{A}\, s(u) + A\,\frac{\mathrm{d}s(u)}{\mathrm{d}t} + \dot{a},\, e_3\rangle = \langle A\,\frac{\mathrm{d}s(u)}{\mathrm{d}t},\, e_3\rangle = \langle\frac{\mathrm{d}s(u)}{\mathrm{d}t},\, u\rangle = 0.$$

Here the second equality follows from (29). The third equality follows from the orthogonality of A and the fact that $A^{-1}\,e_3 = u$. To prove the fourth equality we observe that $\frac{\mathrm{d}s(u)}{\mathrm{d}t}$ belongs to the tangent space of the surface

$\mathcal{S}$ at the point $s = s(u)$, and that $s = s(u)$ is equivalent to the condition that $u = n(s) =$ the interior normal to $\mathcal{S}$ at the point s. This shows that $D \subseteq \ker \mathrm{d}f$.

To prove the second statement in the lemma we choose $X_j \in \mathrm{so}(3)$, for $j = 1, 2$, that is, the X_j are anti-symmetric 3×3-matrices. Define the vector fields Y_j on $\mathrm{E}(3) = \mathrm{SO}(3) \times \mathbb{R}^3$ by $Y_j(A, a) = (A\, X_j, -A\, X_j\, s(u))$, where $u = A^{-1}\, e_3$ and $(A, a) \in \mathrm{SO}(3) \times \mathbb{R}^3$. Then the vector fields Y_j are tangent to D and

$$[Y_1, Y_2] = (A\,[X_1, X_2], -A\,[X_1, X_2]\; s(u)) \\ + (0, A\,(X_2\,\mathrm{D}\,s(u)\,X_1\,u - X_1\,\mathrm{D}\,s(u)\,X_2\,u)).$$

The first term belongs to D. Therefore it is sufficent to show that the vectors

$$X_2\,\mathrm{D}s(u)\,X_1\,u - X_1\,\mathrm{D}s(u)\,X_2\,u \tag{32}$$

span a two-dimensional linear subspace of $\mathbb{R}^3$.

The tangent at u of the mapping $s : S^2 \to \mathcal{S}$ is a linear mapping from $T_u S^2$ to $T_{s(u)}\mathcal{S}$. Because $u = n(s(u))$ is equal to the interior normal of $\mathcal{S}$ at $s(u)$, both tangent spaces can be identified with the orthogonal complement $u^\perp$ of u in $\mathbb{R}^3$. Moreover, $\mathrm{D}s(u)$ is equal to the tangent mapping of s, viewed as a linear transformation in $u^\perp$. Because s is equal to the inverse of the Gauss mapping n, $\mathrm{D}s(u)$ is equal to the inverse of the derivative of the Gauss mapping, which is a positive definite self-adjoint transformation, since $\mathcal{S}$ is a strongly convex surface. From this it follows that $\mathrm{D}s(u)$ is a positive definite self-adjoint transformation of $u^\perp$ as well. Note that the antisymmetry of X_j implies that $\langle X_j\, u, u\rangle = 0$, which means that $X_j\, u \in u^\perp$.

Now define X_2 by $u^\times$. Then (32) is equal to $X_2\,\mathrm{D}\,s(u)\,X_1\,u$. The space of all $X_1\,u$ with X_1 ranging over $\mathrm{so}(3)$, is equal to $u^\perp$. In addition, the mappings $\mathrm{D}s(u)$ and X_2 are bijective linear transformations of $u^\perp$. Therefore the space of vectors (32) where X_1 ranges over $\mathrm{so}(3)$ contains the two-dimensional linear space $A(u^\perp)$. This proves the lemma. □

The fact $D \subseteq \ker \mathrm{d}f$ implies that for each $n \in N = f^{-1}(0)$ we have $D_n \subseteq T_n N$. In other words, our phase space, which consists of all $(q, \dot{q}) \in T\mathrm{E}(3)$ such that $q \in N$ and $\dot{q} \in D_n$, is equal to the smooth vector subbundle

$$D_N = \bigcup_{n \in N} D_n \tag{33}$$

of TN.

Corollary 6.3.4. *If for each $n \in N$ the tangent space T_nN is equal to the span of D_n and the Lie brackets of smooth vector fields on N which take values in D, then N is the accessible set of the distribution D_N on N. Moreover, if we view D_N as a submanifold of TN and let $\tau : TN \to N$ be the tangent bundle projection map, then D_N is the unique accessible set of the distribution $H = (T\tau)^{-1}(D_N) \cap T(D_N)$ in D_N.*

Proof. The first statement in the corollary follows immediately from lemma 6.3.3.

To prove the second statement we use the classical result that if the Lie bracket of vector fields on a connected manifold with values in a distribution span the tangent space at every point, then an accessible set of the distribution is the entire manifold. This follows from Chow's theorem [26], because our assumption implies that the tangent space at every point is spanned by the vector fields with values in the distribution and the transports of X by means of the flow of Y, where X and Y are two such vector fields.

The second statement implies the third one, see proposition 1.10.45 in chapter 1. □

From corollary 6.3.4 it follows that every position of the body lying on the horizontal plane can be reached from any other such position by means of a rolling motion without slipping. Thus the nonholonomic constraint does not lead to any holonomic ones.

Alternatively, we can prove lemma 6.3.3 as follows. The group E(3) has the vector subgroup $\mathbb{R}^3$ as a closed normal subgroup. The quotient is canonically isomorphic to SO(3). In other words, we have the principal $\mathbb{R}^3$-fibration $\mathrm{E}(3) \to \mathrm{SO}(3)$. The $\mathbb{R}^3$-valued 1-form φ (31) is a connection 1-form for this principal fibration for which $D = \ker \varphi$. Because $\mathbb{R}^3$ is abelian, the curvature of φ is equal to its exterior derivative

$$\mathrm{d}\varphi = \mathrm{d}((\mathrm{d}A)\, s + \mathrm{d}a) = \mathrm{d}A \wedge \mathrm{d}s, \quad \text{where } s = s(u).$$

In turn, $\mathrm{d}\varphi$ is equal to the pull-back under the bundle projection map from E(3) onto SO(3) of a closed 2-form on SO(3). Evaluated at $(A\,X_1,\, A\,X_2)$, this 2-form is equal to the A-image of (31). □

For any distribution D on a smooth manifold M, a smooth submanifold I of M is called an *integral manifold* of D if $T_mI \subseteq D_m$ for every $m \in I$. A distribution D in M is called *integrable* if for each $m \in M$ there exists an integral manifold I of D through m such that $T_mI = D_m$. The following proposition shows that the distribution D (29) on E(3) is non-integrable in a very strong sense.

Proposition 6.3.5. *If I is an integral manifold of D, then* $\dim I \leq 1$.

Proof. For any distribution D on a smooth manifold M and each $m \in M$ there is an antisymmetric bilinear mapping $\mu_m : D_m \times D_m \to T_m M / D_m$ such that for each pair of smooth vector fields Y_1, Y_2 on M which take values in D, one has

$$[Y_1, Y_2](m) + D_m = \mu_m(Y_1(m), Y_2(m)), \quad m \in M.$$

If I is an integral manifold of D, then $[Y_1, Y_2](m) \in T_m I$ for any pair of smooth vector fields Y_1, Y_2 in I and any $m \in I$. Using local extensions of these vector fields to smooth vector fields on N with values in D, we find that $\mu_m(u, v) = 0$ for every pair of vectors $u, v \in T_m I$ and every $m \in I$.

In the case of the distribution D (29) on $\mathrm{E}(3) = \mathrm{SO}(3) \times \mathbb{R}^3$, the mapping $v \mapsto (0, A\,v) + D_{(A,\,a)}$ leads to an isomorphism $\widetilde{\mu}$ from $\mathbb{R}^3$ onto $T_{(A,\,a)}\mathrm{E}(3)/D_{(A,\,a)}$. We use $\widetilde{\mu}$ to identify $T_{(A,\,a)}\mathrm{E}(3)/D_{(A,\,a)}$ with $\mathbb{R}^3$. Also we use the mapping $X \mapsto (A\,X, -A\,X\,s(u))$ to identify $D_{(A,\,a)}$ with so(3). With these identifications, the proof of lemma 6.3.3 shows that μ is equal to the antisymmetric bilinear mapping which assigns to $(X_1, X_2) \in \mathrm{so}(3) \times \mathrm{so}(3)$ the vector in $\mathbb{R}^3$ given by (32).

Choose an orthonormal basis f_1, f_2, f_3 in $\mathbb{R}^3$ such that $f_3 = u$ and f_1, f_2 are eigenvectors of $\mathrm{D}s(u)$ with eigenvalues s_1, s_2, respectively. Let X_j^{kl} denote the antisymmetic matrix of X_j with respect to this basis. Then the vector given by (32) is equal to

$$\left(X_2^{12}\,s_2\,X_1^{23} - X_1^{12}\,s_2\,X_2^{23},\, X_2^{21}\,s_1\,X_1^{13} - X_1^{21}\,s_1\,X_2^{13},\, 0\right). \tag{34}$$

The third component of the vector above is zero because

$$\begin{aligned}&\langle X_2\,\mathrm{D}s(u)\,X_1\,u - X_1\,\mathrm{D}s(u)\,X_2\,u,\, u\rangle\\ &\quad = -\langle \mathrm{D}s(u)\,X_1\,u,\, X_2\,u\rangle + \langle \mathrm{D}s(u)\,X_2\,u,\, X_1\,u\rangle = 0.\end{aligned}$$

To obtain the first equality we have used the antisymmetry of the X_j's; while the second follows from the symmetry of $\mathrm{D}s(u)$. Because $s_1 > 0$ and $s_2 > 0$, it follows that the vector (34) is equal to zero if and only if

$$X_2^{12}\,X_1^{23} = X_1^{12}\,X_2^{23} \quad \text{and} \quad X_2^{12}\,X_1^{13} = X_1^{12}\,X_2^{13}. \tag{35}$$

If $X_2^{12} \neq 0$, then (35) implies that $X_1 = (X_1^{12}/X_2^{12})\,X_2$; whereas if $X_1^{12} \neq 0$, then $X_2 = (X_2^{12}/X_1^{12})\,X_1$. On the other hand, if $X_1^{12} = X_2^{12} = 0$, then the vector (34) is equal to zero. Therefore for every $(A, a) \in \mathrm{E}(3)$ there is exactly one linear subspace of $D_{(A,\,a)}$ of dimension > 1 on which $\mu_{(A,\,a)}$ is equal to zero, namely, the two-dimensional space of all vectors of the form $(A\,X,\, -A\,X\,s(u))$ such that $X = A^{-1}\,\dot{A} \in \mathrm{so}(3)$ and

$$\alpha_A(\dot{A}) = \mathrm{tr}((u^{\times}) \circ (A^{-1}\,\dot{A})) = 0. \tag{36}$$

Here $u = A^{-1}\,e_3$ and $u^{\times} \in \mathrm{so}(3)$.

Equation (36) defines a smooth 1-form α on SO(3). The argument of the preceding paragraph implies that for any integral manifold I of D the restriction to I of the bundle projection map from E(3) to SO(3) is a local diffeomorphism onto an integral manifold of the distribution $\ker\alpha$ on SO(3). Now for every $x \in \mathbb{R}^3$ we have $(A^{-1}\,e_3) \times (A^{-1}\,\dot{A}\,x) = A^{-1}\,(e_3 \times (\dot{A}\,x))$, which shows that

$$\mathrm{tr}((u^{\times}) \circ (A^{-1}\,\dot{A})) = \mathrm{tr}(A^{-1} \circ (e_3^{\times}) \circ \dot{A}) = \mathrm{tr}(e_3^{\times} \circ (\dot{A}\,A^{-1})).$$

Thus the 1-form α is invariant under right multiplication by SO(3). Therefore the distribution $\ker\alpha$ is invariant under right multiplication. Because the mapping $X \mapsto X(e)$ is an isomorphism from the Lie algebra of all right invariant vector fields on SO(3) onto so(3), this implies the condition that I, and thus $\ker\alpha$, is integrable at any point is equivalent to the condition that $\ker\alpha(e)$ is a Lie subalgebra of so(3). However, so(3) has no two-dimensional Lie subalgebras. This proves the proposition. □

6.4 Constrained equations of motion

In this section we derive the Lagrange-d'Alembert equations of motion for a smooth compact strongly convex rigid body which rolls without slipping on a horizontal plane. We also give the distributional Hamiltonian form of these equations.

6.4.1 *Vector field on D*

The motion of the rolling rigid body is governed by the *Lagrange-d'Alem bert principle* theorem 1.3.4 §3.2 of chapter 1, which states that at each point p of the constraint manifold D_N, the Lagrangian derivative (20) of the unconstrained Lagrangian (17) annihilates the linear subspace D_n of T_nN. The subspace $D_{(A,a)}$ (29) is the image under the map $\lambda_{(A,a)}$ (16) of $\{(\overline{\omega}, \overline{b}) \in \mathbb{R}^3 \times \mathbb{R}^3 \,\big|\, 0 = A(\overline{\omega} \times s + \overline{b})\}$ or equivalently $\{(\overline{\omega}, -\overline{\omega} \times s) \in \mathbb{R}^3 \times \mathbb{R}^3 \,\big|\, \overline{\omega} \in \mathbb{R}^3\}$. Using $b = -\omega \times s$ and formula (20) for the Lagrange derivative, it follows that the Lagrange-d'Alembert principle is equivalent to

$$0 = \langle I\big(\frac{\mathrm{d}\omega}{\mathrm{d}t}\big) - I(\omega) \times \omega, \overline{\omega}\rangle + \\ \big\langle -\mathrm{m}\,\frac{\mathrm{d}}{\mathrm{d}t}(\omega \times s) + \mathrm{m}\,((\omega \times s) \times \omega) + \mathrm{mg}\,u, s \times \overline{\omega}\big\rangle, \qquad (37)$$

for every $\overline{\omega} \in \mathbb{R}^3$. Here $s = s(u)$. Since Grassmann's identity implies $((\omega \times s) \times \omega) \times s = -(s \times (\omega \times s)) \times \omega$, equation (37) is equivalent to

$$0 = \big\langle I\big(\frac{\mathrm{d}\omega}{\mathrm{d}t}\big) - I(\omega) \times \omega - \langle \mathrm{m}\,\frac{\mathrm{d}}{\mathrm{d}t}(\omega \times s) \times s - \mathrm{m}\,((s \times \omega) \times s) \times \omega \\ + \mathrm{mg}\,u \times s, \overline{\omega}\big\rangle$$

for every $\overline{\omega} \in \mathbb{R}^3$, that is,

$$I\frac{\mathrm{d}\omega}{\mathrm{d}t} = I\omega \times \omega + \mathrm{m}\,\frac{\mathrm{d}}{\mathrm{d}t}(\omega \times s) \times s + \mathrm{m}\,(s \times (\omega \times s)) \times \omega - \mathrm{mg}\,u \times s. \quad (38)$$

Using dot to denote derivative with respect to t and differentiating the second term on the right hand side, equation (38) may be written as

$$\mathcal{Y}(s)(\dot{\omega}) = (\mathcal{Y}(s)\omega) \times \omega + \mathrm{m}\,(\omega \times \dot{s}) \times s - \mathrm{mg}\,u \times s. \qquad (39)$$

Here $\mathcal{Y}(s) : \mathbb{R}^3 \to \mathbb{R}^3$ is the linear map defined by

$$\mathcal{Y}(s)v = I(v) + \mathrm{m}\,(s \times (v \times s)). \qquad (40)$$

Physically, $\mathcal{Y}(s)$ is the moment of inertia of the body about its point of contact with the horizontal plane.

Fact 6.4.1.6. *$\mathcal{Y}(s)$ is invertible and symmetric.*

Proof. From the definition of $\mathcal{Y}(s)$, it follows that for every $v \in \mathbb{R}^3$

$$\langle \mathcal{Y}(s)v, v\rangle = \langle Iv, v\rangle + \mathrm{m}\,\langle s \times (v \times s), v\rangle = \langle Iv, v\rangle + \mathrm{m}\,\langle v \times s, v \times s\rangle. \quad (41)$$

Hence $\mathcal{Y}(s)$ is symmetric. Since the moment of inertia tensor I is positive definite, it follows that $\langle \mathcal{Y}(s)v, v\rangle \geq 0$. If $\langle \mathcal{Y}(s)v, v\rangle = 0$, then $\langle Iv, v\rangle = 0$ which implies that $v = 0$. Thus $\mathcal{Y}(s)$ is positive definite and hence is invertible. □

Using (27) we obtain

$$b = A^{-1}\dot{a} = -A^{-1}\dot{A}s(u) = -\omega \times s,$$

which substituted into (18) gives $T = \frac{1}{2}\,\langle \mathcal{Y}(s)\omega, \omega\rangle$. In other words, $\mathcal{Y}(s)$ is a position dependent symmetric tensor which defines the kinetic energy as a quadratic function of the angular velocity in body coordinates.

From the above discussion we see that the motion of the rolling rigid body is governed by a vector field V on the constraint manifold D, whose integral curves satisfy

$$\begin{aligned} \dot{A} &= A\omega^{\times} \\ \dot{a} &= -A\omega^{\times}s = -A(\omega \times s) \\ \dot{\omega} &= \mathcal{Y}(s)^{-1}\big((\mathcal{Y}(s)\omega) \times \omega + \mathrm{m}\,((\omega \times \dot{s}) \times s) - \mathrm{mg}\, u \times s\big). \end{aligned} \tag{42}$$

Here $s = s(u)$. Moreover,

$$\dot{u} = -A^{-1}\dot{A}A^{-1}e_3 = -\omega^{\times}u = u \times \omega \quad \text{and} \quad \dot{s} = Ds(u)\dot{u} = Ds(u)(u \times \omega).$$

The total energy of the rolling rigid body is

$$E : D \to \mathbb{R} : c \to (T+V)(c) = \tfrac{1}{2}\langle I(\omega), \omega\rangle + \tfrac{1}{2}\,\mathrm{m}\,\langle b, b\rangle + \mathrm{mg}\,\langle a, e_3\rangle. \tag{43}$$

Fact 6.4.1.7. *E is an integral of the vector field V.*

Proof. This follows from corollary 1.3.5 of chapter 1. We give a direct computation of the Lie derivative of E with respect to V. At $(A, a, \omega, b) \in D$ we have

$$\begin{aligned} \dot{E} &= L_V E \\ &= \langle I(\dot{\omega}), \omega\rangle + \mathrm{m}\,\langle \omega \times s, \dot{\omega} \times s\rangle + +\mathrm{m}\langle \omega \times s, \omega \times \dot{s}\rangle + \mathrm{mg}\,\langle \dot{a}, e_3\rangle \\ &= \langle \mathcal{Y}(s)\dot{\omega}, \omega\rangle + \mathrm{m}\,\langle \omega \times s, \omega \times \dot{s}\rangle + \mathrm{mg}\,\langle \dot{a}, e_3\rangle \\ &= \langle \mathcal{Y}(s)(\omega) \times \omega, \omega\rangle + \mathrm{m}\,\langle (\omega \times \dot{s}) \times s, \omega\rangle - \mathrm{mg}\,\langle u \times s, \omega\rangle \\ &\qquad + \mathrm{m}\,\langle \omega \times s, \omega \times \dot{s}\rangle - \mathrm{mg}\,\langle A(\omega \times s), e_3\rangle \\ &= 0. \end{aligned}$$

To obtain the right hand side of the first equality above, we have used $b = -\omega \times s$; to obtain the second, the definition of $\mathcal{Y}(s)$; to obtain the third, the third equation in (42), and to obtain the fourth, the fact that $\langle \dot{a}, e_3\rangle = \langle b, u\rangle = -\langle \omega \times s, u\rangle$. □

Proposition 6.4.1.8. *The integral curves of the vector field V on the constraint manifold D are defined for all time.*

Proof. Let

$$\pi : D \to \Pi = \mathbb{R}^2 : (A, a, \dot{A}, \dot{a}) \mapsto As + a$$

be the projection from the constraint manifold to the horizontal plane Π. Suppose that γ is an integral curve of V (42) which does not exist for all time. Since the energy E (43) is conserved, the image of γ lies in $E^{-1}(e)$ for some $e \in \mathbb{R}$. Because γ does not exist for all time, it runs out of every compact subset of $E^{-1}(e)$ in finite time and hence out of every compact subset of Π in finite time. However, this cannot happen because $\frac{d}{dt}\pi(\gamma(t))$ remains bounded. To see this, note that

$$\begin{aligned} e &= \tfrac{1}{2}\,\langle I\omega,\omega\rangle + \tfrac{1}{2}\,\mathrm{m}\,\langle \omega\times s,\omega\times s\rangle + \mathrm{mg}\,\langle a,e_3\rangle \\ &\geq \tfrac{1}{2}\,\langle I\omega,\omega\rangle, \quad \text{since } \langle a,e_3\rangle > 0 \\ &\geq c\,\langle \omega,\omega\rangle, \quad \text{for some } c>0, \text{ since } I \text{ is positive definite.} \end{aligned}$$

Because the body rolls without slipping,

$$\frac{\mathrm{d}}{\mathrm{d}t}(\pi(\gamma(t))) = \frac{\mathrm{d}}{\mathrm{d}t}(As+a) \;=\; \dot{A}s + A\dot{s} + \dot{a} \;=\; ADs(u)(u\times\omega).$$

Consequently,

$$\|\frac{\mathrm{d}}{\mathrm{d}t}(\pi(\gamma(t))\| \leq \|A\|\,\|Ds(u)\|\,\|\omega\|,$$

which is bounded on $E^{-1}(e)$ since $u \in S^2$, $Ds(u)$ is continuous, and $\|\omega\| \leq \sqrt{\frac{e}{c}}$. □

Note that the vector field V (42) governing the motion of the rolling rigid body also can be derived by applying the Hamilton-d'Alembert principle 1.5.12 of chapter 1 to the Hamiltonian E (43) on the symplectic manifold $(\mathrm{E}(3)\times\mathrm{e}(3),\widetilde{\omega})$. Here $\widetilde{\omega}$ is the pull back of the symplectic form ω on $T\mathrm{E}(3)$ by the map λ (16).

6.4.2 *Computation of H and ϖ in a trivialization*

In this subsection we calculate the distribution H and its symplectic form ϖ_H in a trivialization, see §6.2 of chapter 1.

We start by finding an expression for the 2-form ϖ_{D_N} on the constraint distribution D_N. We will use the inverse of the trivialization

$$\begin{aligned} \lambda_{D_N} &: \mathrm{E}(3)\times\mathbb{R}^3 \to D_N \subseteq TE(3): \\ &\big((A,a),\omega\big) \mapsto \big((A,a),(A\omega^{\times}, -A(\omega\times s(u)))\big). \end{aligned} \tag{44}$$

Since $\lambda_{D_N}((A,a),e_i)$ is the vector field X_i on $\mathrm{E}(3)$ given by $X_i(A,a) = (Ae_i^{\times}, -A(e_i\times s))$, it follows that for every $\eta = \sum_{i=1}^{3}\eta_i\,e_i$ in $\mathbb{R}^3$ the vector field $\eta^{\mathrm{E}(3)}$ on $\mathrm{E}(3)$ is given by

$$\eta^{\mathrm{E}(3)}(A,a) = \sum_{i=1}^{3}\eta_i\,X_i(A,a) = \big(A\eta^{\times}, -A(\eta\times s(u))\big). \tag{45}$$

We now calculate ϖ_{D_N} in the trivialization (44) by evaluating

$$\varpi_{Q\times\mathbb{R}^d}\big(\eta^Q_{\rightarrow}+\zeta^Q_{\uparrow},\widetilde{\eta}^Q_{\rightarrow}+\widetilde{\zeta}^Q_{\uparrow}\big)=\mathrm{k}(q)(\eta^Q(q),\widetilde{\zeta}^Q(q))$$

$$-\,\mathrm{k}(q)(\widetilde{\eta}^Q(q),\zeta^Q(q))-\mathrm{d}\mathrm{k}^\flat(c^Q)(q,c)([\eta^Q(q),\widetilde{\eta}^Q(q)]),\qquad (46)$$

where $Q=\mathrm{E}(3)$, $c=\omega$, and $d=3$, see (52) of §6.2 in chapter 1. The *kinetic energy metric* k on E(3) of the strongly convex rolling rigid body is

$$\mathrm{k}(A,a)\big((A\omega^\times,\eta),(A\mu^\times,\zeta)\big)=\langle I\omega,\mu\rangle+\mathrm{m}\,\langle\eta,\zeta\rangle.\qquad (47)$$

Therefore

$$\begin{aligned}&\mathrm{k}(A,a)\big(\eta^{\mathrm{E}(3)}(A,a),\zeta^{\mathrm{E}(3)}(A,a)\big)\\&\quad=\mathrm{k}(A,a)\big((A\eta^\times,-A(\eta\times s(u))),(A\zeta^\times,-A(\zeta\times s(u)))\big)\\&\quad=\langle I\eta,\zeta\rangle+\mathrm{m}\,\langle A(\eta\times s),A(\zeta\times s)\rangle=\langle\eta,\zeta\rangle+\mathrm{m}\,\langle\eta\times s,\zeta\times s\rangle\\&\quad=\langle I\eta,\zeta\rangle+\mathrm{m}\,\langle s\times(\eta\times s),\zeta\rangle=\langle\mathcal{Y}(s)\eta,\zeta\rangle.\end{aligned}\qquad (48)$$

Now

$$L_{\eta^{\mathrm{E}(3)}}\big(\zeta^{\mathrm{E}(3)}\,\lrcorner\,\mathrm{k}^\flat(\omega^{\mathrm{E}(3)})\big)(A,a)=D\langle\mathcal{Y}(s(u))\omega,\zeta\rangle\,\eta^{\mathrm{E}(3)}(A,a),\quad\text{using (48)}$$

$$=D_1\langle\mathcal{Y}(s(u))\omega,\zeta\rangle\,A\eta^\times,\ \text{because } u=A^{-1}e_3 \text{ and therefore } \langle\mathcal{Y}(s(u))\omega,\zeta\rangle \text{ depends only on } A$$

$$=\frac{\mathrm{d}}{\mathrm{d}t}\Big|_{t=0}\langle\mathcal{Y}(s(\mathrm{e}^{-t\eta^\times}u))\omega,\zeta\rangle,\ \text{because the flow of the vector field } A\eta^\times \text{ on SO(3) is } t\mapsto A\mathrm{e}^{t\eta^\times} \text{ and } (A\mathrm{e}^{t\eta^\times})^{-1}e_3=\mathrm{e}^{-t\eta^\times}A^{-1}e_3=\mathrm{e}^{-t\eta^\times}u$$

$$=m\,\frac{\mathrm{d}}{\mathrm{d}t}\Big|_{t=0}\langle\omega\times s(\mathrm{e}^{-t\eta^\times}u),\zeta\times s(\mathrm{e}^{-t\eta^\times}u)\rangle,\quad\text{using (40)}$$

$$=\mathrm{m}\,\langle\omega\times Ds(u)(u\times\eta),\zeta\times s(u)\rangle+\mathrm{m}\,\langle\omega\times s(u),\zeta\times Ds(u)(u\times\eta)\rangle.\qquad (49)$$

In local coordinates we have

$$[\eta^{\mathrm{E}(3)},\zeta^{\mathrm{E}(3)}](A,a)=D\zeta^{\mathrm{E}(3)}(A,a)\eta^{\mathrm{E}(3)}(A,a)-D\eta^{\mathrm{E}(3)}(A,a)\zeta^{\mathrm{E}(3)}(A,a).$$

But $\eta^{\mathrm{E}(3)}$ and $\zeta^{\mathrm{E}(3)}$ depend only on A, see (45). So

$$\begin{aligned}D\zeta^{\mathrm{E}(3)}(A,a)\eta^{\mathrm{E}(3)}(A,a)&=D_1\zeta^{\mathrm{E}(3)}(A,a)A\eta^\times\\&=\frac{\mathrm{d}}{\mathrm{d}t}\Big|_{t=0}\big((A\mathrm{e}^{t\eta^\times})\zeta^{\eta^\times},-(A\mathrm{e}^{t\eta^\times})\zeta^{\eta^\times}s(\mathrm{e}^{-t\eta^\times}u)\big)\\&=\big(A\eta^\times\zeta^\times,-A\eta^\times\zeta^\times s(u)-A\zeta^\times Ds(u)(u\times\eta)\big).\end{aligned}$$

Consequently,

$$\begin{aligned}[\eta^{\mathrm{E}(3)},\zeta^{\mathrm{E}(3)}](A,a)&=\big(A(\eta^\times\zeta^\times-\zeta^\times\eta^\times),-A(\eta^\times\zeta^\times-\zeta^\times\eta^\times)s(u)\big)\\&\quad+\big(0,A(\eta\times Ds(u)(u\times\zeta)-\zeta\times Ds(u)(u\times\eta))\big)\\&=(\eta\times\zeta)^{\mathrm{E}(3)}(A,a)+\big(0,A(\eta\times Ds(u)(u\times\zeta)-\zeta\times Ds(u)(u\times\eta))\big),\end{aligned}\qquad (50)$$

because $[\eta^\times,\zeta^\times]=\eta^\times\zeta^\times-\zeta^\times\eta^\times=(\eta\times\zeta)^\times$.

Using (48) and (50) we get

$$\begin{aligned}&\big([\eta^{\mathrm{E}(3)},\widetilde{\eta}^{\mathrm{E}(3)}]\,\lrcorner\,\mathrm{k}^{\flat}(\omega^{\mathrm{E}(3)})\big)(A,a)\\&=\langle\mathcal{Y}(s)\omega,\eta\times\widetilde{\eta}\rangle+\mathrm{m}\,\langle\omega\times s,\widetilde{\eta}\times Ds(u)(u\times\eta)-\eta\times Ds(u)(u\times\widetilde{\eta})\rangle\qquad(51)\end{aligned}$$

Therefore

$$\begin{aligned}&\mathrm{d}(\mathrm{k}^{\flat}(\omega^{\mathrm{E}(3)}))(A,a)\big(\eta^{\mathrm{E}(3)}(A,a),\widetilde{\eta}^{\mathrm{E}(3)}(A,a)\big)\\&\qquad=L_{\eta^{\mathrm{E}(3)}}\big(\widetilde{\eta}^{\mathrm{E}(3)}\,\lrcorner\,\mathrm{k}^{\flat}(\omega^{\mathrm{E}(3)})\big)-L_{\widetilde{\eta}^{\mathrm{E}(3)}}\big(\eta^{\mathrm{E}(3)}\,\lrcorner\,\mathrm{k}^{\flat}(\omega^{\mathrm{E}(3)})\big)\\&\qquad\qquad-[\eta^{\mathrm{E}(3)},\widetilde{\eta}^{\mathrm{E}(3)}]\,\lrcorner\,\mathrm{k}^{\flat}(\omega^{\mathrm{E}(3)}),\quad\text{using (51) of chapter 1}\\&\qquad=\mathrm{m}\,\langle\omega\times Ds(u)(u\times\eta),\widetilde{\eta}\times s\rangle-\mathrm{m}\,\langle\omega\times Ds(u)(u\times\widetilde{\eta}),\eta\times s\rangle\\&\qquad\qquad-\langle\mathcal{Y}(s)\omega,\eta\times\widetilde{\eta}\rangle.\qquad(52)\end{aligned}$$

Using (46) and the abbreviations $\eta_{\rightarrow}$ and $\zeta_{\uparrow}$ for $(\eta^{\mathrm{E}(3)}(A,a),0)$ and $(0,\zeta)$, respectively, we obtain

$$\begin{aligned}&\varpi_{\mathrm{E}(3)\times\mathbb{R}^3}((A,a),\omega)(\eta_{\rightarrow}+\zeta_{\uparrow},\widetilde{\eta}_{\rightarrow}+\widetilde{\zeta}_{\uparrow})\\&=\mathrm{k}(A,a)\big(\eta^{\mathrm{E}(3)}(A,a),\widetilde{\zeta}^{\mathrm{E}(3)}(A,a)\big)-\mathrm{k}(A,a)\big(\widetilde{\eta}^{\mathrm{E}(3)}(A,a),\zeta^{\mathrm{E}(3)}(A,a)\big)\\&\quad-\mathrm{d}(\mathrm{k}^{\flat}(\omega^{\mathrm{E}(3)}))(A,a)\big(\eta^{\mathrm{E}(3)}(A,a),\widetilde{\eta}^{\mathrm{E}(3)}(A,a)\big)\\&=\langle\mathcal{Y}(s)\eta,\widetilde{\zeta}\rangle-\langle\mathcal{Y}(s)\widetilde{\eta},\zeta\rangle+\langle\mathcal{Y}(s)\omega,\eta\times\widetilde{\eta}\rangle\\&\quad+\mathrm{m}\,\langle\omega\times Ds(u)(u\times\widetilde{\eta}),\eta\times s\rangle-\mathrm{m}\,\langle\omega\times Ds(u)(u\times\eta),\widetilde{\eta}\times s\rangle\\&=\langle\mathcal{Y}(s)\eta,\widetilde{\zeta}\rangle-\langle\mathcal{Y}(s)\widetilde{\eta},\zeta\rangle+\langle\mathcal{Y}(s)\omega,\eta\times\widetilde{\eta}\rangle\\&\quad-\mathrm{m}\,\langle u\times\widetilde{\eta},Ds(u)(\omega\times(\eta\times s))\rangle-\mathrm{m}\,\langle s\times(\omega\times Ds(u)(u\times\eta)),\widetilde{\eta}\rangle,\\&\qquad\qquad\text{since } Ds(u)\text{ is self adjoint on }\mathrm{span}\{u\}^{\perp}\\&=\langle\mathcal{Y}(s)\eta,\widetilde{\zeta}\rangle-\langle\mathcal{Y}(s)\widetilde{\eta},\zeta\rangle+\langle\mathcal{Y}(s)\omega,\eta\times\widetilde{\eta}\rangle\\&\quad+\mathrm{m}\,\langle u\times Ds(u)(\omega\times(\eta\times s)),\widetilde{\eta}\rangle-\mathrm{m}\,\langle s\times(\omega\times Ds(u)(u\times\eta)),\widetilde{\eta}\rangle.\end{aligned}$$

Thus we have proved

Proposition 6.4.2.9. *In the trivialization $\lambda_{D_N}^{-1}$ the distribution H is $H_{\mathrm{E}(3)\times\mathbb{R}^3}$ where*

$$(H_{\mathrm{E}(3)\times\mathbb{R}^3})_{(A,a)}=(D_N)_{(A,a)}\times\mathbb{R}^3,\qquad(53)$$

see (42) of §6 of chapter 1. Using the abbreviations $\eta_{\rightarrow}$ for $\eta_{\rightarrow}^{\mathrm{E}(3)}(A,a)=(A\eta^{\times},-A(\eta\times s))$ and $\zeta_{\uparrow}$ for $\zeta_{\uparrow}^{\mathrm{E}(3)}(A,a)=(0,\zeta)$, see (45) of chapter 1, the nondegenerate 2-form $\varpi_{\mathrm{E}(3)\times\mathbb{R}^3}$ in the trivialization $\lambda_{D_N}^{-1}$ is given by

$$\begin{aligned}\varpi_{\mathrm{E}(3)\times\mathbb{R}^3}((A,a),\omega)(\eta_{\rightarrow}+\zeta_{\uparrow},\widetilde{\eta}_{\rightarrow}+\widetilde{\zeta}_{\uparrow})&=\langle\mathcal{Y}(s)\eta,\widetilde{\zeta}\rangle-\langle\mathcal{Y}(s)\widetilde{\eta},\zeta\rangle\\+\langle\mathcal{Y}(s)\omega,\eta\times\widetilde{\eta}\rangle&+\mathrm{m}\,\langle u\times Ds(u)(\omega\times(\eta\times s)),\widetilde{\eta}\rangle\\&-\mathrm{m}\,\langle s\times(\omega\times Ds(u)(u\times\eta)),\widetilde{\eta}\rangle.\qquad(54)\end{aligned}$$

6.4.3 Distributional vector field in a trivialization

In the trivialization $\lambda_{D_N}^{-1}$ (44) $\eta_{\rightarrow} = \eta_{\rightarrow}^{\mathrm{E}(3)}(A,a)$ and $\zeta_{\uparrow} = \zeta_{\uparrow}^{\mathrm{E}(3)}(A,a)$ are vector fields on $\mathrm{E}(3)\times\mathbb{R}^3$ with values in $(H_{\mathrm{E}(3)\times\mathbb{R}^3})_{(A,a)}$. By definition $Y_{E_{\mathrm{E}(3)\times\mathbb{R}^3}}((A,a),\omega) = \eta_{\rightarrow} + \zeta_{\uparrow}$, is the distributional Hamiltonian vector field, which corresponds to the energy function

$$
\begin{aligned}
E_{\mathrm{E}(3)\times\mathbb{R}^3}((A,a),\omega) &= E(\lambda_{D_N}((A,a),\omega)) = E(A,a)(A\omega^{\times}, -A(\omega\times s)),\\
&\qquad\text{where } E \text{ is the energy function (43) on } D_N\\
&= \tfrac{1}{2}\langle \mathcal{Y}(s(u))\omega,\omega\rangle - \mathrm{mg}\,\langle s(u),u\rangle,
\end{aligned}
\tag{55}
$$

since

$$
\langle a, e_3\rangle = \langle As(u), e_3\rangle = -\langle s(u), A^{-1}e_3\rangle = -\langle s(u),u\rangle.
$$

Using (55) we find that for any $\widetilde{\eta},\widetilde{\zeta}\in\mathbb{R}^3$

$$
\begin{aligned}
\mathrm{d}E_{\mathrm{E}(3)\times\mathbb{R}^3}((A,a),\omega)(\widetilde{\eta}_{\rightarrow}+\widetilde{\zeta}_{\uparrow}) &= D_1E_{\mathrm{E}(3)\times\mathbb{R}^3}((A,a),\omega)A\eta^{\times} + \langle \mathcal{Y}(s)\omega,\widetilde{\zeta}\rangle,\\
&\qquad\text{since } E_{\mathrm{E}(3)\times\mathbb{R}^3} \text{ does not depend on } a\\
&= \frac{\mathrm{d}}{\mathrm{d}t}\Big|_{t=0}\big(\tfrac{1}{2}\langle \mathcal{Y}(s(\mathrm{e}^{-t\widetilde{\eta}^{\times}}u))\omega,\omega\rangle - \mathrm{mg}\,\langle s(\mathrm{e}^{-t\widetilde{\eta}^{\times}}u), \mathrm{e}^{-t\widetilde{\eta}^{\times}}u\rangle\big) + \langle \mathcal{Y}(s)\omega,\widetilde{\zeta}\rangle\\
&= \mathrm{m}\,\langle \omega\times Ds(u)(u\times\widetilde{\eta}), \omega\times s(u)\rangle - \mathrm{mg}\,\langle Ds(u)(u\times\widetilde{\eta}),u\rangle\\
&\quad - \mathrm{mg}\,\langle s(u), u\times\widetilde{\eta}\rangle + \langle \mathcal{Y}(s)\omega,\widetilde{\zeta}\rangle, \ \text{since } \mathcal{Y}(s)\omega = I\omega + \mathrm{m}\, s\times(\omega\times s)\\
&= \mathrm{m}\,\langle \omega\times Ds(u)(u\times\widetilde{\eta}),\omega\times s\rangle - \mathrm{mg}\,\langle s, u\times\widetilde{\eta}\rangle + \langle \mathcal{Y}(s)\omega,\widetilde{\zeta}\rangle.
\end{aligned}
\tag{56}
$$

The last equality above follows because $Ds(u)(u\times\widetilde{\eta})\in T_{s(u)}\mathcal{S}$ is orthogonal to u. The distributional Hamiltonian vector field $Y_{E_{\mathrm{E}(3)\times\mathbb{R}^3}}$ on $\mathrm{E}(3)\times\mathbb{R}^3$ satisfies

$$
\big(Y_{E_{\mathrm{E}(3)\times\mathbb{R}^3}} \,\lrcorner\, \varpi_{\mathrm{E}(3)\times\mathbb{R}^3}\big)((A,a),\omega) = \mathrm{d}E_{\mathrm{E}(3)\times\mathbb{R}^3}((A,a),\omega), \tag{57}
$$

which evaluated at $\widetilde{\eta}_{\rightarrow}+\widetilde{\zeta}_{\uparrow}$ gives

$$
\begin{aligned}
&\varpi_{\mathrm{E}(3)\times\mathbb{R}^3}((A,a),\omega)(\eta_{\rightarrow}+\zeta_{\uparrow}, \widetilde{\eta}_{\rightarrow}+\widetilde{\zeta}_{\uparrow})\\
&\qquad = \mathrm{d}E_{\mathrm{E}(3)\times\mathbb{R}^3}((A,a),\omega)(\widetilde{\eta}_{\rightarrow}+\widetilde{\zeta}_{\uparrow}).
\end{aligned}
\tag{58}
$$

Using (54) and (56), we see that (58) is equivalent to

$$
\begin{aligned}
&\langle \mathcal{Y}(s)\eta,\widetilde{\zeta}\rangle - \langle \mathcal{Y}(s)\widetilde{\eta},\zeta\rangle + \langle \mathcal{Y}(s)\omega,\eta\times\widetilde{\eta}\rangle\\
&+ \mathrm{m}\,\langle \eta\times s, \omega\times Ds(u)(u\times\widetilde{\eta})\rangle - \mathrm{m}\,\langle \widetilde{\eta}\times s, \omega\times Ds(u)(u\times\eta)\rangle\\
&= \langle \mathcal{Y}(s)\omega,\widetilde{\zeta}\rangle + \mathrm{m}\,\langle \omega\times Ds(u)(u\times\widetilde{\eta}), \omega\times s)\rangle + \mathrm{mg}\,\langle s, u\times\widetilde{\eta}\rangle,
\end{aligned}
\tag{59}
$$

that is, for every $\widetilde{\eta}, \widetilde{\zeta} \in \mathbb{R}^3$ we have

$$\begin{aligned}
&\langle \mathcal{Y}(s)\eta, \widetilde{\zeta}\rangle - \langle \mathcal{Y}(s)\zeta, \widetilde{\eta}\rangle + \langle (\mathcal{Y}(s)\omega) \times \eta, \widetilde{\eta}\rangle \\
&\quad + \mathrm{m}\,\langle u \times Ds(u)(\omega \times (\eta \times s)), \widetilde{\eta}\rangle - \mathrm{m}\,\langle s \times (\omega \times Ds(u)(u \times \eta)), \widetilde{\eta}\rangle \\
&\quad = \langle \mathcal{Y}(s)\omega, \widetilde{\zeta}\rangle + \mathrm{m}\,\langle u \times Ds(u)(\omega \times (\eta \times s)), \widetilde{\eta}\rangle + \mathrm{mg}\,\langle u \times s, \widetilde{\eta}\rangle. \qquad (60)
\end{aligned}$$

Setting $\widetilde{\eta} = 0$, equation (60) becomes

$$\langle \mathcal{Y}(s)\eta, \widetilde{\zeta}\rangle = \langle \mathcal{Y}(s)\omega, \widetilde{\zeta}\rangle, \quad \text{for every } \widetilde{\zeta} \in \mathbb{R}^3.$$

Hence $\mathcal{Y}(s)\eta = \mathcal{Y}(s)\omega$, that is,

$$\eta = \omega, \qquad (61)$$

since $\mathcal{Y}(s)$ is invertible. Setting $\widetilde{\zeta} = 0$ and using (61), equation (60) reads

$$\begin{aligned}
&-\langle \mathcal{Y}(s)\zeta, \widetilde{\eta}\rangle + \langle (\mathcal{Y}(s)\omega) \times \omega, \widetilde{\eta}\rangle - \mathrm{m}\,\langle s \times (\omega \times Ds(u)(u \times \omega)), \widetilde{\eta}\rangle \\
&\qquad = \mathrm{mg}\,\langle u \times s, \widetilde{\eta}\rangle,
\end{aligned}$$

for every $\widetilde{\eta} \in \mathbb{R}^3$. In other words,

$$-\mathcal{Y}(s)\zeta + (\mathcal{Y}(s)\omega) \times \omega - \mathrm{m}\, s \times (\omega \times Ds(u)(u \times \omega)) = \mathrm{mg}\, u \times s.$$

So

$$\zeta = \mathcal{Y}(s)^{-1}\left[(\mathcal{Y}(s)\omega) \times \omega + \mathrm{m}\,(\omega \times Ds(u)(u \times \omega)) \times s - \mathrm{mg}\, u \times s\right]. \qquad (62)$$

Therefore

$$Y_{E_{\mathrm{E}(3)\times\mathbb{R}^3}}((A,a),\omega) = \left(A\omega^{\times}, -A(\omega \times s)\right) + (0,\zeta), \qquad (63)$$

where ζ is given by (62). In other words, the integral curves of $Y_{E_{\mathrm{E}(3)\times\mathbb{R}^3}}$ on D_N satisfy

$$\begin{aligned}
\dot{A} &= A\omega^{\times} \\
\dot{a} &= -A(\omega \times s) \qquad (64)\\
\dot{\omega} &= \mathcal{Y}(s)^{-1}\left[(\mathcal{Y}(s)\omega) \times \omega + \mathrm{m}\,(\omega \times Ds(u)(u \times \omega)) \times s - \mathrm{mg}\, u \times s\right].
\end{aligned}$$

6.5 Reduction of the translational $\mathbb{R}^2$ symmetry

In this subsection we reduce the translational $\mathbb{R}^2$ symmetry of the strongly convex rolling rigid body.

The subgroup $\mathbb{R}^2 = \mathbb{R}^2 \times \{0\} \subseteq \mathbb{R}^3$ of $\mathrm{E}(3)$ formed by *translations of the horizontal plane* $\mathbb{R}^2 \times \{0\}$ acts on $T\mathrm{E}(3)$ by

$$\widehat{\Phi} : \mathbb{R}^2 \times T\mathrm{E}(3) \to T\mathrm{E}(3) : \left(b, ((A,a),(\dot{A},\dot{a}))\right) \mapsto \left((A,a+b),(\dot{A},\dot{a})\right). \qquad (65)$$

The holonomic constraint $\langle As(u) + a, e_3 \rangle = 0$, the nonholonomic constraint $\dot{A}s(u) + \dot{a} = 0$, the kinetic energy $T = \frac{1}{2} \langle \mathcal{Y}(s)\omega, \omega \rangle$, and potential energy $V = -\mathrm{mg} \langle s(u), u \rangle$ with $u = A^{-1}e_3$ are all invariant under the $\mathbb{R}^2$ action $\widehat{\Phi}$. Therefore the phase space D_N given by

$$\{((A,a),(\dot{A},\dot{a})) \in T\mathrm{E}(3) \big| \langle As(u) + a, e_3 \rangle = 0, \text{ and } \dot{A}s(u) + \dot{a} = 0\}, \tag{66}$$

which is diffeomorphic to $T\,\mathrm{SO}(3) \times \mathbb{R}^2$, and the equations of motion

$$\begin{aligned} \dot{A} &= A\omega^{\times} \\ \dot{a} &= -A(\omega \times s) \\ \dot{\omega} &= \mathcal{Y}(s)^{-1} \left[(\mathcal{Y}(s)\omega) \times \omega + \mathrm{m} \left((\omega \times Ds(u)(u \times \omega)) \times s \right) - \mathrm{mg}\, u \times s \right] \end{aligned} \tag{67}$$

are also invariant under $\widehat{\Phi}$. Consequently, the distributional Hamiltonian system $(D_N, H_{D_N}, \varpi_{D_N}, h_{D_N})$, where $D_N = T\,\mathrm{SO}(3) \times \mathbb{R}^2 \subseteq T\mathrm{E}(3) \times \mathbb{R}^3$, $H_{D_N} = \{(H_{\mathrm{E}(3)\times\mathbb{R}^3})_u \big| u = (A,\omega) \in \mathrm{SO}(3) \times \mathbb{R}^3 \subseteq \mathrm{E}(3) \times \mathbb{R}^3\}$, see (53), $\varpi_{D_N} = \varpi_{\mathrm{E}(3)\times\mathbb{R}^3} | (H_{D_N} \times H_{D_N})$, see (54), and $h_{D_N} = (T+V)|D_N$, has an $\mathbb{R}^2$-symmetry.

6.5.1 *The $\mathbb{R}^2$-reduced equations of motion*

Consider the action of $\mathbb{R}^2$ on $T\,\mathrm{SO}(3) \times \mathbb{R}^2$ defined by

$$\begin{aligned} \widehat{\varphi} : \mathbb{R}^2 \times (T\,\mathrm{SO}(3) \times \mathbb{R}^2) &\to T\,\mathrm{SO}(3) \times \mathbb{R}^2 : \\ \big(b, ((A,\dot{A}),c)\big) &\mapsto ((A,\dot{A}), c+b). \end{aligned} \tag{68}$$

This action is transitive. Its orbit space $\overline{T\,\mathrm{SO}(3) \times \mathbb{R}^3} = (T\,\mathrm{SO}(3) \times \mathbb{R}^2)/\mathbb{R}^2$ may be identified with $T\,\mathrm{SO}(3)$, being the $\mathbb{R}^2$-orbit through $T\,\mathrm{SO}(3) \times \{0\}$. The orbit map of this $\mathbb{R}^2$-action is

$$\pi : T\,\mathrm{SO}(3) \times \mathbb{R}^2 \to T\,\mathrm{SO}(3) : ((A,\dot{A}),c) \mapsto (A,\dot{A}). \tag{69}$$

Now

$$\widehat{\psi} : T\,\mathrm{SO}(3) \times \mathbb{R}^2 \to D_N : ((A,\dot{A}),c) \mapsto \Big(\big(A, a = \big(\begin{smallmatrix} c \\ -(As)_3 \end{smallmatrix}\big)\big), (\dot{A}, \dot{a} = -\dot{A}s) \Big)$$

is a diffeomorphism, which intertwines the $\mathbb{R}^2$-action $\widehat{\varphi}$ (68) with the $\mathbb{R}^2$-action $\widehat{\Phi}|(\mathbb{R}^2 \times D_N)$ (65). In other words, for every $b \in \mathbb{R}^2$ we have $\widehat{\psi} \circ \widehat{\varphi}_b = \widehat{\Phi}_b \circ \widehat{\psi}$ on $T\,\mathrm{SO}(3) \times \mathbb{R}^2$. Consequently, $\widehat{\psi}$ induces a smooth mapping

$$\begin{aligned} \overline{\psi} : T\,\mathrm{SO}(3) = \overline{T\,\mathrm{SO}(3) \times \mathbb{R}^2} &\to \overline{D}_N = D_N/\mathbb{R}^2 : \\ (A,\dot{A}) = \overline{((A,\dot{A}),c)} &\mapsto \overline{\widehat{\psi}((A,\dot{A}),c)}, \end{aligned} \tag{70}$$

where $\overline{((A,\dot{A}),c)} = \{\widehat{\psi}_b((A,\dot{A}),c) \in T\,\mathrm{SO}(3) \times \mathbb{R}^2 \big| b \in \mathbb{R}^2\}$ is the orbit of the $\mathbb{R}^2$-action $\widehat{\varphi}$ on $T\,\mathrm{SO}(3) \times \mathbb{R}^2$ through $((A,\dot{A}),c)$ and $\overline{\widehat{\psi}}((A,\dot{A}),c) =$

$\{\Phi_b(\psi((A,\dot{A}),c)) \in D_N \big| b \in \mathbb{R}^2\}$ is the orbit of the $\mathbb{R}^2$-action $\Phi|(\mathbb{R}^2 \times D_N)$ on D_N through $\psi((A,\dot{A}),c)$. To verify that $\overline{\psi}$ is a diffeomorphism it suffices to show that it is injective, because then the map $\overline{(\psi)^{-1}}$ induced by ψ^{-1} is equal to $(\overline{\psi})^{-1}$. Suppose that $\overline{\widehat{\psi}((A,\dot{A}),c)} = \overline{\widehat{\psi}((B,\dot{B}),d)}$ for some $((A,\dot{A}),c)$ and $((B,\dot{B}),d)$ in $T\,\mathrm{SO}(3)\times\mathbb{R}^2$. Then $\widehat{\psi}((A,\dot{A}),c)$ and $\widehat{\psi}((B,\dot{B}),d)$ lie in the same $\mathbb{R}^2$-orbit on D_N, that is, there is a $b \in \mathbb{R}^2$ such that

$$\psi((B,\dot{B}),d) = \widehat{\Phi}_b(\psi((A,\dot{A}),c)) = \psi(\widehat{\varphi}_b((A,\dot{A}),c)) = \psi((A,\dot{A}),b+c).$$

But ψ is injective, so $((B,\dot{B}),d) = ((A,\dot{A}),b+c)$. Consequently, $(B,\dot{B}) = (A,\dot{A})$, that is, the map $\overline{\psi}$ is injective. Let

$$\lambda : \mathrm{SO}(3)\times\mathbb{R}^3 \to T\,\mathrm{SO}(3) : (A,\omega) \mapsto (A, A\omega^{\times}) \tag{71}$$

be the inverse of a trivialization of $T\,\mathrm{SO}(3)$. Since π (69) is the orbit map of the $\mathbb{R}^2$ action on $T\,\mathrm{SO}(3)\times\mathbb{R}^2$, the $\mathbb{R}^2$-*reduced vector field* X on the $\mathbb{R}^2$-*orbit space* $T\,\mathrm{SO}(3)$ is given by

$$\begin{aligned} T_{((A,\dot{A}),c)}\pi Y_{E_{\mathrm{E}(3)\times\mathbb{R}^3}}&|(T\,\mathrm{SO}(3)\times\mathbb{R}^2)((A,\dot{A}=A\omega^{\times}),c) \\ &= X(\pi((A,A\omega^{\times}),c)) = X(A,\omega) \end{aligned}$$

whose integral curves satisfy

$$\begin{aligned} \dot{A} &= A\omega^{\times} \\ \mathcal{Y}(s)\dot{\omega} &= (\mathcal{Y}(s)\omega)\times\omega + \mathrm{m}\,((\omega\times Ds(u)(u\times\omega))\times s) - \mathrm{mg}\,u\times s. \end{aligned} \tag{72}$$

Thus we have shown

Proposition 6.5.1.10. *The distributional Hamiltonian system* $(D_N, H_{D_N}, \varpi_{D_N}, h_{D_N})$ *is a Chaplygin system* (D,ℓ,G)*, where* $D = D_N$*,* $G=\mathbb{R}^2$*, and* $\ell = (T-V)|D_N$.[3]

6.5.2 *Comparison with the Euler-Lagrange equations*

It is tempting to view the kinetic $T = \frac{1}{2}\langle\mathcal{Y}(s)\omega,\omega\rangle$ and potential $V = -\mathrm{mg}\,\langle s(u),u\rangle$ energies as functions on $T\,\mathrm{SO}(3)$ with coordinates $(A,\dot{A} = A\omega^{\times})$ and $u = A^{-1}e_3$. Using the trivialization λ^{-1} (71), we compute the Euler-Lagrange equations of motion $\delta\mathcal{L}=0$, when $\mathcal{L} = \ell\circ\lambda$ and $\ell = T-V$ are thought of as functions on $T\,\mathrm{SO}(3)$. In other words,

$$\mathcal{L} : \mathrm{SO}(3)\times\mathbb{R}^3 \to \mathbb{R} : (A,\omega) \mapsto \tfrac{1}{2}\langle\mathcal{Y}(s)\omega,\omega\rangle + \mathrm{mg}\,\langle s(u),u\rangle. \tag{73}$$

[3]See §4 of chapter 3. The proof of proposition 6.5.1.10 shows that $Q = N = \{(A,a) \in \mathrm{SO}(3)\times\mathbb{R}^2 \big| \langle As(u)+a, e_3\rangle = 0\}$ and $R = \overline{Q} = Q/\mathbb{R}^2 = \mathrm{SO}(3)$.

Using equation (17) of chapter 1 for every $a \in \mathbb{R}^3$ the Lagrange derivative $\langle \delta\mathcal{L}(A(t),\omega(t)) | \, a\rangle$ is equal to

$$\langle \tfrac{\mathrm{d}}{\mathrm{d}t}\Big(\tfrac{\partial \mathcal{L}}{\partial \omega}\Big)(A(t),\omega)\Big|_{\omega=\omega(t)} | \, a\rangle - \tfrac{\partial L}{\partial \dot{A}}(A(t),\dot{A})\Big|_{\dot{A}=\frac{\mathrm{d}A(t)}{\mathrm{d}t}} | \, [\omega(t)^{\mathrm{SO}(3)}, a^{\mathrm{SO}(3)}]\rangle$$
$$- \langle \tfrac{\partial \mathcal{L}}{\partial A}(A,\omega(t))\Big|_{A=A(t)} | \, a^{\mathrm{SO}(3)}\rangle. \qquad (74)$$

But

$$\langle \frac{\mathrm{d}}{\mathrm{d}t}\Big(\frac{\partial \mathcal{L}}{\partial \omega}\Big) | \, a\rangle = \langle \frac{\mathrm{d}(\mathcal{Y}(s)\omega)}{\mathrm{d}t}, a\rangle = \langle \mathcal{Y}(s)\dot{\omega} + \mathrm{m}\, Ds(u)(u\times\omega)\times(\omega\times s)$$
$$+ \mathrm{m}\, s\times(\omega\times Ds(u)(u\times\omega)), a\rangle. \qquad (75)$$

Also

$$\langle \frac{\partial \mathcal{L}}{\partial A}(A,\omega(t)) \, | \, Aa^{\times}\rangle = \frac{\mathrm{d}}{\mathrm{d}t}\Big|_{t=0} \mathcal{L}(A\mathrm{e}^{ta^{\times}},\omega)$$
$$= \frac{\mathrm{d}}{\mathrm{d}t}\Big|_{t=0} \Big[\tfrac{1}{2}\langle\omega,\omega\rangle + \tfrac{1}{2}\langle\omega\times s(\mathrm{e}^{-ta^{\times}}u), \omega\times s(\mathrm{e}^{-ta^{\times}}u)$$
$$+\mathrm{mg}\,\langle s(\mathrm{e}^{-ta^{\times}}u), \mathrm{e}^{-ta^{\times}}u\rangle\Big]$$
$$= \mathrm{m}\,\langle \omega\times Ds(u)(u\times a), \omega\times s\rangle + \mathrm{mg}\,\langle s(u),u\rangle$$
$$\text{since } Ds(u)(u\times a)\in T_{s(u)}\mathcal{S} \text{ is perpendicular to } u$$
$$= \mathrm{m}\,\langle u\times Ds(u)(\omega\times(\omega\times s)), a\rangle - \mathrm{mg}\,\langle u\times s, a\rangle, \qquad (76)$$

because $Ds(u)$ is self adjoint. From

$$L(A, \dot{A} = A\omega^{\times}) = \tfrac{1}{2}\langle\langle(\mathcal{Y}(s)\omega)^{\times}, \omega^{\times}\rangle\rangle + \mathrm{mg}\,\langle s(A^{-1}e_3), A^{-1}e_3\rangle,$$

where $\langle\langle X, Y\rangle\rangle = \frac{1}{2}X^tY$ for X, $Y \in \mathrm{so}(3)$, we find that

$$\langle \frac{\partial L}{\partial \dot{A}} \, | \, [\omega^{\mathrm{SO}(3)}, a^{\mathrm{SO}(3)}]\rangle = \langle\langle(\mathcal{Y}(s)\omega)^{\times}, (\omega\times a)^{\times}\rangle\rangle$$
$$= \langle \mathcal{Y}(s)\omega, \omega\times a\rangle = \langle(\mathcal{Y}(s)\omega)\times\omega, a\rangle. \qquad (77)$$

Substituting (75), (76), and (77) into (74) we get

$$\langle\delta\mathcal{L}\,|\,a\rangle = \langle \mathcal{Y}(s)\dot{\omega} + \mathrm{m}\, Ds(u)(u\times\omega)\times(\omega\times s) + \mathrm{m}\, s\times(\omega\times Ds(u)(u\times a))$$
$$- (\mathcal{Y}(s)\omega)\times\omega - \mathrm{m}\, u\times Ds(u)(\omega\times(\omega\times s)) + \mathrm{mg}\, u\times s, a\rangle. \qquad (78)$$

Consequently, the Euler-Lagrange equations of motion $\delta\mathcal{L} = 0$ are

$$\begin{aligned} \dot{A} &= A\omega^{\times} \\ \dot{\omega} &= \mathcal{Y}(s)^{-1}[(\mathcal{Y}(s)\omega)\times\omega - \mathrm{m}\, Ds(u)(u\times\omega)\times(\omega\times s) \\ &\quad -\mathrm{m}\, s\times(\omega\times Ds(u)(u\times\omega)) + \mathrm{m}\, u\times Ds(u)(\omega\times(\omega\times s)) \\ &\quad -\mathrm{mg}\, u\times s] \end{aligned} \qquad (79)$$

The second equation in (79) differs from the second equation in (72) by

$$\mathrm{m}\,(\omega\times s)\times Ds(u)(u\times\omega) + \mathrm{m}\, u\times Ds(u)(\omega\times(\omega\times s)). \qquad (80)$$

Taking the inner product with η, the expression in (80) becomes

$$\mathrm{m}\,\langle \omega\times s, \omega\times Ds(u)(u\times\eta) - \eta\times Ds(u)(u\times\omega)\rangle, \qquad (81)$$

which is the correction term for a Chaplygin system, see (20) of chapter 3.

6.5.3 The $\mathbb{R}^2$-reduced distribution $\overline{H}_{\overline{D}_N}$ and the 2-form $\varpi_{\overline{D}_N}$

In this subsection we prove

Proposition 6.5.3.11. *On the orbit space* $\overline{D_N} = T\,\mathrm{SO}(3) = \mathrm{SO}(3)\times\mathbb{R}^3$ *of the* $\mathbb{R}^2$*-action* $\widehat{\Phi}|(\mathbb{R}^2\times D_N)$ (65) *the* $\mathbb{R}^2$-reduced distribution

$$\overline{H}_{\overline{D}_N} = \overline{H}_{\mathrm{SO}(3)\times\mathbb{R}^3} = T\,\mathrm{SO}(3)\times\mathbb{R}^3. \tag{82}$$

The $\mathbb{R}^2$-reduced 2-form $\varpi_{\overline{D}_N}$ *on* $\overline{H}_{\overline{D}_N}$ *is the 2-form* $\varpi_{\mathrm{E}(3)\times\mathbb{R}^3}$ (54) *restricted to* $\mathrm{SO}(3)\times\mathbb{R}^3$. *More explicitly, for every* $(A,\omega)\in\mathrm{SO}(3)\times\mathbb{R}^3$ *and every* η, ζ, $\widetilde{\eta}$, $\widetilde{\zeta}\in\mathbb{R}^3$ *we have*

$$\begin{aligned}\varpi_{\overline{D}_N}(A,\omega)\big(\eta_{\rightarrow}+\zeta_{\uparrow},\widetilde{\eta}_{\rightarrow}+\widetilde{\zeta}_{\uparrow}\big) &= \langle\mathcal{Y}(s)\eta,\widetilde{\zeta}\rangle - \langle\mathcal{Y}(s)\zeta,\widetilde{\eta}\rangle\\ &+\langle\mathcal{Y}(s)\omega,\eta\times\widetilde{\eta}\rangle - \mathrm{m}\,\langle s\times(\omega\times Ds(u)(u\times\eta)),\widetilde{\eta}\rangle\\ &+\mathrm{m}\,\langle u\times(Ds(u)(\omega\times(\eta\times s)),\widetilde{\eta}\rangle.\end{aligned} \tag{83}$$

Here $\eta_{\rightarrow} = \eta_{\rightarrow}^{\mathrm{SO}(3)}$ *and* $\zeta_{\uparrow} = \zeta_{\uparrow}^{\mathrm{SO}(3)}$ *are vector fields on* $\mathrm{SO}(3)\subseteq\mathrm{E}(3)$ *with values in* $T\,\mathrm{SO}(3)\times\mathbb{R}^3$ *given by*

$$\eta_{\rightarrow}^{\mathrm{SO}(3)}(A) = (A\eta^{\times},0) \text{ and } \zeta_{\uparrow}^{\mathrm{SO}(3)}(A) = (0,\zeta). \tag{84}$$

Proof. Recall that $N = \{(A,a)\in\mathrm{SO}(3)\times\mathbb{R}^3 \,\big|\, \langle As+a,e_3\rangle = 0\}$ and

$$\begin{aligned}D_N = \Big\{&(n,v)\in TN = T(\mathrm{SO}(3)\times\mathbb{R}^3)\,\big|\, n\in N \text{ and}\\ &v = (\dot{A},\dot{a})\in T_{(n,v)}(\mathrm{SO}(3)\times\mathbb{R}^3) \text{ with } \dot{A}s+\dot{a} = 0\Big\}.\end{aligned}$$

Fix $p = (n,v)\in D_N$ and let $\mathcal{O}_p = \{\widehat{\Phi}_c(p)\in D_N\,\big|\,\text{for all } c\in\mathbb{R}^2\}$ be the orbit through p of the $\mathbb{R}^2$-action $\widehat{\Phi}|(\mathbb{R}^2\times D_N)$ (65). Then

$$\begin{aligned}T_p\mathcal{O}_p &= \{T_e\widehat{\Phi}_p d\in T_pD_N\,\big|\,\text{for every } d\in\mathbb{R}^2\}\\ &= \{\frac{\mathrm{d}}{\mathrm{d}t}\Big|_{t=0}\widehat{\Phi}_{c+td}(p)\in T_pD_N\,\big|\,\text{for every } d\in\mathbb{R}^2\}\\ &= \{\frac{\mathrm{d}}{\mathrm{d}t}\Big|_{t=0}\big((A,\begin{pmatrix} c+td\\ -(As)_3\end{pmatrix}),(\dot{A},-\dot{A}s)\big)\in T_pD_N\,\big|\text{for every } d\in\mathbb{R}^2\}\\ &= \{\big((0,\begin{pmatrix} d\\ 0\end{pmatrix}),(0,0)\big)\in T_pD_N\,\big|\,\text{for every } d\in\mathbb{R}^2\}.\end{aligned} \tag{85}$$

From equation (41) of chapter 1 we know that

$$(H_{D_N})_p = T_{(n,v)}\widetilde{\lambda}(D_n\times\mathbb{R}^3), \tag{86}$$

where $\widetilde{\lambda}$ is the map given by

$$\widetilde{\lambda}: N \times \mathbb{R}^3 \to D_N : (n,c) \mapsto \big(n, (\dot{A} = Ac^{\times}, \dot{a} = -Ac^{\times}s)\big), \tag{87}$$

which is the inverse of a trivialization of D_N. For $\big((Ac^{\times}, -Ac^{\times}s(u)), f\big) \in D_n \times \mathbb{R}^3$ we have

$$\begin{aligned}
T_{(n,c)}\widetilde{\lambda}\big((Ac^{\times}, -Ac^{\times}s(u)), f\big) \\
&= \frac{\mathrm{d}}{\mathrm{d}t}\Big|_{t=0} \widetilde{\lambda}\Big((Ae^{tc^{\times}}, \begin{pmatrix} b \\ -(Ae^{tc^{\times}}s(u))_3 \end{pmatrix}\Big), c+tf) \\
&= \frac{\mathrm{d}}{\mathrm{d}t}\Big|_{t=0} \Big((Ae^{tc^{\times}}, \begin{pmatrix} b \\ -(Ae^{tc^{\times}}s(u))_3 \end{pmatrix}\Big), (A(c+tf)^{\times}, -A(c+tf)^{\times}s(u))\Big) \\
&= \Big((Ac^{\times}, \begin{pmatrix} 0 \\ -(Ac^{\times}s(u))_3 \end{pmatrix}\Big), (Af^{\times}, -Af^{\times}s(u))\Big) \in (H_{D_N})_p,
\end{aligned} \tag{88}$$

where $p = \widetilde{\lambda}(n,c)$. From (86) and (88) we see that $T_p\mathcal{O}_p \cap (H_{D_N})_p = \{0\}$ and $T_pD_N = (H_{D_N})_p \oplus T_p\mathcal{O}_p$. Let

$$\rho : D_N \to \overline{D}_N = T\,\mathrm{SO}(3) : \big((A,a),(\dot{A},\dot{a})\big) \mapsto (A, \dot{A})$$

be the orbit map of the $\mathbb{R}^2$-action $\widehat{\Phi}|(\mathbb{R}^2 \times D_N)$ (65) on D_N. On D_N the distribution $U : D_N \to TD_N : u \mapsto U_u$ defined by

$$\begin{aligned} U_p = \big\{ w_p \in (H_{D_N})_p \,\big|\, \varpi_{D_N}(p)(v_p, w_p) = 0, \\ \text{for every } w_p \in \ker T_p\rho \cap (H_{D_N})_p \big\}, \end{aligned}$$

see proposition 3.3.25 of chapter 3, equals $(H_{D_N})_p$, because $\ker T_p\rho = T_p\mathcal{O}_p$ and $T_p\mathcal{O}_p \cap (H_{D_N})_p = \{0\}$. By definition the $\mathbb{R}^2$-reduced distribution

$$(\overline{H}_{\overline{D}_N})_{\overline{p}} = T_p\rho(p)(H_{D_N})_p = T_{\overline{p}=(A,\dot{A})}(T\,\mathrm{SO}(3)),$$

since $\ker T_p\rho = T_p\mathcal{O}_p$, $T_pD_N = \ker T_p\mathcal{O}_p \oplus (H_{D_N})_p$, and the map ρ is a surjective submersion. This proves (82).

To prove (83) we note (84). So (83) follows immediately from (54). □

Let $\mathrm{SO}(2) = \{R \in \mathrm{SO}(3) \,|\, Re_3 = e_3\}$. We have an $\mathrm{SO}(2)$-action on $T\,\mathrm{SO}(3)$ defined by

$$\varphi : \mathrm{SO}(2) \times T\,\mathrm{SO}(3) \to T\,\mathrm{SO}(3) : (R, (A, \dot{A})) \mapsto (RA, R\dot{A}). \tag{89}$$

Identifying $T\,\mathrm{SO}(3)$ with $\mathrm{SO}(3) \times \mathbb{R}^3$ using the map

$$\lambda : \mathrm{SO}(3) \times \mathbb{R}^3 \to T\,\mathrm{SO}(3) : (A, \omega) \mapsto (A, A\omega^{\times}),$$

the $\mathrm{SO}(2)$-action (89) becomes the action

$$\widetilde{\varphi} : \mathrm{SO}(2) \times (\mathrm{SO}(3) \times \mathbb{R}^3) : (R, (A, \omega)) \mapsto (RA, \omega). \tag{90}$$

We now prove

Lemma 6.5.3.12. *The* $\mathbb{R}^2$*-reduced 2-form* $\varpi_{\mathrm{SO}(3)\times\mathbb{R}^3}$ (83) *on the* $\mathbb{R}^2$*-reduced space* $T\,\mathrm{SO}(3) = \mathrm{SO}(3)\times\mathbb{R}^3$ *is invariant under the* SO(2)*-action* (90).

Proof. Because

$$(RA)^{-1}e_3 = A^{-1}R^{-1}e_3 = A^{-1}e_3 = u,$$

the point $u = A^{-1}e_3$ on the 2-sphere S^2 is fixed by the SO(2)-action (90). From (83) we get $\varphi_R^*\varpi_{\mathrm{SO}(3)\times\mathbb{R}^3} = \varpi_{\mathrm{SO}(3)\times\mathbb{R}^3}$ for every $R \in \mathrm{SO}(2)$. □

Corresponding to $e_3 \in \mathbb{R}^3$ there is the vector field $e_3^{\mathrm{SO}(3)}$ on SO(3) defined by $e_3^{\mathrm{SO}(3)}(A) = e_3^\times A$. Because

$$e_3^\times A = A(A^{-1}e_3^\times A) = A(A^{-1}e_3)^\times = Au^\times,$$

we see that $e_3^{\mathrm{SO}(3)} = u_{\rightarrow}^{\mathrm{SO}(3)} = u_{\rightarrow}$. From the invariance of $\varpi_{\mathrm{SO}(3)\times\mathbb{R}^3}$ under the SO(2)-action (90) it follows that

$$0 = L_{u_\rightarrow}\varpi_{\mathrm{SO}(3)\times\mathbb{R}^3} = u_\rightarrow \,\lrcorner\, (\mathrm{d}\varpi_{\mathrm{SO}(3)\times\mathbb{R}^3}) + \mathrm{d}(u_\rightarrow \,\lrcorner\, \varpi_{\mathrm{SO}(3)\times\mathbb{R}^3}),$$

that is,

$$u_\rightarrow \,\lrcorner\, (\mathrm{d}\varpi_{\mathrm{SO}(3)\times\mathbb{R}^3}) = -\,\mathrm{d}(u_\rightarrow \,\lrcorner\, \varpi_{\mathrm{SO}(3)\times\mathbb{R}^3}). \qquad (91)$$

We now prove

Proposition 6.5.3.13. *At each point of* $\mathrm{SO}(3)\times\mathbb{R}^3$ *where* $\langle s(u), u\rangle \neq 0$ *the kernel of the 2-form* $u_\rightarrow \,\lrcorner\, (\mathrm{d}\varpi_{\mathrm{SO}(3)\times\mathbb{R}^3})$ *is a 2-dimensional vector subspace of the tangent space of* $T\,\mathrm{SO}(3)$*, which contains the vector* $u_\rightarrow$*; otherwise it is at least 4-dimensional.*

Proof. Consider the kinetic energy metric k on SO(3) defined by

$$\mathrm{k}(A)(A\eta^\times, A\zeta^\times) = \langle \mathcal{Y}(s)\eta, \zeta\rangle.$$

Clearly k is invariant under the SO(2)-action (89) on $T\,\mathrm{SO}(3)$. Let $\omega_{\mathrm{SO}(3)}$ be the canonical symplectic form on $T^*\,\mathrm{SO}(3)$. The 2-form $\omega = (\lambda\circ \mathrm{k}^\flat)^*\omega_{\mathrm{SO}(3)}$ on $\mathrm{SO}(3)\times\mathbb{R}^3$ is invariant under the SO(2)-action (90) because λ, k, and $\omega_{\mathrm{SO}(3)}$ are. Therefore

$$0 = L_{u_\rightarrow}\omega = \mathrm{d}(u_\rightarrow \,\lrcorner\, \omega),$$

because $u_\rightarrow \,\lrcorner\, \mathrm{d}\omega = u_\rightarrow \,\lrcorner\, (\lambda\circ\mathrm{k}^\flat)^*\,\mathrm{d}\omega_{\mathrm{SO}(3)} = 0$. So

$$\mathrm{d}(u_\rightarrow \,\lrcorner\, \varpi_{\mathrm{SO}(3)\times\mathbb{R}^3}) = \mathrm{d}\left(u_\rightarrow \,\lrcorner\, (\varpi_{\mathrm{SO}(3)\times\mathbb{R}^3} - \omega)\right).$$

Using (54) we obtain

$$\begin{aligned}
&\omega(A,\omega)\big(\eta_{\rightarrow}+\zeta_{\uparrow},\widetilde{\eta}_{\rightarrow}+\widetilde{\zeta}_{\uparrow}\big)\\
&\quad=\langle\mathcal{Y}(s)\eta,\widetilde{\zeta}\rangle-\langle\mathcal{Y}(s)\zeta,\widetilde{\eta}\rangle+\langle\mathcal{Y}(s)\omega,\eta\times\widetilde{\eta}\rangle\\
&\qquad-\mathrm{m}\,\langle\omega\times Ds(u)(u\times\eta),\widetilde{\eta}\times s\rangle-\mathrm{m}\,\langle\omega\times s,\widetilde{\eta}\times Ds(u)(u\times\eta)\rangle\\
&\qquad\quad+\mathrm{m}\,\langle\omega\times Ds(u)(u\times\widetilde{\eta}),\eta\times s\rangle+\langle\omega\times s,\widetilde{\eta}\times Ds(u)(u\times\widetilde{\eta})\rangle.
\end{aligned}$$

Using (83) we get

$$\begin{aligned}
&(\overline{\varpi}_{\mathrm{SO}(3)\times\mathbb{R}^3}-\omega)(A,a)\big(\eta_{\rightarrow}+\zeta_{\uparrow},\widetilde{\eta}_{\rightarrow}+\widetilde{\zeta}_{\uparrow}\big)\\
&=\langle\mathcal{Y}(s)\eta,\widetilde{\zeta}\rangle-\langle\mathcal{Y}(s)\zeta,\widetilde{\eta}\rangle+\langle\mathcal{Y}(s)\omega,\eta\times\widetilde{\eta}\rangle\\
&-\mathrm{m}\,\langle s\times Ds(u)(u\times\eta),\widetilde{\eta}\rangle-\mathrm{m}\,\langle Ds(u)(u\times\eta),\widetilde{\eta}\rangle\\
&+\mathrm{m}\,\langle u\times(Ds(u)(\omega\times(\eta\times s))),\widetilde{\eta}\rangle-\mathrm{m}\,\langle u\times(Ds(u)((\omega\times s)\times\eta),\widetilde{\eta}\rangle\\
&\quad-\big(\langle\mathcal{Y}(s)\eta,\widetilde{\zeta}\rangle-\langle\mathcal{Y}(s)\zeta,\widetilde{\eta}\rangle+\langle\mathcal{Y}(s)\omega,\eta\times\widetilde{\eta}\rangle\\
&\quad-\mathrm{m}\,\langle s\times Ds(u)(u\times\eta),\widetilde{\eta}\rangle+\mathrm{m}\,\langle u\times Ds(u)(\omega\times(\eta\times s)),\widetilde{\eta}\rangle\big)\\
&=\mathrm{m}\,\langle u\times Ds(u)(\eta\times(\omega\times s)),\widetilde{\eta}\rangle-\mathrm{m}\,\langle Ds(u)(u\times\eta)\times(\omega\times s),\widetilde{\eta}\rangle,
\end{aligned}\tag{92}$$

which does not depend on either ζ or $\widetilde{\zeta}$. Therefore

$$\begin{aligned}
\Omega&=\big(u_{\rightarrow}\lrcorner\,(\overline{\varpi}_{\mathrm{SO}(3)\times\mathbb{R}^3}-\omega)\big)(A,a)\big(\eta_{\rightarrow}+\zeta_{\uparrow},\widetilde{\eta}_{\rightarrow}+\widetilde{\zeta}_{\uparrow}\big)\\
&=-\mathrm{m}\,\langle\omega\times s,u\times Ds(u)(u\times\widetilde{\eta})\rangle,
\end{aligned}\tag{93}$$

setting $\widetilde{\eta}=u$ in (92) and using $u\times u=0$. Consequently, the matrix of $\mathrm{d}(u_{\rightarrow}\lrcorner\,\overline{\varpi}_{\mathrm{SO}(3)\times\mathbb{R}^3})$ with respect to the basis $\{(e_i,0),(0,e_i)\}_{i=1}^3$ of the space $T_{(A,\omega)}(\mathrm{SO}(3)\ \times\mathbb{R}^3)=\mathbb{R}^3\times\mathbb{R}^3$, equals the matrix of $\mathrm{d}\Omega$, which is given by $\left(\begin{smallmatrix}B & -\mathrm{m}\,C^t\\ \mathrm{m}\,C & 0\end{smallmatrix}\right)$. Here C is the matrix of the linear transformation

$$\begin{aligned}
\mathbb{R}^3\rightarrow\mathbb{R}^3:\widetilde{\eta}&\mapsto s\times(u\times Ds(u)(u\times\widetilde{\eta}))\\
&=\langle s,Ds(u)(u\times\widetilde{\eta})\rangle\,u-\langle s,u\rangle\,Ds(u)(u\times\widetilde{\eta}).
\end{aligned}\tag{94}$$

If $C\widetilde{\eta}=0$ and $\langle s,u\rangle\neq0$, then (94) implies that $Ds(u)(u\times\widetilde{\eta})$ is a multiple μ of u. Because $Ds(u)(u\times\widetilde{\eta})\in T_{s(u)}\mathcal{S}$, which is perpendicular to u, it follows that $\mu=0$. Hence $u\times\widetilde{\eta}=0$, since $Ds(u)$ is invertible, being symmetric and positive definite. Therefore $\widetilde{\eta}=\widetilde{\mu}u$. Consequently, $\ker C$ is 1-dimensional, which implies that $\ker C^t$ is 1-dimensional also. Therefore for any $(A,\omega)\in\mathrm{SO}(3)\times\mathbb{R}^3$ the kernel of $-\mathrm{d}(u_{\rightarrow}\lrcorner\,\overline{\varpi}_{\mathrm{SO}(3)\times\mathbb{R}^3}(A,\omega)$ is at most 2-dimensional. Because $\mathrm{d}\overline{\varpi}_{\mathrm{SO}(3)\times\mathbb{R}^3}(A,\omega)$ is skew symmetric, it follows that $u_{\rightarrow}\lrcorner\,\big(u_{\rightarrow}\lrcorner\,\mathrm{d}\overline{\varpi}_{\mathrm{SO}(3)\times\mathbb{R}^3}\big)(A,\omega)=0$. In other words, $u_{\rightarrow}\in\ker\big(u_{\rightarrow}\lrcorner\,\mathrm{d}\overline{\varpi}_{\mathrm{SO}(3)\times\mathbb{R}^3}\big)(A,\omega)$. Since the kernel of a 2-form is even dimensional, we find that the kernel of $u_{\rightarrow}\lrcorner\,\mathrm{d}\overline{\varpi}_{\mathrm{SO}(3)\times\mathbb{R}^3}$ is 2-dimensional at any $(A, \omega) \in \mathrm{SO}(3) \times \mathbb{R}^3$ where $\langle s(u), u \rangle \neq 0$. This proves the first conclusion of the proposition.

If $\langle s(u), u \rangle = 0$, then C is the linear map

$$C : \mathbb{R}^3 \to \mathbb{R}^3 : \widetilde{\eta} \mapsto \langle s, Ds(u)(u \times \widetilde{\eta}) \rangle \, u.$$

Hence $\ker C$ is the kernel of the linear form

$$\mathbb{R}^3 \to \mathbb{R} : \widetilde{\eta} \mapsto \langle s, Ds(u)(u \times \widetilde{\eta}) \rangle,$$

which is at least 2-dimensional. So $\dim \ker C^t \geq 2$. Because $u_{\rightarrow}$ lies in $\ker \big(u_{\rightarrow} \lrcorner \, \mathrm{d}\overline{\varpi}_{\mathrm{SO}(3) \times \mathbb{R}^3}\big)$, it follows that the kernel of the matrix M at (A, ω) contains the vector space $\mathrm{span}\{u_{\rightarrow}, \ker C\}$, whose dimension is at least 3. Because $\mathrm{d}(u_{\rightarrow} \lrcorner \, \mathrm{d}\overline{\varpi}_{\mathrm{SO}(3) \times \mathbb{R}^3})$ is skew symmetric, its kernel is even dimensional and hence has dimension greater than or equal to 4. This proves the second assertion and thus the proposition. □

If the center of mass of the body $\mathcal{B}$, which is assumed to be at the origin, is in the interior of its bounding surface $\mathcal{S}$, then $\langle s(u), u \rangle < 0$ at every point of $\mathcal{S}$, because we assumed that u is an *inward* pointing normal to $\mathcal{S}$. If the center of mass of $\mathcal{B}$ is on $\mathcal{S}$, then the locus of points on $\mathcal{S}$ where $\langle s(u), u \rangle = 0$ is a codimension 2 submanifold defined by $s = s(u) = 0$. If the center of mass of $\mathcal{B}$ is outside of $\mathcal{S}$, then the locus of points on $\mathcal{S}$, where $\langle s(u), u \rangle = 0$ is a codimension 1 submanifold traced out by all normals to $\mathcal{S}$ which pass through the origin, which is the center of mass of $\mathcal{B}$.

6.6 Reduction of E(2) symmetry

In this section we remove the symmetry group E(2) of Euclidean motions of the plane from the smooth strongly convex rigid body rolling without slipping on a horizontal plane by applying the reduction process described in chapter 3.

E(2) is the subgroup of E(3) consisting of all $(R, x) \in \mathrm{SO}(3) \times \mathbb{R}^3 = \mathrm{E}(3)$ such that R is a rotation about the e_3-axis in $\mathbb{R}^3$ and $x \in \mathbb{R}^2 \times \{0\} \subseteq \mathbb{R}^3$. E(2) acts on $\mathbb{R}^3$ by sending $((R, x), y)$ to $Ry + x$. We let E(2) act on E(3) by

$$\mathrm{E}(2) \times \mathrm{E}(3) \to \mathrm{E}(3) : \big((R, x), (A, a)\big) \mapsto (RA, Ra + x).$$

This induces an E(2) action on $T\mathrm{E}(3)$ given by

$$\mathrm{E}(2) \times T\mathrm{E}(3) \to T\mathrm{E}(3) : \big((R, x), ((A, a), (\dot{A}, \dot{a}))\big) \mapsto \big((Ra, Ra + x), (R\dot{A}, R\dot{a})\big).$$

Finally, using the trivialization

$$TE(3) \to (\mathrm{SO}(3)\times\mathbb{R}^3)\times\mathbb{R}^3\times\mathbb{R}^3 = \mathrm{E}(3)\times\mathbb{R}^6 : \big((A,a),(\dot{A},\dot{a})\big) \mapsto (A,a,\omega,b),$$

where $\dot{A} = A\omega^{\times}$ and $\dot{a} = Ab$, leads to the action

$$\begin{aligned}\Phi : \mathrm{E}(2)\times(\mathrm{E}(3)\times\mathbb{R}^6) \to \mathrm{E}(3)\times\mathbb{R}^6 :&\\ \big((R,x),(A,a,\omega,b)\big) \mapsto (RA, Ra+x, \omega, b).&\end{aligned} \tag{95}$$

6.6.1 E(2) *symmetry*

We now explain what we mean by E(2) being a symmetry group of the strongly convex rolling rigid body.

The smooth strongly convex body rolling on a horizontal plane is a distributional Hamiltonian system $(D_N, H_{D_N}, \varpi_{D_N}, E_{D_N})$, where

1. D_N is the accessible set of the constraint distribution D on E(3) defined by

$$D_N = \{(A,a,\omega,b) \in \mathrm{E}(3)\times\mathrm{e}(3) \,\big|\, b = -\omega\times s(u), \langle a, e_3\rangle = -\langle s(u), u\rangle\}, \tag{96}$$

see lemma 6.3.3;

2. H_{D_N} is the ϖ_{D_N}-symplectic distribution on D_N, which at $v \in D_N$ is

$$\begin{aligned}&(H_{D_N})_v \\ &= \{(Ay^{\times}, -A(y\times s(u)), z^{\times}, -z\times(s(u)) + \omega\times Ds(u)(u\times y)) \in T_vD_N \,\big|\, y,z\in\mathbb{R}^3\}.\end{aligned} \tag{97}$$

3. ϖ_{D_N} is the symplectic form

$$\begin{aligned}&\varpi_{D_N}(v)\big((Ay^{\times}, -A(y\times s), z^{\times}, -z\times s + \omega\times Ds(u)(u\times y),\\ &\quad (A\widetilde{y}^{\times}, -A(\widetilde{y}\times s, \widetilde{z}^{\times}, -\widetilde{z}\times s + \omega Ds(u)(u\times\widetilde{y})\big)\\ &\quad = -\langle \mathcal{Y}(s)(z), \widetilde{y}\rangle + \langle \mathcal{Y}(s)(\widetilde{z}, y\rangle + \langle \mathcal{Y}(s)\omega, y\times\widetilde{y}\rangle\\ &\qquad - \mathrm{m}\,\langle(\omega\times Ds(u)(u\times y))\times s, \widetilde{y}\rangle + \mathrm{m}\,\langle(\omega\times Ds(u)(u\times\widetilde{y}))\times s, y\rangle.\end{aligned} \tag{98}$$

4. E_{D_N} is the Hamiltonian

$$E_{D_N} : D_N \subseteq D \to \mathbb{R} : v \to \tfrac{1}{2}\,\langle I(\omega), \omega\rangle + \tfrac{1}{2}\,\mathrm{m}\,\langle b, b\rangle + \mathrm{mg}\,\langle a, e_3\rangle. \tag{99}$$

We now prove

Proposition 6.6.1.14. E(2) *is a symmetry of the distributional Hamiltonian system* $(D_N, H_{D_N}, \varpi_{D_N}, E_{D_N})$.

Proof. 1. From the definition of Φ (95) it follows the Φ leaves b and ω unchanged. However, the SO(2) action $R \cdot u$ on u leaves u fixed, since

$$(RA)^{-1}e_3 = A^{-1}R^{-1}e_3 = A^{-1}e_3.$$

Because the action Φ maps $\dot{A}$ to $R\dot{A}$ and $\dot{a}$ to $R\dot{a}$, it maps the defining equation $\dot{A}s(u) + \dot{a} = 0$ of D_N to $R\dot{A}s(u) + R\dot{a} = 0$. Therefore D_N is invariant under Φ.

2. Differentiating the map $\Phi_{(R,x)}$ gives

$$T_v\Phi_{(R,x)}(\dot{A}, \dot{a}, \dot{\omega}, \dot{b}) = (R\dot{A}, R\dot{a}, \dot{\omega}, \dot{b}), \tag{100}$$

for every $v \in D_N$, every $(R, x) \in \mathrm{E}(2)$, and every $(\dot{A}, \dot{a}, \dot{\omega}, \dot{b}) \in T_vD_N$. Applying $T_v\Phi_{(R_\theta,x)}$ to

$$w_v = (Ay^\times, -A(y \times s(u)), z^\times, -z \times s(u) + \omega \times Ds(u)(u \times y) \in H_v$$

gives

$$\begin{aligned} T_v\Phi_{(R,x)}w_v = \big(&RAy^\times, -RA(y \times s(R \cdot u)), z^\times, -z \times s(R \cdot u) \\ &+ \omega \times Ds(R \cdot u)(R \cdot u \times y)\big). \end{aligned} \tag{101}$$

From the definition of the distribution H_{D_N} (97) it follows that $T_v\Phi_{(R,x)}w_v \in H_{D_N}$. Thus H_{D_N} is Φ-invariant.

3. From the definition of the 2-form $\varpi_{D_N}(v)$ (98) we see that

$$\begin{aligned} &\varpi_{D_N}(\Phi_{(R,x)}(v))(T_v\Phi_{(R,x)}w_v, T_v\Phi_{(R,x)}w_v') = \\ &\quad -\langle \mathcal{Y}(s(R \cdot u))(z), \widetilde{y}\rangle + \langle \mathcal{Y}(s(R \cdot u))(\widetilde{z}, y\rangle + \langle \mathcal{Y}(s(R \cdot u))\omega, y \times \widetilde{y}\rangle \\ &\qquad - \mathrm{m}\,\langle(\omega \times Ds(R \cdot u)(R \cdot u \times y) \times s(R \cdot u), \widetilde{y}\rangle \\ &\qquad\quad + \mathrm{m}\,\langle(\omega \times Ds(R \cdot u)(R \cdot u \times \widetilde{y})) \times s(R \cdot u), y\rangle \\ &\quad = \varpi_{D_N}(v)(w_v, w_v'), \end{aligned}$$

since $R \cdot u = u$. Thus ϖ_{D_N} is Φ-invariant.

4. We compute

$$\begin{aligned} E_{D_N}(RA, Ra + x, \omega, b) &= \tfrac{1}{2}\langle I(\omega), \omega\rangle + \tfrac{1}{2}\mathrm{m}\,\langle b, b\rangle + \mathrm{mg}\,\langle Ra + x, e_3\rangle \\ &= E_{D_N}(A, a, \omega, b). \end{aligned} \tag{102}$$

Thus the Hamiltonian E_{D_N} is Φ-invariant.

Consequently, E(2) is a symmetry of the distributional Hamiltonian system $(D_N, H_{D_N}, \varpi_{D_N}, E_{D_N})$ which describes the strongly convex rolling rigid body. □

6.6.2 E(2)-*orbit space*

In this subsection we reduce the E(2)-symmetry of the rolling rigid body.

The first step is to construct the space of E(2)-orbits $\overline{D_N} = D_N/\mathrm{E}(2)$ of the E(2)-action Φ restricted to D_N. Towards this goal we prove

Lemma 6.6.2.15. *The* E(2)*-action*

$$\widetilde{\Phi} : \mathrm{E}(2) \times D_N \to D_N : \big((R,x),(A,a,\omega,b)\big) \mapsto (RA, Ra + x, \omega, b)$$

is free and proper.

Proof. Suppose that $(RA, Ra + x, \omega, b) = (A, a, \omega, b)$. Then $R = e$ and $x = 0$. Thus the action $\widetilde{\Phi}$ is free.

Next we need to show that the E(2)-action $\widetilde{\Phi}$ on D_N is proper. We will use the criterion given in paragraph 4 of §2 of chapter 2. Suppose that the sequences $\{(A_n, a_n, \omega_n, b_n)\}$ and $\{(R_nA_n, R_na_n + x_n, \omega_n, b_n)\}$ in D_N converge. We must show that there is a subsequence of the sequence $\{(R_n, x_n)\}$ in E(2) which converges. Towards this goal, we observe that the sequence $\{R_n = (R_nA_n)A_n^{-1}\}$ in $\mathrm{SO}(3,\mathbb{R})$ converges, since by hypothesis the sequence $\{R_nA_n\}$ in $\mathrm{SO}(3,\mathbb{R})$ converges and the sequence $\{A_n^{-1}\}$ in $\mathrm{SO}(3,\mathbb{R})$ converges because by hypothesis the sequence $\{A_n\}$ in $\mathrm{SO}(3,\mathbb{R})$ converges. Next the sequence $\{x_n = (R_na_n + x_n) - R_na_n\}$ in $\mathbb{R}^3$ converges, since the sequences $\{R_na_n + x_n\}$ and $\{a_n\}$ converge in $\mathbb{R}^3$ by hypothesis. So the sequence $\{R_na_n\}$ in $\mathbb{R}^3$ converges. Therefore the sequence $\{(R_n, x_n)\}$ in E(2) converges. This shows that the E(2)-action $\widetilde{\Phi}$ is proper. □

From lemma 6.6.2.15 we see that the orbit space $\overline{D_N}$ is a smooth manifold such that the canonical projection map $\rho : D_N \to D_N/\mathrm{E}(2) = \overline{D_N}$ defines an E(2) principal bundle. More precisely, we have

Lemma 6.6.2.16. *Let*

$$\pi : D_N \to S^2 \times \mathbb{R}^3 : (A, a, \omega, b) \mapsto (u, \omega), \tag{103}$$

where $u = A^{-1}e_3$*, and let* $\rho : D_N \to D_N/\mathrm{E}(2) = \overline{D}_N$ *be the orbit map of the* E(2)*-action* Φ (95). *Then there is a unique map* $\psi : \overline{D_N} \to S^2 \times \mathbb{R}^3$ *such that* $\pi = \psi \circ \rho$. *Moreover,* ψ *is a diffeomorphism.*

Proof. The map π is constant on E(2)-orbits because

$$\begin{aligned}\pi(RA, Ra + x, \omega, b) &= ((RA)^{-1}e_3, \omega) = (A^{-1}e_3, \omega) \\ &= (u, \omega) = \pi(A, a, \omega, b).\end{aligned}$$

This proves the existence of the map ψ. Also ψ is unique because ρ is surjective.

Next we show that each fiber of π is a single E(2)-orbit on D_N, which implies that ψ is injective. Suppose that $(A, a, \omega, b), (A', a', \omega', b') \in D_N$ and that

$$\begin{aligned}(u, \omega) &= (A^{-1}e_3, \omega) = \pi(A, a, \omega, b)\\ &= \pi(A', a', \omega', b') = ((A')^{-1}e_3, \omega') = (u', \omega').\end{aligned}$$

Then $\omega = \omega'$ and $u = u'$. Written out the second equation reads

$$A^{-1}e_3 = u = u' = (A')^{-1}e_3,$$

which implies that $Re_3 = e_3$, where $R = A(A')^{-1}$. Because A and A' lie in $\mathrm{SO}(3, \mathbb{R})$, the matrix of R with respect to the orthonormal basis $\{e_1, e_2, e_3\}$ of $\mathbb{R}^3$ is $\left(\begin{smallmatrix}\widetilde{R} & 0\\ 0 & 1\end{smallmatrix}\right)$, where $\widetilde{R} \in \mathrm{SO}(2, \mathbb{R})$.

Since (A, a, ω, b) and (A', a', ω', b') lie in D_N, it follows that

$$b = -\omega \times s(u) = -\omega' \times s(u') = b'$$

and

$$\begin{aligned}\langle a, e_3\rangle + \langle s(u), u\rangle &= 0 = \langle a', e_3\rangle + \langle s(u'), u'\rangle\\ &= \langle a', e_3\rangle + \langle s(u), u\rangle, \quad \text{since } u = u'.\end{aligned}$$

Therefore $\langle a, e_3\rangle = \langle a', e_3\rangle$. Now

$$\langle Ra, e_3\rangle = \langle a, R^{-1}e_3\rangle = \langle a, e_3\rangle = \langle a', e_3\rangle,$$

which implies $\langle a' - Ra, e_3\rangle = 0$. So there is a vector $x \in \mathrm{span}\{e_1, e_2\}$ such that $a' = Ra + x$.

The above argument shows that there is a rotation $R = \left(\begin{smallmatrix}\widetilde{R} & 0\\ 0 & 1\end{smallmatrix}\right)$, where $\widetilde{R} \in \mathrm{SO}(2, \mathbb{R})$, and a vector $x \in \mathrm{span}\{e_1, e_2\}$ such that $(A', a'\omega', b') = (RA, Ra + x, \omega, b)$. In other words, (A, a, ω, b) and (A', a', ω', b') lie in the same E(2)-orbit on D_N.

Because the mapping

$$\mathrm{SO}(3) \to S^2 \subseteq \mathbb{R}^3 : A \mapsto A^{-1}e_3$$

is a submersion, the smooth map π (103) is a submersion. Let $\sigma : U \subseteq \overline{D}_N \to D_N$ be a smooth local section of ρ, defined on an open neighborhood U of a point $\overline{v} \in \overline{D}_N$. Then on U we have $\psi = \psi \circ \rho \circ \sigma = \pi \circ \sigma$, which shows that the mapping ψ is smooth. Using a local section of π, a similar argument shows that ψ^{-1} is smooth as well. Consequently, ψ is a diffeomorphism. □

Another way to find the E(2)-reduced space $D_N/\mathrm{E}(2)$ is first to note that $\mathbb{R}^2$ is a normal subgroup of E(2). Then the E(2)-action on D_N induces an action of the quotient group $\mathrm{E}(2)/\mathbb{R}^2$, which is isomorphic to SO(2), on the $\mathbb{R}^2$-orbit space $D_N/\mathbb{R}^2 = T\,\mathrm{SO}(3)$ (70). In particular, the induced SO(2)-action on $T\,\mathrm{SO}(3)$ is given by

$$\mathrm{SO}(2) \times T\,\mathrm{SO}(3) \to T\,\mathrm{SO}(3) : (R, (A, \dot{A})) \mapsto (RA, R\dot{A}),$$

which using

$$\mathrm{SO}(3) \times \mathbb{R}^3 \to T\,\mathrm{SO}(3) : (A, \omega) \mapsto (A, A\omega^{\times}) = (A, \dot{A}),$$

becomes the SO(2) action

$$\mathrm{SO}(2) \times (\mathrm{SO}(3) \times \mathbb{R}^3) \to \mathrm{SO}(3) \times \mathbb{R}^3 : (R, (A, \omega)) \mapsto (RA, \omega). \tag{104}$$

Therefore the orbit space $(\mathrm{SO}(3) \times \mathbb{R}^3)/\,\mathrm{SO}(2)$ of the SO(2)-action (104) is $S^2 \times \mathbb{R}^3$, since

$$\widetilde{\rho} : \mathrm{SO}(3) \times \mathbb{R}^3 \to S^2 \times \mathbb{R}^3 : (A, \omega) \mapsto (A^{-1}e_3 = u, \omega) \tag{105}$$

is the orbit map of the action (104). Hence the E(2)-orbit space $D_N/\mathrm{E}(2)$ is $\big(D_N/\mathbb{R}^2\big)/\,\mathrm{SO}(2)$, which in turn is $(\mathrm{SO}(3) \times \mathbb{R}^3)/\,\mathrm{SO}(2) = S^2 \times \mathbb{R}^3$.

6.6.3 E(2)-*reduced distribution and 2-form*

In this subsection we determine the E(2)-reduced distribution $\overline{H}_{S^2\times\mathbb{R}^3}$ and 2-form $\overline{\varpi}_{S^2\times\mathbb{R}^3}$. We show that the accessible set of $\overline{H}_{S^2\times\mathbb{R}^3}$ is $S^2 \times \mathbb{R}^3$ and give conditions when $S^2 \times \mathbb{R}^3$ has a contact structure.

Because each point $u = A^{-1}e_3$ of the 2-sphere is left fixed by the SO(2) action (104) the total energy

$$E = \tfrac{1}{2}\,\langle \mathcal{Y}(s(u))\omega, \omega\rangle + \mathrm{mg}\,\langle s(u), u\rangle \tag{106}$$

on $\mathrm{SO}(3) \times \mathbb{R}^3$, the 2-form

$$\begin{aligned}\varpi_{\mathrm{SO}(3)\times\mathbb{R}^3}(A,\omega)(\eta_{\rightarrow} + \zeta_{\uparrow}, \widetilde{\eta}_{\rightarrow} + \widetilde{\zeta}_{\uparrow}) &= \langle \mathcal{Y}(s(u))\eta, \widetilde{\zeta}\rangle - \langle \mathcal{Y}(s(u))\zeta, \widetilde{\eta}\rangle \\ &+ \langle \mathcal{Y}(s(u))\omega, \eta \times \widetilde{\eta}\rangle - \mathrm{m}\,\langle s \times (\omega \times Ds(u)(u \times \eta)), \widetilde{\eta}\rangle \\ &\qquad + \mathrm{m}\,\langle u \times Ds(u)(\omega \times (\eta \times s)), \widetilde{\eta}\rangle,\end{aligned} \tag{107}$$

see (83), and the distribution $H_{\mathrm{SO}(3)\times\mathbb{R}^3}$ (82) on $\mathrm{SO}(3) \times \mathbb{R}^3$ are invariant under the SO(2) action. Thus we obtain the E(2)-reduced distributional Hamiltonian system $(\overline{H}_{S^2\times\mathbb{R}^3}, T(S^2 \times \mathbb{R}^2), \overline{\varpi}_{S^2\times\mathbb{R}^3}, E_{S^2\times\mathbb{R}^3} = E)$ by reducing the SO(2)-symmetry (104) on $\mathrm{SO}(3) \times \mathbb{R}^3$.

We start by finding the SO(2)-reduced distribution $\overline{H}_{S^2\times\mathbb{R}^3}$. Let γ be the 1-form on $\mathbb{R}^9$ with coordinates (u,s,ω) given by

$$\langle \mathcal{Y}(s)u, \mathrm{d}\omega\rangle + \langle \mathcal{Y}(s)\omega, \mathrm{d}u\rangle + \mathrm{m}\,\langle u,\omega\rangle\,\langle s, \mathrm{d}s\rangle. \tag{108}$$

Lemma 6.6.3.17. *Consider the mapping*

$$\mu : S^2\times\mathbb{R}^3 \to \mathbb{R}^9 : (u,\omega)\mapsto (u, s(u), \omega).$$

Let α be the 1-form on $S^2\times\mathbb{R}^3$ given by $\mu^*\gamma$. Then

$$\widetilde{\rho}^{\,*}\alpha = u_{\rightarrow} \lrcorner\, \varpi_{\mathrm{SO}(3)\times\mathbb{R}^3}, \tag{109}$$

where $\widetilde{\rho}$ is the SO(2) *orbit map* (105) *on* $\mathrm{SO}(3)\times\mathbb{R}^3$.

Proof. Using (107) we find that

$$\begin{aligned}
&(u_{\rightarrow} \lrcorner\, \varpi_{\mathrm{SO}(3)\times\mathbb{R}^3}(A,\omega)(\widetilde{\eta}_{\rightarrow} + \widetilde{\zeta}_{\uparrow})\\
&\quad = \langle \mathcal{Y}(s)u, \widetilde{\zeta}\rangle + \langle \mathcal{Y}(s)\omega, u\times\widetilde{\eta}\rangle + \mathrm{m}\,\langle u\times Ds(u)(\omega\times(u\times s)), \widetilde{\eta}\rangle,\\
&\qquad\qquad \text{since } u_{\rightarrow} = (Au^{\times},0), \text{ which implies } \eta = u \text{ and}\\
&\qquad\qquad \zeta = 0 \text{ in (107)}\\
&\quad = \langle \mathcal{Y}(s)u, \widetilde{\zeta}\rangle + \langle \mathcal{Y}(s)\omega, u\times\widetilde{\eta}\rangle - \mathrm{m}\,\langle \omega\times(u\times s), Ds(u)(u\times\widetilde{\eta})\rangle\\
&\quad = \langle \mathcal{Y}(s)u, \widetilde{\zeta}\rangle + \langle \mathcal{Y}(s)\omega, u\times\widetilde{\eta}\rangle - \mathrm{m}\,\langle \omega, s\rangle\,\langle u, Ds(u)(u\times\widetilde{\eta})\rangle\\
&\qquad + \mathrm{m}\,\langle u,\omega\rangle\,\langle s, Ds(u)(u\times\widetilde{\eta})\rangle\\
&\quad = \langle \mathcal{Y}(s)u, \widetilde{\zeta}\rangle + \langle \mathcal{Y}(s)\omega, u\times\widetilde{\eta}\rangle + \mathrm{m}\,\langle u,\omega\rangle\,\langle s, Ds(u)(u\times\widetilde{\eta})\rangle, \qquad (110)\\
&\qquad\qquad \text{since } Ds(u)(u\times\widetilde{\eta}) \in T_{s(u)}\mathcal{S} \text{ and } u \perp T_{s(u)}\mathcal{S}.
\end{aligned}$$

Under the tangent $T\widetilde{\rho}$ of the SO(2)-orbit map $\widetilde{\rho}$ (105), the image of the vector field $\widetilde{\eta}_{\rightarrow} + \widetilde{\zeta}_{\uparrow}$ given by $(A,\omega)\mapsto (A\widetilde{\eta}^{\times}, \widetilde{\zeta})$ is the vector field

$$\begin{aligned}
(u,\omega) \mapsto \frac{\mathrm{d}}{\mathrm{d}t}\Big|_{t=0} \widetilde{\rho}\big(A\mathrm{e}^{t\widetilde{\eta}^{\times}}, \omega + t\widetilde{\zeta}\big) &= \frac{\mathrm{d}}{\mathrm{d}t}\Big|_{t=0} \big((A\mathrm{e}^{t\widetilde{\eta}^{\times}})^{-1}e_3, \omega + t\widetilde{\zeta}\big)\\
&= (-\widetilde{\eta}^{\times}u, \widetilde{\zeta}) = (u\times\widetilde{\eta}, \widetilde{\zeta})
\end{aligned} \tag{111}$$

on $S^2\times\mathbb{R}^3$. Note that $\widetilde{\eta}_{\rightarrow} + \widetilde{\zeta}_{\uparrow} \in \ker T\widetilde{\rho}$ if and only if $u\times\widetilde{\eta} = 0 = \widetilde{\zeta}$, that is, $\ker T\widetilde{\rho} = \mathrm{span}\{u_{\rightarrow}\}$. Let $\widetilde{\zeta}^t = (\mu\circ\widetilde{\rho})^*\,\mathrm{d}\omega$, $(u\times\widetilde{\eta})^t = (\mu\circ\widetilde{\rho})^*\,\mathrm{d}u$, and $(\mu\circ\widetilde{\rho})^*\,\mathrm{d}s = \big(Ds(u)(u\times\widetilde{\eta})\big)^t$. Then

$$\begin{aligned}
\widetilde{\rho}^{\,*}\alpha &= (\mu\circ\widetilde{\rho})^*\gamma\\
&= \langle \mathcal{Y}(s)u, \widetilde{\zeta}\rangle + \langle \mathcal{Y}(s)\omega, u\times\widetilde{\eta}\rangle + \langle u,\omega\rangle\,\langle s, Ds(u)(u\times\widetilde{\eta})\rangle.
\end{aligned} \tag{112}$$

Since the right hand sides of (110) and (112) are equal, we have proved (109). □

Corollary 6.6.3.18. *The* E(2)*-reduced distribution* $\overline{H}_{S^2\times\mathbb{R}^3}$ *is* $\ker\alpha$.

Proof. Since $\ker T_{(A,\omega)}\widetilde{\rho} = \mathrm{span}\{u_{\rightarrow}(A,\omega)\}$, the distribution U on $\mathrm{SO}(3)\times\mathbb{R}^3$ is defined by

$$\begin{aligned}U_{(A,\omega)} &= \{v_{(A,\omega)} \in \big(H_{\mathrm{SO}(3)\times\mathbb{R}^3}\big)_{(A,\omega)} \,\big|\, \varpi_{\mathrm{SO}(3)\times\mathbb{R}^3}(A,\omega)\big(v_{(A,\omega)}, w_{(A,\omega)}\big) = 0,\\ &\qquad\qquad \text{for all } w_{(A,\omega)} \in \ker T_{(A,\omega)}\widetilde{\rho} \cap \big(H_{\mathrm{SO}(3)\times\mathbb{R}^3}\big)_{(A,\omega)}\}\\ &= \mathrm{span}\{u_{\rightarrow}(A,\omega)\} = \ker\big(u_{\rightarrow} \mathbin{\lrcorner} \varpi_{\mathrm{SO}(3)\times\mathbb{R}^3}\big)(A,\omega).\end{aligned}$$

From (109) it follows that

$$\begin{aligned}\alpha(A,\omega)T_{(A,\omega)}\widetilde{\rho}\,U_{(A,\omega)} &= \widetilde{\rho}^{\,*}\alpha|U_{(A,\omega)}\\ &= \big(u_{\rightarrow} \mathbin{\lrcorner} \varpi_{\mathrm{SO}(3)\times\mathbb{R}^3}\big)(A,\omega)|U_{(A,\omega)} = 0. \qquad (113)\end{aligned}$$

Using proposition 3.3.25 of chapter 3, we see that the SO(2)*-reduced distribution* $\big(\overline{H}_{S^2\times\mathbb{R}^3}\big)_{(A,\omega)}$ equals $T_{(A,\omega)}\widetilde{\rho}\,U_{(A,\omega)}$. From (113) we deduce that $\overline{H}_{S^2\times\mathbb{R}^3} = \ker\alpha$, which is a codimension 1 subbundle of $T(S^2\times\mathbb{R}^3)$. □

By proposition 3.3.26 of chapter 3 the SO(2)*-reduced* 2*-form* $\overline{\varpi}_{S^2\times\mathbb{R}^3}$ on $\overline{H}_{S^2\times\mathbb{R}^3} = \ker\alpha$ is

$$\overline{\varpi}_{S^2\times\mathbb{R}^3}(u,\omega)\big((v,w),(\widetilde{v},\widetilde{w})\big) = \varpi_{\mathrm{SO}(3)\times\mathbb{R}^3}\big(\eta_{\rightarrow}+\zeta_{\uparrow}, \widetilde{\eta}_{\rightarrow}+\widetilde{\zeta}_{\uparrow}\big),$$

where $\widetilde{\rho}(A,\omega) = (u,\omega)$. In addition, $\eta_{\rightarrow}+\zeta_{\uparrow} = (A\eta^{\times},\zeta)$ and $\widetilde{\eta}_{\rightarrow}+\widetilde{\zeta}_{\uparrow} = (A\widetilde{\eta}^{\times},\widetilde{\zeta})$ lie in $U_{(A,\omega)} = \ker(u_{\rightarrow} \mathbin{\lrcorner} \varpi_{\mathrm{SO}(3)\times\mathbb{R}^3})$. Also $(v,w) = T_{(A,\omega)}\widetilde{\rho}(A\eta^{\times},\zeta) = (u\times\eta,\zeta)$ and $(\widetilde{v},\widetilde{w}) = T_{(A,\omega)}\widetilde{\rho}(A\widetilde{\eta}^{\times},\widetilde{\zeta}) = (u\times\widetilde{\eta},\widetilde{\zeta})$ lie in $\big(\overline{H}_{S^2\times\mathbb{R}^3}\big)_{(u,\omega)}$. Using (107) and (110) we have proved

Lemma 6.6.3.19. *The* SO(2)*-reduced* 2*-form* $\overline{\varpi}_{S^2\times\mathbb{R}^3}$ *on* $\overline{H}_{S^2\times\mathbb{R}^3}$ *is given by*

$$\begin{aligned}&\overline{\varpi}_{S^2\times\mathbb{R}^3}(u,\omega)\big((u\times\eta,\zeta),(u\times\widetilde{\eta},\widetilde{\zeta})\big)\\ &\quad = \langle\mathcal{Y}(s)\eta,\widetilde{\zeta}\rangle + \langle\mathcal{Y}(s)\zeta,\widetilde{\eta}\rangle + \langle\mathcal{Y}(s)\omega,\eta\times\widetilde{\eta}\rangle \qquad (114)\\ &\quad - \mathrm{m}\,\langle s\times(\omega\times Ds(u)(u\times\eta)),\widetilde{\eta}\rangle + \mathrm{m}\,\langle u\times Ds(u)(\omega\times(\eta\times s)),\widetilde{\eta}\rangle,\end{aligned}$$

where

$$0 = \langle\mathcal{Y}(s(u))u,\zeta\rangle + \langle\mathcal{Y}(s(u))\omega,u\times\eta\rangle + \mathrm{m}\,\langle u,\omega\rangle\,\langle s(u),Ds(u)(u\times\eta)\rangle$$

and

$$0 = \langle\mathcal{Y}(s(u))u,\widetilde{\zeta}\rangle + \langle\mathcal{Y}(s(u))\omega,u\times\widetilde{\eta}\rangle + \mathrm{m}\,\langle u,\omega\rangle\,\langle s(u),Ds(u)(u\times\widetilde{\eta})\rangle.$$

Until the end of this subsection we will simplify notation by writing H for $\overline{H}_{S^2 \times \mathbb{R}^3}$ and ϖ for $\overline{\varpi}_{S^2 \times \mathbb{R}^3}$.

Next we prove

Proposition 6.6.3.20. *The accessible set of the distribution H is all of $S^2 \times \mathbb{R}^3$.*

Proof. First assume that we are at a point $\widetilde{\rho}(A, \omega) = (u, \omega) \in S^2 \times \mathbb{R}^3$ where $\langle s(u), u \rangle \neq 0$. Then

$$\ker \mathrm{d}(u_{\rightarrow} \lrcorner\, \varpi) = \mathrm{d}(\widetilde{\rho}^* \alpha) = \widetilde{\rho}^* (\mathrm{d}\alpha)$$

is a 2-dimensional subspace of $T_{(A,\omega)}(\mathrm{SO}(3) \times \mathbb{R}^3)$, which contains the subspace $\mathrm{span}\{u_{\rightarrow}(A, \omega)\}$. Let $\beta = \mathrm{d}\alpha$. Because $\ker T\widetilde{\rho} = \mathrm{span}\{u_{\rightarrow}\}$, it follows that $Z = \ker \beta$ is 1-dimensional. Let $K \subseteq H$ be the kernel of the 2-form $\mathrm{d}\alpha|(H \times H)$. If $Z \cap H = \{0\}$ and $v \in K$, then v is β perpendicular to $H + Z = T(S^2 \times \mathbb{R}^3)$. Therefore $v \in Z \cap H = \{0\}$, that is, $v = 0$. Hence $K = \{0\}$. In other words, β is nondegenerate. If $Z \subseteq H$, then choose a vector $w \notin H$. This implies $w \notin Z$. Consequently, $L = \mathrm{span}\{w\}^{\beta}$ is a codimension 1 subspace of H which contains Z. Because $K \cap L \subseteq Z$, we conclude that $\dim K \leq 2$. Since $Z \subseteq K$ and $\dim K$ is even, we find that $\dim K = 2$. But $\dim H = 4$. Thus the 2-form β is nonzero.

Since

$$\mathrm{d}\alpha(X, Y) = X(Y \lrcorner\, \alpha) - Y(X \lrcorner\, \alpha) + \alpha([X, Y]),$$

we see that at points of $S^2 \times \mathbb{R}^3$ where $\langle s(u), u \rangle \neq 0$, the tangent space $T(S^2 \times \mathbb{R}^3)$ is spanned by H and vector fields on $S^2 \times \mathbb{R}^3$ whose Lie brackets take values in H. By Chow's theorem it follows that each connected component of the subset of points of $S^2 \times \mathbb{R}^3$, where $\langle s(u), u \rangle \neq 0$, is an accessible set of H.

If the center of mass of the strongly convex body $\mathcal{B}$ is in the interior of its boundary $\mathcal{S}$, then $\langle s(u), u \rangle < 0$ on $\mathcal{S}$. Consequently, the accessible set of the distribution H is all of $S^2 \times \mathbb{R}^3$. If the center of mass of $\mathcal{B}$ is on $\mathcal{S}$, then $M = \{ s \in \mathcal{S} \big| \, \langle s, u \rangle = 0 \}$ is a smooth codimension 2 submanifold of $\mathcal{S}$ defined by $s = 0$. Its complement $\mathcal{S} \setminus M$ is connected, and thus is contained in an accessible set of H in view of Chow's theorem. At every point $m \in M$ the codimension 1 subspace H_m is *not* contained in the codimension 2 subspace $T_m M$ of $T_m(S^2 \times \mathbb{R}^3)$. Therefore we can move out of M at m by using a motion which is tangent to H_m. Thus the accessible set of H is all of $S^2 \times \mathbb{R}^3$. Finally, if the center of mass of $\mathcal{B}$ is in the exterior of $\mathcal{S}$, then $\mathcal{C} = \{s \in \mathcal{S} \big| \langle s, u \rangle = 0\}$ is a smooth codimension 1 submanifold of $\mathcal{S}$. By Chow's theorem each connected component of $\mathcal{S} \setminus \mathcal{C}$ is an accessible set. For each $c \in \mathcal{C}$ we see that both $T_c\mathcal{C}$ and H_c are codimension 1 subspaces of $T_c(S^2 \times \mathbb{R}^3)$. So $\dim H_c = \dim T_c\mathcal{C}$. If $H_c \subseteq T_c\mathcal{C}$, then $H_c = T_c\mathcal{C}$. From (109) it follows that belonging to $H = \ker \alpha$ is a nontrivial restriction on the $\mathbb{R}^3$ component of a tangent vector to $S^2 \times \mathbb{R}^3$; whereas belonging to $T_c\mathcal{C}$ is a restriction only on the S^2 component. Therefore for every $c \in \mathcal{C}$, we have $H_c \neq T_c\mathcal{C}$. So H_c is not contained in $T_c\mathcal{C}$. Consequently, from $c \in \mathcal{C}$ we can move off $\mathcal{C}$ into each of the connected components of $\mathcal{S} \setminus \mathcal{C}$ by a motion which is tangent to H_c. Thus the accessible set of H is all of $S^2 \times \mathbb{R}^3$. □

One may wonder if the codimension 1 distribution $H = \ker \alpha$ is a *contact structure* on $S^2 \times \mathbb{R}^3$, that is, if the restriction to $H \times H$ of $\mathrm{d}\alpha$ is nondegenerate.

From lemma 6.6.3.17 it follows that

$$\widetilde{\rho}^*(\mathrm{d}\alpha) = \mathrm{d}(\widetilde{\rho}^*\alpha) = \mathrm{d}(u_{\rightarrow} \,\lrcorner\, \varpi).$$

From proposition 6.6.3.20 we know that if $\langle s, u \rangle = 0$ then $\dim \ker \mathrm{d}(u_{\rightarrow} \,\lrcorner\, \varpi) \geq 4$ and contains the vector $u_{\rightarrow}$. Because $\ker T\widetilde{\rho} = \mathrm{span}\{u_{\rightarrow}\}$, we see that $\dim \ker \mathrm{d}\alpha \geq 3$ and $\dim(H \cap \ker \mathrm{d}\alpha) \geq 2$. Consequently, H is *not* a contact structure at any point of $S^2 \times \mathbb{R}^3$ where $\langle s, u \rangle = 0$.

Now suppose that $\langle s, u \rangle \neq 0$. The fact that $u_{\rightarrow} \in \ker \mathrm{d}(u_{\rightarrow} \,\lrcorner\, \varpi)$ implies that $(u, 0) \in \ker \mathrm{d}(u_{\rightarrow} \,\lrcorner\, \varpi)$, that is, $Cu = 0 = Bu$, where C and B are the linear maps defined in (94) and the matrix M above this equation. Conversely, $\ker \mathrm{d}(u_{\rightarrow} \,\lrcorner\, \varpi) = \mathrm{span}\{u\} \times \ker C^t$. Now $\ker C^t = (\mathrm{im}\, C)^{\perp}$. So $C^t\widetilde{\zeta} = 0$ if and only if

$$\begin{aligned} 0 &= \langle \langle s, Ds(u)(u \times \widetilde{\eta}) \rangle u - \langle s, u \rangle\, Ds(u)(u \times \widetilde{\eta}), \widetilde{\zeta} \rangle \\ &= \langle Ds(u)(u \times \widetilde{\eta}), \langle u, \widetilde{\zeta} \rangle s - \langle s, u \rangle \widetilde{\zeta} \rangle \end{aligned} \tag{115}$$

for every $\widetilde{\eta} \in \mathbb{R}^3$. Because $Ds(u)$ is a positive definite symmetric linear transformation of $\mathrm{span}\{u\}^{\perp}$ into itself, condition (115) is equivalent to

$$\langle u, \widetilde{\zeta} \rangle s - \langle s, u \rangle \widetilde{\zeta} = \mu\, u \tag{116}$$

for some $\mu \in \mathbb{R}$. Taking the inner product on both sides of (116) with u gives $\mu = 0$. Because $\langle s, u \rangle \neq 0$, we deduce that $\widetilde{\zeta}$ is a multiple of s. Therefore $\ker \mathrm{d}\alpha = \mathrm{span}\{(0, s)\}$. From the proof of proposition 6.6.3.20 it follows that H is *not* a contact structure at a given point of $S^2 \times \mathbb{R}^3$ if at that point $\ker d\alpha \subseteq H = \ker \alpha$. In view of the definition of α (109) this condition amounts to

$$0 = \langle \mathcal{Y}(s)u, s\rangle = \langle u, \mathcal{Y}(s)s\rangle = \langle u, Is + \mathrm{m}\, s \times (s \times s)\rangle = \langle u, Is\rangle.$$

Thus we have proved

Proposition 6.6.3.21. *H is a contact structure on all of $S^2 \times \mathbb{R}^3$ if and only if $\langle s, u\rangle \neq 0$ and $\langle u, Is\rangle \neq 0$ at every point of $S^2 \times \mathbb{R}^3$.*

Assume that the center of mass of $\mathcal{B}$ is in interior of $\mathcal{S}$. Then $\langle s, u\rangle < 0$ at $S^2 \times \mathbb{R}^3$. The condition $\langle u, Is\rangle \neq 0$ means that the function $u \mapsto \langle s(u), Iu\rangle$ on S^2 has no zeroes. For $u = e_j$ we have $Iu = I_j e_j = I_j u$. Therefore $\langle s(u), Iu\rangle = I_j\langle s(u), u\rangle < 0$, when $u = e_j$. Thus the condition $\langle u, s\rangle \neq 0$ on $S^2 \times \mathbb{R}^3$ means that $\langle Is, u\rangle < 0$ everywhere $S^2 \times \mathbb{R}^3$.

The hypotheses of proposition 6.6.3.21 are satisfied for the Chaplygin sphere, where $\mathcal{S}$ is a sphere with its center of mass at its geometric center. In this case $s(u) = -ru$, where r is the radius of sphere. Therefore $\langle Is, u\rangle = -r\langle Iu, u\rangle < 0$. This situation also occurs when the moment of inertia tensor of $\mathcal{S}$ is a multiple of the identity and the center of mass is in the interior of $\mathcal{S}$.

The collection of strongly convex rigid bodies rolling without slipping on a horizontal plane for which H is a contact structure for all $S^2 \times \mathbb{R}^3$ is an open set in the sense that a small perturbation of the surface $\mathcal{S}$ and its moment of inertia tensor I does not destroy this property.

6.6.4 *Reduced distributional system*

In this section we complete the reduction of the E(2) symmetry of the rigid body rolling on a horizontal plane, by computing the E(2)-reduced distributional Hamiltonian system on the E(2)-orbit space $S^2 \times \mathbb{R}^3$.

First we prove

Fact 6.6.4.22. *On the accessible set D_N the vector field V (42) induces the vector field $\overline{V}$ on $S^2 \times \mathbb{R}^3$ whose integral curves satisfy*

$$\dot{u} = u \times \omega \tag{117a}$$

$$\begin{aligned}\mathcal{Y}(s)\dot{\omega} = {} & \mathcal{Y}(s)\omega \times \omega + \mathrm{m}\,(\omega \times Ds(u)(u \times \omega)) \times s \\ & - \mathrm{mg}\, u \times s.\end{aligned} \tag{117b}$$

Proof. The E(2) action

$$\begin{aligned}\Phi : \mathrm{E}(2) \times (\mathrm{E}(3) \times \mathrm{e}(3)) &\rightarrow \mathrm{E}(3) \times \mathrm{e}(3) : \\ \big((R,x),(A,a,\omega,b)\big) &\mapsto (RA, Ra+x, \omega, b)\end{aligned} \tag{118}$$

given by (95) leaves the constraint manifold D_N (96) invariant. Thus Φ restricts to an E(2)-action on D_N. The vector field V on D_N, which governs the motion of the rolling rigid body, has integral curves which satisfy (42)

$$\begin{aligned}\dot{A} &= A\omega^{\times} \\ \dot{a} &= -A\omega^{\times}s \;=\; -A(\omega \times s) \\ \dot{\omega} &= \mathcal{Y}(s)^{-1}\big((\mathcal{Y}(s)\omega) \times \omega + \mathrm{m}\,((\omega \times \dot{s}) \times s) - \mathrm{mg}\, u \times s\big).\end{aligned} \tag{119}$$

Note that V preserves D_N because $b = A^{-1}\dot{a} = -\omega \times s$, using the second equation in (119); while

$$\frac{\mathrm{d}}{\mathrm{d}t}\big(\langle a, e_3\rangle + \langle s(u), u\rangle\big) = \langle \dot{a}, e_3\rangle + \langle Ds(u)(u \times \omega), u\rangle + \langle s(u), u \times \omega\rangle,$$

$$\text{since } \dot{u} = \tfrac{\mathrm{d}}{\mathrm{d}t}(A^{-1})e_3 = -A^{-1}\dot{A}A^{-1}e_3 = -\omega^{\times}u = u \times \omega$$

$$= \langle A^{-1}\dot{a}, A^{-1}e_3\rangle + \langle s(u), u\rangle,$$

$$\text{since } Ds(u)(u \times \omega) \in T_{s(u)}\mathcal{S}, \text{ which is perpendicular to } u$$

$$= \langle b, u\rangle + \langle \omega \times s, u\rangle = 0, \text{ since } b = -\omega \times s.$$

The vector field V on D_N is invariant under the E(2)-action $\Phi|(\mathrm{E}(2) \times D_N)$, because for $((A,a),\omega) \in \mathrm{E}(3) \times \mathbb{R}^3$, which parametrizes D_N, we have

$$\begin{aligned}T_{(A,a,\omega)}\Phi_{(R,x)}V(A,a,\omega) &= (R\dot{A}, R\dot{a}, \dot{\omega}) \\ &= V(RA, Ra+x, \omega) = V(\Phi_{(R,x)}(A,a,\omega)),\end{aligned}$$

since $s(R \cdot u) = s(u)$. Now

$$\rho : D_N \rightarrow \overline{D}_N = S^2 \times \mathbb{R}^3 : (A,a,\omega) \mapsto (u = A^{-1}e_3, \omega)$$

is the orbit map of the E(2)-action $\Phi|(\mathrm{E}(2) \times D_N)$. So $T_{(A,a,\omega)}\rho(\dot{A}, \dot{a}, \dot{\omega}) = (\dot{u}, \dot{\omega})$. Thus the E(2)-reduced vector field $\overline{V}$ on $\overline{D}_N = S^2 \times \mathbb{R}^3$ is given by

$$\overline{V}(u,\omega) = \overline{V}(\rho(A,a,\omega)) = T_{(A,a,\omega)}\rho V(A,a,\omega),$$

whose integral curves satisfy (117a) and (117b). □

Proposition 6.6.4.23. *The* E(2) *reduced vector field* $\overline{V}$ *on* $S^2 \times \mathbb{R}^3$ *is the distributional Hamiltonian vector field* $Y_{\overline{E}}$ *associated to the* E(2) *reduced Hamiltonian*

$$\begin{aligned}\overline{E} = \overline{E}_{S^2 \times \mathbb{R}^3} &: S^2 \times \mathbb{R}^3 \rightarrow \mathbb{R} : \\ (u,\omega) &\mapsto \tfrac{1}{2}\langle \mathcal{Y}(s)\omega, \omega\rangle - \mathrm{mg}\,\langle s(u), u\rangle\end{aligned} \tag{120}$$

In other words,

$$\overline{V} \mathbin{\lrcorner} \varpi_{\overline{H}} = \partial_{\overline{H}}\overline{E}, \tag{121}$$

where $\overline{H} = \overline{H}_{S^2 \times \mathbb{R}^3}$.

Proof. Instead of appealing to proposition 3.3.27 of chapter 3, we will verify (121) directly. Differentiating

$$\overline{E} : S^2 \times \mathbb{R}^3 \to \mathbb{R} : (u, \omega) \mapsto \tfrac{1}{2} \langle I\omega, \omega \rangle + \mathrm{m} \langle \omega \times s, \omega \times s \rangle - \mathrm{mg} \langle s, u \rangle \quad (122)$$

in the direction $(u \times w, v) \in \overline{H}_{(u,\omega)}$ gives

$$\begin{aligned} \partial_{\overline{H}} \overline{E}(u, \omega)(u \times w, v) &= \langle I(\omega), v \rangle + \mathrm{m} \langle s \times (\omega \times s), v \rangle \\ &\quad - \mathrm{mg} \langle s \times u, w \rangle - \mathrm{m} \langle u \times Ds(u)((\omega \times s) \times \omega), w \rangle, \end{aligned} \quad (123)$$

using (117b). Since $(u \times w, v) \in \overline{H}_{(u,\omega)}$, from lemma 6.6.3.19 it follows that

$$0 = \langle u, \mathcal{Y}(s)v \rangle + \langle u \times w, \mathcal{Y}(s)\omega + \mathrm{m} (u \times Ds(u)(\omega \times (s \times u))) \rangle. \quad (124)$$

Let

$$\begin{aligned} \mathcal{Y}(s)v &= \mathcal{Y}(s)\omega \times \omega + \mathrm{m} (\omega \times Ds(u)(u \times \omega)) \times s - \mathrm{mg}\, u \times s \\ w &= v \end{aligned} \quad (125)$$

A straightforward calculation shows that if v and w are defined by (125), then equation (124) holds. Thus $\overline{V}(u, \omega) = (u \times w, v) \in \overline{H}_{(u,\omega)}$.

From lemma 6.6.3.19 for every $(u \times w', v') \in \overline{H}_{(u,\omega)}$ we get

$$\begin{aligned} (\overline{V}(u, \omega) \lrcorner\, \varpi_{\overline{H}})(u, \omega)(u \times w', v') &= \langle \mathcal{Y}(s)v', w \rangle - \langle \mathcal{Y}(s)v, w' \rangle \\ &+ \langle \mathcal{Y}(s)\omega, w \times w' \rangle - \mathrm{m} (\omega \times Ds(u)(w \times u)) \times s, w' \rangle \\ &\quad + \mathrm{m} \langle (\omega \times Ds(u)(w' \times u)) \times s, w \rangle \\ &= \langle \mathcal{Y}(s)v', \omega \rangle - \langle (\mathcal{Y}(s)\omega) \times \omega, w' \rangle - \mathrm{m} \langle (\omega \times Ds(u)(u \times \omega)) \times s, w' \rangle \\ &+ \mathrm{mg} \langle u \times s, w' \rangle + \langle (\mathcal{Y}(s)\omega) \times \omega, w' \rangle - \mathrm{m} (\omega \times Ds(u)(\omega \times u)) \times s, w' \rangle \\ &\quad + \mathrm{m} \langle (\omega \times Ds(u)(w' \times u)) \times s, \omega \rangle, \quad \text{using (125)} \\ &= \langle \mathcal{Y}(s)\omega, v' \rangle - \mathrm{m} \langle u \times Ds(u)((\omega \times s) \times \omega), w' \rangle - \mathrm{mg} \langle s \times u, w' \rangle \\ &= \partial_{\overline{H}} \overline{E}(u, \omega)(u \times w', v'), \end{aligned}$$

using (123). □

After removal of the E(2) symmetry, the rolling strongly convex rigid body is the distributional Hamiltonian system $\big(S^2 \times \mathbb{R}^3, \overline{H}_{S^2 \times \mathbb{R}^3}, \varpi_{\overline{H}_{S^2 \times \mathbb{R}^3}}, \overline{E}_{S^2 \times \mathbb{R}^3}\big)$.

6.7 Body of revolution

In this section we discuss the motion of a strongly convex body of revolution which rolls without slipping on a horizontal plane under the influence of a constant vertical gravitational force.

In body coordinates a rotation $R \in \mathrm{SO}(3)$ sends the reference position of the strongly convex rigid body $\mathcal{B}$ to its actual position acted on by the rotation R^{-1}. Under this action of R the vector $u = A^{-1} e_3$ is sent to $(A R^{-1})^{-1} e_3 = R A^{-1} e_3 = R u$. Also the vectors $A\, s(u)$ and $\dot{A}\, s(u)$ are sent to $A R^{-1} s(R u)$ and $\dot{A} R^{-1} s(R u)$, respectively. It follows that the nonholonomic constraint (28) is invariant if and only if $s(R u) = R\, s(u)$ for every $u \in S^2$. This means that the surface $\mathcal{S}$ of $\mathcal{B}$ is invariant under R. If this is the case, then the holonomic constraint (27) is invariant as well and we will say the strongly convex rigid body is *geometrically symmetric* under R.

Since the action of R maps the rotation A to $A R^{-1}$, it follows that $\dot{A}$ is mapped to $\dot{A} R^{-1}$. Therefore $\omega^{\times} = A^{-1} \dot{A}$ is mapped to $R A^{-1} \dot{A} R^{-1} = R(\omega^{\times})R^{-1} = (R\omega)^{\times}$. This implies that ω is mapped to $R\omega$. Therefore from the geometric invariance of the body it follows that the first two equations defining the vector field V (42), which governs its motion, are invariant.

If $IR = RI$, then from the definition of $\mathcal{Y}(s)$ (40) it follows that

$$\begin{aligned} \mathcal{Y}(R s)\, R v &= IRv + \mathrm{m}\, Rs \times (Rv \times Rs) = RIv + \mathrm{m}\, Rs \times R(v \times s) \\ &= RIv + \mathrm{m}\, R(s \times (v \times s)) = R(Iv + \mathrm{m}\, s \times (v \times s)) \\ &= R\mathcal{Y}(s)v \end{aligned}$$

for every $v \in \mathbb{R}^3$, that is, $\mathcal{Y}(Rs) = R\mathcal{Y}(s)R^{-1}$. Therefore the third equation defining the vector field V (42) is invariant under the substitutions $\omega \mapsto R\omega$, $s \mapsto Rs$, and $u \mapsto Ru$.

The body $\mathcal{B}$ is *dynamically symmetric* under R if $IR = RI$. The body $\mathcal{B}$ is *symmetric* under R if it is both geometrically and dynamically symmetric under R. When this is the case, the holonomic constraint (27), the nonholonomic constraint (28), and the equations of motion (42) are invariant under the action of R.

The body $\mathcal{B}$ is a *body of revolution* if it is both geometrically and dynamically symmetric under all rotations about a given axis. We may choose this axis to be the e_3-axis. Then dynamic symmetry means that in the principal axis frame its moment of inertia tensor I is diagonal, with eigenvalues I_1,

I_2, I_3, and $I_1 = I_2$. Note that the inequalities (13) satisfied by the principal moments of inertia imply that $I_3 \le 2\, I_1$.

6.7.1 *Geometric and dynamic symmetry*

Let $\mathcal{B}$ be the reference body with reference frame $\{e_1, e_2, e_3\}$ having its origin at the center of mass. Since $\mathcal{S}$, the boundary of $\mathcal{B}$, is a smooth strictly convex surface of revolution, it can be parametrized by

$$\mathbf{x} : S^1 \times [-1,1] \to \mathbb{R}^3 : (\theta, x_3) \mapsto (f(x_3)\cos\theta, f(x_3)\sin\theta, x_3), \tag{126}$$

where $f : [-1,1] \to \mathbb{R}_{\ge 0}$ is continuous with $f(\pm 1) = 0$ and is smooth on $(-1,1)$ with $f'' < 0$. Let $g_{\pm}$ be a local inverse to f with $g_{\pm}(0) = \pm 1$, which in a neighborhood of 0 extends to a smooth *even* function. Then $\lim_{x_3 \to \pm 1^{\mp}} f'(x_3) = \mp\infty$.

Let $\mathbf{n}(\mathbf{x})$ be the *inward* pointing unit normal to $\mathcal{S}$ at $\mathbf{x} = \mathbf{x}(\theta, x_3)$. A short calculation shows that

$$\mathbf{n}(\mathbf{x}(\theta, x_3)) = \frac{1}{\sqrt{1 + (f'(x_3))^2}}\,(\cos\theta, \sin\theta, -f'(x_3)). \tag{127}$$

Thus the *Gauss map* of $\mathcal{S}$ is

$$\mathbf{n} : \mathcal{S} \subseteq \mathbb{R}^3 \to S^2 \subseteq \mathbb{R}^3 : \mathbf{x} \mapsto \mathbf{n}(\mathbf{x}) = (u_1, u_2, u_3) \;=\; u. \tag{128}$$

Its inverse is the map

$$s : S^2 \to \mathcal{S} : u \mapsto (\rho(u_3)\, u_1, \rho(u_3)\, u_2, \zeta(u_3)), \tag{129}$$

where $\rho : (-1,1) \to \mathbb{R}$ and $\zeta : (-1,1) \to \mathbb{R}$ are smooth functions defined by

$$f'(\zeta(u_3)) = -u_3\,(1 - u_3^2)^{-1/2} \tag{130a}$$

$$\rho(u_3) = f(\zeta(u_3))\,(1 - u_3^2)^{-1/2}. \tag{130b}$$

Proof. Using the fact that $\mathbf{x} = s(\mathbf{n}(\mathbf{x}))$, we see that the components of (126) and (129) are equal. This gives

$$f^2(x_3) = \rho^2(u_3)(1 - u_3^2),$$

since $u_1^2 + u_2^2 = 1 - u_3^2$. Because $f \ge 0$ and $x_3 = \zeta(u_3)$, we obtain (130b). Since $u = \mathbf{n}(\mathbf{x})$ we get

$$u_3 = -\frac{f'(x_3)}{\sqrt{1 + (f'(x_3))^2}} \tag{131}$$

from which (130a) follows.

Because $f'' < 0$ on $(-1,1)$, the function f' is strictly decreasing on $(-1,1)$. Thus the function ζ in (130a) is well defined and smooth on $(-1,1)$. Consequently, the function ρ in (130b) is smooth on $(-1,1)$. Moreover, differentiating (130a) and using the fact that $f'' < 0$ shows that $\zeta' > 0$ on $(-1,1)$. Thus ζ is monotonically increasing on $(-1,1)$. □

Although the right hand sides of (130a) and (130b) are not defined at $u_3 = \pm 1$, the functions ζ and ρ extend to smooth functions in a neighborhood of ± 1.

Proof. Because $y \mapsto g_\pm(y) = x_3$ is a smooth local inverse of f with $g_\pm(0) = \pm 1$, using the chain rule we get $g'_\pm(y) = \frac{1}{f'(x_3)}$. From (131) it follows that

$$u_3 = -(1 + (g'_\pm(y))^2)^{-1/2},$$

which is equivalent to

$$g'_\pm(y) = -\frac{(1-u_3^2)^{1/2}}{u_3}. \tag{132}$$

The inverse of $g'_\pm$ extends to a smooth *odd* function in a neighborhood of 0. Since $y = g_\pm^{-1}(x_3) = g_\pm^{-1}(\zeta(u_3))$, equation (132) reads

$$-\frac{(1-u_3^2)^{1/2}}{u_3} = \frac{1}{f'(\zeta(u_3))} = (g'_\pm \circ g_\pm^{-1})(\zeta(u_3)),$$

that is,

$$\zeta(u_3) = g_\pm \circ (g'_\pm)^{-1}\left(-\frac{(1-u_3^2)^{1/2}}{u_3}\right). \tag{133}$$

Thus ζ is a smooth function near $u_3 = \pm 1$. Composing both sides of (133) with f and using (130b), we obtain

$$\rho(u_3) = f(\zeta(u_3))(1-u_3^2)^{-1/2} = (g'_\pm)^{-1}\left(-\frac{(1-u_3^2)^{1/2}}{u_3}\right)(1-u_3^2)^{-1/2}.$$

Because $(g'_\pm)^{-1}$ is a smooth *odd* function near ± 1, it follows that ρ is smooth near ± 1. □

The following argument shows that the functions ρ and ζ satisfy

$$(1-u_3^2)\,\rho'(u_3) - u_3\,\rho(u_3) = -u_3\,\zeta'(u_3). \tag{134}$$

Proof. Multiplying both sides of (130a) by $\zeta'(u_3)$ gives

$$(f\circ\zeta)'(u_3) = -u_3\,(1-u_3^2)^{-1/2}\zeta'(u_3).$$

Differentiating (130b) we obtain

$$\begin{aligned}\rho'(u_3) &= (f\circ\zeta)'(u_3)\,(1-u_3^2)^{-1/2} + (f\circ\zeta)(u_3)\,u_3(1-u_3^2)^{-3/2}\\ &= (-\zeta'(u_3)+\rho(u_3))\,u_3(1-u_3^2)^{-1}.\end{aligned}$$

□

Since $\mathcal{S}$ is a surface of revolution, it is invariant under the *axial symmetry*

$$S^1\times\mathbb{R}^3\to\mathbb{R}^3 : (\psi,\mathbf{x})\mapsto R\mathbf{x} = \begin{pmatrix}\cos\psi & -\sin\psi & 0\\ \sin\psi & \cos\psi & 0\\ 0 & 0 & 1\end{pmatrix} \tag{135}$$

of rotation about the e_3-axis. From (127) it is easy to see that the Gauss map $\mathbf{n}$ is *equivariant* under this axial symmetry, that is,

$$\mathbf{n}(R(\mathbf{x})) = R(\mathbf{n}(\mathbf{x})),$$

for every rotation R about e_3. Consequently, the inverse s of the Gauss map of $\mathcal{S}$ is also equivariant, that is, $s(R(u)) = R(s(u))$ for every $u\in S^2$ and every $\psi\in S^1$. Thus a body of revolution is geometricially symmetric.

Next to ensure the dynamic symmetry of the rolling body of revolution $\mathcal{B}$, we assume that in the principal axis frame $\{e_1,e_2,e_3\}$ the moment of inertia tensor I of $\mathcal{B}$ is $\mathrm{diag}(I_1,I_1,I_3)$. Here $0<I_3\le 2I_1$. Thus, for every $R\in S^1$ we have

$$R\circ I = I\circ R. \tag{136}$$

From equation (96) of §6 we know that the motion of $\mathcal{B}$ takes place on the constraint manifold

$$D_N = \{(A,a,\omega,b)\in \mathrm{E}(2)\times\mathrm{e}(2)\,\big|\, b=\omega\times s(u),\ \langle a,e_3\rangle = -\langle s(u),u\rangle\}. \tag{137}$$

In §6.2 we have seen that E(2) acts on D_N by

$$\widetilde{\Phi} : \mathrm{E}(2)\times D_N\to D_N : ((A',a'),(A,a,\omega,b))\mapsto (A'A, A'a+a',\omega,b). \tag{138}$$

On D_N define an *axial action* of S^1 by

$$\begin{aligned}&\Phi : D_N\times S^1\to D_N :\\ &((A,a,\omega,b),R)\mapsto (AR^{-1},Ra,R\omega,Rb).\end{aligned} \tag{139}$$

From

$$Rb = -(R\omega)^{\times}Rs(u) = -(R\omega)^{\times}s(Ru)$$

and

$$\langle Ra, Re_3 \rangle = \langle a, e_3 \rangle = -\langle s(u), u \rangle = -\langle Rs(u), Ru \rangle,$$

it follows that the axial S^1-action Φ preserves D_N. Because $\mathrm{E}(2)$ acts on D_N on the left, while S^1 acts on D_N on the right, the actions commute. Thus Φ induces an S^1 action on the $\mathrm{E}(2)$-orbit space $D_N/\mathrm{E}(2) = S^2 \times \mathbb{R}^3$ (see §6.2) defined by

$$\Psi : S^1 \times (S^2 \times \mathbb{R}^3) \rightarrow S^2 \times \mathbb{R}^3 : (\psi, u, \omega) \mapsto (Ru, R\omega). \tag{140}$$

We call the action Ψ the *induced axial symmetry*. Ψ is the restriction to $S^2 \times \mathbb{R}^3$ of the lift of the axial symmetry (135) to $T\mathbb{R}^3$.

From §6.4 we know that the motion of the body $\mathcal{B}$ on the horizontal plane after removing its $\mathrm{E}(2)$-symmetry is governed by a vector field $\overline{V}$ on $S^2 \times \mathbb{R}^3$, whose integral curves satisfy

$$\begin{aligned} \dot{u} &= u \times \omega \\ \mathcal{Y}(s)\dot{\omega} &= \mathcal{Y}(s)\omega \times \omega + \mathrm{m}\,(\omega \times Ds(u)(u \times \omega)) \times s - \mathrm{mg}\, u \times s. \end{aligned} \tag{141}$$

Here $\mathcal{Y}(s)v = Iv + \mathrm{m}\,(s(u) \times (v \times s(u)))$ for every $v \in \mathbb{R}^3$.

Claim 6.7.1.24. *The* $\mathrm{E}(2)$*-reduced vector field* $\overline{V}$ *on* $S^2 \times \mathbb{R}^3$ *is invariant under the induced axial symmetry* Ψ (140).

Proof. This follows because the axial symmetry of $\mathcal{B}$ implies the invariance of (42) under Ψ. □

From proposition 6.6.4.23 we know that the function

$$\overline{E} : S^2 \times \mathbb{R}^3 \rightarrow \mathbb{R} : (u, \omega) \mapsto \tfrac{1}{2}\,\langle \mathcal{Y}(s)\omega, \omega \rangle - \mathrm{mg}\,\langle s(u), u \rangle \tag{142}$$

is an integral of the vector field $\overline{V}$.

Claim 6.7.1.25. *The* $\mathrm{E}(2)$*-reduced energy* $\overline{E}$ *is invariant under the induced axial symmetry* (140).

Proof. We compute

$$\begin{aligned} \overline{E}(Ru, R\omega) &= \tfrac{1}{2}\,\langle \mathcal{Y}(s(Ru))R\omega, \omega \rangle - \mathrm{mg}\,\langle s(Ru), Ru \rangle \\ &= \tfrac{1}{2}\,\langle R(\mathcal{Y}(s)\omega), R\omega \rangle - \mathrm{mg}\,\langle R(s(u)), Ru \rangle \\ &= \overline{E}(u, \omega). \end{aligned}$$

□

6.7.2 Reduction of the induced axial symmetry

Here we reduce the induced axial symmetry (140) on $S^2 \times \mathbb{R}^3$.

To carry out this reduction we will use *invariant theory*. The algebra of polynomials on $\mathbb{R}^3 \times \mathbb{R}^3$, which are invariant under the $S^1 = \{R \in \mathrm{SO}(3) \big| \, Re_3 = e_3\}$-action

$$\widehat{\Psi} : S^1 \times (\mathbb{R}^3 \times \mathbb{R}^3) \to \mathbb{R}^3 \times \mathbb{R}^3 : (R, (u, \omega)) \to (Ru, R\omega), \tag{143}$$

is generated by

$$\begin{array}{lll} \sigma_1 = u_3 & \sigma_2 = u_1\omega_2 - u_2\omega_1 & \sigma_3 = u_1\omega_1 + u_2\omega_2 \\ \sigma_4 = \omega_3 & \sigma_5 = \omega_1^2 + \omega_2^2 & \sigma_6 = u_1^2 + u_2^2 \end{array} \tag{144}$$

subject to the relation

$$\sigma_2^2 + \sigma_3^2 = \sigma_5\sigma_6, \quad \sigma_5 \ge 0, \; \sigma_6 \ge 0. \tag{145}$$

Restricting the action $\widehat{\Psi}$ to $S^2 \times \mathbb{R}^3$ gives the induced axial action Ψ (140).

Its algebra of invariant polynomials is generated by restricting the polynomials $\sigma_1, \ldots, \sigma_6$ to $S^2 \times \mathbb{R}^3$ (which we again denote by $\sigma_1, \ldots, \sigma_6$). In addition to (145), these polynomials satisfy the relation

$$\sigma_6 + \sigma_1^2 = 1, \quad \sigma_6 \ge 0. \tag{146}$$

Equation (146) is just the defining equation of $S^2 \times \mathbb{R}^3$ expressed in terms of invariants. Eliminating σ_6 from (145) and (146) shows that the space $(S^2 \times \mathbb{R}^3)/S^1$ of orbits of the induced axial action is the semialgebraic variety $\mathcal{M}$ in $\mathbb{R}^5$ (with coordinates $(\sigma_1, \ldots, \sigma_5)$) defined by

$$\sigma_2^2 + \sigma_3^2 = (1 - \sigma_1^2)\sigma_5, \quad |\sigma_1| \le 1, \; \sigma_5 \ge 0. \tag{147}$$

$\mathcal{M}$ is singular with

$$\mathcal{M}_{\text{sing}} = \{(\pm 1, 0, 0, \sigma_4, 0) \in \mathcal{M} \big| \sigma_4 \in \mathbb{R}\} \tag{148}$$

being its subvariety of singular points. $\mathcal{M}_{\text{sing}}$ is the image of the submanifold

$$(S^2 \times \mathbb{R}^3)_{S^1} = \{(0, 0, \pm 1, 0, 0, \omega_3) \in S^2 \times \mathbb{R}^3 \big| \; \omega_3 \in \mathbb{R}\}$$

of points of $S^2 \times \mathbb{R}^3$ of symmetry type (=isotropy type, see §3 of chapter 2) S^1 of the action Ψ under its S^1-orbit map

$$\pi : S^2 \times \mathbb{R}^3 \to \mathcal{M} \subseteq \mathbb{R}^5 : (u, \omega) \mapsto \big(\sigma_1(u, \omega), \ldots, \sigma_5(u, \omega)\big). \tag{149}$$

If we disregard the dummy variable σ_4, then $\mathcal{M}$ has two singular points, near each of which $\mathcal{M}$ looks like a quadratic cone. Since $(S^2 \times \mathbb{R}^3) \setminus (S^2 \times$

$\mathbb{R}^3)_{S^1}$ is the submanifold of $S^2 \times \mathbb{R}^3$ of symmetry type $\{e\}$, its image under π is the smooth manifold $\mathcal{M}_{\mathrm{reg}} = \mathcal{M} \setminus \mathcal{M}_{\mathrm{sing}}$. In the domain where $-1 < \sigma_1 < 1$, $\mathcal{M}_{\mathrm{reg}}$ is equal to the graph of the smooth function

$$\sigma_5 = \frac{\sigma_2^2 + \sigma_3^2}{1 - \sigma_1^2}, \quad -1 < \sigma_1 < 1. \tag{150}$$

The invariants σ_j, $1 \le j \le 4$ form a system of coordinates in this domain. However, $\mathcal{M}_{\mathrm{reg}}$ contains the points where $\sigma_5 > 0$ and $\sigma_1 = \pm 1$. The graphs of the smooth functions

$$\sigma_1 = \pm(1 - (\sigma_2^2 + \sigma_3^2)/\sigma_5)^{1/2}, \quad \sigma_5 > \sigma_2^2 + \sigma_3^2, \tag{151}$$

give an open neighborhood of each of these additional points. The invariants σ_j, $2 \le j \le 5$ form a regular system of coordinates on each of these two open subsets.

6.7.3 *Axially reduced equations of motion*

In this subsection we study the motion of the axially reduced system in terms of the invariants σ_j in (144).

6.7.3.1 *Chaplygin's equations*

The next lemma leads to Chaplygin's equations for a surface of revolution.

Lemma 6.7.3.1.26. *Using dot to denote derivative with respect to time, the following equations hold*

$$\dot{\sigma}_1 = \sigma_2 \tag{152a}$$

$$\rho\, I_1\, \dot{\sigma}_3 + \zeta\, I_3\, \dot{\sigma}_4 + \rho\, I_3\, \sigma_4\, \dot{\sigma}_1 = 0. \tag{152b}$$

Here $\rho = \rho(u_3) = \rho(\sigma_1)$ *and* $\zeta = \zeta(u_3) = \zeta(\sigma_1)$, *given by* (130a) *and* (130b), *respectively, are viewed as functions of* σ_1.

Proof. The third component of $\dot{u} = u \times \omega$ yields (152a) From (130a) and (130b) it follows that

$$s = \rho\, u - \rho\, u_3\, e_3 + \zeta\, e_3 = \rho\, u + \tau\, e_3, \quad \text{where } \tau = \zeta - \rho\, \sigma_1.$$

Therefore

$$\langle I\dot{\omega} - (I\omega) \times \omega,\, s \rangle = \rho\, \langle I\dot{\omega} - (I\omega) \times \omega,\, u \rangle + \tau\, \langle I\dot{\omega} - (I\omega) \times \omega,\, e_3 \rangle.$$

Now $-\langle (I\omega) \times \omega, u \rangle = \langle I\omega, u \times \omega \rangle = \langle I\omega, \dot{u} \rangle$ implies that

$$\langle I\dot{\omega} - (I\omega) \times \omega, u \rangle = \langle I\dot{\omega},\, u \rangle + \langle I\omega, \dot{u} \rangle = (\langle I\omega, u \rangle)^{\cdot},$$

where we have used $(\ )^{\cdot}$ to denote $\frac{\mathrm{d}}{\mathrm{d}t}(\ )$. In view of (144) and $I_1 = I_2$ we see that $\langle I\omega, u\rangle = I_1\,\sigma_3 + I_3\,\sigma_4\,\sigma_1$. But

$$\langle (I\omega)\times\omega, e_3\rangle = I_1\,\omega_1\,\omega_2 - I_2\,\omega_2\,\omega_1 = 0. \tag{153}$$

So we arrive at

$$\begin{aligned}\langle I\,\dot\omega - (I\,\omega)\times\omega,\ s\rangle &= \rho\,(I_1\,\sigma_3 + I_3\,\sigma_4\,\sigma_1)^{\cdot} + (\zeta - \rho\,\sigma_1)\,I_3\,\dot\sigma_4\\ &= \rho\,I_1\,\dot\sigma_3 + \zeta\,I_3\,\dot\sigma_4 + \rho\,I_3\,\sigma_4\,\dot\sigma_1.\end{aligned}$$

Therefore taking the inner product of (38) with s gives (152b). □

Lemma 6.7.3.1.27. *We have*

$$\begin{aligned}-\mathrm{m}\rho\zeta\,\dot\sigma_3 + (I_3 + \mathrm{m}\rho^2\,(1-\sigma_1^2))\,\dot\sigma_4&\\ -\ \mathrm{m}\rho\,((\zeta'-\rho)\,\sigma_3 + \zeta'\,\sigma_1\,\sigma_4)\,\dot\sigma_1 &= 0.\end{aligned} \tag{154}$$

Here ρ' and ζ' are the derivatives of ρ and ζ with respect to σ_1.

Proof. Let i_k be the inner product of e_3 with the kth term of

$$I\dot\omega - I(\omega)\times\omega - \mathrm{m}\,(\omega\times s)^{\cdot}\times s - \mathrm{m}\,\langle s,\omega\rangle\,\omega\times s + \mathrm{mg}\,u\times s. \tag{155}$$

Then $i_2 = \langle I(\omega)\times\omega, e_3\rangle = 0$ because of (153), and $i_5 = \langle u\times s, e_3\rangle = u_1\,s_2 - u_2\,s_1 = u_1\,\rho\,u_2 - u_2\,\rho\,u_1 = 0$. For $i_3 = \langle(\omega\times s)^{\cdot}\times s, e_3\rangle$ we write

$$\langle(\omega\times s)^{\cdot}\times s,\ e_3\rangle = (\omega\times s)^{\cdot}_1\,s_2 - (\omega\times s)^{\cdot}_2\,s_1.$$

Substituting $s_1 = \rho\,u_1$ and $s_2 = \rho\,u_2$ gives

$$(\omega\times s)_1 = \omega_2\,s_3 - \omega_3\,s_2 = \omega_2\,\zeta - \sigma_4\,\rho\,u_2,$$

and

$$(\omega\times s)_2 = \omega_3\,s_1 - \omega_1\,s_3 = \sigma_4\,\rho\,u_1 - \omega_1\,\zeta.$$

Therefore i_3 is equal to $\mathrm{m}\,\rho$ times

$$\begin{aligned}-(\omega_2\,\zeta - \sigma_4\,\rho\,u_2)^{\cdot}\,u_2 &+ (\sigma_4\,\rho\,u_1 - \omega_1\,\zeta)^{\cdot}\,u_1\\ &= -\zeta\,(\dot\omega_2\,u_2 + \dot\omega_1\,u_1) - \dot\zeta\,(\omega_2\,u_2 + \omega_1\,u_1)\\ &\qquad + \sigma_4\,\rho(\dot u_2\,u_2 + \dot u_1\,u_1) + (\sigma_4\,\rho)^{\cdot}\,({u_2}^2 + {u_1}^2)\\ &= -\zeta\,(\dot\omega_2\,u_2 + \dot\omega_1\,u_1) - \dot\zeta\,\sigma_3 + \tfrac{1}{2}\,\sigma_4\,\rho\,\dot\sigma_6 + (\sigma_4\,\rho)^{\cdot}\,\sigma_6.\end{aligned}$$

Because

$$\begin{aligned}\dot\omega_2\,u_2 + \dot\omega_1\,u_1 &= \langle\dot\omega,\ u\rangle - \dot\omega_3\,u_3 = (\langle\omega,\ u\rangle)^{\cdot} - \langle\omega,\ \dot u\rangle - \dot\sigma_4\,\sigma_1\\ &= (\sigma_3 + \sigma_4\,\sigma_1)^{\cdot} - \langle\omega,\ u\times\omega\rangle - \dot\sigma_4\,\sigma_1 = \dot\sigma_3 + \sigma_4\,\dot\sigma_1,\end{aligned}$$

we see that $i_3 = \langle (\omega \times s)^{\cdot} \times s, e_3 \rangle$ is equal to m ρ times

$$-\zeta\,(\dot\sigma_3 + \sigma_4\,\dot\sigma_1) - \dot\zeta\,\sigma_3 + \tfrac{1}{2}\,\sigma_4\,\rho\,\dot\sigma_6 + (\sigma_4\,\rho)^{\cdot}\,\sigma_6.$$

For $i_4 = \langle s, \omega\rangle\,\langle \omega \times s, e_3\rangle$ we write

$$\langle \omega,\, s\rangle = \omega_1\,\rho\,u_1 + \omega_2\,\rho\,u_2 + \omega_3\,\zeta = \rho\,\sigma_3 + \zeta\,\sigma_4, \tag{156}$$

and

$$(\omega \times s)_3 = \omega_1\,s_2 - \omega_2\,s_1 = -\rho\sigma_2,$$

which implies that

$$i_4 = \mathrm{m}\rho\,(\rho\,\sigma_3 + \zeta\,\sigma_4)\,\sigma_2.$$

Because $\dot\sigma_1 = \sigma_2$, the term $\mathrm{m}\rho\zeta\,\sigma_4\,\dot\sigma_1$ in i_3 cancels the term $\mathrm{m}\rho\zeta\,\sigma_4\,\sigma_2$ in i_4. Collecting all the terms, we see that the inner product of (155) with e_3 is

$$I_3\,\dot\sigma_4 + \mathrm{m}\rho\big(-\zeta\,\dot\sigma_3 - \dot\zeta\,\sigma_3 + \tfrac{1}{2}\,\sigma_4\,\rho\dot\sigma_6 + (\sigma_4\,\rho)^{\cdot}\,\sigma_6 + \rho\,\sigma_3\,\dot\sigma_1\big),$$

which equals 0, because of (38).

We conclude the proof of (154) by observing that $\rho = \rho(u_3) = \rho(\sigma_1)$, $\zeta = \zeta(u_3) = \zeta(\sigma_1)$. Therefore $\dot\rho = \rho'(\sigma_1)\,\dot\sigma_1$ and $\dot\zeta = \zeta'(\sigma_1)\,\dot\sigma_1$. Furthermore $\sigma_6 = 1 - \sigma_1^2$, see (146). Hence $\dot\sigma_6 = -2\sigma_1\,\dot\sigma_1$. Consequently,

$$\begin{aligned}\tfrac{1}{2}\,\sigma_4\,\rho\dot\sigma_6 + (\sigma_4\,\rho)^{\cdot}\,\sigma_6 &= -\sigma_4\,\rho\,\sigma_1\dot\sigma_1 + \sigma_4\,\rho'\,\dot\sigma_1\,(1-\sigma_1^2) + \dot\sigma_4\,\rho\,(1-\sigma_1^2)\\ &= -\sigma_4\,\zeta'\sigma_1\,\dot\sigma_1 + \dot\sigma_4\,\rho\,(1-\sigma_1^2).\end{aligned}$$

To obtain the last equation above we used $-\rho\,\sigma_1 + \rho'\,(1-\sigma_1^2) = -\zeta'\,\sigma_1$, see (134). This proves (154). □

Writing $\frac{\dot\sigma_j}{\dot\sigma_1} = \frac{\mathrm{d}\sigma_j}{\mathrm{d}t} \big/ \frac{\mathrm{d}\sigma_1}{\mathrm{d}t} = \frac{\mathrm{d}\sigma_j}{\mathrm{d}\sigma_1}$, equations (152b) and (154) imply

$$\begin{aligned}0 &= \rho\,I_1\,\tfrac{\mathrm{d}\sigma_3}{\mathrm{d}\sigma_1} + \zeta\,I_3\,\tfrac{\mathrm{d}\sigma_4}{\mathrm{d}\sigma_1} + \rho\,I_3\,\sigma_4\\ 0 &= -\mathrm{m}\rho\,\zeta\,\tfrac{\mathrm{d}\sigma_3}{\mathrm{d}\sigma_1} + (I_3 + \mathrm{m}\,\rho^2\,(1-{\sigma_1}^2))\,\tfrac{\mathrm{d}\sigma_4}{\mathrm{d}\sigma_1}\\ &\quad - \mathrm{m}\rho\,\left((\zeta' - \rho)\,\sigma_3 + \zeta'\,\sigma_1\,\sigma_4\right).\end{aligned} \tag{157}$$

Solving for $\frac{\mathrm{d}\sigma_3}{\mathrm{d}\sigma_1}$ and $\frac{\mathrm{d}\sigma_4}{\mathrm{d}\sigma_1}$ from (157), we arrive at the following system of ordinary differential equations, which we call *Chaplygin's equations*,

$$\begin{aligned}\frac{\mathrm{d}\sigma_3}{\mathrm{d}\sigma_1} &= a(\sigma_1)\,\sigma_3 + b(\sigma_1)\,\sigma_4,\\ \frac{\mathrm{d}\sigma_4}{\mathrm{d}\sigma_1} &= c(\sigma_1)\,\sigma_3 + d(\sigma_1)\,\sigma_4,\end{aligned} \tag{158}$$

where

$$\begin{aligned} a(\sigma_1)\,\Delta &= -\mathrm{m}I_3\,\zeta\,(\zeta' - \rho), \\ b(\sigma_1)\,\Delta &= -I_3\,\left(I_3 + \mathrm{m}\,\rho^2\,(1-\sigma_1^2) + \mathrm{m}\,\zeta\,\zeta'\,\sigma_1\right), \\ c(\sigma_1)\,\Delta &= \mathrm{m}I_1\,\rho\,(\zeta' - \rho), \\ d(\sigma_1)\,\Delta &= \mathrm{m}\rho(I_1\,\zeta'\,\sigma_1 - I_3\,\zeta), \end{aligned} \tag{159}$$

and

$$\Delta = I_1\,I_3 + m\,I_1\,\rho^2\,(1-\sigma_1^2) + m\,I_3\,\zeta^2. \tag{160}$$

Note that Chaplygin's system is a linear system of ordinary differential equations with coefficients which smoothly depend on σ_1. Here σ_1 is treated as the independent variable; whereas σ_3 and σ_4 are treated as the dependent variables.

In §7.1 we have observed that the functions ρ and ζ extend to smooth functions on a neighborhood of the closed segment $[-1, 1]$. From this it follows that the coefficients a, b, c and d in Chaplygin's equations (158) are smooth functions on $[-1, 1]$ for the strongly convex body of revolution. This is quite remarkable, because as $\sigma_1 \to \pm 1$ we leave the part of the reduced phase space where σ_1, σ_2, σ_3, σ_4 is a system of coordinates.

6.7.3.2 *Two additional constants of motion*

From the theory of linear systems of ordinary differential equations, see chapter 3 of [27], it follows that for each $(\overline{\sigma}_3, \overline{\sigma}_4) \in \mathbb{R}^2$ there is a unique solution

$$\sigma_1 \mapsto \left(\sigma_3(\sigma_1;\,\overline{\sigma}_3,\,\overline{\sigma}_4),\,\sigma_4(\sigma_1;\,\overline{\sigma}_3,\,\overline{\sigma}_4)\right), \tag{161}$$

of Chaplygin's system (158) on $[-1,\,1]$ such that

$$(\sigma_3(0;\,\overline{\sigma}_3,\,\overline{\sigma}_4),\,\sigma_4(0;\,\overline{\sigma}_3,\,\overline{\sigma}_4)) = (\overline{\sigma}_3,\,\overline{\sigma}_4).$$

Moreover, the solution (161) is a smooth function of $\sigma_1 \in [-1,\,1]$ and depends linearly on the initial vector $(\overline{\sigma}_3,\,\overline{\sigma}_4) \in \mathbb{R}^2$. If for $j = 3, 4$ we write

$$\sigma_j^{1,\,0}(\sigma_1) = \sigma_j(\sigma_1;\,1,\,0), \quad \sigma_j^{0,\,1}(\sigma_1) = \sigma_j(\sigma_1;\,0,\,1), \tag{162}$$

which are smooth functions of $\sigma_1 \in [-1,\,1]$, then the linear dependence on the initial vector takes the form

$$\sigma_j(\sigma_1;\,\overline{\sigma}_3,\,\overline{\sigma}_4) = \overline{\sigma}_3\,\sigma_j^{1,\,0}(\sigma_1) + \overline{\sigma}_4\,\sigma_j^{0,\,1}(\sigma_1), \tag{163}$$

for $j = 3, 4$. Geometrically this means that the solution curves of the Chaplygin system define a smooth fibration of $(\sigma_1,\,\sigma_3,\,\sigma_4)$-space $[-1, 1] \times \mathbb{R}^2$, with the inverse of the diffeomorphism

$$\psi : (\sigma_1,\,(\overline{\sigma}_3,\,\overline{\sigma}_4)) \mapsto \left(\sigma_1,\,\sigma_3(\sigma_1;\,\overline{\sigma}_3,\,\overline{\sigma}_4),\,\sigma_4(\sigma_1;\,\overline{\sigma}_3,\,\overline{\sigma}_4)\right) \tag{164}$$

of $[-1,\,1]\times\mathbb{R}^2$ being a trivialization of this fibration. If

$$\pi_1 : [-1,\,1]\times\mathbb{R}^2 \to \mathbb{R}^2 : (\sigma_1,\,(\overline{\sigma}_3,\overline{\sigma}_4)) \mapsto (\overline{\sigma}_3,\,\overline{\sigma}_4)$$

denotes the projection map which forgets σ_1, then

$$\pi = \pi_1\circ\psi^{-1} : [-1,\,1]\times\mathbb{R}^2 \to \mathbb{R}^2$$

is the projection map of a fibration where the fiber over $(\overline{\sigma}_3,\,\overline{\sigma}_4)$ is the solution curve of Chaplygin equations with initial condition $(\overline{\sigma}_3,\,\overline{\sigma}_4)$.

From the fact that the equations of motion of the $\mathrm{E}(2)\times S^1$- reduced system imply Chaplygin's equations, it follows that the image of each solution of the $\mathrm{E}(2)\times S^1$-reduced system under the projection map

$$P : [-1,1]\times\mathbb{R}^3 \to [-1,1]\times\mathbb{R}^2 : (\sigma_1,\sigma_2,\sigma_3,\sigma_4) \mapsto (\sigma_1,\sigma_3,\sigma_4)$$

is contained in a fiber of P. Equivalently, the two components of the vector valued function $\pi\circ P$ are *constants of motion*. With a slight abuse of notation, we shall write

$$\pi\circ P = (\overline{\sigma}_3,\,\overline{\sigma}_4), \tag{165}$$

where $\overline{\sigma}_3$ and $\overline{\sigma}_4$ are viewed as smooth functions on the reduced phase space.

Note that these constants of motion, although smooth, are not given by algebraic formulæ involving the shape of the body of revolution. Instead the constants of motion are obtained by solving a linear system of ordinary diffferential equations with quite general variable coefficients.

6.7.3.3 *The total energy*

We now combine the constants of motion defined by the Chaplygin's equations with the $\mathrm{E}(2)\times S^1$-reduced total energy. Recall that total energy is a constant of motion for any system which satisfies the Lagrange-d'Alembert principle in §3 of chapter 1.

Lemma 6.7.3.3.28. *In terms of the invariants σ_j (144), the $\mathrm{E}(2)\times S^1$-reduced total energy $\overline{E}$ (142) is given by*

$$\begin{aligned}\overline{E} = \tfrac{1}{2}\,(I_1\,\sigma_5 + I_3\,\sigma_4^2) + \tfrac{1}{2}\,\mathrm{m}\,\big(\rho^2\,\sigma_2^2 + \zeta^2\,\sigma_5 + \rho^2\,(1-\sigma_1^2)\,\sigma_4^2 \\ - 2\rho\zeta\,\sigma_3\,\sigma_4\big) - \mathrm{mg}\,\big(\rho\,(1-\sigma_1^2) + \zeta\,\sigma_1\big),\end{aligned} \tag{166}$$

where $\rho=\rho(\sigma_1)$ and $\zeta=\zeta(\sigma_1)$ are smooth functions of σ_1 defined by (130a) *and* (130b).

Proof. We use (122) for the E(2) $\times$ S^1-reduced total energy. In it, we substitute $\langle I\,\omega,\,\omega\rangle = I_1\,\sigma_5 + I_3\,\sigma_4^2$ and

$$\begin{aligned}\langle \omega\times s,\,\omega\times s\rangle &= \langle\omega,\,\omega\rangle\,\langle s,s\rangle - \langle\omega,s\rangle^2\\ &= (\sigma_5+\sigma_4^2)\,(\rho^2\,\sigma_6+\zeta^2) - (\rho\,\sigma_3+\zeta\,\sigma_4)^2\\ &= \sigma_5\,\rho^2\,\sigma_6 + \sigma_5\,\zeta^2 + \sigma_4^2\,\rho^2\,\sigma_6 + \sigma_4^2\,\zeta^2 - \rho^2\,\sigma_3^2 - \zeta^2\,\sigma_4^2 - 2\rho\,\zeta\,\sigma_3\,\sigma_4\\ &= \rho^2\,\sigma_2^2 + \zeta^2\,\sigma_5 + \rho^2\,(1-\sigma_1^2)\sigma_4^2 - 2\rho\,\zeta\,\sigma_3\,\sigma_4.\end{aligned}$$

To obtain the second equality above we use (156) and for the fourth equality equation $\sigma_6 = 1-\sigma_1^2$. Furthermore, $\langle s,\,u\rangle = \rho\,\sigma_6 + \zeta\,\sigma_1 = \rho(1-\sigma_1^2)+\zeta\,\sigma_1$. Collecting terms and using the definition of the invariants (144) we get (166). □

The E(2) $\times$ S^1-reduced total energy $\overline{E}$ (166) is a smooth function of σ_j, $1\le j\le 5$. In the domain $0\le\sigma_1^2<1$, where the invariants σ_j, $1\le j\le 4$, form a system of coordinates, substituting $\sigma_5 = (\sigma_2^2+\sigma_3^2)(1-\sigma_1^2)^{-1}$ gives

$$\begin{aligned}\overline{E} &= \tfrac12\,\frac{I_1+\mathrm{m}\rho\,(1-\sigma_1^2)+\mathrm{m}\,\zeta^2}{1-\sigma_1^2}\,\sigma_2^2 + \tfrac12\,\frac{I_1+\mathrm{m}\,\zeta^2}{1-\sigma_1^2}\,\sigma_3^2 - \mathrm{m}\,\rho\,\zeta\,\sigma_3\,\sigma_4\\ &\quad+\tfrac12\,(I_3+\mathrm{m}\,\rho^2\,(1-\sigma_1^2))\,\sigma_4^2 - \mathrm{mg}\,\big(\rho\,(1-\sigma_1^2)+\zeta\,\sigma_1\big)\,.\end{aligned}\tag{167}$$

In the domain $\sigma_5>\sigma_2^2+\sigma_3^2$, where the invariants σ_j, $2\le j\le 5$, form a system of coordinates, we substitute $\sigma_1 = \pm\big(1-(\sigma_2^2+\sigma_3^2)/\sigma_5\big)^{1/2}$ and obtain $\overline{E}$ is a smooth function of $\sigma_2,\ldots,\sigma_5$.

6.7.3.4 *A conservative Newtonian system*

If we fix the values of the constants of motion $\overline{\sigma}_3$ and $\overline{\sigma}_4$ given by Chaplygin's equations (158), then the fiber $(\pi\circ P)^{-1}(\overline{\sigma}_3,\overline{\sigma}_4)$, which lies in the domain $-1<\sigma_1<1$, is a smooth surface $\mathcal{N}$ in the E(2)$\times S^1$-reduced phase space, which is parametrized by $(\sigma_1,\,\sigma_2)\in(-1,\,1)\times\mathbb{R}$. The surface $\mathcal{N}$ is invariant under the flow of the E(2) $\times$ S^1-*reduced vector field*, for which the E(2) $\times$ S^1-reduced total energy $\overline{E}$ is a constant of motion. In terms of the parameters $(\sigma_1,\,\sigma_2)$, the restriction to $\mathcal{N}$ of the E(2) $\times$ S^1-*reduced total energy* (167) is

$$\overline{E} = \overline{T}(\sigma_1,\,\sigma_2) + \overline{V}_{\overline{\sigma}_3,\overline{\sigma}_4}(\sigma_1),\tag{168}$$

where

$$\overline{T}(\sigma_1,\,\sigma_2) = \tfrac12\,\frac{I_1+\mathrm{m}\,\rho\,(1-\sigma_1^2)+\mathrm{m}\,\zeta^2}{1-\sigma_1^2}\,\sigma_2^2\tag{169}$$

plays the role of a *kinetic energy* and

$$\overline{V}_{\overline{\sigma}_3,\,\overline{\sigma}_4}(\sigma_1) = \tfrac{1}{2}\,\frac{I_1 + \mathrm{m}\,\zeta^2}{1-\sigma_1^2}\,\sigma_3^2 + \tfrac{1}{2}\,(I_3 + \mathrm{m}\,\rho^2\,(1-\sigma_1^2))\,\sigma_4^2 - \mathrm{m}\,\rho\,\zeta\,\sigma_3\,\sigma_4$$
$$- \,\mathrm{mg}\,\big(\rho\,(1-\sigma_1^2) + \zeta\,\sigma_1\big) \tag{170}$$

is viewed as a *potential energy* function. Here $\rho = \rho(\sigma_1)$, $\zeta = \zeta(\sigma_1)$, $\sigma_3 = \sigma_3(\sigma_1;\,\overline{\sigma}_3,\,\overline{\sigma}_4)$ and $\sigma_4 = \sigma_4(\sigma_1;\,\overline{\sigma}_3,\,\overline{\sigma}_4)$ are solutions of Chaplygin's equations (158) viewed as functions of σ_1 with the initial conditions $\overline{\sigma}_3$ and $\overline{\sigma}_4$ being viewed as parameters.

Because $\dot{\sigma}_1 = \sigma_2$, we have a one degree of freedom classical mechanical system of the form kinetic plus potential energy on a configuration space parametrized by σ_1. The kinetic energy defined by the Riemannian structure $(I_1 + \mathrm{m}\,\rho\,(1-\sigma_1^2) + \mathrm{m}\,\zeta^2)/(1-\sigma_1^2)$ times the standard one and the potential energy is given as a function of σ_1 by (169). The kinetic energy seems to be singular at $\sigma_1 = \pm 1$. However, it can be regularized by means of the change of variable $\sigma_1 = \sin\varphi$ with $-\pi/2 < \varphi < \pi/2$, for then

$$(1-\sigma_1^2)^{-1/2}\,\dot{\sigma}_1 = (\cos\varphi)^{-1}\,(\cos\varphi)\,\dot{\varphi} = \dot{\varphi}. \tag{171}$$

Therefore

$$\overline{E} = \tfrac{1}{2}\,k(\varphi)\,\dot{\varphi}^2 + \overline{V}_{\overline{\sigma}_3,\,\overline{\sigma}_4}(\sin\varphi), \tag{172}$$

where

$$k(\varphi) = I_1 + \mathrm{m}\,\rho(\sin\varphi)\,\cos^2\varphi + \mathrm{m}\,\zeta(\sin\varphi)^2. \tag{173}$$

Conservation of $\mathrm{E}(2)\times S^1$-reduced energy yields

$$\tfrac{1}{2}\,k'(\varphi)\,\dot{\varphi}^2 + k(\varphi)\,\ddot{\varphi} + \frac{\partial\,\overline{V}_{\overline{\sigma}_3,\,\overline{\sigma}_4}(\sin\varphi)}{\partial\varphi} = 0. \tag{174}$$

Equation (174) describes a *conservative Newtonian system* with kinetic energy $\tfrac{1}{2}\,k(\varphi)\,\dot{\varphi}^2$ subject to a force $-\frac{\partial\,\overline{V}_{\overline{\sigma}_3,\,\overline{\sigma}_4}(\sin\varphi)}{\partial\varphi}$. Actually, the Lie derivative $\dot{\overline{E}}$ of $\overline{E}$ with respect to the $\mathrm{E}(2)\times S^1$-reduced vector field is equal to $\dot{\varphi}$ times the left hand side of (174). Therefore (174) follows from $\dot{\overline{E}} = 0$ if $\dot{\varphi} \neq 0$. However, if $\dot{\varphi} = 0$, then (174) follows by continuity, because from §1.2 of chapter 3 we know that the motion on the smooth part of the orbit space is defined by a smooth vector field. Therefore $\ddot{\varphi}$ is a smooth function of φ and $\dot{\varphi}$ if $-\pi/2 < \varphi < \pi/2$.[4]

[4]From

$$\sin\varphi = \sigma_1 = \langle u,\, e_3\rangle = \langle A^{-1}\,e_3,\, e_3\rangle = \langle e_3,\, A\,e_3\rangle$$

it follows that φ is equal to the angle between the oriented axis of symmetry of the body of revolution in its actual position in space, and the horizontal plane. The angle is taken to be positive if $A\,e_3$ points upward.

Note that if $\varphi(t)$ is known, then we know $\sigma_1(t) = \sin\varphi(t)$, $\sigma_2(t) = \dot{\sigma}_1(t) = \dot{\varphi}(t)\cos\varphi(t)$, $\sigma_3(t) = \sigma_3(\sigma_1(t); \overline{\sigma}_3, \overline{\sigma}_4)$, and $\sigma_4(t) = \sigma_4(\sigma_1(t); \overline{\sigma}_3, \overline{\sigma}_4)$. In other words, we know the solution of the $\mathrm{E}(2) \times S^1$-reduced system on the part $-1 < \sigma_1 < 1$ of the $\mathrm{E}(2) \times S^1$-orbit space $\mathcal{M}$.

6.7.3.5 *Solutions of the one degree of freedom system*

Conservative one degreee of freedom classical mechanical systems are the best understood mechanical systems. We recall the classical description of its solutions. To simplify notation we write $\overline{V}(\varphi) = \overline{V}_{\overline{\sigma}_3,\overline{\sigma}_4}(\sin\varphi)$.

Choose a value E of the value of the constant of motion given by the $\mathrm{E}(2) \times S^1$ reduced energy (172), which is larger than the infimum of the potential function $\overline{V}$. Let I be a connected component of the open set of all $\varphi \in (-\pi/2, \pi/2)$ such that $\overline{V}(\varphi) < E$. Then I is an open interval (φ_-, φ_+), where $\pi/2 \le \varphi_- < \varphi_+ \le \pi/2$. Note that $\overline{V}(\varphi_-) = E$ and $\overline{V}(\varphi_+) = E$ if $-\pi/2 < \varphi_- < \pi/2$ and $-\pi/2 < \varphi_+ < \pi/2$, respectively. Depending on the sign of $\dot{\varphi}$, we have

$$\dot{\varphi} = \pm \big(\frac{2}{k(\varphi)}\,(E - \overline{V}(\varphi))\big)^{1/2}, \tag{175}$$

as long as $\varphi_- < \varphi < \varphi_+$. Separating variables and integrating, (175) gives

$$t_1 - t_0 = \pm \int_{\varphi(t_0)}^{\varphi(t_1)} \big(\frac{2}{k(\varphi)}\,(E - \overline{V}(\varphi))\big)^{-1/2}\, \mathrm{d}\varphi \tag{176}$$

for the time needed to go from $\varphi(t_0)$ to $\varphi(t_1)$. The solution $t_1 \mapsto \varphi(t_1)$ of (175) is obtained by inverting the function $\varphi(t_1) \mapsto t_1$ given by the right hand side of (176).

Take the plus sign in (176). If $\phi_+ < \pi/2$ and $\overline{V}'(\phi_+) \neq 0$, then $\overline{V}'(\phi_+) > 0$ and the right hand side of (176) has a finite positive limit as $\varphi(t_1) \uparrow \varphi_+$. This means that $\varphi(t)$ reaches φ_+ at a finite time $t = T$. From (174) it follows that $\ddot{\varphi} < 0$ when $t = T$, which means that the continuation of the solution for $t > T$ is determined by taking the minus sign in (176). Therefore $\varphi(t) = \varphi(2T - t)$ for all t. If also $\varphi_- > -\pi/2$ and $\overline{V}'(\varphi_-) \neq 0$, then $\overline{V}'(\varphi_-) < 0$ and for $t > T$ the solution will first reach φ_- at the finite time $t = S > T$. Hence $\varphi(t) = \varphi(2S - t)$ for all t. It follows that

$$\varphi(t) = \varphi(2T - t) = \varphi(2S - (2T - t)) = \varphi(2(S - T) + t)$$

for all t, that is, the solution is periodic with minimal positive period equal to $2(S - T)$. Here $\varphi(t)$ oscillates back and forth between φ_- and φ_+, moving monotonically during each traversal with the time reversed motion during the succeeding traversal. The number $S-T$ is equal to the right hand side of (176) with the plus sign and $\varphi(t_0)$ and $\varphi(t_1)$ replaced by φ_- and φ_+, respectively. Therefore the minimal positive period τ of the motion is given by the convergent improper integral

$$\tau = 2\int_{\varphi_-}^{\varphi_+} \Big(\frac{2}{k(\varphi)}\,(E-\overline{V}(\varphi))\Big)^{-1/2}\,\mathrm{d}\varphi. \tag{177}$$

If $\overline{V}'(\varphi_+) = 0$, then $(\varphi, \dot{\varphi}) = (\varphi_+, 0)$ is an equilibrium point in the phase plane of (174), which corresponds to a relative equilibrium of the rolling body of revolution. Because

$$E - \overline{V}(\varphi) = \overline{V}(\varphi_+) - \overline{V}(\varphi) = \mathrm{O}((\varphi-\varphi_+)^2) \quad \text{as } \varphi \uparrow \varphi_+,$$

the right hand side of (176) diverges as $\varphi(t_1) \uparrow \varphi_+$. We conclude that $(\varphi(t), \dot{\varphi}(t)) \rightarrow (\varphi_+, 0)$ as $t \rightarrow \infty$. With the choice of the minus sign in (176), we get $(\varphi(t), \dot{\varphi}(t)) \rightarrow (\varphi_+, 0)$ as $t \rightarrow \infty$. Similar conclusions hold with φ_+ replaced by φ_-.

Now assume that $\sigma_3(\pm 1; \overline{\sigma}_3, \overline{\sigma}_4) \neq 0$. From (170) it follows that $\overline{V}(\varphi) \rightarrow \infty$ if $\varphi \rightarrow \pm\pi/2$. If the parameters $(\overline{\sigma}_3, \overline{\sigma}_4) \in \mathbb{R}^2$ avoid the two straight lines through the origin in $\mathbb{R}^2$ where $\sigma_3(\pm 1; \overline{\sigma}_3, \overline{\sigma}_4) = 0$, then $\overline{V}(\varphi) \rightarrow \infty$ as $\varphi \downarrow -\pi/2$ and as $\varphi \uparrow -\pi/2$. So $\varphi(t)$ moves in a potential well with infinitely steep walls at $\varphi = \pm\pi/2$. Therefore $-\pi/2 < \varphi_- < \varphi_+ < \pi/2$. Consequently, $\varphi(t)$ oscillates back and forth between φ_- and φ_+, unless the solution converges to an unstable relative equilibrium.

6.7.3.6 *Quasi-periodic solutions*

The above analysis of the conservative one degree of freedom Newtonian system (174) shows that, *except* for initial values on a submanifold of codimension one, the solutions in the $\mathrm{E}(2) \times S^1$-orbit space are periodic.

Let $((A, a), B) \in \mathrm{E}(2) \times S^1$ be the shift element, see lemma 4.3.1.41 of chapter 4, where $A \in \mathrm{SO}(2)$. If $A \neq I$, then from the second item in example 4.3.6.1.53 of chapter 4 it follows that (A, a) is a regular and stably elliptic element of $\mathrm{E}(2)$ with a circle subgroup C of $\mathrm{E}(2)$ as its centralizer. Consequently, $((A, a), B)$ is a regular and stably elliptic element of $\mathrm{E}(2) \times S^1$ with the two-dimensional torus $C \times S^1$ as its centralizer. From proposition 4.3.6.2.58 of chapter 4 we see that over the open subset of the periodic solutions in the $\mathrm{E}(2) \times S^1$-orbit space $\mathcal{M}$ where $A \neq I$, the motion of the strongly convex rolling body of revolution is quasi-periodic on three-dimensional tori in phase space.

The strongly convex body of revolution differs from the rolling disk because for the disk the $\mathrm{E}(2) \times S^1$-reduced equations of motion are truly singular when $\sigma_1 \to \pm 1$. These motions correspond to the disk falling flat. In chapter 7 we will treat the motion of the rolling disk in much more detail.

6.7.3.7 *Appendix. The* $\mathrm{E}(2) \times S^1$*-reduced equations of motion*

In the domain where $-1 < \sigma_1 < 1$ of the $\mathrm{E}(2) \times S^1$-reduced space $\mathcal{M}_{\mathrm{reg}}$, where the invariants σ_j, $1 \le j \le 4$, form a regular system of coordinates, the equations of motion for the $\mathrm{E}(2) \times S^1$-reduced system expressed in terms of the invariants σ_j, $1 \le j \le 4$, can be obtained in the following way. Start with $\sigma_2 = \dot{\sigma}_1$ and

$$\dot{\sigma}_3 = \big(a(\sigma_1)\,\sigma_3 + b(\sigma_1)\,\sigma_4\big)\,\sigma_2, \tag{178}$$

$$\dot{\sigma}_4 = \big(c(\sigma_1)\,\sigma_3 + d(\sigma_1)\,\sigma_4\big)\,\sigma_2, \tag{179}$$

which follow from Chaplygin's equations (158) and $\sigma_2 = \dot{\sigma}_1$. To find an expression for $\dot{\sigma}_2$ in terms of the σ_j, $1 \le j \le 4$, we use the fact that the time derivative of the $\mathrm{E}(2) \times S^1$-reduced total energy $\overline{E}$ (167) is equal to zero. In working out the time derivative of (167), we use

$$\dot{\rho} = \rho'(\sigma_1)\,\dot{\sigma}_1 = \rho'(\sigma_1)\,\sigma_2, \quad \text{and} \quad \dot{\zeta} = \zeta'(\sigma_1)\,\dot{\sigma}_1 = \zeta'(\sigma_1)\,\sigma_2.$$

Because $(\sigma_2^2)^{\cdot} = 2\sigma_2\,\dot{\sigma}_2$, all the terms in $\frac{\mathrm{d}\overline{E}}{\mathrm{d}t}$ have a factor σ_2. When $\sigma_2 \neq 0$, we can divide this factor out, which leads to the desired equation for $\dot{\sigma}_2$. By continuity, this equation remains valid when $\sigma_2 = 0$, because we know *a priori* that in the smooth part of the $\mathrm{E}(2) \times S^1$-orbit space the motion is determined by a smooth vector field, see §1.2 of chapter 3. The formula for $\dot{\sigma}_2$ in terms of σ_j, $1 \le j \le 4$, obtained this way is explicit, but quite unwieldy. Moreover, it does not show in an obvious way that the $\mathrm{E}(2) \times S^1$-reduced equations of motion are integral curves of a smooth vector field on the $\mathrm{E}(2) \times S^1$-orbit space including those points where $\sigma_1 = \pm 1$. Below we present an alternative calculation of smooth vector field in the reduced phase space.

We now remove the axial symmetry from the $\mathrm{E}(2)$-reduced vector field $\overline{V}$ (141). Because $\overline{V}$ is invariant under the induced axial symmetry Ψ (140), it induces a derivation $\widehat{\mathcal{V}}$ on $(C^\infty(\mathcal{M}), \{\,,\,\})$. From the theory of nonholonomic reduction in §2 of chapter 3, we expect that the integral curves of $\widehat{\mathcal{V}}$ satisfy almost Poisson bracket equations of the form

$$\frac{d\sigma_i}{dt} = \{\sigma_i, \mathcal{E}\} \quad \text{for } i = 1, \ldots, 5, \tag{180}$$

where $\mathcal{E}$ is the $\mathrm{E}(2) \times S^1$ reduced energy given by

$$\tfrac{1}{2}\,(I_1 + \mathrm{m}(\rho^2(1-\sigma_1^2)+\zeta^2)\sigma_5 + \tfrac{1}{2}\,(I_3 + \mathrm{m}\,\rho^2(1-\sigma_1^2))\sigma_4^2 \\ - \tfrac{1}{2}\,\mathrm{m}\rho^2\sigma_3^2 - \mathrm{m}\rho\zeta\,\sigma_3\sigma_4 - \mathrm{mg}\,(\rho(1-\sigma_1^2)\zeta\sigma_1).$$

A direct calculation of the nonholonomic bracket on the right hand side of (180) is quite involved. So we use the following indirect method.

Consider the functions

$$\tau_1 = \langle \mathcal{Y}(s)\omega, e_3\rangle \tag{181a}$$

$$\tau_2 = \langle \mathcal{Y}(s)\omega, s\rangle \tag{181b}$$

$$\tau_3 = \langle \mathcal{Y}(s)\omega, u \times e_3\rangle \tag{181c}$$

$$\tau_4 = \langle \mathcal{Y}(s)\omega, \omega\rangle, \tag{181d}$$

each of which is invariant under the induced axial action on $S^2 \times \mathbb{R}^3$. Hence each τ_i is a smooth function of the invariants σ_j (144) for $j = 1, \ldots, 5$. A straightforward calculation shows that

$$\tau_1 = -\mathrm{m}\,\rho\zeta\,\sigma_3 + (I_3 + \mathrm{m}\rho^2(1-\sigma_1^2))\sigma_4 \tag{182a}$$

$$\tau_2 = I_1\rho\,\sigma_3 + I_3\zeta\,\sigma_4 \tag{182b}$$

$$\tau_3 = -(I_1 + \mathrm{m}\rho^2(1-\sigma_1^2) + \mathrm{m}\,\zeta^2)\sigma_2 \tag{182c}$$

$$\tau_4 = (I_1 + \mathrm{m}\zeta^2)\sigma_5 + (I_3 + \mathrm{m}\rho^2(1-\sigma_1^2))\sigma_4^2 \\ - 2\,\mathrm{m}\rho\zeta\,\sigma_3\sigma_4 + \mathrm{m}\rho^2\,\sigma_2^2 \tag{182d}$$

We have

Proposition 6.7.3.7.29. *The* $\mathrm{E}(2) \times S^1$ *reduced equations of motion of a dynamically symmetric smooth strongly convex surface of revolution* $\mathcal{S}$ *which rolls without slipping on a horizontal plane under the influence of a constant vertical gravitational force are*

$$\dot\sigma_1 = \sigma_2 \tag{183a}$$

$$-\mathrm{m}\rho\zeta\,\dot\sigma_3 + (I_3 + \mathrm{m}\rho^2(1-\sigma_1^2))\dot\sigma_4 = \mathrm{m}\rho(\zeta' - \rho)\sigma_2\sigma_3 + \mathrm{m}\rho\zeta'\,\sigma_1\sigma_2\sigma_4 \tag{183b}$$

$$I_1\rho\,\dot\sigma_3 + I_3\zeta\,\dot\sigma_4 = -I_3\rho\,\sigma_2\sigma_4 \tag{183c}$$

$$(I_1 + \mathrm{m}\rho^2(1-\sigma_1^2) + \mathrm{m}\,\zeta^2)\dot\sigma_2 = (I_3 + \mathrm{m}\rho(1-\sigma_1^2) + \mathrm{m}\rho\zeta\,\sigma_1)\sigma_3\sigma_4 \\ -\mathrm{m}\rho\zeta\,\sigma_3^2 - (I_1 + \mathrm{m}\,\zeta^2)\sigma_1\sigma_5 + \mathrm{mg}\,(\zeta - \rho\,\sigma_1)(1-\sigma_1^2) \\ -\mathrm{m}\big(\zeta\zeta' + \rho\rho'\,(1-\sigma_1^2)\big)\sigma_2^2 \tag{183d}$$

$$\tfrac{1}{2}\,(I_1 + \mathrm{m}\zeta^2)\,\dot\sigma_5 + \big[(I_3 + \mathrm{m}\rho^2(1-\sigma_1^2))\sigma_4 - \mathrm{m}\rho\zeta\,\sigma_3\big]\,\dot\sigma_4 \\ -\mathrm{m}\rho\zeta\,\sigma_4\dot\sigma_3 + \mathrm{m}\rho^2\sigma_2\dot\sigma_2 = \big[\mathrm{m}(\rho'\zeta + \rho\zeta'\sigma_3\sigma_4 - \rho\rho'\sigma_2^2) \\ +\ \mathrm{mg}\,(\zeta - \rho\,\sigma_1) - \mathrm{m}\zeta\zeta'\,\sigma_5 - \mathrm{m}\rho\,\big(\rho'(1-\sigma_1^2) - \rho\,\sigma_1)\big)\sigma_4^2\big]\,\sigma_2. \tag{183e}$$

Proof. Computing the derivative of (181a) and (181b) and equating the result with the derivative of (182a) and (182b) gives (183b) and (183c), respectively. These equations can be obtained from (157) by multiplying both sides by $\frac{\mathrm{d}\sigma_1}{\mathrm{d}t} = \sigma_2$ (183a).

We repeat this process for τ_3 and τ_4, but give all the details. We begin by differentiating (181c), which gives

$$\begin{aligned}\dot{\tau}_3 &= \langle \frac{\mathrm{d}(\mathcal{Y}(s)\omega)}{\mathrm{d}t}, u \times e_3\rangle + \langle \mathcal{Y}(s)\omega, \dot{u} \times e_3\rangle \\ &= \langle (\mathcal{Y}(s)\omega) \times \omega + \mathrm{m}(\dot{s} \times (\omega \times s)) - \mathrm{mg}\, u \times s, u \times e_3\rangle \\ &\quad + \langle \mathcal{Y}(s)\omega, (u \times \omega) \times e_3\rangle, \end{aligned} \tag{184}$$

since

$$\begin{aligned}\frac{\mathrm{d}\mathcal{Y}(s)\omega)}{\mathrm{d}t} &= I\dot{\omega} + \mathrm{m}\, \dot{s} \times (\omega \times s) + \mathrm{m}\, s \times (\dot{\omega} \times s) + \mathrm{m}\,(s \times (\omega \times \dot{s})) \\ &= I\omega \times \omega + \mathrm{m}\,\langle s, \omega\rangle\,(\omega \times s) + \mathrm{m}\,\dot{s} \times (\omega \times s) - \mathrm{mg}\, u \times s \\ &= (\mathcal{Y}(s)\omega) \times \omega + \mathrm{m}\,\dot{s} \times (\omega \times s) - \mathrm{mg}\, u \times s. \end{aligned} \tag{185}$$

Now using Grassmann's identity we get

$$\begin{aligned}\omega \times (u \times e_3) &= u\langle \omega, e_3\rangle - e_3\langle u, \omega\rangle \\ (\omega \times s) \times (u \times e_3) &= u\langle \omega \times s, e_3\rangle - e_3\langle \omega \times s, u\rangle \\ \langle u \times s, u \times e_3\rangle &= \langle u, s \times (u \times e_3)\rangle = \langle s, e_3\rangle - \langle u, e_3\rangle\,\langle s, u\rangle \\ e_3 \times (u \times \omega) &= u\langle e_3, \omega\rangle - \omega\langle e_3, u\rangle. \end{aligned}$$

So

$$\begin{aligned}\dot{\tau}_3 &= -\langle \mathcal{Y}(s)\omega, e_3\rangle\langle u, \omega\rangle + \mathrm{m}\,\langle \omega \times s, e_3\rangle\langle \dot{s}, u\rangle - \mathrm{m}\,\langle \omega \times s, u\rangle\langle \dot{s}, e_3\rangle \\ &\quad - \mathrm{mg}\,\big(\langle s, e_3\rangle - \langle u, e_3\rangle\langle s, u\rangle\big) + \langle u, e_3\rangle\langle \mathcal{Y}(s)\omega, \omega\rangle. \end{aligned} \tag{186}$$

We now express each term in (186) in terms of invariants. We find that the first term is

$$\begin{aligned}&-\langle u, \omega\rangle\langle \mathcal{Y}(s)\omega, e_3\rangle \\ &\qquad = (u_1\omega_1 + u_2\omega_2 + u_3\omega_3)(I_3\omega_3 + \mathrm{m}\,\omega_3\langle s, s\rangle - \mathrm{m}\, s_3\langle \omega, s\rangle) \\ &\qquad = -(\sigma_3 + \sigma_1\sigma_4)\big((I_3 + \mathrm{m}\rho^2(1 - \sigma_1^2))\sigma_4 - \mathrm{m}\zeta\rho\,\sigma_3\big); \end{aligned}$$

the second term is

$$\begin{aligned}&\mathrm{m}\,\langle \omega \times s, e_3\rangle\langle \dot{s}, u\rangle \\ &\quad = -\mathrm{m}\rho(u_1\omega_2 - u_2\omega_1)^2(-\rho\, u_3 + \rho'\,(u_1^2 + u_2^2) + \zeta'\, u_3), \\ &\qquad\qquad \text{since } \omega \times s = (\zeta\omega_2 - \rho u_2\omega_3, \rho u_1\omega_3 - \zeta\omega_1, -\rho(u_1\omega_2 - u_2\omega_1)) \\ &\qquad\qquad \text{and } Ds(u)(u \times \omega) = \begin{pmatrix} \rho(u_2\omega_3 - u_3\omega_2) + \rho' u_1(u_1\omega_2 - u_2\omega_1) \\ \rho(u_3\omega_1 - u_1\omega_3) + \rho' u_2(u_1\omega_2 - u_2\omega_1) \\ \zeta'(u_1\omega_2 - u_2\omega_2) \end{pmatrix} \\ &\quad = -\mathrm{m}\rho\,\big((\zeta' - \rho)\sigma_1 + \rho'(1 - \sigma^2)\big)\sigma_2^2; \end{aligned}$$

the third term is

$$-\mathrm{m}\,\langle \omega \times s, u\rangle\langle \dot{s}, e_3\rangle = -\mathrm{m}\,\zeta'\,(\zeta - \rho\, u_3)(u_1\omega_2 - u_2\omega_1)^2 = -\mathrm{m}\,\zeta'\,(\zeta - \rho\,\sigma_1)\sigma_2^2;$$

the fourth term is

$$-\mathrm{mg}\,\big(s_3 - u_3\langle s, u\rangle\big) = -\mathrm{mg}\,(\zeta - \rho\,\sigma_1)(1-\sigma_1^2);$$

and the fifth term is

$$\begin{aligned}&\langle u, e_3\rangle\langle \mathcal{Y}(s)\omega, \omega\rangle = \sigma_1\tau_4\\ &= \sigma_1\big((I_1 + \mathrm{m}\,\zeta^2)\sigma_5 + (I_3 + \mathrm{m}\,\rho^2(1-\sigma_1^2))\sigma_4^2 - 2\mathrm{m}\,\rho\zeta\,\sigma_3\sigma_4 + \mathrm{m}\,\rho^2\,\sigma_2^2\big).\end{aligned}$$

Therefore

$$\begin{aligned}\dot{\tau}_3 &= -(I_3 + \mathrm{m}\,\rho^2(1-\sigma_1^2)\sigma_3\sigma_4 + \mathrm{m}\,\rho\zeta\,\sigma_3^2\\ &\quad - (I_3 + \mathrm{m}\rho^2(1-\sigma_1^2))\sigma_1\sigma_4 + \mathrm{m}\,\rho\zeta\,\sigma_3\sigma_1\sigma_4 - \mathrm{m}\,\rho\,\big((\zeta'-\rho)\sigma_1 + \rho(1-\sigma_1^2)\big)\sigma_2^2\\ &\qquad - \mathrm{m}\,\zeta'(\zeta - \rho\sigma_1)\sigma_2^2 - \mathrm{mg}\,(\zeta - \rho\sigma_1)(1-\sigma_1^2)\\ &\quad + (I_1 + \mathrm{m}\,\zeta^2)\sigma_1\sigma_5 + (I_3 + \mathrm{m}\,\rho^2(1-\sigma_1^2))\sigma_1\sigma_4^2 - 2\mathrm{m}\,\rho\zeta\,\sigma_1\sigma_3\sigma_4 + \mathrm{m}\,\rho^2\,\sigma_1\sigma_2^2\\ &= -\big(I_3 + \mathrm{m}\,\rho\zeta\,\sigma_1 + \mathrm{m}\,\rho^2(1-\sigma_1^2)\big)\sigma_3\sigma_4 + \mathrm{m}\,\rho\zeta\,\sigma_3^2 + \big(2\mathrm{m}\,\rho^2\,\sigma_1 - \mathrm{m}\zeta\zeta'\\ &\quad - \mathrm{m}\,\rho\rho'(1-\sigma_1^2)\big)\sigma_2^2 + (I_1 + \mathrm{m}\,\zeta^2)\sigma_1\sigma_5 - \mathrm{mg}\,(\zeta - \rho\,\sigma_1)(1-\sigma_1^2). \quad (187)\end{aligned}$$

Differentiating $\tau_3 = -\big(I_1 + \mathrm{m}\,\zeta^2 + \mathrm{m}\rho^2(1-\sigma_1^2)\big)\sigma_2$ gives

$$\begin{aligned}\dot{\tau}_3 &= -2\mathrm{m}\,\big(\zeta\zeta' + \rho\rho'(1-\sigma_1^2) - \rho^2\sigma_1\big)\sigma_2\\ &\quad - \big(I_1 + \mathrm{m}\,\zeta^2 + \mathrm{m}\,\rho^2(1-\sigma_1^2)\big)\dot{\sigma}_2. \qquad (188)\end{aligned}$$

Equating (187) and (188) gives (183e).

Differentiating the expression (181d) for τ_4 and using $\sigma_2 = \dot{\sigma}_1$ gives

$$\begin{aligned}\dot{\tau}_4 &= 2\mathrm{m}\,\zeta\zeta'\,\sigma_2\sigma_5 + (I_1 + \mathrm{m}\,\zeta^2)\dot{\sigma}_5\\ &\quad + 2\mathrm{m}\,\rho\big(\rho'\,(1-\sigma_1^2) - \rho\,\sigma_1\big)\sigma_2\sigma_4^2 + 2\big(I_3 + \mathrm{m}\,\rho^2(1-\sigma_1^2)\big)\sigma_4\dot{\sigma}_4\\ &\qquad - 2\mathrm{m}\,\big((\rho'\zeta + \rho\zeta')\sigma_2\sigma_3\sigma_4 + \rho\zeta(\dot{\sigma}_3\sigma_4 + \sigma_3\dot{\sigma}_4)\big)\\ &\qquad\quad + 2\mathrm{m}\,\rho(\rho'\,\sigma_2^2 + \rho\,\dot{\sigma}_2)\sigma_2. \qquad (189)\end{aligned}$$

Next differentiating the expression (182d) for τ_4 gives

$$\begin{aligned}\dot{\tau}_4 &= \frac{\mathrm{d}}{\mathrm{d}t}\langle \mathcal{Y}(s)\omega, \omega\rangle = \frac{\mathrm{d}}{\mathrm{d}t}\,[\langle I\omega, \omega\rangle + \mathrm{m}\,\langle \omega \times s, \omega \times s\rangle]\\ &= 2\langle I\dot{\omega}, \omega\rangle + 2\mathrm{m}\,\langle \frac{\mathrm{d}}{\mathrm{d}t}(\omega \times s), \omega \times s\rangle.\end{aligned}$$

Substituting

$$I\dot{\omega} = I\omega \times \omega + \mathrm{m}\,\frac{\mathrm{d}}{\mathrm{d}t}((\omega \times s) \times s) + \mathrm{m}\,\langle s, \omega\rangle(\omega \times s) - \mathrm{mg}\,u \times s$$

into the above expression for $\dot{\tau}_4$ we get

$$\begin{aligned}\dot{\tau}_4 &= -2\mathrm{m}\,\big(\langle\frac{\mathrm{d}}{\mathrm{d}t}((\omega\times s)\times s) - \mathrm{g}\,\langle u\times s,\omega\rangle + \langle\frac{\mathrm{d}}{\mathrm{d}t}(\omega\times s),\omega\times s\rangle\big)\\ &= 2\mathrm{mg}\,\langle u,\omega\times s\rangle = 2\mathrm{mg}\,(u_1\omega_2 - u_2\omega_1)(\zeta - \rho\,u_3)\\ &= 2\mathrm{mg}\,(\zeta - \rho\,\sigma_1)\sigma_2. \qquad (190)\end{aligned}$$

Equating (189) and (190) gives (183e). This proves proposition 6.7.3.7.29. □

6.8 Notes

The approach to the dynamics of a rigid body used here follows the techniques developed by Hermans [53]. The fact that every position of the body lying on the horizontal plane can be reached from any other such position by means of a rolling motion without slipping was verified for the rolling disk in §1 of chapter 1 of Neĭmark and Fufaev [82]. For a classical treatment see [95]. As was mentioned in the notes to chapter 1, taking the Euler-Lagrange equations on the tangent bundle of the orbit space of an abelian $\mathbb{R}^n$ action to find the equations of motion of a nonholonomically constrained system with an $\mathbb{R}^n$ symmetry was quite common at the end of the 19th century. Comparing the mathematical model obtained by this method with the one obtained by applying the Lagrange-d'Alembert principle in the case of a rigid body rolling without slipping on a horizontal plane, experimental evidence favors the latter, see Lewis and Murray [67]. For instance, the *rattleback* has spinning modes which are asymptotically stable on a level set of the energy, see Walker [115, 116]. On the other hand, this contradicts the Hamiltonian character of the Euler-Lagrange system. In general the 2-form $\overline{\varpi}_{\overline{D}_N}$ on $\overline{D}_N = T\,\mathrm{SO}(3)$ is *not* closed. For if it were, then the $\mathbb{R}^2$-reduced total energy $\overline{h}_{\overline{D}_N}$ on $\overline{D}_N$ and $\overline{\varpi}_{\overline{D}_N}$ would form a Hamiltonian system on $\overline{D}_N$ and therefore would preserve the volume element on $\overline{D}_N$. But in the case of the rattleback there are asymptotically stable motions. This precludes volume preservation. Equations (158) are Chaplgyin's equations in [23] except that we have taken $s = 0$ in his equation (15) because we do not have a little gyroscope on the top of our body of revolution. Our analysis of the rolling body of revolution grew out of attempts to understand Chaplygin's equations. In [23] Chaplygin was focused on finding the motion of the rolling body of revolution by means of quadratures. He did not pursue the analysis of the motion of the general body of revolution any further than deriving Chaplygin's equations (158).

The potential energy function $\overline{V}_{\overline{\sigma}_3,\overline{\sigma_4}}$ (170) is *not* equal to the potential energy mg $a_3(t)$ of the rolling body of revolution. However, it plays the role of a potential energy function in the description of the motion of $\sigma_1(t)$. So we allow ourselves to use the term "potential energy function" for $\overline{V}_{\overline{\sigma}_3,\overline{\sigma_4}}$. The E(2) $\times$ S^1-reduced equations of motion in proposition 6.7.3.7.29 are new.

Chapter 7

The rolling disk

Summary

In this chapter we discuss the motion of a *rolling disk*, which is a body of revolution with its edge rolling on a horizontal plane under the influence of a constant vertical gravitational field. The rim of the disk is a planar circle of radius r with its center at the center of mass. We assume that during the motion the lowest point of the rim remains in contact with the horizontal plane, thus preventing the disk from taking off into space.

The rolling disk has a *symmetry group* $\mathrm{E}(2) \times S^1$. Here $\mathrm{E}(2)$ is the *Euclidean motion group of the plane*, which consists of horizontal translations and rotations about the vertical. The group of *internal rotational symmetries* of the disk is S^1. The group $\mathrm{E}(2) \times S^1$ acts on positions of the unconstrained rolling disk in Euclidean 3-space. The fully reduced space $\mathcal{M}$ of $\mathrm{E}(2) \times S^1$-orbits on the constraint manifold $\mathcal{C}$ of the rolling disk is parametrized by four invariants σ_j, $1 \le j \le 4$. Here $\sigma_1 = \sin\varphi$, where φ is the angle between the oriented plane of the rim of the disk and the vertical direction. When the position of the disk is flat, which is *not* part of the fully reduced phase space $\mathcal{M}$, we have $\sigma_1 = \pm 1$. In fact $\mathcal{M} = (-1,1) \times \mathbb{R}^3$, see (10). Because every *motion* of the disk on the constraint manifold $\mathcal{C}$ is an integral curve of a vector field V (4) which commutes with the action of the symmetry group $\mathrm{E}(2) \times S^1$ on $\mathcal{C}$, there is an induced flow on $\mathcal{M}$, which is generated by the *fully reduced vector field* $\mathcal{V}$ (13). In §2 we derive the *equations of motion* satisfied by the integral curves of $\mathcal{V}$. In §3 we describe how motions on $\mathcal{C}$ can be *reconstructed* from the fully reduced motions on $\mathcal{M}$. The simplest fully reduced motions are equilibrium points. The corresponding reconstructed motions on the constraint manifold are *relative equilibria*, which we describe in §4.

For any body of revolution rolling on a horizontal plane, Chaplygin discovered a system of ordinary differential equations of the form

$$\frac{d\sigma_3}{d\sigma_1} = a(\sigma_1)\,\sigma_3 + b(\sigma_1)\,\sigma_4 \quad \text{and} \quad \frac{d\sigma_4}{d\sigma_1} = c(\sigma_1)\,\sigma_3 + d(\sigma_1)\,\sigma_4,$$

called *Chaplygin's equations*. The image of each integral curve of the fully reduced vector field $\mathcal{V}$ under the projection map $(\sigma_1, \sigma_2, \sigma_3, \sigma_4) \mapsto (\sigma_1, \sigma_3, \sigma_4)$ is a solution of Chaplygin's equations, see (76). Because Chaplygin's equations are a system of linear differential equations, $(\sigma_1, \sigma_3, \sigma_4)$-space is fibered by its solution curves, each of which is parametrized by $\sigma_1 \in (-1, 1)$. Each fiber of the fibration defined by the above projection map is determined by its intersection point $(0, \overline{\sigma}_3, \overline{\sigma}_4)$ with the plane $\{\sigma_1 = 0\}$. Viewing the parameters $(\overline{\sigma}_3, \overline{\sigma}_4)$ as constant functions on the solution curves of Chaplygin's equations, they become *constants of motion* (or integrals) of the fully reduced vector field $\mathcal{V}$. Using the *fully reduced energy* $\overline{E}$, see (19) as a third constant of motion, the fully reduced system in (σ_1, σ_2)-space, namely $(-1, 1) \times \mathbb{R}$, becomes a *conservative one degree of freedom Newtonian system* of the form kinetic plus potential energy. The potential function $\varphi \mapsto V_{\overline{\sigma}_3, \overline{\sigma}_4}(\varphi)$ depends on the parameters $(\overline{\sigma}_3, \overline{\sigma}_4)$, see §5.

For most values of $(\overline{\sigma}_3, \overline{\sigma}_4)$, the potential function $\sigma_1 \mapsto V_{\overline{\sigma}_3, \overline{\sigma}_4}(\sigma_1)$ tends to ∞ as $\sigma_1 \to \pm 1$. Moreover, it has either one or three nondegenerate critical points, which are either one or two local minima and a local maximum. Therefore most of the solutions of the one degree of freedom Newtonian system are periodic, where the angle $\varphi(t)$ oscillates in a well of the potential $V_{\overline{\sigma}_3, \overline{\sigma}_4}$. At exceptional values of $(\overline{\sigma}_3, \overline{\sigma}_4)$, which lie on two lines $\ell_\perp$ that pass through the origin, the potential does not tend to ∞ at $\sigma_1 = \pm 1$. Here the disk *falls flat* in finite time. There are also exceptional values of the parameters $(\overline{\sigma}_3, \overline{\sigma}_4)$ where the potential $V_{\overline{\sigma}_3, \overline{\sigma}_4}$ has a degenerate critical points. These values define the *degeneracy locus* Δ. In §9 we show that Δ is a closed curve in the $(\overline{\sigma}_3, \overline{\sigma}_4)$-plane, which bounds an open set R containing the origin. Δ is analytic except at four points on the coordinate axes where it has a *cusp*. At the cusp points all three critical points of the potential have merged to form a fourth order degenerate minimum. The region R looks like a diamond, see figure 7.2. At each point in R the potential has three critical points. At each point of the complement of the closure of R, the potential has only one critical point which is a nondegenerate minimum, see §9 and §12.

The set Σ of all $(\overline{\sigma}_3, \overline{\sigma}_4, \mathrm{v})$, where v is a critical value of the potential

$V_{\overline{\sigma}_3,\overline{\sigma}_4}$ is called the *critical surface*. It has a cusped edge over the smooth part of the degeneracy locus Δ, a transversal self-intersection over each part of the coordinate axes which lie in R, and a cusped edge along two lines $\ell_\pm$ which pass through the origin. Otherwise it is a smooth analytic surface, see figure 7.13. The critical value surface is also the set of *critical values* of the integral map

$$\mathcal{I} : \mathcal{M} \rightarrow \mathbb{R}^3 : (\sigma_1, \sigma_2, \sigma_3, \sigma_4) \rightarrow \big(\overline{\sigma}_3, \overline{\sigma}_4, E_{\overline{\sigma}_3,\overline{\sigma}_4}(\sigma_1, \sigma_2)\big).$$

The various components of $\mathbb{R}^3 \setminus \Sigma$ correspond to different qualitative behaviors of the rolling disk. Local maxima of the potential, which correspond to *hyperbolic relative equilibria*, exist only over R. Because there is an upper bound to all such critical values, we obtain a global *gyroscopic stabilization principle*, namely, *relative equilibria* are *stable* (= *elliptic*) if their energy is larger than a fixed number, see §13.

When the parameters $(\overline{\sigma}_3, \overline{\sigma}_4)$ lie on one of the *exceptional lines* $\ell_\pm$, where the potential $V_{\overline{\sigma}_3,\overline{\sigma}_4}$ has a finite limit as $\sigma_1 \rightarrow \pm 1$, the disk falls flat in finite time. In §10 we give an asymptotic analysis of the situation when the disk *near falls flat* and then rises up again. A surprising result of this analysis is the existence of a universal constant $\Delta\chi = (1 + \frac{\mathrm{mr}^2}{I_1})^{1/2}\pi$ for the change in the angle of the point of contact.

This asymptotic analysis is taken up once more in §14 to show that the critical surface has a downward pointing cusped edge over $\ell_\pm$ and again in §15 to investigate the *spatial rotational shift* for motions of the disk which remain nearly flat.

7.1 General set up

Consider a *reference disk* D with a rim ∂D, which is a planar circle of radius r. Suppose that the total mass of the disk is m. Assume that the reference frame $\{e_1, e_2, e_3\}$ for the reference disk is the *principal axis frame* with origin at the center of mass O. In this frame the *moment of inertia tensor* I of the disk is $\mathrm{diag}(I_1, I_2, I_3)$. We assume that $I_1 = I_2$. We make no further assumptions on the mass distribution. For example, if the mass is distributed uniformly, then

$$I_1 = I_2 = \tfrac{1}{4}\,\mathrm{mr}^2 \quad \text{and} \quad I_3 = \tfrac{1}{2}\,\mathrm{mr}^2; \tag{1}$$

and the disk is called the *uniform disk* ; whereas, if the mass is distributed uniformly on the rim of the disk, then

$$I_1 = I_2 = \tfrac{1}{2}\,\mathrm{mr}^2 \quad \text{and} \quad I_3 = \mathrm{mr}^2 \tag{2}$$

and the disk is called a *hoop*.

Proof. Recall that in the principal axis frame $I_1 = M_2 + M_3$, $I_2 = M_3 + M_1$ and $I_3 = M_1 + M_2$. Here M_i is the integral of the function x_i^2 with respect

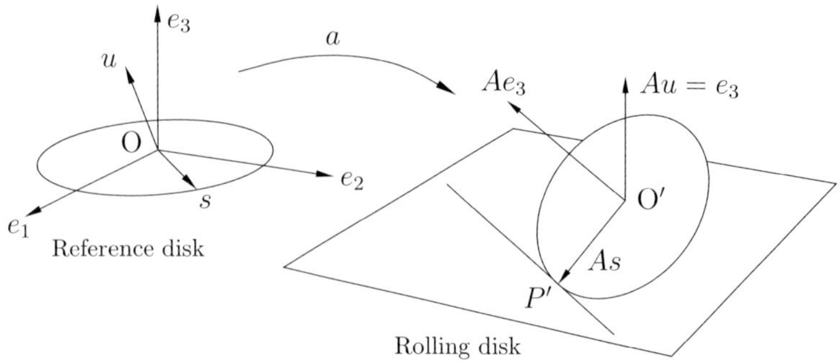

Fig. 7.1 The reference disk (left) and the rolling disk (right). $(A, a) \in \mathrm{E}(3)$. The points O and O' are the centers of the reference disk and moving disk, respectively. P' is the point of contact of the moving disk with the horizontal plane.

to the mass distribution. Note that $0 \le I_3 \le I_1 + I_2 = 2I_1$. If the mass distribution is uniform with density μ, we use polar coordinates to obtain

$$M_1 = M_2 = \mu \int_0^{2\pi} \int_0^r (\rho \cos\theta)^2 \rho \, \mathrm{d}\rho \, \mathrm{d}\theta = \tfrac{1}{4}\mu \mathrm{r}^4 \int_0^{2\pi} (\cos\theta)^2 \, \mathrm{d}\theta = \tfrac{1}{4}\mu\pi \, \mathrm{r}^4;$$

whereas $M_3 = 0$, because the reference disk lies in the e_1-e_2 plane. Since the total mass of the disk is $\mathrm{m} = \mu\pi\,\mathrm{r}^2$, we get $I_1 = I_2 = \frac{1}{4}\mathrm{m}\,\mathrm{r}^2$, which proves (1).

In the case of the hoop with angular mass density μ on the rim we have $\mathrm{m} = 2\pi\,\mu$ and

$$M_1 = M_2 = \mu \int_0^{2\pi} (r\cos\theta)^2 \, \mathrm{d}\theta = \mu\pi\,\mathrm{r}^2 = \tfrac{1}{2}\mathrm{m}\,\mathrm{r}^2$$

and $M_3 = 0$. This yields (2). □

The *position* of the rolling disk in Euclidean 3-space is given by an element (A, a) of the 3-dimensional Euclidean group $\mathrm{E}(3) \subseteq \mathrm{SO}(3) \times \mathbb{R}^3$. As for any smooth convex surface rolling on a horizontal plane, the motion of the disk takes place on the *constraint manifold* $\mathcal{C}$ given by

$$\big\{ \big((A, a), (\omega, \dot{a})\big) \in \mathrm{E}(3) \times \mathbb{R}^6 \,\big|\, Ae_3 \ne \pm e_3, \\ \dot{a} = -A(\omega \times s(u)), \ \langle a, e_3 \rangle = -\langle s(u), u \rangle \big\}, \qquad (3)$$

see (30) of chapter 6. The reason for the first condition $\dot{a} = -A(\omega \times s(u))$ in (3) is that it is just the *nonholonomic constraint* $0 = \dot{a} + \dot{A}s$ that the disk rolls without slipping, see (28) of chapter 6. The *holonomic constraint* that the disk moves on a horizontal plane is given by $\langle a, e_3 \rangle = -\langle s(u), u \rangle$.

The *motion* of the rolling disk is governed by a vector field V on $\mathcal{C}$, whose integral curves satisfy

$$\begin{aligned} \dot{A} &= A\omega^{\times}, \quad \text{where } \omega^{\times} : \mathbb{R}^3 \to \mathbb{R}^3 : x \mapsto \omega \times x. \text{ So } \omega^{\times} \in \mathrm{so}(3) \\ \dot{a} &= -A(\omega \times s) \\ \dot{\omega} &= \mathcal{Y}(s)^{-1} \left[(\mathcal{Y}(s)\omega) \times \omega + \mathrm{m}(\omega \times Ds(u)(u \times \omega)) \times s - \mathrm{mg}\, u \times s \right], \end{aligned} \tag{4}$$

see (42) of chapter 6.

Here s is the inverse of the *Gauss map* of the disk, which we calculate as follows. Parametrize the rim ∂D of the reference disk by

$$\mathbf{x} : S^1 \subseteq \mathbb{R}^2 \to \mathbb{R}^3 : \theta \mapsto \mathrm{r}\,(\cos\theta, \sin\theta, 0).$$

Since the *inward* unit normal $n(\mathbf{x})$ at $\mathbf{x} \in \partial D$ is $-(\cos\theta, \sin\theta, 0)$, the *inverse of the Gauss map* is

$$\begin{aligned} &s : (S^2 \setminus \{\pm e_3\}) \subseteq \mathbb{R}^3 :\to \mathbb{R}^3 : \\ &u = (u_1, u_2, u_3) \mapsto -\tfrac{\mathrm{r}}{\sqrt{1-u_3^2}}(u_1, u_2, 0). \end{aligned} \tag{5}$$

To relate this to the basic set up of the body of revolution treated in §7 of chapter 6, we observe that the functions ρ and ζ introduced in §7.1 in (130a) and (130b) of chapter 6 are

$$\rho(u_3) = -\mathrm{r}\,(1 - u_3^2)^{-1/2} \text{ and } \zeta(u_3) = 0,$$

respectively, for the rolling disk.

Note that s is *not* defined when $u = \pm e_3$, that is, when the disk is lying flat. This is the origin of the condition $Ae_3 \neq \pm e_3$ in the definition of the constraint manifold $\mathcal{C}$. The physical reason for this is that the point of contact with the horizontal plane of a flat lying disk is not defined. This gives rise to apparent singularities in the motion of the disk.

7.2 Reduction of the E(2) × S^1 symmetry

As is shown for a smooth convex surface of revolution in §6 of chapter 6, the 2-dimensional Euclidean group E(2), consisting of translations b_{hor} of the horizontal plane and rotations B around the vertical, and the internal rotational symmetry S^1, consisting of rotations C about the e_3-axis, together form a symmetry group for the rolling disk. In this section we remove this $\mathrm{E}(2) \times S^1$ *symmetry* and obtain the *fully reduced equations of motion*. We do this in two ways: first, by removing the E(2) symmetry and then the S^1 symmetry, and second, by removing the S^1 symmetry first and then the E(2) symmetry. Both of these reductions lead to the same fully reduced vector field and both are useful in carrying out the *reconstruction* of the full motion from the fully reduced motion.

7.2.1 *First* E(2), *then* S^1

Here we carry out the reduction of the $\mathrm{E}(2) \times S^1$ symmetry of the rolling disk, by first reducing the E(2) symmetry and then reducing the induced S^1 symmetry.

The argument used in §6 of chapter 6 applies to the disk. It shows that the E(2)-*reduced motions* of the disk are governed by a vector field $\overline{\mathcal{V}}$ on $M = (S^2 \setminus \{\pm e_3\}) \times \mathbb{R}^3$ (with coordinates (u, ω)), whose integral curves satisfy

$$\begin{aligned} \dot{u} &= u \times \omega \\ \dot{\omega} &= \mathcal{Y}(s)^{-1}[\mathcal{Y}(s)\omega \times \omega + \mathrm{m}(\omega \times Ds(u)(u \times \omega)) \times s - \mathrm{mg}\, u \times s]. \end{aligned} \tag{6}$$

Here $\mathcal{Y}(s)v = Iv + \mathrm{m}(s \times (v \times s))$ for every $v \in \mathbb{R}^3$. As usual, g is the strength of the gravitational field. The E(2)-reduced energy of the disk is

$$\begin{aligned} \overline{E}(u, \omega) &= \tfrac{1}{2} \langle I\omega, \omega \rangle + \tfrac{1}{2} \mathrm{m} \langle \omega \times s, \omega \times s \rangle - \mathrm{mg} \langle s, u \rangle \\ &= \tfrac{1}{2} I_1(\omega_1^2 + \omega_2^2) + \tfrac{1}{2} (I_3 + \mathrm{mr}^2)\omega_3^2 + \tfrac{1}{2} \mathrm{mr}^2 \frac{(u_1\omega_2 - u_2\omega_1)^2}{1 - u_3^2} \\ &\qquad + \mathrm{mgr} \sqrt{1 - u_3^2}, \end{aligned} \tag{7}$$

which is an integral of $\overline{\mathcal{V}}$, see (117a) and (117b) of chapter 6.

The E(2)-*reduced vector field* $\overline{\mathcal{V}}$ (6) and the E(2)-reduced energy $\overline{E}$ (7) are invariant under the proper free S^1-action

$$\Psi : S^1 \times M \to M : (C, (u, \omega)) \mapsto (Cu, C\omega), \tag{8}$$

where $S^1 = \{C \in \mathrm{SO}(3) \,\big|\, Ce_3 = e_3\}$. This S^1 symmetry arises from the rotational symmetry of the reference disk about its e_3-principal axis and the fact that the disk is dynamically symmetric because its moment of inertia tensor I commutes with rotations about e_3. Thus the reduction of the axial S^1 symmetry (8) follows along the same lines as the *reduction of the axial symmetry* of a smooth convex surface of revolution rolling on a horizontal plane given in §7 of chapter 6. We give the details.

We first construct the space $\mathcal{M}$ of orbits of the S^1-action Ψ (8) using *invariant theory*. Note that

$$\begin{array}{lll} \sigma_1 = u_3 & \sigma_2 = u_2\omega_1 - u_1\omega_2 & \sigma_3 = u_1\omega_1 + u_2\omega_2 \\ \sigma_4 = \omega_3 & \sigma_5 = \omega_1^2 + \omega_2^2 & \sigma_6 = u_1^2 + u_2^2 \end{array} \tag{9}$$

generate the algebra of polynomials on M which are invariant under the action Ψ. They satisfy the relations

$$\sigma_2^2 + \sigma_3^2 = \sigma_5\sigma_6, \;\; \sigma_5 \geq 0, \;\; \sigma_6 \geq 0 \text{ and } \sigma_6 + \sigma_1^2 = 1, \;\; |\sigma_1| < 1.$$

Eliminating σ_6 gives

$$\sigma_2^2 + \sigma_3^2 = \sigma_5(1 - \sigma_1^2), \;\; \sigma_5 \geq 0, \;\; |\sigma_1| < 1, \tag{10}$$

which defines the space $\mathcal{M}$ of S^1-orbits on M. In other words, $\mathcal{M}$ is the image of the *orbit map*

$$\widetilde{\pi} : M = (S^2 \setminus \{\pm e_3\}) \times \mathbb{R}^3 \subseteq \mathbb{R}^3 \times \mathbb{R}^3 \to \mathbb{R}^5 : (u, \omega) \mapsto (\sigma_1, \ldots, \sigma_5). \tag{11}$$

Since $|\sigma_1| < 1$, the orbit space $\mathcal{M}$ is the graph of the function

$$\sigma_5 = \frac{\sigma_2^2 + \sigma_3^2}{1 - \sigma_1^2}, \quad |\sigma_1| < 1 \tag{12}$$

and thus is diffeomorphic to $(-1, 1) \times \mathbb{R}^3$ (with coordinates $(\sigma_1, \sigma_2, \sigma_3, \sigma_4)$).

Because the E(2)-reduced vector field $\overline{\mathcal{V}}$ (6) is invariant under the S^1-action Ψ, it induces the *fully reduced vector field* $\mathcal{V}$ on the orbit space $\mathcal{M}$.

Claim 7.2.1.1. *The integral curves of $\mathcal{V}$ on $\mathcal{M}$ satisfy*

$$\begin{aligned} \frac{d\sigma_1}{dt} &= \sigma_2 \\ \frac{d\sigma_2}{dt} &= -\frac{\sigma_1\sigma_2^2}{1-\sigma_1^2} - \frac{I_1}{I_1 + \mathrm{mr}^2}\frac{\sigma_1\sigma_3^2}{1-\sigma_1^2} + \frac{I_3 + \mathrm{mr}^2}{I_1 + \mathrm{mr}^2}\sigma_3\sigma_4 \\ &\quad + \frac{\mathrm{mgr}}{I_1 + \mathrm{mr}^2}\sigma_1\sqrt{1-\sigma_1^2} \\ \frac{d\sigma_3}{dt} &= -\frac{I_1}{I_3}\sigma_2\sigma_4 \\ \frac{d\sigma_4}{dt} &= -\frac{\mathrm{mr}^2}{I_3 + \mathrm{mr}^2}\frac{\sigma_2\sigma_3}{1-\sigma_1^2}. \end{aligned} \tag{13}$$

Proof. We give the details how we obtained (13).

Let $u_{\text{hor}} = (u_1, u_2, 0) = u - u_3 e_3$. Then the inverse of the Gauss map $s = s(u) = -\frac{\mathrm{r}}{\|u_{\text{hor}}\|} u_{\text{hor}}$. Hence $\langle s, s\rangle = \mathrm{r}^2$ and $\langle s, e_3\rangle = 0$. In what follows we will use the formulæ

$$\langle I\omega \times \omega, e_3\rangle = 0 \qquad \langle \omega \times s, u_{\text{hor}}\rangle = 0$$
$$\langle s \times u, e_3\rangle = 0 \qquad \langle u \times s, u_{\text{hor}}\rangle = 0$$
$$\langle \frac{\mathrm{d}s}{\mathrm{d}t}, e_3\rangle = 0 \qquad \langle \frac{\mathrm{d}s}{\mathrm{d}t}, u_{\text{hor}}\rangle = 0$$
$$\|u_{\text{hor}}\| \, \langle \frac{\mathrm{d}s}{\mathrm{d}t}, u \times e_3\rangle = \|u_{\text{hor}}\| \, \langle s \times \omega, u \times e_3\rangle - \mathrm{r}\, u_3 \, \langle e_3 \times \omega, u \times e_3\rangle$$

without explicit mention. They are easily verified. Also we will need

$$\begin{aligned}\frac{\mathrm{d}}{\mathrm{d}t}(I\omega) = I\omega \times \omega - \mathrm{m}\mathrm{r}^2 \frac{\mathrm{d}\omega}{\mathrm{d}t} + \mathrm{m}\, \langle \frac{\mathrm{d}\omega}{\mathrm{d}t}, s\rangle\, s + \mathrm{m}\, \langle s, \omega\rangle \frac{\mathrm{d}s}{\mathrm{d}t} \\ + \mathrm{m}\, \langle \omega, s\rangle\, \omega \times s - \mathrm{m}\mathrm{g}\, u \times s,\end{aligned} \tag{14}$$

which follows from the second equation in (6).

We now begin our derivation of (13). From the definition of σ_2 and the first equation in (6) we obtain

$$\sigma_2 = \langle u \times \omega, e_3\rangle = \langle \frac{\mathrm{d}u}{\mathrm{d}t}, e_3\rangle = \frac{\mathrm{d}}{\mathrm{d}t}\langle u, e_3\rangle = \frac{\mathrm{d}\sigma_1}{\mathrm{d}t}. \tag{15}$$

Differentiating $I_1\sigma_3 = \langle I\omega, u_{\text{hor}}\rangle$ and using (14) gives

$$\begin{aligned}I_1 \frac{\mathrm{d}\sigma_3}{\mathrm{d}t} &= \langle \frac{\mathrm{d}(I\omega)}{\mathrm{d}t}, u_{\text{hor}}\rangle + \langle I\omega, \frac{\mathrm{d}u_{\text{hor}}}{\mathrm{d}t}\rangle \\ &= \langle I\omega \times \omega, u_{\text{hor}}\rangle - \mathrm{m}\mathrm{r}^2 \langle \frac{\mathrm{d}\omega}{\mathrm{d}t}, u_{\text{hor}}\rangle + \mathrm{m}\, \langle \frac{\mathrm{d}\omega}{\mathrm{d}t}, s\rangle \langle s, u_{\text{hor}}\rangle \\ &\quad + \mathrm{m}\, \langle s, \omega\rangle \langle \frac{\mathrm{d}s}{\mathrm{d}t}, u_{\text{hor}}\rangle + \mathrm{m}\, \langle \omega, s\rangle \langle \omega \times s, u_{\text{hor}}\rangle \\ &\quad - \mathrm{m}\mathrm{g}\, \langle u \times s, u_{\text{hor}}\rangle + \langle I\omega, \frac{\mathrm{d}u}{\mathrm{d}t} - \dot{u}_3 e_3\rangle \\ &= \langle I\omega \times \omega, u_{\text{hor}}\rangle + \langle I\omega, u \times \omega\rangle - \dot{u}_3 \, \langle I\omega, e_3\rangle \\ &= -I_3\, \omega_3 \dot{u}_3 = -I_3\, \sigma_2\sigma_4.\end{aligned} \tag{16}$$

Multiplying the derivative of $I_3\sigma_4 = \langle I\omega, e_3\rangle$ by $1-u_3^2$ gives

$$I_3(1-u_3^2)\frac{\mathrm{d}\sigma_4}{\mathrm{d}t} = (1-u_3^2)\langle\frac{\mathrm{d}(I\omega)}{\mathrm{d}t}, e_3\rangle$$
$$= (1-u_3^2)\Big[\langle I\omega\times\omega, e_3\rangle - \mathrm{mr}^2\langle\frac{\mathrm{d}\omega}{\mathrm{d}t}, e_3\rangle + \mathrm{m}\langle\frac{\mathrm{d}\omega}{\mathrm{d}t}, s\rangle\langle s, e_3\rangle$$
$$+\mathrm{m}\,\langle s,\omega\rangle\langle\frac{\mathrm{d}s}{\mathrm{d}t}, e_3\rangle + \mathrm{m}\,\langle\omega, s\rangle\langle\omega\times s, e_3\rangle - \mathrm{mg}\,\langle u\times s, e_3\rangle\Big]$$
$$= -\mathrm{mr}^2(1-u_3^2)\dot{\omega}_3 - \mathrm{mr}^2\,\langle\omega, u_{\mathrm{hor}}\rangle\langle\omega\times u_{\mathrm{hor}}, e_3\rangle$$
$$= -\mathrm{mr}^2(1-u_3^2)\dot{\omega}_3 - \mathrm{mr}^2\,\langle\omega, u_{\mathrm{hor}}\rangle\dot{u}_3.$$

In other words,

$$(I_3+\mathrm{mr}^2)(1-\sigma_1^2)\frac{\mathrm{d}\sigma_4}{\mathrm{d}t} = -\mathrm{mr}^2\,\sigma_2\sigma_3. \tag{17}$$

Multiplying the derivative of $I_1\sigma_2 = I_1\langle u\times\omega, e_3\rangle = \langle u\times I\omega, e_3\rangle$ by $1-u_3^2$ gives

$$I_1(1-u_3^2)\frac{\mathrm{d}\sigma_2}{\mathrm{d}t} = (1-u_3^2)\Big[\langle\frac{\mathrm{d}u}{\mathrm{d}t}\times I\omega, e_3\rangle + \langle u\times\frac{\mathrm{d}(I\omega)}{\mathrm{d}t}, e_3\rangle\Big]$$
$$= \Big[\langle(u\times\omega)\times I\omega, e_3\rangle + \langle u\times(I\omega\times\omega), e_3\rangle - \mathrm{mr}^2\,\langle u\times\frac{\mathrm{d}\omega}{\mathrm{d}t}, e_3\rangle$$
$$+\mathrm{m}\,\langle s,\omega\rangle\langle u\times\frac{\mathrm{d}s}{\mathrm{d}t}, e_3\rangle + \mathrm{m}\,\langle\omega, s\rangle\langle u\times(\omega\times s), e_3\rangle$$
$$-\mathrm{mg}\,\langle u\times(u\times s), e_3\rangle\Big]\,(1-u_3^2)$$
$$= (1-u_3^2)\Big[-\langle I\omega,\omega\rangle\,u_3 + \langle\omega, u\rangle I_3\,\omega_3 - \mathrm{mr}^2\frac{\mathrm{d}}{\mathrm{d}t}\langle u\times\omega, e_3\rangle$$
$$+\mathrm{mr}^2\langle\frac{\mathrm{d}u}{\mathrm{d}t}\times\omega, e_3\rangle - \mathrm{m}\,\langle s,\omega\rangle\langle s\times\omega, u\times e_3\rangle + \mathrm{m}\,\langle\omega, s\rangle\langle s\times\omega, u\times e_3\rangle\Big]$$
$$+\mathrm{mr}\,u_3\,\langle s,\omega\rangle\langle e_3\times\omega, u\times e_3\rangle\,\|u_{\mathrm{hor}}\| + \mathrm{mgr}\,u_3\,\|e_3\times u_{\mathrm{hor}}\|^2\,\|u_{\mathrm{hor}}\|.$$

Thus

$$(I_1+\mathrm{mr}^2)(1-\sigma_1^2)\frac{\mathrm{d}\sigma_2}{\mathrm{d}t}$$
$$= [(I_3+\mathrm{mr}^2)\,\omega_3\langle u_{\mathrm{hor}},\omega\rangle - (I_1+\mathrm{mr}^2)\,u_3(\omega_1^2+\omega_2^2)](1-\sigma_1^2)$$
$$+\mathrm{mr}^2\,u_3\,\langle u_{\mathrm{hor}},\omega\rangle^2 + \mathrm{mgr}\,u_3(1-u_3^2)^{3/2}$$
$$= -(I_1+\mathrm{mr}^2)\,\sigma_1\sigma_5(1-\sigma_1^2) + (I_3+\mathrm{mr}^2)\sigma_3\sigma_4(1-\sigma_1^2) + \mathrm{mr}^2\,\sigma_1\,\sigma_3^2$$
$$+\mathrm{mgr}\,\sigma_1(1-\sigma_1^2)^{3/2}$$
$$= -(I_1+\mathrm{mr}^2)\,\sigma_1\,\sigma_2^2 - I_1\,\sigma_1\,\sigma_3^2 + (I_3+\mathrm{mr}^2)\,\sigma_3\sigma_4(1-\sigma_1^2)$$
$$+\mathrm{mgr}\,\sigma_1(1-\sigma_1^2)^{3/2}, \tag{18}$$

since $\sigma_5(1-\sigma_1^2) = \sigma_2^2+\sigma_3^2$. Thus equations (15)–(18) yield (13). □

The E(2)-*reduced energy* $\overline{E}$ on $S^2 \times \mathbb{R}^3$ is invariant under the S^1 action Ψ (8). Therefore $\overline{E}$ (7) is a function of the invariants σ_j (9). So

$$\overline{E} = \tfrac{1}{2}(I_1 + \mathrm{mr}^2)\,\frac{\sigma_2^2}{1-\sigma_1^2} + \tfrac{1}{2}I_1\,\frac{\sigma_3^2}{1-\sigma_1^2} + \tfrac{1}{2}(I_3 + \mathrm{mr}^2)\sigma_4^2 + \mathrm{mgr}\,(1-\sigma_1^2)^{1/2}. \qquad (19)$$

$\overline{E}$ is an integral of the fully reduced vector field $\mathcal{V}$ on $\mathcal{M}$.

From the definition of the invariants σ_2 and σ_3 (9) we have

$$u_2\omega_1 - u_1\omega_2 = \sigma_2$$

$$u_1\omega_1 + u_2\omega_2 = \sigma_3,$$

which can be solved for ω_1 and ω_2. Together with the definition of σ_6 in (9) we obtain

$$\begin{aligned}
\omega_1 &= \tfrac{1}{u_1^2+u_2^2}(u_2\sigma_2 + u_1\sigma_3) = \tfrac{1}{1-\sigma_1^2}(u_2\sigma_2 + u_1\sigma_3)\\
\omega_2 &= \tfrac{1}{u_1^2+u_2^2}(-u_1\sigma_2 + u_2\sigma_3) = \tfrac{1}{1-\sigma_1^2}(-u_1\sigma_2 + u_2\sigma_3) \qquad (20)\\
\omega_3 &= \sigma_4.
\end{aligned}$$

Therefore the E(2)-*reduced equations of motion* for $\dot{\omega}$, see (6), are a consequence of the fully reduced equations of motion (13), equation (20), and the first equation in (6).

7.2.2 *First S^1, then* E(2)

Here we carry out the reduction of the $\mathrm{E}(2)\times S^1$ symmetry by first reducing the S^1 symmetry and then the induced E(2) symmetry.

Recall that the S^1 *action on the constraint manifold* $\mathcal{C}$ (3) is given by

$$(C, (A, a, \omega, \dot{a})) \mapsto (AC^{-1}, a, C\omega, \dot{a}).$$

Therefore the space $\mathcal{N}$ of S^1 orbits on $\mathcal{C}$ is parametrized by $(v, a_{\mathrm{hor}}, \nu)$, where $v = Ae_3$, $\nu^{\times} = \dot{A}A^{-1}$, and a_{hor} is the projection of a on the horizontal e_1–e_2 plane. Recall that if $\nu = (\nu_1, \nu_2, \nu_3)$, then $\nu^{\times} = \left(\begin{smallmatrix} 0 & -\nu_3 & \nu_2\\ \nu_3 & 0 & -\nu_1\\ -\nu_2 & \nu_1 & 0\end{smallmatrix}\right) \in \mathrm{so}(3)$. In terms of u, ω and A, we have $v = A^2u$ and $\nu = A\omega$. Consequently, $\mathcal{N}$ is diffeomorphic to $S^2 \times \mathbb{R}^2 \times \mathbb{R}^3$.

Next we find the S^1-*reduced equations of motion* on $\mathcal{N}$, which describe the motion of the disk ignoring its internal rotation. Differentiating the definition of the vector v and using the definition of ν we get

$$\dot{v} = \dot{A}e_3 = \nu^{\times}Ae_3 = \nu^{\times}v = \nu \times v. \qquad (21)$$

To find the equation for $\dot{a}$, the velocity of the center of mass, we use the nonholonomic constraint $\dot{a} = -A\omega \times As(u)$. But $\nu = A\omega$ and $s(u) = -\frac{\mathrm{r}}{\sqrt{1-u_3^2}}(u - u_3\, e_3)$. Since

$$v_3 = \langle v, e_3 \rangle = \langle Ae_3, e_3 \rangle = \langle e_3, A^{-1}e_3 \rangle = \langle e_3, u \rangle = u_3,$$

we find that

$$As(u) = -\frac{\mathrm{r}}{\sqrt{1-v_3^2}}(Au - v_3\, Ae_3) = -\frac{\mathrm{r}}{\sqrt{1-v_3^2}}(e_3 - v_3\, v).$$

Therefore

$$\dot{a} = \frac{\mathrm{r}}{\sqrt{1-v_3^2}}(\nu \times (e_3 - v_3\, v)). \tag{22}$$

Taking the horizontal component of (22) gives

$$\dot{a}_{\mathrm{hor}} = \frac{\mathrm{r}}{\sqrt{1-v_3^2}}(\nu \times (e_3 - v_3\, v))_{\mathrm{hor}}. \tag{23}$$

A straightforward calculation of the S^1-reduced equation for $\dot{\nu}$ is not available. Instead we procede indirectly. First we complete the *reduction of the* E(2) *action induced on* $\mathcal{N}$ by the E(2)-action on the constraint manifold $\mathcal{C}$. The E(2)-action on $\mathcal{C}$ is given by

$$((B, b_{\mathrm{hor}}), (A, a, \omega, \dot{a})) \mapsto (BA, Ba + b_{\mathrm{hor}}, \omega, B\dot{a});$$

while the induced E(2) action on $\mathcal{N}$ is given by

$$((B, b_{\mathrm{hor}}), (v, a_{\mathrm{hor}}, \nu)) \mapsto (Bv, Ba_{\mathrm{hor}} + b_{\mathrm{hor}}, B\nu).$$

Since the E(2) action $((B, b_{\mathrm{hor}}), a_{\mathrm{hor}}) \mapsto Ba_{\mathrm{hor}} + b_{\mathrm{hor}}$ on $\mathbb{R}^2$ is transitive, the space of E(2)-orbits on $\mathcal{N}$ is diffeomorphic to the fully reduced space $\mathcal{M}$ (10). The E(2)-invariant vector field on $\mathcal{N}$ induced by the S^1 invariant vector field on $\mathcal{C}$, which governs the motion of the disk, is the same as the E(2) $\times S^1$-reduced vector field $\mathcal{V}$ (13) on the fully reduced space $\mathcal{M}$. We can express the invariants σ_j, $1 \le j \le 4$ (9) in terms of the vectors v and ν as follows

$$\begin{aligned} \sigma_1 &= u_3 = v_3 \\ \sigma_2 &= \dot{\sigma}_1 = \dot{v}_3 = (\nu \times v)_3 = \nu_1 v_2 - \nu_2 v_1 \\ \sigma_3 + \sigma_1\sigma_4 &= \langle u, \omega \rangle = \langle A^{-2}v, A^{-1}\nu \rangle = \langle A^{-1}v, \nu \rangle = \langle e_3, \nu \rangle = \nu_3 \\ \sigma_4 &= \langle \omega, e_3 \rangle = \langle A^{-1}\nu, e_3 \rangle = \langle \nu, v \rangle. \end{aligned} \tag{24}$$

Solving the equations

$$\begin{aligned} v_2\nu_1 - v_1\nu_2 &= \sigma_2 \\ v_1\nu_1 + v_2\nu_2 &= \sigma_4 - v_3\nu_3 = \sigma_4 - \sigma_1(\sigma_3 + \sigma_1\sigma_4) = (1 - \sigma_1^2)\sigma_4 - \sigma_1\sigma_3 \end{aligned}$$

for ν_1 and ν_2, using $v_1^2 + v_2^2 = 1 - v_3^2 = 1 - \sigma_1^2$, combined with the third equation in (24) gives

$$\begin{aligned}
\nu_1 &= (\sigma_4 - \frac{\sigma_1\sigma_3}{1-\sigma_1^2})v_1 + \frac{\sigma_2}{1-\sigma_1^2}v_2 \\
\nu_2 &= -\frac{\sigma_2}{1-\sigma_1^2}\,v_1 + (\sigma_4 - \frac{\sigma_1\sigma_3}{1-\sigma_1^2})v_2 \\
\nu_3 &= \sigma_3 + \sigma_1\sigma_4.
\end{aligned} \tag{25}$$

Therefore the equation for $\dot{\nu}$ of the S^1-reduced vector field on $\mathcal{N}$ is a consequence of the fully reduced equations of motion (13), equation (25), and $\dot{v} = \nu \times v$.

7.3 Reconstruction

In this section we reconstruct the actual motion of the disk in Euclidean 3-space from its fully reduced motion.

7.3.1 *The* **E(2)***-reduced flow*

First we *reconstruct* the flow of the E(2)-reduced vector field $\overline{\mathcal{V}}$ (6) on $M = (S^2 \setminus \{\pm e_3\}) \times \mathbb{R}^3$ from the flow of the fully reduced vector field $\mathcal{V}$ (13) on $\mathcal{M}$ (10).

We can reconstruct the vector $\omega(t)$ from the vector $u(t)$ and the fully reduced motion $(\sigma_1(t), \ldots, \sigma_4(t))$ because

$$\begin{aligned}
\omega_1 &= \frac{1}{1-\sigma_1^2}(u_2\sigma_2 + u_1\sigma_3) \\
\omega_2 &= \frac{1}{1-\sigma_1^2}(-u_1\sigma_2 + u_2\sigma_3) \\
\omega_3 &= \sigma_4,
\end{aligned} \tag{26}$$

see (20).

In order to reconstruct the vector $u(t) = A(t)^{-1}e_3$, we use the equation $\dot{u} = u \times \omega$. This leads to

$$\begin{aligned}
u_1\dot{u}_2 - u_2\dot{u}_1 &= u_1(u_3\omega_1 - u_1\omega_3) - u_2(u_2\omega_3 - u_3\omega_2) \\
&= (u_1\omega_1 + u_2\omega_2)u_3 - (u_1^2 + u_2^2)\omega_3 = \sigma_3\sigma_1 - (1-\sigma_1^2)\sigma_4,
\end{aligned}$$

using the definition (9) of the invariants. It is now convenient to introduce spherical coordinates (φ, ψ) on the unit sphere S^2 so that

$$u = (\cos\varphi\cos\psi, \cos\varphi\sin\psi, \sin\varphi). \tag{27}$$

Here $\sin\varphi = \sigma_1$. Thus φ is the angle between the oriented plane of the rim of the disk and the vertical axis. Note that the condition $Ae_3 \neq \pm e_3$, or equivalently $|\sigma_1| < 1$, implies that $-\pi/2 < \varphi < \pi/2$. Therefore $\cos\varphi > 0$. Here the cosine of the angle φ is positive because the upward vertical direction is on the positive side of the plane of the rim. Thus the angle $\psi \in \mathbb{R}/2\pi\,\mathbb{Z}$ is well defined and depends analytically on u. Using (27) we obtain

$$\begin{aligned} u_1\dot{u}_2 - u_2\dot{u}_1 &= \cos\varphi\cos\psi\left[-(\sin\varphi)\dot{\varphi}\sin\psi + \cos\varphi\cos\psi\,\dot{\psi}\right] \\ &\quad - \cos\varphi\sin\psi\left[-(\sin\varphi)\dot{\varphi}\cos\psi - \cos\varphi(\sin\psi)\dot{\psi}\right] \\ &= (\cos^2\varphi)\dot{\psi} = (1-\sigma_1^2)\dot{\psi}. \end{aligned}$$

Therefore

$$\frac{\mathrm{d}\psi}{\mathrm{d}t} = \frac{\sigma_1\sigma_3}{1-\sigma_1^2} - \sigma_4. \tag{28}$$

So we obtain $\psi(t)$ from $\sigma_j(t)$ by integration. Once $\psi(t)$ has been determined, we can read off $u(t)$ from (27), since $\varphi(t)$ is uniquely determined by $\sin^{-1}\sigma_1(t)$. Therefore $\omega(t)$ follows from (26) and taking its horizontal component. This completes the reconstruction of the E(2)-reduced motion.

7.3.2 *The full motion*

In this subsection we reconstruct the full motion $t \mapsto (A(t), a(t)) \in \mathrm{E}(3)$ of the rolling disk from its E(2)-reduced motion $t \mapsto (u(t), \omega(t)) \in M = (S^2 \setminus \{\pm e_3\}) \times \mathbb{R}^3$.

The rotational motion $t \mapsto A(t)$ of the disk is governed by

$$\dot{A} = A\omega^{\times}. \tag{29}$$

Since $t \mapsto \omega(t)$ is known from (26), integrating (29) with the initial condition $A(0) = I$ gives $t \mapsto A_1(t)$. Therefore $A(t) = A(0)A_1(t)$.

To reconstruct the translational motion $t \mapsto a(t)$ of the center of mass of the disk, we parametrize points on the rim ∂D of the reference disk D by $\mathbf{x}(\theta) = \mathrm{r}(\cos\theta, \sin\theta, 0)$, where $\theta \in \mathbb{R}/2\pi\,\mathbb{Z}$. The point of contact $p(t)$ of the rolling disk at time t occurs at the value of θ where the function

$\theta \mapsto \langle A(t)\mathbf{x}(\theta) + a(t), e_3\rangle$ assumes its minimum value. This value is 0, because the point of contact lies on the horizontal plane. But

$$\langle A(t)\mathbf{x}(\theta) + a(t), e_3\rangle = \langle \mathbf{x}(\theta), u(t)\rangle + a_3(t) \tag{30}$$

attains its minimum value when $\mathbf{x}(\theta)$ lies in the same direction as $-u_{\mathrm{hor}}(t)$. From the fact that $u_3(t) = \sigma_1(t)$, it follows that $\|u_{\mathrm{hor}}(t)\| = \sqrt{1-\sigma_1^2(t)}$. Therefore

$$\mathbf{x}(\theta) = -\mathrm{r}(1-\sigma_1^2(t))^{-1/2}\, u_{\mathrm{hor}}(t) = -\mathrm{r}(\cos\psi(t), \sin\psi(t), 0), \tag{31}$$

using (27). Because the left hand side of (30) is 0 and $\mathbf{x}(\theta)$ is given by (31), it follows from (30) that the height of the center of mass of the rolling disk above the horizontal plane is

$$a_3(t) = \mathrm{r}(1-\sigma_1^2(t))^{1/2}. \tag{32}$$

For fixed θ the condition that the disk rolls without slipping implies that the derivative of the motion of the point of contact $p(t) = A(t)\mathbf{x}(\theta) + a(t)$ is zero. Here $\mathbf{x}(\theta)$ is given by (31). In other words,

$$\begin{aligned}\frac{\mathrm{d}a}{\mathrm{d}t} &= \mathrm{r}(1-\sigma_1^2)^{-1/2}\frac{\mathrm{d}A}{\mathrm{d}t}(u_{\mathrm{hor}}) = \mathrm{r}(1-\sigma_1^2)^{-1/2}A(\omega \times u_{\mathrm{hor}}) \\ &= -\mathrm{r}A(t)(\omega(t) \times (\cos\psi(t), \sin\psi(t), 0)).\end{aligned} \tag{33}$$

Therefore, given the rotational motion $t \mapsto A(t)$ of the disk, the angular velocity $t \mapsto \omega(t)$ and the angle $t \mapsto \psi(t)$ (see (28)), we obtain the horizontal component $t \mapsto a_{\mathrm{hor}}(t)$ of the center of mass of the rolling disk by integrating (33). This completes the reconstruction of the actual motion of the disk.

7.3.3 *The S^1-reduced flow*

In this subsection we *reconstruct* the motion $t \mapsto (v(t), a(t), \nu(t))$ of the S^1-invariant vector field on $\mathcal{N}$ from a motion of the fully reduced vector field on $\mathcal{M}$.

The vector $\nu(t)$ can be reconstructed from the vector $v(t)$ and the $\sigma_j(t)$ of the fully reduced motion because

$$\begin{aligned}\nu_1 &= \Big(\sigma_4 - \frac{\sigma_1\sigma_3}{1-\sigma_1^2}\Big)v_1 + \frac{\sigma_2}{1-\sigma_1^2}\,v_2 \\ \nu_2 &= -\frac{\sigma_1}{1-\sigma_1^2}\,v_1 + \Big(\sigma_4 - \frac{\sigma_1\sigma_3}{1-\sigma_1^2}\Big)v_2 \\ \nu_3 &= \sigma_3 + \sigma_1\sigma_4,\end{aligned} \tag{34}$$

see (25).

To reconstruct the vector $v(t)$, we use spherical coordinates (φ, χ) on S^2 to write

$$v = (\cos\varphi\cos\chi, \cos\varphi\sin\chi, \sin\varphi). \tag{35}$$

Here φ is the same angle as used in (27), because $v_3 = \sigma_1 = u_3 = \sin\varphi$. Using the same argument as for the angle ψ, we see that the angle χ depends analytically on v. An argument similar to the one used to prove (28) gives

$$\begin{aligned}(1-\sigma_1^2)\dot{\chi} &= \cos^2\varphi\,\dot{\chi} = v_1\dot{v}_2 - v_2\dot{v}_1 = (v \times \dot{v})_3 \\ &= (v \times (\nu \times v))_3 = \langle v, v\rangle \nu_3 - \langle v, \nu\rangle v_3 \\ &= \sigma_3 + \sigma_1\sigma_4 - \sigma_4\sigma_1, \quad \text{using (24)} \\ &= \sigma_3.\end{aligned}$$

Therefore

$$\frac{\mathrm{d}\chi}{\mathrm{d}t} = \frac{\sigma_3}{1-\sigma_1^2}. \tag{36}$$

To reconstruct $a(t)$ from $v(t)$ and $\nu(t)$ we use (22). This gives

$$\frac{\mathrm{d}a}{\mathrm{d}t} = \mathrm{r}(1-\sigma_1^2)^{-1/2}(\nu \times (e_3 - \sigma_1 v)),$$

since $v_3 = \sigma_1$. In other words

$$\frac{\mathrm{d}a}{\mathrm{d}t} = \mathrm{r}(\sigma_4\,\sin\chi - \dot{\varphi}\,\cos\varphi\cos\chi, -\sigma_4\,\cos\chi - \dot{\varphi}\,\cos\varphi\sin\chi, -\dot{\varphi}\,\sin\varphi). \tag{37}$$

Equation (37) follows because

$$\begin{aligned}\nu = \Big(&(\sigma_4 - \frac{\sigma_1\sigma_3}{1-\sigma_1^2})\cos\varphi\cos\chi + \frac{\dot{\sigma}_1}{1-\sigma_1^2}\,\cos\varphi\sin\chi \\ &- \frac{\dot{\sigma}_1}{1-\sigma_1^2}\,\cos\varphi\cos\chi + (\sigma_4 - \frac{\sigma_1\sigma_3}{1-\sigma_1^2})\cos\varphi\sin\chi, \sigma_3 + \sigma_1\sigma_4\Big)\end{aligned}$$

and

$$e_3 - \sigma_1 v = (-\sigma_1\cos\varphi\cos\chi, -\sigma_1\cos\varphi\sin\chi, 1 - \sigma_1\sin\varphi).$$

The two equations above were obtained using $\sigma_2 = \dot{\sigma}_1 = \cos\varphi\,\dot{\varphi}$ and equations (34) and (35).

In terms of $(v(t), a(t), \nu(t))$ the point of contact of the rolling disk with the horizontal plane is

$$\begin{aligned}p(t) &= A(t)\mathbf{x}(\theta) + a(t) = -\mathrm{r}(1-\sigma_1^2)^{-1/2}A(t)u_{\mathrm{hor}}(t) + a(t), \quad \text{using (31)} \\ &= -\mathrm{r}(1-\sigma_1^2)^{-1/2}(e_3 - v_3\,v(t)) + a(t), \\ &\qquad \text{because } Au_{\mathrm{hor}} = A(u - u_3\,e_3) = e_3 - v_3\,v \\ &= -a_3(t)\,e_3 + \mathrm{r}(1-\sigma_1^2)^{-1/2}\sigma_1 v_{\mathrm{hor}}(t) + a(t), \\ &\qquad \text{since } e_3 - v_3\,v = e_3 - v_3(v_{\mathrm{hor}} + v_3 e_3) \\ &\qquad = (1-v_3^2)e_3 - v_3\,v_{\mathrm{hor}} \text{ and } a_3 = \mathrm{r}\sqrt{1-\sigma_1^2} \\ &= \mathrm{r}\,\frac{\sigma_1}{\sqrt{1-\sigma_1^2}}\,v_{\mathrm{hor}}(t) + a_{\mathrm{hor}}(t).\end{aligned} \tag{38}$$

7.3.4 *Geometry of the* $\mathrm{E}(2) \times S^1$ *reduction map*

In this subsection we show that the map

$$\pi : \mathcal{C} \to \mathcal{M} : (A, a, \omega, \dot{a}) \mapsto (\sigma_1, \sigma_2, \sigma_3, \sigma_4), \tag{39}$$

which carries out the *reduction of the* $\mathrm{E}(2) \times S^1$ *symmetry* of the rolling disk, defines a trivial fibration with fiber $\mathrm{E}(2) \times S^1$. Note that π is the orbit map of the $\mathrm{E}(2) \times S^1$ action on $\mathcal{C}$.

First we prove

Lemma 7.3.4.2. *If* $A \in \mathrm{SO}(3)$ *with* $Ae_3 \neq e_3$, *then there are unique rotations* B *and* C *about the* e_3 *axis through angles* $\chi - \pi$ *and* ψ, *respectively, and a unique angle* $-\pi/2 < \varphi < \pi/2$ *such that*

$$A = B \begin{pmatrix} \cos(\varphi + \pi/2) & 0 & -\sin(\varphi + \pi/2) \\ 0 & 1 & 0 \\ \sin(\varphi + \pi/2) & 0 & \cos(\varphi + \pi/2) \end{pmatrix} C^{-1}.$$

Proof. Let $\mathcal{A} = B^{-1}AC$. We have

$$\widetilde{v} = \mathcal{A}e_3 = B^{-1}ACe_3 = B^{-1}Ae_3 = B^{-1}v,$$

where $v = (\cos\varphi \cos\chi, \cos\varphi \sin\chi, \sin\varphi)$. Since $Ae_3 \neq \pm e_3$, it follows that $\mathcal{A}e_3 \neq \pm e_3$. Thus there is a unique rotation B about the e_3-axis such that $(B^{-1}v)_{\mathrm{hor}}$ is a negative multiple of e_1. Consequently,

$$\mathcal{A}e_3 = B^{-1}v = -(\cos\varphi)e_1 + (\sin\varphi)e_3. \tag{40}$$

B is a rotation through an angle $\chi - \pi$, since

$$B\widetilde{v} = \begin{pmatrix} \cos(\chi - \pi) & -\sin(\chi - \pi) & 0 \\ \sin(\chi - \pi) & \cos(\chi - \pi) & 0 \\ 0 & 0 & 1 \end{pmatrix} \begin{pmatrix} -\cos\varphi \\ 0 \\ \sin\varphi \end{pmatrix} = \begin{pmatrix} \cos\varphi \cos\chi \\ \cos\varphi \sin\chi \\ \sin\varphi \end{pmatrix} = v.$$

Similarly

$$\widetilde{u} = \mathcal{A}^{-1}e_3 = C^{-1}A^{-1}Be_3 = C^{-1}A^{-1}e_3 = C^{-1}u,$$

where $u = (\cos\varphi \cos\psi, \cos\varphi \sin\psi, \sin\varphi)$. Thus there is a unique rotation C about the e_3-axis such that $(C^{-1}u)_{\mathrm{hor}}$ is a positive multiple of e_1. Consequently,

$$\mathcal{A}^{-1}e_3 = (\cos\varphi)e_1 + (\sin\varphi)e_3. \tag{41}$$

C is a rotation through ψ since

$$C\widetilde{u} = \begin{pmatrix} \cos\psi & -\sin\psi & 0 \\ \sin\psi & \cos\psi & 0 \\ 0 & 0 & 1 \end{pmatrix} \begin{pmatrix} \cos\varphi \\ 0 \\ \sin\varphi \end{pmatrix} = \begin{pmatrix} \cos\varphi \cos\psi \\ \cos\varphi \sin\psi \\ \sin\varphi \end{pmatrix} = u.$$

Applying $\mathcal{A}$ to both sides of (41) and then using (40) gives

$$e_3 = (\cos\varphi)\mathcal{A}e_1 + (\sin\varphi)\mathcal{A}e_3 = (\cos\varphi)\mathcal{A}e_1 - (\sin\varphi\cos\varphi)e_1 + (\sin^2\varphi)e_3,$$

that is,

$$(\cos\varphi)\mathcal{A}e_1 = \cos\varphi\,[(\sin\varphi)e_1 + (\cos\varphi)e_3]\,.$$

But $\cos\varphi > 0$. So

$$\mathcal{A}e_1 = (\sin\varphi)e_1 + (\cos\varphi)e_3. \tag{42}$$

From (42) and (40) it follows that the linear map $\mathcal{A}$ leaves the 2-plane $\Pi = \mathrm{span}\{e_1, e_3\}$ invariant. Since $\mathcal{A} \in \mathrm{SO}(3)$, we deduce that $\mathcal{A}$ leaves $\Pi^\perp = \mathrm{span}\{e_2\}$ invariant. From the fact that $\det\mathcal{A} = 1$ and $\det\mathcal{A}|\Pi = 1$ it follows that $\mathcal{A}e_2 = e_2$. Therefore the matrix of $\mathcal{A}$ with respect to the basis $\{e_1, e_2, e_3\}$ is

$$\begin{pmatrix} \sin\varphi & 0 & -\cos\varphi \\ 0 & 1 & 0 \\ \cos\varphi & 0 & \sin\varphi \end{pmatrix} = \begin{pmatrix} \cos(\varphi+\pi/2) & 0 & -\sin(\varphi+\pi/2) \\ 0 & 1 & 0 \\ \sin(\varphi+\pi/2) & 0 & \cos(\varphi+\pi/2) \end{pmatrix}.$$

This proves the lemma. □

Note that the angles $\varphi + \pi/2$, $\chi - \pi$, and ψ are *Euler angles* for the rotation A.

Corollary 7.3.4.3. *Let* $\mathrm{SO}(3)^* = \mathrm{SO}(3) \setminus \{A \in \mathrm{SO}(3) \big|\, Ae_3 = \pm e_3\}$. *For* $x \in \mathbb{R}^3$ *let* $\mathrm{SO}(3)_x = \{A \in \mathrm{SO}(3) \big|\, Ax = x\}$. *The map*

$$\mathrm{SO}(3)^* \to \mathrm{SO}(3)_{e_3} \times \mathrm{SO}(3)_{e_2} \times \mathrm{SO}(3)_{e_3} : A \to (B, \mathcal{A}, C),$$

is smooth. Here $A = B\mathcal{A}C^{-1}$.

Proof. Just observe that the vectors $v = (\cos\varphi\cos\chi, \cos\varphi\sin\chi, \sin\varphi)$ and $u = (\cos\varphi\sin\psi, \cos\varphi\sin\psi, \sin\varphi)$ depend smoothly on $A \in \mathrm{SO}(3)^*$. Consequently, the Euler angles $\chi - \pi$, $\varphi + \pi/2$, and ψ do also. Thus the rotations B and C about the e_3 axis through the angles $\chi - \pi$ and ψ and the rotation $\mathcal{A}$ about the e_2-axis through $\varphi + \pi/2$ depend smoothly on $A \in \mathrm{SO}(3)^*$. □

To proceed further we note that the constraint manifold $\mathcal{C}$ (3) is parametrized by $(A, \omega, a_{\mathrm{hor}})$, where $A \in \mathrm{SO}(3)^*$, $a_{\mathrm{hor}} = (a_1, a_2, 0)$ with $(a_1, a_2) \in \mathbb{R}^2$ and $\omega \in \mathbb{R}^3$. To verify this, observe that A determines the vector $u = Ae_3$, which is not a multiple of e_3 since $Ae_3 \neq \pm e_3$. Therefore $u_3 \neq \pm 1$. Consequently, $s(u) = -\frac{\mathrm{r}}{\|u_{\mathrm{hor}}\|}\, u_{\mathrm{hor}}$ is defined. The third component of the vector $a = (a_1, a_2, a_3)$ is determined by the holonomic constraint $a_3 = -\mathrm{r}\langle s(u), u\rangle = \mathrm{r}\|u_{\mathrm{hor}}\|$. The velocity of the center of mass of the rolling disk is defined by the nonholonomic constraint $\dot{a} = -A\omega \times As(u)$, which is determined by A, ω, and $s(u)$. Thus $\mathcal{C}$ is parametrized by A, ω, and a_{hor}.

Consider the smooth map

$$\Sigma : \mathcal{M} \to \mathcal{C} : \sigma = (\sigma_1, \sigma_2, \sigma_3, \sigma_4) \mapsto \left(\begin{pmatrix} \sigma_1 & 0 & -\sqrt{1-\sigma_1^2} \\ 0 & 1 & 0 \\ \sqrt{1-\sigma_1^2} & 0 & \sigma_1 \end{pmatrix}, 0, \begin{pmatrix} \frac{\sigma_3}{\sqrt{1-\sigma_1^2}} \\ -\frac{\sigma_2}{\sqrt{1-\sigma_1^2}} \\ 0 \end{pmatrix} \right) = (\mathcal{A}, 0, \widetilde{\omega}) \tag{43}$$

Lemma 7.3.4.4. *The map Σ is a smooth section of the bundle $\pi : \mathcal{C} \to \mathcal{M}$ (39).*

Proof. Now $\widetilde{u} = \mathcal{A}^{-1}e_3 = (\sqrt{1-\sigma_1^2}, 0, \sigma_1)$ and $\widetilde{\omega} = \frac{1}{\sqrt{1-\sigma_1^2}}(\sigma_3, \sigma_2, 0)$. But $(\widetilde{u}, \widetilde{\omega}) = (C^{-1}u, C^{-1}\omega)$ for a rotation C about the e_3-axis through an angle ψ. To verify this write $\widetilde{\omega} = (\cos\varphi \frac{\sigma_3}{1-\sigma_1^2}, -\cos\varphi \frac{\sigma_2}{1-\sigma_1^2}, \sigma_4)$. Then

$$C\widetilde{\omega} = \begin{pmatrix} \cos\varphi\cos\psi \frac{\sigma_3}{1-\sigma_1^2} + \cos\varphi\sin\psi \frac{\sigma_2}{1-\sigma_1^2} \\ \cos\varphi\sin\psi \frac{\sigma_3}{1-\sigma_1^2} - \cos\varphi\cos\psi \frac{\sigma_2}{1-\sigma_1^2} \\ \sigma_4 \end{pmatrix} = \begin{pmatrix} u_1 \frac{\sigma_3}{1-\sigma_1^2} + u_2 \frac{\sigma_2}{1-\sigma_1^2} \\ u_2 \frac{\sigma_3}{1-\sigma_1^2} - u_1 \frac{\sigma_2}{1-\sigma_1^2} \\ \sigma_4 \end{pmatrix},$$

$$\text{using } u = (\cos\varphi\cos\psi, \cos\varphi\sin\psi, \sin\varphi) = C\widetilde{u}$$

$$= \omega, \quad \text{using (26)}.$$

Therefore $\widetilde{u}_3 = u_3 = \sigma_1 = \sin\varphi$ and $\widetilde{\omega}_3 = \omega_3 = \sigma_4$. Moreover, $\widetilde{u}_2\widetilde{\omega}_1 - \widetilde{u}_1\widetilde{\omega}_2 = u_2\omega_1 - u_1\omega_2 = \sigma_2$ and $\widetilde{u}_1\widetilde{\omega}_1 + \widetilde{u}_2\widetilde{\omega}_2 = u_1\omega_1 + u_2\omega_2 = \sigma_3$. So $\pi \circ \Sigma(\sigma) = \sigma$ for every $\sigma \in \mathcal{M}$. □

Next consider the map

$$\Xi : (\mathrm{E}(2) \times S^1) \times \mathcal{M} \to \mathcal{C} : (((B, b_{\mathrm{hor}}), C), \sigma) \mapsto ((B, b_{\mathrm{hor}}), C) \cdot \Sigma(\sigma). \tag{44}$$

Then

$$\Xi(((B, b_{\mathrm{hor}}), C), \sigma) = ((B, b_{\mathrm{hor}}), C) \cdot (\mathcal{A}, 0, \widetilde{\omega}) = (B\mathcal{A}C^{-1}, b_{\mathrm{hor}}, C\widetilde{\omega}) = (A, b_{\mathrm{hor}}, \omega).$$

Proposition 7.3.4.5. *Ξ is a trivialization of the bundle $\pi : \mathcal{C} \to \mathcal{M}$ (39).*

Proof. Because

$$\pi \circ \Xi(((B, b_{\mathrm{hor}}), C), \sigma) = \pi(((B, b_{\mathrm{hor}}), C) \cdot \Sigma(\sigma)) = \pi \circ \Sigma(\sigma) = \sigma = \pi_2(\sigma),$$

where π_2 is projection on the second factor, that is,

$$\pi_2 : (\mathrm{E}(2) \times S^1) \times \mathcal{M} \to \mathcal{M} : (((B, b_{\mathrm{hor}}), C), \sigma) \mapsto \sigma,$$

it suffices to show that Ξ is a diffeomorphism. Clearly Ξ is smooth. Consider the map

$$\xi : \mathcal{C} \to (\mathrm{E}(2) \times S^1) \times \mathcal{M} : (A, b_{\mathrm{hor}}, \omega) \mapsto (((B, b_{\mathrm{hor}}), C), \pi(\mathcal{A}, 0, \widetilde{\omega})).$$

Using corollary 7.3.4.3 it follows that ξ is smooth. Next we verify that $\Xi \circ \xi = \mathrm{id}_{\mathcal{C}}$ and $\xi \circ \Xi = \mathrm{id}_{\mathcal{M}}$. We compute

$$\begin{aligned}\Xi \circ \xi(A, b_{\mathrm{hor}}, \omega) &= \Xi(((B, b_{\mathrm{hor}}), C), \pi(\mathcal{A}, 0, \widetilde{\omega})) = ((B, b_{\mathrm{hor}}), C) \cdot \Sigma \circ \pi(\mathcal{A}, 0, \widetilde{\omega})) \\ &= ((B, b_{\mathrm{hor}}), C) \cdot (\mathcal{A}, 0, \widetilde{\omega}) = (A, b_{\mathrm{hor}}, \omega)\end{aligned}$$

and

$$\begin{aligned}\xi \circ \Xi(((B, b_{\mathrm{hor}}), C), \sigma) &= \xi(((B, b_{\mathrm{hor}}), C) \cdot \Sigma(\sigma)) \\ &= (((B, b_{\mathrm{hor}}), C), \pi \circ \Sigma(\sigma)) = (((B, b_{\mathrm{hor}}), C), \sigma).\end{aligned}$$

This shows that ξ is the inverse of Ξ and completes the proof of the proposition. □

7.4 Relative equilibria

In this section we study the set of equilibrium points of the fully reduced vector field $\mathcal{V}$ (13). These points are the $\mathrm{E}(2) \times S^1$-*relative equilibria* of the rolling disk.

7.4.1 *The manifold of relative equilibria*

Because σ_2 is a factor in the right hand side of the last two equations in (13), it follows that $(\sigma_1, \sigma_2, \sigma_3, \sigma_4)$ is an equilibrium point of $\mathcal{V}$ if and only if $\sigma_2 = \frac{\mathrm{d}\sigma_1}{\mathrm{d}t} = 0$ and $\frac{\mathrm{d}\sigma_2}{\mathrm{d}t} = \frac{\mathrm{d}^2\sigma_1}{\mathrm{d}t^2} = 0$. Substituting $\sigma_2 = 0$ into the second equation in (13) and using $\frac{\mathrm{d}\sigma_2}{\mathrm{d}t} = 0$ gives

$$-I_1 \frac{\sigma_1}{1 - \sigma_1^2} \sigma_3^2 + (I_3 + \mathrm{mr}^2)\sigma_3\sigma_4 + \mathrm{mgr}\, \sigma_1(1 - \sigma_1^2)^{1/2} = 0. \tag{45}$$

When $\sigma_3 \neq 0$, equation (45) defines σ_4 as an analytic function of $(\sigma_1, \sigma_3) \in (-1, 1) \times (\mathbb{R} \setminus \{0\})$. If $\sigma_3 = 0$, then the set of solutions of (45) is the σ_4-axis. At the point $(\sigma_1^0, \sigma_3^0, \sigma_4^0) = (0, 0, \sigma_4)$ the differential of the left hand side of (45) is

$$\Omega = (I_3 + \mathrm{mr}^2)\sigma_4^0 \, \mathrm{d}\sigma_3 + \mathrm{mgr}\, \mathrm{d}\sigma_1,$$

which is nonzero. This shows that the set N of equilibrium points of the fully reduced vector field $\mathcal{V}$ is an analytic surface in $(\sigma_1, \sigma_2, \sigma_3, \sigma_4)$-space. Because $N \subseteq \{\sigma_2 = 0\}$, it has codimension 2. In fact N is connected. This follows because at points of the σ_4-axis, the tangent space to N, which is $\{\sigma_2 = 0\} \cap \ker \Omega$, contains a vector with nonzero third component. Thus the intersection of the σ_4-axis with N is connected. So N is connected being a graph over the half planes $\{(\sigma_1, 0, \sigma_3) \big| \, \sigma_3 \geq 0\}$ and $\{(\sigma_1, 0, \sigma_3) \big| \sigma_3 \leq 0\}$.

7.4.2 One parameter groups

Here we take a closer look at the relative equilibria of the rolling disk.

Let (A, a) be an element of the 3-dimensional Euclidean group E(3), which defines the configuration space of the unconstrained disk. Let $S^1 = \mathrm{SO}(2)$ be the subgroup of SO(3) consisting of rotations, which leave the vertical basis vector e_3 of $\mathbb{R}^3$ fixed. Let E(2) be the 2-dimensional Euclidean subgroup of E(3) whose elements are (B, b_{hor}), where $B \in \mathrm{SO}(2)$ and $b_{\mathrm{hor}} = (b_1, b_2, 0)$.

The group $\mathrm{E}(2) \times S^1$ acts on E(3) by

$$((B, b_{\mathrm{hor}}), C)\bullet(A, a) = (B, b_{\mathrm{hor}}) \cdot (A, a) \cdot (C, 0) = (BAC^{-1}, a + b_{\mathrm{hor}}).$$

Here $\cdot$ is multiplication in E(3). This action induces the $\mathrm{E}(2) \times S^1$ action

$$\begin{aligned} &\Phi : (\mathrm{E}(2) \times S^1) \times \mathcal{C} \to \mathcal{C} : \\ &(((B, b_{\mathrm{hor}}), C), (A, a_{\mathrm{hor}}, \omega)) \mapsto (BAC^{-1}, Ba_{\mathrm{hor}} + b_{\mathrm{hor}}, C\omega) \end{aligned} \tag{46}$$

on the constraint manifold $\mathcal{C}$ (3). The $\mathrm{E}(2) \times S^1$-action Φ is free and proper. Therefore from proposition 2.1.1 of chapter 4 we know that to every equilibrium point $\sigma^0 = (\sigma_1^0, \sigma_2^0, \sigma_3^0, \sigma_4^0)$ of the fully reduced vector field $\mathcal{V}$ on the fully reduced space $\mathcal{M}$ there is a motion $t \mapsto c(t)$ of the disk in the fiber $\pi^{-1}(\sigma^0) \subseteq \mathcal{C}$ of the $\mathrm{E}(2) \times S^1$ reduction map π (39). Also, there is a unique element ξ in the Lie algebra $\mathrm{e}(2) \times \mathbb{R}$ of $\mathrm{E}(2) \times S^1$ such that $c(t) = \Phi_{\exp t\xi}(c(0))$. The choice of $c(0) \in \pi^{-1}(\sigma^0)$ depends on $\dim \mathcal{C} - \dim(\mathrm{E}(2) \times S^1) = 8 - 4 = 4$ parameters: two of which fix the initial horizontal component of the center of mass of the disk; one of which fixes the amount of initial rotation of the disk about the vertical; and the fourth which fixes the amount of initial internal rotation of the disk its symmetry axis.

Let $E_3 = (e_3)^\times = \begin{pmatrix} 0 & -1 & 0 \\ 1 & 0 & 0 \\ 0 & 0 & 0 \end{pmatrix}$. Then every element $\xi \in \mathrm{e}(2) \times \mathbb{R}$ is of the form $((\eta\, E_3, \beta_{\mathrm{hor}}), \zeta\, E_3)$, where $(\eta, \zeta) \in \mathbb{R}^2$ and $\beta_{\mathrm{hor}} = (\beta_1, \beta_2, 0)$. Physically, η is the spatial angular velocity of the disk about the vertical; β_{hor} is the *horizontal translational velocity* of the center of mass of the moving disk; and ζ is the *intrinsic angular velocity* of the disk *about its symmetry axis*. Since $\lambda = (\eta\, E_3, \beta_{\mathrm{hor}}) \in \mathrm{e}(2)$, the corresponding *one parameter subgroup* of E(2) is $t \mapsto \exp t\lambda = g(t)$, where $g(t)$ satisfies the differential equation $g'(t) = \lambda \cdot g(t)$ with $g(0)$ being the identity element of E(2). In other words, $g(t) = (B(t), b_{\mathrm{hor}}(t))$ satisfies

$$B'(t) = \eta\, E_3\, B(t) \quad \text{and} \quad b'_{\mathrm{hor}}(t) = \eta\, E_3\, b_{\mathrm{hor}}(t) + \beta_{\mathrm{hor}}$$

with $B(0) = I$ and $b(0) = 0$. Consequently, $B(t) = \mathrm{e}^{t\eta\, E_3}$ and

$$\begin{aligned} b_{\mathrm{hor}}(t) &= \left(\int_0^t \mathrm{e}^{(t-s)\eta\, E_3}\, \mathrm{d}s \right) \beta_{\mathrm{hor}} \\ &= \begin{cases} \frac{1}{\eta}(1 - \mathrm{e}^{t\eta\, E_3}) E_3 \beta_{\mathrm{hor}}, & \text{if } \eta \neq 0 \\ t\beta_{\mathrm{hor}}, & \text{if } \eta = 0. \end{cases} \end{aligned} \tag{47}$$

Therefore the action of $\exp t\xi$ on $\mathrm{E}(3)$ is given by

$$\begin{aligned} (A(t), a(t)) &= \exp t\xi \cdot (A^0, a^0) = (B(t), b_{\mathrm{hor}}(t)) \cdot (A^0, a^0) \cdot (\mathrm{e}^{-t\zeta\, E_3}, 0) \\ &= (B(t) A^0 \mathrm{e}^{-t\zeta\, E_3}, B(t) a^0 + b_{\mathrm{hor}}) \end{aligned} \tag{48}$$

Thus we have proved

Lemma 7.4.2.6. *For* $\xi = ((\eta\, E_3, \beta_{\mathrm{hor}}), \zeta\, E_3) \in \mathrm{e}(2) \times \mathbb{R}$, *the curve* $t \mapsto c(t) = \Phi_{\exp t\xi}(c(0))$ *on the constaint manifold* $\mathcal{C}$, *which reconstructs the motion of the rolling disk corresponding to a relative equilibrium determined by* $\xi \in \mathrm{e}(2) \times \mathbb{R}$, *is given by*

$$\begin{aligned} c(0) &= (A^0, a^0_{\mathrm{hor}}, \omega^0) \mapsto c(t) = (A(t), a_{\mathrm{hor}}(t), \omega(t)) \\ &= \begin{cases} \left(\mathrm{e}^{t\eta\, E_3} A^0 \mathrm{e}^{-t\zeta\, E_3}, \mathrm{e}^{t\eta\, E_3} a^0_{\mathrm{hor}} + \frac{1}{\eta}(1 - \mathrm{e}^{t\eta\, E_3}) E_3 \beta_{\mathrm{hor}}, \mathrm{e}^{t\eta\, E_3} \omega^0 \right), & \textit{if } \eta \neq 0 \\ (A^0 \mathrm{e}^{-t\zeta\, E_3}, a^0_{\mathrm{hor}} + t\beta_{\mathrm{hor}}, \omega^0), & \textit{if } \eta = 0. \end{cases} \end{aligned} \tag{49}$$

7.4.3 *Angular speeds in terms of invariants*

In this subsection we calculate the angular speeds η and ζ of the disk about the vertical direction and its axis of symmetry, respectively, in terms of the components of the equilibrium point σ^0 of the fully reduced vector field.

Since

$$\begin{aligned} u(t) &= A(t)^{-1} e_3 = \mathrm{e}^{t\zeta\, E_3} (A^0)^{-1} \mathrm{e}^{-t\eta\, E_3} e_3, \quad \text{using (49)} \\ &= \mathrm{e}^{t\zeta\, E_3} u^0, \quad \text{since } \mathrm{e}^{-t\eta\, E_3} \text{ is a rotation about } e_3 \end{aligned}$$

we get

$$\begin{aligned} (\cos t\zeta\, u_1^0 - \sin t\zeta\, u_2^0, \sin t\zeta\, u_1^0 + \cos t\zeta\, u_2^0, u_3^0) &= u(t) \\ &= (\cos\varphi^0 \cos\psi(t), \cos\varphi^0 \sin\psi(t), \sin\varphi^0), \end{aligned} \tag{50}$$

using (27). Setting $t = 0$ in the above equation gives

$$u^0 = (\cos\varphi^0 \cos\psi^0, \cos\varphi^0 \sin\psi^0, u_3^0).$$

Therefore the first component of (50) reads

$$\cos\varphi^0 \cos\psi(t) = \cos\varphi^0 (\cos t\zeta \cos\psi^0 - \sin t\zeta \sin\psi^0) = \cos\varphi^0 \cos(t\zeta + \psi^0). \tag{51}$$

Since $|\sigma^0| < 1$, it follows that $\cos\varphi^0 > 0$ and that $\psi(t)$ is uniquely determined. Therefore

$$\psi(t) = t\zeta + \psi^0. \tag{52}$$

Hence

$$\zeta = \frac{\mathrm{d}\psi}{\mathrm{d}t}(0) = \frac{\sigma_1^0\sigma_3^0}{1-(\sigma_1^0)^2} - \sigma_4^0, \tag{53}$$

using (28).

Since $v(t) = A(t)e_3 = \mathrm{e}^{t\eta\, E_3}v^0$, we get

$$(\cos t\eta\, v_1^0 - \sin t\eta\, v_2^0, \sin t\eta\, v_1^0 + \cos t\eta\, v_2^0, v_3^0) = (\cos\varphi^0 \cos\chi(t), \cos\varphi^0 \sin\chi(t), \sin\varphi^0).$$

An argument similar to the one given in the preceding paragraph shows that

$$\chi(t) = t\eta + \chi^0. \tag{54}$$

Hence

$$\eta = \frac{\mathrm{d}\chi}{\mathrm{d}t}(0) = \frac{\sigma_3^0}{1-(\sigma_1^0)^2}, \tag{55}$$

using (36). Solving (53) and (55) gives

$$\sigma_3^0 = \eta(1-(\sigma_1^0)^2) \text{ and } \sigma_4^0 = \eta\,\sigma_1^0 - \zeta. \tag{56}$$

Substituting both expressions in (56) into (45) shows that the one parameter group (49) generated by $\xi = ((\eta, \beta_{\mathrm{hor}}), \zeta) \in \mathrm{e}(2) \times \mathbb{R}$ is a relative equilibrium if and only if

$$\mathrm{mgr}\,\sigma_1^0(1-(\sigma_1^0)^2)^{-1/2} = \eta^2(I_1 - I_3 - \mathrm{mr}^2)\sigma_1^0 + \eta\zeta\,(I_3 + \mathrm{mr}^2). \tag{57}$$

The above equation expresses the balance between the force of gravity and the centrifugal force acting on the rolling disk.

Observe that the mapping

$$(-1,1)\times\mathbb{R}^2 \to (-1,1)\times\mathbb{R}^2 : (\sigma_1^0, \eta, \zeta) \mapsto (\sigma_1^0, \eta(1-(\sigma_1^0)^2), \eta\,\sigma_1^0 - \zeta)$$

is a diffeomorphism from the solution set of (57) onto the manifold N of relative equilibria defined by (45). Therefore the solution set of (57) is a connected 2-dimensional analytic submanifold of $(-1,1)\times\mathbb{R}^2$. Because the left hand side of (57) tends to $\pm\infty$ as $\sigma_1^0 \to \pm 1$, equation (57) has at least one solution for every $(\eta,\zeta)\in\mathbb{R}^2$. If $I_3 + \mathrm{mr}^2 \ge I_1$ (which holds for the uniform disk and the hoop), then (57) has a unique solution for every $(\eta,\zeta)\in\mathbb{R}^2$, which depends analytically on (η,ζ). If $I_3 + \mathrm{mr}^2 < I_1$, then

(57) can have at most three solutions. When $\sigma_1^0 \to \pm 1$, which means that the disk is almost flat, then η and ζ cannot both remain bounded.

If $\eta \neq 0$, then (57) defines ζ as an analytic function of $(\sigma^0, \eta) \in (-1,1) \times (\mathbb{R} \setminus \{0\})$. Explicitly,

$$\zeta = \left[\mathrm{mgr}\, \sigma_1^0 (1 - (\sigma_1^0))^{-1/2} - \eta^2 (I_1 - I_3 - \mathrm{mr}^2)\right] \left[\eta(I_3 + \mathrm{mr}^2)\right]^{-1}. \qquad (58)$$

If $\eta = 0$, then $\sigma_1^0 = 0$ is the only solution of (57); moreover, the value of ζ is arbitrary.

7.4.4 *Motion of the relative equilibria*

In this subsection we discuss how the disk is moving in space when its motion is a relative equilibrium corresponding to $\xi = ((\eta, \beta_{\mathrm{hor}}), \zeta) \in \mathrm{e}(2) \times \mathbb{R}$.

Since

$$\begin{aligned}\sigma_1(t) &= u_3(t) = (A(t)^{-1} e_3)_3 = (\mathrm{e}^{t\zeta\, E_3} (A^0)^{-1} \mathrm{e}^{-t\eta\, E_3} e_3)_3, \text{ using (49)} \\ &= (\mathrm{e}^{t\zeta\, E_3} u^0)_3 = u_3^0 = \sigma_1^0 = \sin\varphi^0, \qquad (59)\end{aligned}$$

the angle $\varphi(t)$ of inclination of the oriented plane of the rim of the disk with the horizontal plane is constant, namely φ^0. From equation (32) it follows that

$$a_3(t) = \mathrm{r}\sqrt{1 - (\sigma_1^0)^2}, \qquad (60)$$

that is, the height of the center of mass of the rolling disk above the horizontal plane is constant.

For a relative equilibrium, we have

$$v(t) = A(t) e_3 = \mathrm{e}^{t\eta\, E_3} A^0 \mathrm{e}^{-t\zeta\, E_3} e_3 = \mathrm{e}^{t\eta\, E_3} v^0, \qquad (61)$$

which implies that $v_3(t) = v_3^0 = \sigma_1^0$ and $v_{\mathrm{hor}}(t) = \mathrm{e}^{t\eta\, E_3} v_{\mathrm{hor}}^0$. Also from the definition of ν we get

$$\begin{aligned}\nu^\times &= \dot{A} A^{-1} = (\eta\, E_3 A - \zeta\, A E_3) A^{-1}, \\ &\qquad \text{by differentiating } A(t) = \mathrm{e}^{t\eta\, E_3} A^0 \mathrm{e}^{-t\zeta\, E_3} \\ &= \eta\, E_3 - \zeta\, A E_3 A^{-1} = \eta\, (e_3)^\times - \zeta\, (A e_3)^\times \\ &= (\eta\, e_3 - \zeta\, v)^\times.\end{aligned}$$

Therefore

$$\begin{aligned}\nu(t) &= \eta\, e_3 - \zeta\, v(t) &(62)\\ &= \mathrm{e}^{t\eta\, E_3}(\eta\, e_3 - \zeta\, v^0), \quad \text{using (61)}\\ &= \mathrm{e}^{t\eta\, E_3}\nu^0. &(63)\end{aligned}$$

Using (61) and (63), equation (23) for the derivative of the horizontal compontent of the position of the center of mass of the rolling disk becomes

$$\frac{\mathrm{d}a_{\mathrm{hor}}}{\mathrm{d}t} = \mathrm{r}(1-(\sigma_1^0)^2)^{-1/2}\,\mathrm{e}^{t\eta\, E_3}(\nu^0 \times (e_3 - \sigma_1^0\, v^0))_{\mathrm{hor}}. \tag{64}$$

Evaluating (62) at $t = 0$ gives $\nu^0 = \eta\, e_3 - \zeta\, v^0$. So

$$\begin{aligned}\nu^0 \times (e_3 - \sigma_1^0\, v^0) = (\eta\, e_3 - \zeta\, v^0) \times (e_3 - \sigma_1^0\, v^0) = (\zeta - \eta\, \sigma_1^0)e_3 \times v^0\\ = -\sigma_4^0\, e_3 \times v^0, \quad \text{using (56).}\end{aligned}$$

Therefore we may rewrite (64) as

$$\frac{\mathrm{d}a_{\mathrm{hor}}}{\mathrm{d}t} = -\mathrm{r}(1-(\sigma_1^0)^2)^{-1/2}\,\sigma_4^0\,\mathrm{e}^{t\eta\, E_3}E_3 v_{\mathrm{hor}}^0,$$

which integrated gives

$$a_{\mathrm{hor}}(t) - a_{\mathrm{hor}}^0 = \begin{cases} \mathrm{r}(1-(\sigma_1^0)^2)^{-1/2}\,\frac{\sigma_4^0}{\eta}(1-\mathrm{e}^{t\eta\, E_3})v_{\mathrm{hor}}^0, \text{ if } \eta \neq 0\\ -\mathrm{r}(1-(\sigma_1^0)^2)^{-1/2}\,\sigma_4^0(e_3 \times v^0)_{\mathrm{hor}}\, t, \text{ if } \eta = 0. \end{cases} \tag{65}$$

Now suppose that $\eta \neq 0$. From equation (65) it follows that $a_{\mathrm{hor}}(t)$ traces out a circle C_a in the horizontal plane with uniform angular speed η having its center at

$$C^0 = a_{\mathrm{hor}}^0 + \mathrm{r}(1-(\sigma_1^0)^2)^{-1/2}\,\frac{\sigma_4^0}{\eta}\, v_{\mathrm{hor}}^0 \tag{66}$$

and of radius

$$R_a = \mathrm{r}(1-(\sigma_1^0)^2)^{-1/2}\,\Big|\frac{\sigma_4^0}{\eta}\Big|\,\|v_{\mathrm{hor}}^0\| = \mathrm{r}\Big|\frac{\sigma_4^0}{\eta}\Big|, \tag{67}$$

since $v^0 = v_{\mathrm{hor}}^0 + \sigma_1^0\, e_3$. We now find the motion $p(t)$ of the point of contact. From (38) we have

$$\begin{aligned}p(t) &= \mathrm{r}(1-(\sigma_1^0)^2)^{-1/2}\,\sigma_1^0\, v_{\mathrm{hor}}(t) + a_{\mathrm{hor}}(t) &(68)\\ &= \mathrm{r}(1-(\sigma_1^0)^2)^{-1/2}(\sigma_1^0 - \frac{\sigma_4^0}{\eta})\mathrm{e}^{t\eta\, E_3}v_{\mathrm{hor}}^0\\ &\quad + \mathrm{r}(1-(\sigma_1^0)^2)^{-1/2}\,\frac{\sigma_4^0}{\eta}\, v_{\mathrm{hor}}^0 + a_{\mathrm{hor}}^0,\end{aligned}$$

using $v_{\rm hor}(t) = e^{t\eta E_3} v^0_{\rm hor}$ and (65). Therefore the point of contact traces out a circle C_p in the horizontal plane at uniform angular speed η whose center is at C^0 (66). Its radius is

$$R_p = \mathrm{r}(1-(\sigma^0_1)^2)^{-1/2}\,|\frac{\zeta}{\eta}|\,\|v^0_{\rm hor}\| = \mathrm{r}|\frac{\zeta}{\eta}|, \tag{69}$$

because by (56) we have

$$\sigma^0_1 - \frac{\sigma^0_4}{\eta} = \sigma^0_1 - \frac{(\eta\,\sigma^0_1 - \zeta)}{\eta} = \frac{\zeta}{\eta}.$$

Lemma 7.4.4.7. *Suppose that $\eta \neq 0$ and $\zeta \neq 0$. If $I_3 + \mathrm{mr}^2 \geq 2I_1$ then the disk leans* inward *toward the center of the circle C_p traced out by its point of contact. If $I_3 + \mathrm{mr}^2 < 2I_1$, then the disk leans* outward *from the center of C_p if and only if $\sigma^0_1 \neq 0$, $(1-(\sigma^0_1)^2)^{1/2}\eta^2 > \frac{2\mathrm{mgr}}{(2I_1 - I_3 - \mathrm{mr}^2)}$ and (σ^0, η) satisfies equation* (57).

Proof. The disk leans inward if and only if $R_p > R_a$, that is, if and only if

$$\mathrm{r}^2\frac{\zeta^2}{\eta^2} > \mathrm{r}^2(\sigma^0_1 - \frac{\zeta}{\eta})^2. \tag{70}$$

After multiplying both sides by η^2, equation (70) is equivalent to

$$(\sigma^0_1)^2\eta^2 - 2\sigma^0_1\zeta\eta < 0. \tag{71}$$

Using (57) to eliminate $\zeta\eta$, we find that (71) is equivalent to

$$0 > (\sigma^0_1)^2\left[(2I_1 - I_3 - \mathrm{mr}^2)\eta^2 - 2\mathrm{mgr}(1-(\sigma^0_1)^2)^{-1/2}\right]. \tag{72}$$

If $(2I_1 - I_3 - \mathrm{mr}^2) < 0$ and $\sigma^0_1 \neq 0$, then (72) holds for every η. If $\sigma^0_1 = 0$ and $\eta \neq 0$, then (57) implies that $\zeta = 0$. This contradicts the hypothesis of the lemma.

The disk leans outward if and only if $R_p < R_a$, that is, if and only if

$$0 < (\sigma^0_1)^2\left[(2I_1 - I_3 - \mathrm{mr}^2)\eta^2 - 2\mathrm{mgr}(1-(\sigma^0_1)^2)^{-1/2}\right]. \tag{73}$$

Clearly the above inequality does not hold if $(2I_1 - I_3 - \mathrm{mr}^2) \leq 0$. On the other hand, if $(2I_1 - I_3 - \mathrm{mr}^2) > 0$ then (73) holds for an arbitrary nonzero $\sigma^0_1 \in (-1, 1)$. This can be arranged by choosing $|\eta|$ large enough and then making sure that (57) is satisfied by letting ζ be defined by (58). □

For the hoop $2I_1 = I_3 < I_3 + \mathrm{mr}^2$. So when the hoop is moving in a relative equilibrium, which is a circle, it leans inward toward the center of this circle. If the mass of the reference disk is concentrated in two points which lie opposite to each other on the e_3-axis and at a distance greater than $\frac{1}{2}\mathrm{r}$ from the origin, then $I_3 + \mathrm{mr}^2 < 2I_1$. So the disk leans outward, when moving in a relative equilibrium which is a circle.

7.4.5 *Nearly flat relative equilibria*

Interesting relative equilibria occur when the rim of the disk approaches a horizontal position and its fully reduced energy remains bounded.

From equations (56) and (57) it follows that

$$I_1\eta^2\sigma_1 - I_3\eta\sigma_4 = \mathrm{mgr}\,\sigma_1(1-\sigma_1^2)^{-1/2} \tag{74}$$

In view of (19), the boundedness of the fully reduced energy $\overline{E}$ implies that σ_4 remains bounded as $\sigma_1 \to \pm 1$. Then (74) implies that

$$\eta^2 \sim \frac{\mathrm{mgr}}{I_1}(1-\sigma_1^2)^{-1/2} \to \infty \quad \text{as } \sigma_1 \to \pm 1. \tag{75}$$

Moreover, equation (56) implies that $\zeta/\eta = \sigma_1 - \sigma_4/\eta \to \pm 1$ and from (69) it follows that $|p(t)| \to \mathrm{r}$. Therefore the point of contact $p(t)$ traces out a circle of radius approximately r at an angular speed which is approximately $(\mathrm{mgr}/I_1)^{1/2}(1-\sigma_1^2)^{-1/4}$. This speed tends to ∞ as $\sigma_1 \to \pm 1$. At the same time the center of mass of the disk traces out a circle of radius equal to $\mathrm{r}|\sigma_1 - \zeta/\eta| = \mathrm{r}|\sigma_4/\eta|$, see (67) and (56), which is small of order $(1-\sigma_1^2)^{1/4}$ as $\sigma_1 \to \pm 1$. Conversely, for any given σ_3 and σ_4, there is a real solution η of (57). If σ_4 is bounded as $\sigma_1 \to \pm 1$, then (75) holds. Therefore $\sigma_3^2 = \eta^2(1-\sigma_1^2)^2$ is of order $(1-\sigma_1^2)^{3/2}$. From (19) with $\sigma_2 = 0$ it follows that the fully reduced energy $\overline{E}$ remains bounded as $\sigma_1 \to \pm 1$. This shows that relative equilibria always exist with bounded fully reduced energy and almost horizontal rim.

This completes our discussion of the relative equilibria of the rolling disk.

7.5 A potential function on an interval

7.5.1 *Chaplygin's equations*

If in the fully reduced equations of motion (13) we substitute the first equation $\frac{\mathrm{d}\sigma_1}{\mathrm{d}t} = \sigma_2$ into the third and fourth equation and then divide both sides of the resulting equations by $\frac{\mathrm{d}\sigma_1}{\mathrm{d}t}$ we obtain *Chaplygin's equations*

$$\begin{aligned} \frac{\mathrm{d}\sigma_3}{\mathrm{d}\sigma_1} &= -\frac{I_1}{I_3}\sigma_4 \\ \frac{\mathrm{d}\sigma_4}{\mathrm{d}\sigma_1} &= -\frac{\mathrm{mr}^2}{I_3+\mathrm{mr}^2}\frac{1}{1-\sigma_1^2}\sigma_3. \end{aligned} \tag{76}$$

Observe that for the rolling disk, Chaplygin's system has the following *discrete symmetries*

$$(\sigma_1, \sigma_3, \sigma_4) \to (-\sigma_1, -\sigma_3, \sigma_4) \to (-\sigma_1, \sigma_3, -\sigma_4) \to (\sigma_1, -\sigma_3, -\sigma_4). \quad (77)$$

Chaplygin's equations are a homogeneous linear system of differential equations for (σ_3, σ_4) as functions of σ_1, whose coefficients are analytic functions of σ_1 on $(-1, 1)$. From the theory for such equations, see [27], it follows that for each $(\overline{\sigma}_3, \overline{\sigma}_4) \in \mathbb{R}^2$ there is a unique solution

$$(-1, 1) \to \mathbb{R}^2 : \sigma_1 \mapsto \big(\sigma_3(\sigma_1; \overline{\sigma}_3, \overline{\sigma}_4), \sigma_4(\sigma_1; \overline{\sigma}_3, \overline{\sigma}_4)\big) \quad (78)$$

such that $\sigma_3(0; \overline{\sigma}_3, \overline{\sigma}_4) = \overline{\sigma}_3$ and $\sigma_4(0; \overline{\sigma}_3, \overline{\sigma}_4) = \overline{\sigma}_4$. Moreover the solution (78) is an analytic function of σ_1 and depends linearly on the initial condition $(\overline{\sigma}_3, \overline{\sigma}_4) \in \mathbb{R}^2$.

Geometrically this means that the solution curves of Chaplgyin's system define an analytic fibration of $(\sigma_1, \sigma_3, \sigma_4)$-space $(-1, 1) \times \mathbb{R}^2$. A trivialization of this fibration is given by the inverse of the analytic diffeomorphism

$$\begin{aligned} \psi &: (-1, 1) \times \mathbb{R}^2 \to (-1, 1) \times \mathbb{R}^2 : \\ &\big(\sigma_1, (\overline{\sigma}_3, \overline{\sigma}_4)\big) \mapsto \big(\sigma_1, \sigma_3(\sigma_1; \overline{\sigma}_3, \overline{\sigma}_4), \sigma_4(\sigma_1; \overline{\sigma}_3, \overline{\sigma}_4)\big), \end{aligned} \quad (79)$$

which is the *flow* of Chaplygin's system. If

$$\pi_2 : (-1, 1) \times \mathbb{R}^2 \to \mathbb{R}^2 : (\sigma_1, (\overline{\sigma}_3, \overline{\sigma}_4)) \mapsto (\overline{\sigma}_3, \overline{\sigma}_4) \quad (80)$$

denotes the projection onto the second factor, then

$$\pi = \pi_2 \circ \psi^{-1} : (-1, 1) \times \mathbb{R}^2 \to \mathbb{R}^2$$

is a projection map, whose fiber over $(\overline{\sigma}_3, \overline{\sigma}_4)$ is a solution curve of Chaplygin's equations with initial condition $(\overline{\sigma}_3, \overline{\sigma}_4)$. The fact that the fully reduced equations of motion imply the Chaplygin system means that under the projection map

$$P : (-1, 1) \times \mathbb{R}^3 \to (-1, 1) \times \mathbb{R}^2 : (\sigma_1, \sigma_2, \sigma_3, \sigma_3) \mapsto (\sigma_1, \sigma_3, \sigma_4)$$

each solution of the fully reduced system is mapped onto a fiber of the map π. Equivalently, this means that the components of the $\mathbb{R}^2$-valued function $\pi \circ P$ are *constants of motion* of the fully reduced system. We shall write $\pi \circ P$ as $(\overline{\sigma}_3, \overline{\sigma}_4)$, where $\overline{\sigma}_3$ and $\overline{\sigma}_4$ are viewed as analytic functions on the fully reduced phase space $(-1, 1) \times \mathbb{R}^3$.

7.5.2 *A conservative Newtonian system*

In this subsection we use the integrals found in the preceding subsection to obtain a conservative one degree of freedom Newtonian system on $(-1,1)\times\mathbb{R}$, whose energy is the sum of kinetic and potential energies.

If we fix the values of the constants of motion $\overline{\sigma}_3$ and $\overline{\sigma}_4$ of the fully reduced system on $(-1,1)\times\mathbb{R}^3$, then the corresponding fiber of the map $\pi\circ P$ is an analytic surface Σ in the *fully reduced space*, which is parametrized by $(\sigma_1,\sigma_2)\in(-1,1)\times\mathbb{R}$. The surface Σ is invariant under the flow of the fully reduced system. Another constant of motion of the fully reduced system is the *fully reduced energy*. In terms of the parameters (σ_1,σ_2), the restriction to Σ of the fully reduced energy (19) is

$$\overline{E} = T(\sigma_1,\sigma_2) + V_{\overline{\sigma}_3,\overline{\sigma}_4}(\sigma_1), \tag{81}$$

where

$$T(\sigma_1,\sigma_2) = \tfrac{1}{2}\,(I_1+\mathrm{mr}^2)\,\frac{\sigma_2^2}{1-\sigma_1^2} \tag{82}$$

plays the role of *kinetic energy* and

$$\begin{aligned} V_{\overline{\sigma}_3,\overline{\sigma}_4}(\sigma_1) = \tfrac{1}{2}\,I_1\,\frac{1}{1-\sigma_1^2}\,\sigma_3(\sigma_1;\overline{\sigma}_3,\overline{\sigma}_4)^2 \\ + \tfrac{1}{2}\,(I_3+\mathrm{mr}^2)\sigma_4(\sigma_1;\overline{\sigma}_3,\overline{\sigma}_4)^2 + \mathrm{mgr}(1-\sigma_1^2)^{1/2} \end{aligned} \tag{83}$$

is viewed as *potential energy*. The potential function $V_{\overline{\sigma}_3,\overline{\sigma}_4}$ is invariant under the discrete symmetries (77) with the variables (σ_3,σ_4) there being replaced by $(\overline{\sigma}_3,\overline{\sigma}_4)$.

The kinetic energy (82) simplifies if we parametrize the one dimensional configuration space $(-1,1)$ by arc length. This amounts to introducing angle φ by setting $\sigma_1=\sin\varphi$ with $-\pi/2<\varphi<\pi/2$, which is the same as the angle in (27). Then $(1-\sigma_1^2)^{-1/2}\dot{\sigma}_1 = (\cos\varphi)^{-1}(\cos\varphi)\dot{\varphi} = \dot{\varphi}$. So

$$\overline{E} = \tfrac{1}{2}\,(I_1+\mathrm{mr}^2)(\dot{\varphi})^2 + V(\varphi), \tag{84}$$

where $V(\varphi) = V_{\overline{\sigma}_3,\overline{\sigma}_4}(\sin\varphi)$. Conservation of energy $\frac{\mathrm{d}\overline{E}}{\mathrm{d}t}=0$ is equivalent to the *conservative Newtonian system*

$$M\,\frac{\mathrm{d}^2\varphi}{\mathrm{d}t^2} + \frac{\partial V(\varphi)}{\partial\varphi} = 0 \tag{85}$$

with inertial mass $M = I_1+\mathrm{mr}^2$ and force equal to $-\frac{\partial V(\varphi)}{\partial\varphi}$, provided $\frac{\mathrm{d}\varphi}{\mathrm{d}t}\neq 0$. Because the analytic vector field $X_{\overline{\sigma}_3,\overline{\sigma}_4}$ on $(-\pi/2,\pi/2)$ associated to the second order equation (85) has integral curves governed by

$$\begin{aligned}\frac{\mathrm{d}\varphi}{\mathrm{d}t} &= \frac{1}{M}\theta \\ \frac{\mathrm{d}\theta}{\mathrm{d}t} &= -\frac{\partial V(\varphi)}{\partial \varphi},\end{aligned} \tag{86}$$

it follows that (85) holds when $\frac{\mathrm{d}\varphi}{\mathrm{d}t} = 0$.

Recall that if $t \mapsto \varphi(t)$ is a solution of the Newtonian system (85), then $\sigma_2(t) = \frac{\mathrm{d}\sigma_1}{\mathrm{d}t} = \dot{\varphi}(t)\cos\varphi(t)$ is known. Moreover, we see that substituting $t \mapsto \sigma_1(t) = \sin\varphi(t)$ into a solution of Chaplygin's system (76) gives

$$t \mapsto \big(\sigma_3(\sigma_1(t); \overline{\sigma}_3, \overline{\sigma}_4), \sigma_4(\sigma_1(t); \overline{\sigma}_3, \overline{\sigma}_4)\big),$$

which is a solution of the third and fourth equations in the fully reduced system (13). Thus, given a solution of (85), we know a solution $t \mapsto (\sigma_1(t), \sigma_2(t), \sigma_3(t), \sigma_4(t))$ of the *fully reduced equations of motion.* From this solution we can reconstruct the full motion of the disk in space as in §3.2.

7.5.3 *Qualitative behavior*

In this subsection we discuss the *qualitative behavior* of the integral curves of the vector field $X_{\overline{\sigma}_3,\overline{\sigma}_4}$ on $(-\pi/2, \pi/2)\times\mathbb{R}$ (86), or equivalently the second order differential equation (85) on the configuration space $(-\pi/2, \pi/2)$. We now recall the classical description of these motions.

Choose a value E of the energy $\overline{E}$ (84) which is larger than the infimum of the potential function V on $(-\pi/2, \pi/2)$. Let $(\varphi_-, \varphi_+) \subseteq [-\pi/2, \pi/2]$ be a connected component of the open set $\{\varphi \in (-\pi/2, \pi/2) \,\big|\, V(\varphi) < E\}$. Note that $V(\varphi_-) = E$ and $V(\varphi_+) = E$ if $\varphi_\pm$ both lie in $(-\pi/2, \pi/2)$. As long as $\varphi \in (\varphi_-, \varphi_+)$ we have

$$\frac{\mathrm{d}\varphi}{\mathrm{d}t} = \pm\Big(\frac{2}{M}(E - V(\varphi))\Big)^{1/2}.$$

Separating the variables and integrating yields

$$t_1 - t_0 = \pm\int_{\varphi(t_0)}^{\varphi(t_1)} \Big(\frac{2}{M}(E - V(\varphi))\Big)^{-1/2}\,\mathrm{d}\varphi, \tag{87}$$

which is a formula for the time needed to go from $\varphi(t_0)$ to $\varphi(t_1)$. A solution $t \mapsto \varphi(t)$ of (85) is obtained by inverting the function $\varphi(t_1) \mapsto t_1$ defined by (87).

Take the plus sign in (87). If $\varphi_+ < \pi/2$ and $V'(\varphi_+) \neq 0$, then $V'(\varphi_+) > 0$ and the right hand side of (87) has a finite positive limit as $\varphi(t_1) \uparrow \varphi_+$. This means that $\varphi(t)$ reaches φ_+ for the first (finite) time at $t = T$. From (85) it follows that at $t = T$ we have $\frac{\mathrm{d}^2\varphi}{\mathrm{d}t^2} < 0$. Thus the continuation of the solution for $t > T$ is determined by taking the minus sign in (87). So $\varphi(t) = \varphi(2T - t)$ for all t. If in addition $\varphi_- > -\pi/2$ and $V'(\varphi_-) \neq 0$, then $V'(\varphi_-) < 0$ and for $t > T$ the solution will reach φ_- for the first (finite) time $t = S > T$. So we have $\varphi(t) = \varphi(2S - t)$ for all t. It follows that

$$\varphi(t) = \varphi(2T - t) = \varphi(2S - (2T - t)) = \varphi(2(S - T) + t) \quad \text{for all } t.$$

Consequently, the solution is *periodic* with minimal positive period equal to $2(S - T)$. Thus $\varphi(t)$ oscillates in between φ_- and φ_+, monotonically increasing from φ_- to φ_+ and following the time reversed motion when it goes from φ_+ to φ_-. Here $S - T$ is equal to the right hand side of (87) with the limits $\varphi(t_0)$ and $\varphi(t_1)$ replaced by φ_- and φ_+. Therefore the minimal positive period τ of the motion is given by

$$\tau = 2\int_{\varphi_-}^{\varphi_+} \big(\frac{2}{M}(E - V(\varphi))\big)^{-1/2}\,\mathrm{d}\varphi. \tag{88}$$

If $V'(\varphi_+) = 0$, then $(\varphi_+, 0)$ is an equilibrium point of the vector field $X_{\overline{\sigma}_3,\overline{\sigma}_4}$ (86). Because

$$E - V(\varphi) = V(\varphi_+) - V(\varphi) = \mathrm{O}((\varphi - \varphi_+)^2) \quad \text{as } \varphi \uparrow \varphi_+.$$

The right hand side of (87) diverges as $\varphi(t_1) \uparrow \varphi_+$. Therefore the integral curve $t \mapsto (\varphi(t), \dot{\varphi}(t))$ of $X_{\overline{\sigma}_3,\overline{\sigma}_4}$ *converges to the equilibrium point* $(\varphi_+, 0)$ as $t \to \infty$. Choosing the minus sign in the right hand side of (87), a similar argument shows that the integral curve $t \mapsto (\varphi(t), \dot{\varphi}(t))$ converges to the equilibrium point $(\varphi_+, 0)$ as $t \to -\infty$. Similar conclusions hold when φ_+ is replaced by φ_-.

Now assume that $\varphi_+ = \pi/2$ and that $\lim_{\varphi\uparrow\pi/2} V(\varphi) < E$. Then the right hand side of (87) has a finite limit as $\varphi(t_1) \uparrow \pi/2$. Thus there is a finite time T such that $\varphi(t) \uparrow \pi/2$ as $t \uparrow T$, provided we take the plus sign in (87). In this case the planar vector field $X_{\overline{\sigma}_3,\overline{\sigma}_4}$ is *incomplete*, that is, it has integral curves which leave every compact subset of $(-\pi/2, \pi/2) \times \mathbb{R}$ in finite time. Because φ is the angle between the oriented plane of the rim of the disk and the vertical direction, the plane of the rim of the disk for such solutions reaches the horizontal position in finite time. In other words, the disk falls *flat* in finite time. The choice of minus sign in (87) leads to an integral curve of $X_{\overline{\sigma}_3,\overline{\sigma}_4}$ where the plane of the disk *rises* in a finite time from the horizontal position.

In §12 we will see that there are no other cases to be considered for the potential $V_{\overline{\sigma}_3,\overline{\sigma}_4}$ (83). This concludes our review of the properties of the conservative Newtonian system (85).

7.5.4 *A special case of falling flat*

A simple special case of an integral curve of the vector field $X_{\overline{\sigma}_3,\overline{\sigma}_4}$ (86) occurs when $\overline{\sigma}_3 = \overline{\sigma}_4 = 0$. Because this implies that the solutions of the Chaplygin system (76) are linearly dependent, we see that $\sigma_3(t) = \sigma_4(t) = 0$. Therefore (26) implies that at all time the angular velocity vector ω of the disk is horizontal and is perpendicular to the vector $u(t)$, see (9).

From the definition of the potential function $V_{\overline{\sigma}_3,\overline{\sigma}_4}$ (83) it follows that $V_{0,0} = \mathrm{mgr}(1-\sigma_1^2)^{1/2}$, or, using the substitution $\sigma_1 = \sin\varphi$, that $V(\varphi) = \mathrm{mgr}\cos\varphi$ for $\varphi \in (-\pi/2, \pi/2)$. Therefore the conservative Newtonian system (85) is equal to the *mathematical pendulum* on the upper semicircle of radius r. Because all of its solutions, except for the unstable equilibrium point $(0,0)$ and its asymptotic stable manifold reach $\varphi = \pm\pi/2$ in finite time, the disk falls flat in finite time. Thus the planar vector field $X_{\overline{\sigma}_3,\overline{\sigma}_4}$ (86) on $(-\pi/2, \pi/2) \times \mathbb{R}$ is *incomplete* in the sense that not all of its solutions are defined for all time.

We now reconstruct this motion of the disk. The differential equation (28) satisfied by the angle ψ is $\frac{\mathrm{d}\psi}{\mathrm{d}t} = 0$, using the fact that $\sigma_3(t) = \sigma_4(t) = 0$ for all time. Therefore the angle ψ is constant. Applying a rotation in body coordinates, which is an element of the S^1 symmetry group, we can arrange that $\psi = \pi$. In other words,

$$A(t)^{-1}e_3 = u(t) = -\cos\varphi(t)e_1 + \sin\varphi(t)e_3. \tag{89}$$

From equations (33) with $\psi(t) = \pi$ and (89) we obtain $\dot{a} = -\mathrm{r}\,\dot{A}e_1$, which integrated gives

$$a(t) = a(0) - \mathrm{r}(A(t) - A(0))e_1.$$

In view of equation (38) for the point of contact, the above equation implies that the point of contact is *constant*, namely, $p = -\mathrm{r}A(0)e_1 + a(0)$. Using an element of the translation subgroup of the symmetry goup E(2), we can arrange that $p = 0$. Having done this the position of the center of mass of the disk in space is given by $a(t) = \mathrm{r}A(t)e_1$. From the differential equation (36) governing the angle χ, it follows that $\chi(t)$ is constant. Using a suitable rotation B about the vertical axis, which leaves the vector $u(t)$ fixed and takes $v(t)$ to $Bv(t)$, we can arrange that $\chi = 0$. This means that

$$A(t)e_3 = v(t) = \cos\varphi(t)e_1 + \sin\varphi(t)e_3. \tag{90}$$

Equations (89) and (90) imply that $A(t)$ is a rotation about the e_2-axis through an angle of $\varphi(t) + \pi/2$, see lemma 7.3.4.2. Thus the action of the element $(A(t), a(t))$ of E(2) on a material point x on the rim of the disk is $A(t)x + a(t) = A(t)(x + \mathrm{r}e_1)$. The above argument shows that after bringing the rim of the disk into the vertical position with its point of contact at the origin and its rim to lie in the e_2–e_3 plane, then the disk moves so that at time t the plane of the rim of the disk has rotated about the e_2-axis through an angle $\varphi(t)$.

The above motion of the disk falling flat has the full motion of the mathematical pendulum as its analytic continuation. Here the disk is attached to the horizontal plane at its fixed point of contact and is allowed to pass *through* the horizontal plane.

In §11.1 we will investigate motions when the disk comes close to the flat position but then rises up. In the *fully reduced phase space* these nearby motions converge to a motion $\varphi(t)$ where at the horizontal position we have an elastic reflection. This means that $\frac{\mathrm{d}\varphi}{\mathrm{d}t}$ changes sign, after which the motion continues as before but in the opposite direction, see lemma 7.11.1.25. For the motion of the disk in space this limiting behavior is subtler, see §11.2 and §11.3.

7.6 Scaling

In this section we rescale Chaplygin's equations (76) to eliminate superfluous parameters. This will simplify the discussion later on of the asymptotic behavior of its solutions.

Consider the scaling defined by

$$\sigma_3 = -\frac{I_3}{I_1}\widetilde{u}\,x, \quad \sigma_4 = \widetilde{u}\,y, \text{ and } \sigma_1 = z, \tag{91}$$

where x, y, and z are new variables and $\widetilde{u} > 0$ is a positive constant. Then Chaplygin's equations (76) become the *rescaled Chaplygin equations*

$$\begin{aligned} x'(z) &= y(z) \\ y'(z) &= c\,\frac{1}{1-z^2}\,x(z), \end{aligned} \tag{92}$$

where $c = \frac{\mathrm{mr}^2}{I_3+\mathrm{mr}^2}\frac{I_3}{I_1}$. Note that the rescaled Chaplygin equations imply the second order linear differential equation

$$x''(z) = c\,\frac{1}{1-z^2}\,x(z). \tag{93}$$

Consider the function

$$\widetilde{V}(\sigma_1,\sigma_3,\sigma_4) = \tfrac{1}{2}\,I_1\,\frac{\sigma_3^2}{1-\sigma_1^2} + \tfrac{1}{2}\,(I_3+\mathrm{mr}^2)\sigma_4^2 + \mathrm{mgr}(1-\sigma_1^2)^{1/2}$$

on $(-1,1)\times\mathbb{R}^2$. Then $\widetilde{V}\circ\psi = V_{\overline{\sigma}_3,\overline{\sigma}_4}$. Here ψ is the map defined in (79). If we now rescale $\widetilde{V}$ using (91) and set $\widetilde{u} = \sqrt{\frac{\mathrm{mgr}}{I_3+\mathrm{mr}^2}}$, we obtain $\widetilde{V}(\sigma_1,\sigma_3,\sigma_4) = \mathrm{mgr}\,\widetilde{W}(x,y,z)$, where

$$\widetilde{W}(x,y,z) = \tfrac{1}{2}\,d\,\frac{1}{1-z^2}\,x^2 + \tfrac{1}{2}\,y^2 + (1-z^2)^{1/2} \tag{94}$$

and $d = \frac{I_3^2}{I_1(I_3+\mathrm{mr}^2)}$. Let

$$W_{X,Y}(z) = \tfrac{1}{2}\,d\,\frac{1}{1-z^2}\,x(z;X,Y)^2 + \tfrac{1}{2}\,y(z;X,Y)^2 + (1-z^2)^{1/2} \tag{95}$$

be the *rescaled potential function*. Here $z \mapsto (x(z;X,Y), y(z;X,Y))$ is the solution of the rescaled Chaplygin equations (92) with $(X,Y) = (x(0;X,Y), y(0;X,Y))$. Thus $\widetilde{W}\circ\psi = W_{X,Y}$.

We wish to compute the derivative of $W_{X,Y}$ with respect to z. To do this we evaluate the partial differential operator

$$\mathfrak{C} = \frac{\partial}{\partial z} + y\frac{\partial}{\partial x} + \frac{c}{1-z^2}\frac{\partial}{\partial y}, \tag{96}$$

on the function $\widetilde{W}$. The operator $\mathfrak{C}$ represents the vector field on $(-1,1)\times\mathbb{R}^2$ whose integral curves are governed by the rescaled Chaplygin equations (92). This gives

$$\mathfrak{C}\widetilde{W} = d\,\frac{z}{1-z^2}\,x^2 + (c+d)\,\frac{1}{1-z^2}\,x\,y - \frac{z}{(1-z^2)^{1/2}}. \tag{97}$$

Note that $\mathfrak{C}\widetilde{W} = \frac{\mathrm{d}W_{X,Y}}{\mathrm{d}z}$. If we take

$$\overline{\sigma}_3 = -\frac{I_3}{I_1}\widetilde{u}\,X \quad\text{and}\quad \overline{\sigma}_4 = \widetilde{u}\,Y \tag{98}$$

with $\widetilde{u} = \sqrt{\frac{\mathrm{mgr}}{I_3+\mathrm{mr}^2}}$, then we obtain $V_{\overline{\sigma}_3,\overline{\sigma}_4}(\sigma_1) = \mathrm{mgr}\,W_{X,Y}(z)$.

As a result of the rescaling, we have only two parameters c and d in our *rescaled Newtonian equations* of motion $\ddot{\varphi} = -\frac{\partial W_{X,Y}}{\partial\varphi}(\sin\varphi)$. The parameter d appears in the rescaled potential function and c in the rescaled

Chaplygin equations. For the uniform disk we have $c = \frac{4}{3}$ and $d = \frac{2}{3}$, while for the hoop $c = 1$ and $d = 1$. In general, the parameters c and d are subject to the inequalities

$$c > 0, \quad d > 0, \quad \text{and} \quad c + d \le 2. \tag{99}$$

The last inequality above follows because

$$c + d = \frac{I_3}{I_1}\left(\frac{\mathrm{mr}^2}{I_3 + \mathrm{mr}^2} + \frac{I_3}{I_3 + \mathrm{mr}^2}\right) = \frac{I_3}{I_1} \le 2.$$

Equality holds above if all the mass of the disk is in the plane of the rim.

7.7 Solutions of the rescaled Chaplygin equations

In this section we give a detailed description of the solutions of the rescaled Chaplygin equations (92) as $z \to \pm 1$. We begin by finding the *recessive solution* of the rescaled Chaplygin equations and then use the Wronskian method to finish solving the Chaplygin system.

7.7.1 *The recessive solution*

If we rewrite equation (93) as

$$2(1+z)x''(z) = cx(z) + (1+z)^2 x''(z)$$

and then substitute the power series

$$r(z) = \sum_{n=1}^{\infty} r_n (1+z)^n, \tag{100}$$

we obtain the recurrcnce relation

$$r_{n+1} = \frac{c + n(n-1)}{2(n+1)n} r_n \quad \text{with } r_1 = 1. \tag{101}$$

The ratio test shows that the radius of convergence of (100) is 2. This agrees with the fact that the other singular point of equation (93) is $z = 1$, which is at a distance 2 from the singular point at $z = -1$. Thus the power series (100) for $r(z)$ with coefficients satisfying (101) is a solution of (93) for all $z \in (-1, 1)$. Because $c > 0$, from (101) we see that $r_n > 0$ for every $n \ge 1$. In particular it follows that $r(z) > 0$ and $r'(z) > 0$ for every $z \in (-1, 1)$. A formula for the constants $r(0)$ and $r'(0)$ as an entire analytic function of c will be given in §7.4.

For later use we prove

Lemma 7.7.1.8. *There is a function $u(z)$, which is analytic at $z = -1$ such that*

$$\frac{1}{r(z)^2} = \frac{1}{(1+z)^2} - \tfrac{1}{2}\,c\,\frac{1}{1+z} + u(z). \tag{102}$$

Proof. Since $r(-1) = 0$, it follows that $r(z) = (1+z)s(z)$ where $s(z)$ is analytic at $z = -1$ and $s(-1) = 1$. Therefore $r(z)^{-2} = (1+z)^2 t(z)$ where $t(z) = s(z)^{-2}$ is analytic at $z = -1$ with $t(-1) = 1$ and

$$t'(-1) = -2s'(-1) = -2r_2 = -\tfrac{1}{2}\,c,$$

using (101). From this equation (102) follows. □

7.7.2 *Asymptotics*

In this subsection we study the *asymptotic behavior* of the solutions of the rescaled Chaplygin equations as $z \downarrow -1$. To do this we first prove

Lemma 7.7.2.9. *There is a function $w(z)$, which is analytic on $(-1,1)$ with $w(-1) = 1$ such that the solution $x(z)$ of the rescaled Chaplygin equations (93) with $x(0) = X$ and $x'(0) = Y$ is given by*

$$x(z) = \frac{X}{r(0)}\,r(z) + (Xr'(0) - Yr(0))[\tfrac{1}{2}\,c\,r(z)\ln(1+z) + w(z)]. \tag{103}$$

Proof. We use the Wronskian method to find a second solution of (93). Because

$$(x'r - xr')' = x''r - xr'' = c\,\frac{xr - rx}{1-z^2} = 0,$$

the Wronski determinant $w = x'(z)r(z) - x(z)r'(z)$ is constant, say, $Yr(0) - Xr'(0)$. Integrating the identity $(x/r)' = w/r^2$ from 0 to z gives

$$x(z) = \frac{X}{r(0)}\,r(z) + wr(z)\int_0^z \frac{1}{r(\zeta)^2}\,\mathrm{d}\zeta \quad \text{for } z \in (-1,1). \tag{104}$$

Inserting (102) into the above equation and integrating gives

$$x(z) = \frac{X}{r(0)}\,r(z) + w\,r(z)\left(-\frac{1}{1+z} - \tfrac{1}{2}\,c\,\ln(1+z) + v(z)\right),$$

where $v(z) = \int_0^z u(\zeta)\,\mathrm{d}\zeta$. Note that $v(z)$ is analytic at $z = -1$. Using the fact that $r(-1) = 0$ and $r'(-1) = 1$, we obtain (103) where $w(z) = r(z)((1+z)^{-1} - v(z))$. Observe that $w(z)$ in analytic in $(-1,1)$ and $w(-1) = r'(-1) = 1$. □

We now draw some conclusions about the behavior of the solution (103). From (103) we see that $x(z)$ has a finite limit $Xr'(0) - Yr(0)$ as $z \downarrow -1$. This limit is equal to zero if and only if the Wronskian determinant equals zero, which implies that $x(z)$ is a constant multiple of $r(z)$. For this reason $r(z)$ is called the *normalized recessive solution* of (93) at $z = -1$. Every solution of (93) which has a nonzero limit as $z \downarrow -1$ is called a *dominant solution* at $z = -1$.

Differentiating (103) gives

$$x'(z) = \frac{X}{r(0)}\, r'(z) + (Xr'(0) - Yr(0))\big(\tfrac{1}{2}\, c[r(z)(1+z)^{-1} \\ + r'(z)\ln(1+z)] + w'(z)\big). \tag{105}$$

This shows that

$$x'(z) \sim \tfrac{1}{2}\, c(Xr'(0) - Yr(0))\ln(1+z) \quad \text{as } z \downarrow -1. \tag{106}$$

Therefore $x'(z)$ has an infinite limit as $z \downarrow -1$ provided $w_- = Xr'(0) - Yr(0) \neq 0$. The sign of this limit is opposite to the sign of w_-. If $w_- = 0$ then $x'(z) = \frac{X}{r(0)}\, r'(z)$, which converges to $\frac{X}{r(0)}$ as $z \downarrow -1$.

Because equation (93) has the discrete symmetry $z \mapsto -z$, the function $z \mapsto x(-z)$ is another solution with $x(0) = X$ and $x'(0) = -Y$. Equation (103) remains valid if we first replace $x(z)$ and Y by $x(-z)$ and $-Y$, respectively, and then replace z by $-z$. This gives

$$x(z) = \frac{X}{r(0)} r(-z) + (Xr'(0) + Yr(0))[\tfrac{1}{2}\, c\, r(-z)\ln(1-z) + w(-z)]. \tag{107}$$

Recall that the function $w(z)$ is analytic at $z = -1$ with $w(-1) = 1$. This implies that $\lim_{z \uparrow 1} x(z) = Xr'(0) + Yr(0)$. Similarly, we obtain

$$x'(z) = -\frac{X}{r(0)}\, r'(-z) - (Xr'(0) + Yr(0))\big(\tfrac{1}{2}\, c[r(-z)(1-z)^{-1} \\ + r'(-z)\ln(1-z)] + w'(-z)\big),$$

which gives

$$x'(z) \sim -\tfrac{1}{2}\, c(Xr'(0) + Yr(0))\ln(1-z) \quad \text{as } z \uparrow 1, \tag{108}$$

because $r(-1) = 0$ and $r'(-1) = 1$. Therefore $x'(z)$ has an infinite limit as $z \uparrow 1$, provided $w_+ = Xr'(0) + Yr(0) \neq 0$. Moreover the limit has the same sign as w_+. If $w_+ = 0$, then $x'(z) = -\frac{X}{r(0)}\, r'(-z)$, which converges to $-\frac{X}{r(0)}$ as $z \uparrow 1$.

7.7.3 The normalized even and odd solutions

In this subsection we define the normalized even and odd solutions of the rescaled Chaplygin equations.

Let $e(z)$ and $o(z)$ be solutions of

$$x''(z) = \frac{c}{1-z^2}\, x(z) \tag{109}$$

such that $e(-z) = e(z)$ with $e(0) = 1$, $e'(0) = 0$ and $o(z) = -o(-z)$ with $o(0) = 0$, $o'(0) = 1$. Then $e(z)$ and $o(z)$ are the *normalized even* and *odd* solutions of (109), respectively. Every solution of the rescaled Chaplygin equations (92) can be written as

$$\begin{aligned} x(z) &= X\, e(z) + Y\, o(z) \\ y(z) = x'(z) &= X\, e'(z) + Y\, o'(z) \end{aligned} \tag{110}$$

Now $e(z) = \sum_{k=0}^{\infty} e_k\, z^{2k}$ with $e_0 = 1$, $e_1 = c/2$, and $e_k > 0$ for every $k > 1$. The Taylor expansion of $e(z)$ at $z = 0$ is

$$e(z) = 1 + \tfrac{1}{2} c\, z^2 + \tfrac{1}{24} c(c+2)\, z^4 + \mathrm{O}(z^6), \tag{111}$$

which implies

$$e'(z) = c\, z + \tfrac{1}{6} c(c+2)\, z^3 + \mathrm{O}(z^5). \tag{112}$$

Moreover, $e(z) > 1$ and $\frac{e'(z)}{z} > 1$, when $z > 0$. Also $o(z) = \sum_{k=0}^{\infty} o_k\, z^{2k+1}$ with $o_1 = 1$ and $o_k > 0$ for every $k > 1$. Thus the Taylor expansion of $o(z)$ at $z = 0$ is

$$o(z) = z + \tfrac{1}{6} c\, z^3 + \tfrac{1}{120} c(c+6)\, z^5 + \mathrm{O}(z^7) \tag{113}$$

and

$$o'(z) = 1 + \tfrac{1}{2} c\, z^2 + \tfrac{1}{24} c(c+6)\, z^4 + \mathrm{O}(z^6). \tag{114}$$

Moreover, $\frac{o(z)}{z} > 1$ and $o'(z) > 1$, when $z > 0$.

7.7.4 Computation of $r(0)$ and $r'(0)$

In this subsection we give an explicit formula for the constants $r(0)$ and $r'(0)$ as a function of c.

In order to calculate $r(0)$ and $r'(0)$ we need to understand the relation of equation (109) with the hypergeometric equation. Because e is an even analytic function, $e(z) = f(z^2)$ for some analytic function $f(\zeta)$ of $\zeta = z^2$ in a neighborhood of $\zeta = 0$ which satisfies $f(0) = 1$. Since

$$e'(z) = 2z\, f'(z^2) \quad \text{and} \quad e''(z) = 2\, f'(z^2) + 4z^2\, f''(z^2),$$

we see that (109) is equivalent to

$$\zeta(1-\zeta)f''(\zeta) + \tfrac{1}{2}\,(1-\zeta)f'(\zeta) - \tfrac{1}{4}cf(\zeta) = 0. \tag{115}$$

Equation (115) is the *hypergeometric equation*

$$\zeta(1-\zeta)f''(\zeta) + (\gamma - (\alpha+\beta+1)\zeta)f'(\zeta) - \alpha\beta f(\zeta) = 0, \tag{116}$$

with $\gamma = \alpha + \beta + 1 = \frac{1}{2}$ and $\alpha\beta = \frac{1}{4}c$. In other words,

$$\alpha = -\tfrac{1}{4} + \tfrac{1}{4}\sqrt{1-4c}, \quad \beta = -\tfrac{1}{4} - \tfrac{1}{4}\sqrt{1-4c}, \quad \gamma = \tfrac{1}{2}\,. \tag{117}$$

The parameters α and β are nonreal complex conjugate numbers if $c > \frac{1}{4}$.

For arbitrary values of α, β, and γ, the solution $f(\zeta)$ of (116), which is analytic at $\zeta = 0$ and satisfies $f(0) = 1$, is *Gauss' hypergeometric function*

$$F(\alpha, \beta, \gamma; \zeta) = \sum_{n=0}^{\infty} \frac{(\alpha)_n(\beta)_n}{(\gamma)_n}\,\frac{\zeta^n}{n!}. \tag{118}$$

Here $(\delta)_0 = 1$ and $(\delta)_n = \delta(\delta+1)\cdots(\delta+n-1)$ when $n \geq 1$. It is known that

$$\lim_{\zeta \uparrow 1} F(\alpha, \beta, \gamma; \zeta) = \frac{\Gamma(\gamma)\Gamma(\gamma-\alpha-\beta)}{\Gamma(\gamma-\alpha)\Gamma(\gamma-\beta)}, \tag{119}$$

if $\mathrm{Re}\,(\gamma - \alpha - \beta) > 0$, see pp. 249–251 in [28]. From (116) and (117) it follows that

$$e(z) = F(-\tfrac{1}{4} + \tfrac{1}{4}\sqrt{1-4c}, -\tfrac{1}{4} - \tfrac{1}{4}\sqrt{1-4c}, \tfrac{1}{2}\,; z^2). \tag{120}$$

Because $\gamma - \alpha - \beta = 1 > 0$, we can combine (119) with $\lim_{z\uparrow 1} x(z) = X\, r'(0) + Y\, r(0)$, see (103), and $(X, Y) = (1, 0)$ to obtain

$$r'(0) = \lim_{z\uparrow 1} e(z) = \frac{\Gamma(\frac{1}{2})}{\Gamma(\frac{3}{4} - \frac{1}{4}\sqrt{1-4c})\Gamma(\frac{3}{4} + \frac{1}{4}\sqrt{1-4c})}. \tag{121}$$

Similarly the normalized odd solution $o(z)$ of (109) can be written as $o(z) = zg(z^2)$, where $g(\zeta)$ is an analytic function of $\zeta = z^2$ in a neighborhood of 0, which satisfies $g(0) = 1$. Since

$$o'(z) = g(z^2) + 2z^2\, g'(z^2) \quad \text{and} \quad o''(z) = 6z\, g'(z^2) + 4z^3\, g''(z^2),$$

we see that (109) is equivalent to the hypergeometric equation

$$\zeta(1-\zeta)g''(\zeta) + \tfrac{3}{2}(1-\zeta)g'(\zeta) - \tfrac{1}{4}c\, g(\zeta) = 0, \tag{122}$$

with $\gamma = \alpha + \beta + 1 = \frac{3}{2}$ and $\alpha\beta = \frac{1}{4}c$. In other words,

$$\alpha = \tfrac{1}{4} + \tfrac{1}{4}\sqrt{1-4c}, \quad \beta = \tfrac{1}{4} - \tfrac{1}{4}\sqrt{1-4c}, \quad \text{and} \quad \gamma = \tfrac{3}{2}.$$

Therefore

$$o(z) = z\,F(\tfrac{1}{4} + \tfrac{1}{4}\sqrt{1-4c}, \tfrac{1}{4} - \tfrac{1}{4}\sqrt{1-4c}, \tfrac{3}{2}; z^2). \tag{123}$$

Again, because $\gamma - \alpha - \beta = 1$, we can combine (119) with (103) and $(X, Y) = (0, 1)$ to obtain

$$r(0) = \lim_{z\uparrow 1} o(z) = \frac{\Gamma(\frac{3}{2})}{\Gamma(\frac{5}{4} - \frac{1}{4}\sqrt{1-4c})\Gamma(\frac{5}{4} + \frac{1}{4}\sqrt{1-4c})}. \tag{124}$$

7.8 Bifurcations of a vertical disk

In this section we discuss the *bifurcations that a disk in a vertical position* undergoes, when it is either spinning about a fixed point of contact or rolling in a straight line.

7.8.1 *Degenerate equilibria*

Suppose that p is an equilibrium point of a smooth vector field X on a smooth manifold P. Then its linearization $DX(p)$ at p is a well defined linear map of the tangent space of P at p into itself. The equilibrium points of the *fully reduced vector field* $\mathcal{V}$ (13) on the fully reduced phase space $\mathcal{M} = (-1, 1) \times \mathbb{R}^3$ form a connected two dimensional analytic submanifold N of relative equilibria described in §4. Because $(\overline{\sigma}_3, \overline{\sigma}_4)$ can be used as coordinates on N, the rank of $D\mathcal{V}(\sigma^0)$ at $\sigma^0 \in N$ is at most 2. From the first equation in (13) we see that its rank is at least 1.

From the discussion in §5 it follows that an equilibrium point σ^0 of $\mathcal{V}$ corresponds to an equilibrium point $(\varphi^0, 0)$ of the vector field $X_{\overline{\sigma}_3,\overline{\sigma}_4}$ (86) on $(-\pi/2, \pi/2) \times \mathbb{R}$. In particular, φ^0 corresponds to a point where the derivative of the potential function $V(\varphi) = V_{\overline{\sigma}_3,\overline{\sigma}_4}(\sin\varphi) = V_{\overline{\sigma}_3,\overline{\sigma}_4}(\upsilon_1)$ (83) vanishes, that is, $\varphi^0 \in (-\pi/2, \pi/2)$ is a *critical point* of V. The equilibrium point $(\varphi^0, 0)$ of the vector field $X_{\overline{\sigma}_3,\overline{\sigma}_4}$ is *stable* (of *elliptic* type) or *unstable* (of *hyperbolic type*) if the second derivative $\frac{d^2 V}{d\varphi^2}$ at φ^0 is positive or negative, respectively. We will use the same terms for the corresponding relative equilibria.

The relative equilibrium σ^0 of the fully reduced vector field $\mathcal{V}$ is *degenerate* if the rank of $D\mathcal{V}(\sigma^0)$ is 1. This is equivalent to saying that at the equilibrium point $(\varphi^0, 0)$ of $X_{\overline{\sigma}_3,\overline{\sigma}_4}$ the potential V has a *degenerate critical point* at φ^0, that is,

$$\frac{\mathrm{d}V}{\mathrm{d}\varphi}(\varphi^0) = \frac{\mathrm{d}^2V}{\mathrm{d}\varphi^2}(\varphi^0) = 0.$$

In other words, $\sigma_1^0 = \sin\varphi^0$ is a degenerate critical point of the potential $V_{\overline{\sigma}_3,\overline{\sigma}_4}$, that is,

$$V'_{\overline{\sigma}_3,\overline{\sigma}_4}(\sigma_1^0) = V''_{\overline{\sigma}_3,\overline{\sigma}_4}(\sigma_1^0) = 0.$$

We call the set Δ of parameter values $(\overline{\sigma}_3, \overline{\sigma}_4)$ for which the potential $V_{\overline{\sigma}_3,\overline{\sigma}_4}$ has a degenerate critical point in $(-1, 1)$ the *degeneracy locus* of $V_{\overline{\sigma}_3,\overline{\sigma}_4}$. In the remainder of this section we will study the geometry of Δ.

7.8.2 *Vertical degenerate relative equilibria*

In this subsection we investigate the bifurcation of the relative equilibria of the disk for which $\sigma_1 = 0$. From §4 we see that these relative equilibria correspond to the disk being in a vertical position either spinning about a fixed point of contact or rolling along a straight line.

Since $0 = \sigma_1 = z$, we look at the rescaled potential function $W_{X,Y}$ (95) near 0. Recall that the solutions $z \mapsto \big(x(z; X, Y), y(z; X, Y)\big)$ of the rescaled Chapygin equations (92) with initial conditions $\big(x(0; X, Y), y(0; X, Y)\big) = (X, Y)$ are

$$x(z; X, Y) = X\, e(z) + Y\, o(z) \ \text{ and } \ y(z; X, Y) = X\, e'(z) + Y\, o'(z). \quad (125)$$

Here $e(z)$ and $o(z)$ are the normalized even and odd solutions, whose Taylor series about $z = 0$ to fourth order are given in equations (111) and (113). Consequently, the Taylor expansion of $W_{X,Y}$ about $z = 0$ to fourth order terms is

$$W_{X,Y}(z) = \sum_{j=1}^{4} w_j(X, Y)\, z^j + \mathrm{O}(z^5), \quad (126)$$

where

$$w_0(X, Y) = 1 + \tfrac{1}{2} d\, X + \tfrac{1}{2}\, Y^2 \quad (127)$$

$$w_1(X, Y) = (c + d)XY \quad (128)$$

$$w_2(X, Y) = -\tfrac{1}{2} + \tfrac{1}{2}\,(d + cd + c^2)X^2 + \tfrac{1}{2}(c + d)Y^2 \quad (129)$$

$$w_3(X, Y) = \tfrac{1}{3}(c + 3d + 2cd + 2c^2)XY \quad (130)$$

$$w_4(X, Y) = -\tfrac{1}{8} + \tfrac{1}{6}(3d + 14cd + 2c^2 + c^2 d + c^3)X^2 + \tfrac{1}{12}(3c + 6d + 2cd + 2c^2)Y^2. \quad (131)$$

From equations (127)–(131) we see that the rescaled potential function $W_{X,Y}$ has a critical point at $z = 0$ if and only if $0 = w_1(X, Y) = XY$. There are two cases.

CASE 1. Suppose that $X = 0$. Since $\sigma_1(t) = \sigma_1^0 = z = 0$, we obtain $\sigma_2(t) = \frac{\mathrm{d}\sigma_1}{\mathrm{d}t} = 0$. Then the last two equations in (13) read $\frac{\mathrm{d}\sigma_3}{\mathrm{d}t} = 0$ and $\frac{\mathrm{d}\sigma_4}{\mathrm{d}t} = 0$. Therefore $\sigma_3(t) = \overline{\sigma}_3 = 0$, because $\overline{\sigma}_3 = -\frac{I_1}{I_3}\widetilde{u}\,X$, see (98). In view of the first equation in (56), this means that $\eta = 0$. Thus the disk is in a vertical position and is rolling in a straight line with constant speed. From (129) we see that if $w_2(0, Y) < 0$, that is, $|Y| < (c+d)^{-1/2}$, then this relative equilibrium is unstable; whereas if $w_2(0, Y) > 0$, that is, $|Y| > (c+d)^{-1/2}$, then it is stable. Because $\sigma_4(t) = \overline{\sigma}_4$ at $z = 0$, from the second equation in (56), $\eta = 0$ and (98), we obtain $Y = \frac{\overline{\sigma}_4}{\widetilde{u}} = \frac{\zeta}{\widetilde{u}}$. Since $\widetilde{u} = \sqrt{\frac{\mathrm{mgr}}{I_3 + \mathrm{mr}^2}}$, we see that the relative equilibrium is unstable, respectively, stable, if the angular speed $|\zeta|$ of rolling is less than, respectively, greater than, the *critical speed* $|\zeta_{\mathrm{crit}}| = \frac{1}{\sqrt{c+d}}\sqrt{\frac{\mathrm{mgr}}{I_3+\mathrm{mr}^2}} = \overline{\sigma}_4$. Thus the vertical rolling disk undergoes *gyroscopic stabilization* when $|\zeta|$ increases through $|\zeta_{\mathrm{crit}}|$. At $\big(0, \pm\frac{1}{\sqrt{c+d}}\big) = (0, Y_0)$, we have $w_1(0, Y_0) = w_2(0, Y_0) = w_3(0, Y_0) = 0$, but

$$w_4(0, Y_0) = \tfrac{1}{12}(3c + 6d + 2cd + 2c^2)(c + d)^{-1},$$

which is positive. Thus the rescaled potential W_{0,Y_0} has a stable degenerate equilibrium point at $z = 0$. In other words, the points

$$(\overline{\sigma}_3, \overline{\sigma}_4) = \left(0, \pm\frac{1}{\sqrt{c+d}}\sqrt{\frac{\mathrm{mgr}}{I_3 + \mathrm{mr}^2}}\right) \tag{132}$$

lie on the dengeneracy locus Δ of the potential $V_{\overline{\sigma}_3,\overline{\sigma}_4}$.

CASE 2. Now consider the case when $Y = 0$. Because $\sigma_4(t) = \overline{\sigma}_4 = \widetilde{u}\,Y = 0$, from (56) it follows that $\zeta = 0$. Thus the disk is spinning in a vertical position with its point of contact fixed, see (69). From (129) we see that if $w_2(X, 0) < 0$, that is, $|X| < (d + cd + c^2)^{-1/2}$, then the relative equilibrium is unstable; whereas, if $w_2(X, 0) > 0$, that is, $|X| > (d + cd + c^2)^{-1/2}$ then it is stable. Because $\sigma_3(t) = \overline{\sigma}_3$ at $z = 0$, we have $X = -\frac{\overline{\sigma}_3}{\widetilde{u}}\frac{I_1}{I_3} = -\frac{\eta}{\widetilde{u}}\frac{I_1}{I_3}$, using (98). Since $\widetilde{u} = \sqrt{\frac{\mathrm{mgr}}{I_3+\mathrm{mr}^2}}$, we see that the relative equilibrium is unstable, respectively, stable if the speed of spinning $|\eta|$ is less than, respectively, greater than the critical speed $|\eta_{\mathrm{crit}}| = \frac{I_3}{I_1}\frac{1}{\sqrt{d+cd+c^2}}\sqrt{\frac{\mathrm{mgr}}{I_3+\mathrm{mr}^2}}$. Thus the vertical spinning motion of the disk undergoes gyroscopic stabilization when $|\eta|$ increases through $|\eta_{\mathrm{crit}}|$. At $\big(\pm\frac{1}{\sqrt{d+cd+c^2}}, 0\big) = (X_0, 0)$, we have $w_1(X_0, 0) = w_2(X_0, 0) = w_3(X_0, 0) = 0$, but

$$w_4(X_0, 0) = \tfrac{1}{24}(9d + 5c^2 + 11cd + 4c^2d + 4c^3)(d + cd + c^2)^{-1},$$

which is positive. Therefore $W_{X_0,0}$ has a stable degenerate equilibrium point at $z = 0$. In other words, the points

$$(\overline{\sigma}_3, \overline{\sigma}_4) = \left(\pm\frac{I_3}{I_1}\frac{1}{\sqrt{d+cd+c^2}}\sqrt{\frac{\mathrm{mgr}}{I_3+\mathrm{mr}^2}}, 0\right) \tag{133}$$

lie on the dengeneracy locus Δ of the potential $V_{\overline{\sigma}_3,\overline{\sigma}_4}$.

7.8.3 *Normal form of the potential*

In order to describe the *qualitative properties of the degeneracy locus* Δ in a neighborhood of the four points $(\overline{\sigma}_3, \overline{\sigma}_4)$ given in (132) and (133), we need to describe the *qualitative properties of the family of rescaled potential functions*

$$W : (X, Y) \mapsto W_{X,Y} \tag{134}$$

when the parameters (X, Y) are close to $(0, Y_0)$ or $(X_0, 0)$. To do this we use the *theory of singularities of mappings*, see [8] or [40].

The key concept in this theory is the notion of equivalence of two analytic families f and g of analytic functions. For f to be an *analytic family of analytic functions* we mean that $f : \mathbb{R}^n \times \mathbb{R}^k \to \mathbb{R} : (x, \alpha) \mapsto f(x, \alpha)$ is an analytic function near (x_0, α_0), which we think of as an n-parameter analytic family of analytic functions of k-variables given by $x \mapsto f_x$. Here f_x is the analytic function on $\mathbb{R}^k$ defined by holding the variable x in the function f fixed, that is, $\alpha \mapsto f_x(\alpha) = f(x, \alpha)$. Also x varies in a neighborhood of x_0 with f_x being defined in a neighborhood of α_0. Let $g : \mathbb{R}^n \times \mathbb{R}^k \to \mathbb{R} : (y, \beta) \mapsto g(y, \beta)$ define an analytic family near (y_0, β_0). Informally we say that the analytic families f and g are equivalent if for each y near y_0 the function $\beta \mapsto g(y, \beta)$, defined for β near β_0, can be transformed into the function $\alpha \mapsto f(x, \alpha)$, defined for α near α_0, plus a constant $h(x)$ for x near x_0 by a substitution of variables of the form $(x, \alpha) \mapsto (y(x), \beta(x, \alpha))$, which is defined near (x_0, α_0). More precisely, we say that two analytic families f and g given above are *equivalent* if there is a local analytic diffeomorphism

$$H : \mathbb{R}^n \times \mathbb{R}^k \to \mathbb{R}^n \times \mathbb{R}^k : (x, \alpha) \mapsto \big(y(x), \beta(x, \alpha)\big)$$

such that $H(x_0, \alpha_0) = (y_0, \beta_0)$ and

$$g(y(x), \beta(x, \alpha)) = f(x, \alpha) + h(x) \tag{135}$$

in some open neighborhood of (x_0, α_0) in $\mathbb{R}^n \times \mathbb{R}^k$. Here $h : \mathbb{R}^n \to \mathbb{R}$ is an analytic function near x_0.

The following proposition is a consequence of results from the theory of singularities of mappings.

Proposition 7.8.3.10. *Suppose that we are given a two parameter analytic family g of analytic functions near (y_0, β_0) in $\mathbb{R}^2 \times \mathbb{R}$. In other words we are given a function*

$$g : \mathbb{R}^2 \times \mathbb{R} \to \mathbb{R} : (y, \beta) \mapsto g(y, \beta),$$

which is analytic near (y_0, β_0). In addition suppose that the function g satisfies the following conditions.

1. *There is a positive integer p such that the partial derivatives*

$$\frac{\partial g}{\partial \beta} = \frac{\partial^2 g}{\partial \beta^2} = \cdots = \frac{\partial^{p+1} g}{\partial \beta^{p+1}} = 0$$

at (y_0, β_0).
2. *At (y_0, β_0) the partial derivative $\frac{\partial^{p+2} g}{\partial \beta^{p+2}}$ is nonzero. If p is even then the sign in equation (136) below is equal to the sign of $\frac{\partial^{p+2} g}{\partial \beta^{p+2}}$ at (y_0, β_0). Otherwise the plus sign holds.*
3. *Let D be the $2 \times p$ matrix*

$$\begin{pmatrix} \frac{\partial^2 g}{\partial y_1 \, \partial \beta}(y_0, \beta) & \cdots & \frac{\partial^{p+1} g}{\partial y_1 \, \partial \beta^p}(y_0, \beta) \\ \frac{\partial^2 g}{\partial y_2 \, \partial \beta}(y_0, \beta) & \cdots & \frac{\partial^{p+1} g}{\partial y_2 \, \partial \beta^p}(y_0, \beta) \end{pmatrix}_{\beta=\beta_0}.$$

We suppose that $\operatorname{rank} D = 2$.

Then the analytic family g near (y_0, β_0) is equivalent to the analytic family

$$\begin{aligned} &f_{p+2} : \mathbb{R}^p \times \mathbb{R} \to \mathbb{R} : \\ &\qquad ((x_1, \ldots, x_p), \alpha) \mapsto \pm\alpha^{p+2} + x_1 \, \alpha + x_2 \, \alpha^2 + \cdots + x_p \, \alpha^p \end{aligned} \tag{136}$$

near $(0, 0)$.

Proof. See [8]. □

The family (136) is the *normal form* of the *A_{p+1}-singularity*.

We now show that we can apply proposition 7.8.3.10 to the family W of rescaled potential functions (134) near $((X, Y), z) = ((0, Y_0), 0)$ with $p = 2$. From (126) and (127)–(131) it follows that at $((0, Y_0), 0)$ the partial derivatives

$$\frac{\partial W_{X,Y}}{\partial z} = \frac{\partial^2 W_{X,Y}}{\partial z^2} = \frac{\partial^3 W_{X,Y}}{\partial z^3} = 0$$

and $\frac{\partial^4 W_{X,Y}}{\partial z^4} > 0$. Moreover, using (127)–(131) we find that at $((0, Y_0), z)$ we have

$$\frac{\partial^2 W_{X,Y}}{\partial X \partial z} = (c+d)Y_0 + \mathrm{O}(z) \qquad \frac{\partial^3 W_{X,Y}}{\partial X \partial z^2} = (c+d)Y_0^2 + \mathrm{O}(z)$$

$$\frac{\partial^2 W_{X,Y}}{\partial Y \partial z} = 2(c+d)Y_0 z + \mathrm{O}(z^2) \qquad \frac{\partial^3 W_{X,Y}}{\partial Y \partial z^2} = (c+d)Y_0 + \mathrm{O}(z^2).$$

So at $((0, Y_0), 0)$ the matrix $D = \begin{pmatrix} (c+d)Y_0 & (c+d)Y_0^2 \\ 0 & (c+d)Y_0 \end{pmatrix}$. But then $\det D = (c+d)^2 Y_0^2 = (c+d) > 0$. Therefore D has rank equal to 2. Thus all the hypotheses of proposition 7.8.3.10 with $p = 2$ are satisfied. Hence the analytic family W of rescaled potential functions near $(0, Y_0)$ is equivalent to the normal form family f_4 (136) near $(0, 0)$ for the A_3-singularity.

Corollary 7.8.3.11. *The derivative of the local diffeomorphism*

$$\mathbb{R}^2 \to \mathbb{R}^2 : (x_1, x_2) \mapsto \big(X(x_1, x_2), Y(x_1, x_2)\big), \tag{137}$$

which is defined by the equivalence between the normal form family near $(0, 0)$ *and the family* W *of rescaled potential functions near* $(0, Y_0)$*, maps the standard basis vector* e_2 *into a positive multiple of* $Y_0\, e_2$.

Proof. Taking the partial derivative $\frac{\partial^2}{\partial x_2 \partial \alpha}$ on both sides of the equation

$$W\big((X(x_1,x_2), Y(x_1,x_2)), z((x_1,x_2),\alpha)\big) = f((x_1,x_2),\alpha) + h(x_1,x_2), \tag{138}$$

which defines the equivalence between the famlies f and W gives

$$\frac{\partial^2 f}{\partial x_2 \partial \alpha} = \frac{\partial^2 W}{\partial z \partial X}\frac{\partial X}{\partial x_2}\frac{\partial z}{\partial \alpha} + \frac{\partial^2 W}{\partial z \partial Y}\frac{\partial Y}{\partial x_2}\frac{\partial z}{\partial \alpha} + \frac{\partial W}{\partial z}\frac{\partial^2 z}{\partial x_2 \partial \alpha}.$$

Evaluating both sides of the above equation at $(x, \alpha) = (0, 0)$ and $((X, Y), z) = ((0, Y_0), 0)$ gives

$$0 = \frac{\partial^2 W}{\partial z \partial X}\frac{\partial X}{\partial x_2}\frac{\partial z}{\partial \alpha}, \tag{139}$$

because $\frac{\partial^2 f}{\partial x_2 \partial \alpha} = 0$ at $(0, 0)$ and $\frac{\partial^2 W}{\partial z \partial Y} = \frac{\partial W}{\partial z} = 0$ at $((0, Y_0), 0)$. Since $\frac{\partial^2 W}{\partial z \partial X} = (c+d)Y_0 \neq 0$ at $((0, Y_0), 0)$ and $\frac{\partial z}{\partial \alpha} \neq 0$ at $(0, 0)$ by hypothesis, from (139) we obtain $\frac{\partial X}{\partial x_2} = 0$ at $(0, 0)$. Taking the partial derivative $\frac{\partial^3}{\partial x_2 \partial \alpha^2}$ on both sides of (138) gives

$$\frac{\partial^3 W}{\partial Y \partial z^2}\frac{\partial Y}{\partial x_2}\left(\frac{\partial z}{\partial \alpha}\right)^2 = 2. \tag{140}$$

Since $\frac{\partial^3 W}{\partial Y \partial z^2} = 2(c+d)Y_0$ at $((0, Y_0), z)$, equation (140) implies that $\frac{\partial Y}{\partial x_2}$ at $(0, 0)$ has the same sign as Y_0. Consequently, the derivative of the map (137) at $(0, 0)$ has the form $\begin{pmatrix} \frac{\partial X}{\partial x_1} & 0 \\ \frac{\partial Y}{\partial x_1} & \frac{\partial Y}{\partial x_2} \end{pmatrix}$. Therefore it maps the standard basis vector e_2 into a positive multiple of $Y_0 e_2$. □

Taking the partial derivative of the equivalence equation (138) with respect to x_2 and evaluating at $(0,0)$, we get $Y_0 \frac{\partial Y}{\partial x_2} = \frac{\partial h}{\partial x_2}$. Using corollary 7.8.3.11, which states that $\frac{\partial Y}{\partial x_2}$ at $(0,0)$ has the same sign as Y_0, it follows that $\frac{\partial h}{\partial x_2} > 0$ at $(0,0)$. Therefore the function h *cannot* be eliminated from the normal form family by adjusting the equivalence change of variables.

For the parameter values $(X,Y) = (X_0, 0)$, the fact that

$$\frac{\partial^2 W_{X,Y}}{\partial X \partial z} = (d + cd + c^2)X_0 + O(z^2) \qquad \frac{\partial^3 W_{X,Y}}{\partial X \partial z^2} = (d + cd + c^2)X_0 + O(z)$$

$$\frac{\partial^2 W_{X,Y}}{\partial Y \partial z} = (c + d)X_0 + O(z^2) \quad \frac{\partial^3 W_{X,Y}}{\partial Y \partial z^2} = (c + 3d + 2cd + 2c^2)X_0 z + O(z^2)$$

shows that at $z = 0$ the matrix D in hypothesis 3 of proposition 7.8.3.10 is $\begin{pmatrix} 0 & (d+cd+c^2)X_0 \\ (c+d)X_0 & 0 \end{pmatrix}$. Thus D has rank 2, since its determinant is positive. Moreover,

$$\frac{\partial W_{X,Y}}{\partial z} = \frac{\partial^2 W_{X,Y}}{\partial z^2} = \frac{\partial^3 W_{X,Y}}{\partial z^3} = 0$$

at $((X_0), 0)$ and

$$\frac{\partial^4 W_{X,Y}}{\partial z^4} = (12d + 56cd + 8c^2 + 4c^2 d + 4c^3)X_0^2 > 0.$$

Therefore the hypotheses of proposition 7.8.3.10 hold and the family W of rescaled potential functions at $(X_0, 0)$ is equivalent to the normal form family f_4 (136) for the A_3-singularity.

The following proposition summarizes our results about the normal form of the potential $V_{\overline{\sigma}_3,\overline{\sigma}_4}$ (83).

Proposition 7.8.3.12. *The potential $V_{\overline{\sigma}_3,\overline{\sigma}_4}$ on $(-1,1)$ has a degenerate critcal point at $\sigma_1 = 0$ if and only if $(\overline{\sigma}_3^*, \overline{\sigma}_4^*)$ is equal to either $(\overline{\sigma}_3, 0)$ or $(0, \sigma_4)$, see (133) or (132). Moreover, there is*
1. *an open neighborhood $U \times A \subseteq \mathbb{R}^2 \times \mathbb{R}$ of $((0,0),0)$;*
2. *an analytic substitution of variables $A \rightarrow (-1,1) : \alpha \mapsto \sigma_1(x,\alpha)$, depending analytically on $x \in U$, which maps $(0,0)$ to 0,*
3. *a local analytic diffeomorphism $\mathbb{R}^2 \rightarrow \mathbb{R}^2 : x \mapsto (\overline{\sigma}_3(x), \overline{\sigma}_4(x))$, which maps $(0,0)$ to $(\overline{\sigma}_3^*, \overline{\sigma}_4^*)$, such that for every $(x,\alpha) \in U \times A$*

$$V_{\overline{\sigma}_3(x),\overline{\sigma}_4(x)}(\sigma_1(x,\alpha)) = \alpha^4 + x_1\,\alpha + x_2\,\alpha^2 + h(x), \tag{141}$$

where $h : U \subseteq \mathbb{R}^2 \rightarrow \mathbb{R}$ is an analytic function on U.

7.8.4 *Cusps of the degeneracy locus*

In this subsection we discuss the geometry of the normal form family

$$f : \mathbb{R}^2 \times \mathbb{R} \to \mathbb{R} : ((x_1, x_2), \alpha) \mapsto \alpha^4 + x_1\,\alpha + x_2\,\alpha^2 + h(x_1, x_2) \qquad (142)$$

for the A_3 singularity.

The function $f_{(x_1,x_2)}$ has a degenerate critical point if and only if

$$f'_{(x_1,x_2)}(\alpha) = 4\alpha^3 + x_1 + 2x_2\alpha = 0 \qquad (143)$$

and

$$f''_{(x_1,x_2)}(\alpha) = 12\alpha^2 + 2x_2 = 0 \qquad (144)$$

Equations (143) and (144) are equivalent to

$$x_1 = 8\alpha^3 \text{ and } x_2 = -6\alpha^2. \qquad (145)$$

Equations (145) parametrize the *standard cusp*

$$\Gamma_0 = \{(x_1, x_2) \in \mathbb{R}^2 \big|\, 27x_1^2 + 8x_2^3 = 0\} \qquad (146)$$

in the (x_1, x_2)-plane. The curve Γ_0, which is contained in the half plane $\{x_2 \le 0\}$, has the origin as its cusp point with branches which straddle the x_2-axis, and the cusp points vertically upward. The branches of the cusp have contact of order $\frac{3}{2}$ with the x_2-axis at the origin.

The image Γ of the standard cusp Γ_0 under the local diffeomorphism $(x_1, x_2) \to \big(X(x_1, x_2), Y(x_1, x_2)\big)$, which maps $(0, 0)$ to $(0, Y_0)$ $((X_0, 0))$, is equal to the set of parameter values (X, Y) near $(0, Y_0)$ $((X_0, 0))$ where the rescaled potential $W_{X,Y}$ has a degenerate critical point at $z = 0$. It follows from proposition 7.8.3.10 and corollary 7.8.3.11 that Γ is an analytic nondegenerate cusp with cusp point at $(0, Y_0)$ (or $(X_0, 0)$) pointing away from the origin in parameter space. The image of the x_2-axis under the equivalence diffeomorphism is a smooth curve γ, which is tangent to the Y-axis at $(0, Y_0)$ (or the X-axis at $(X_0, 0)$). Therefore γ has contact of order at least 2 with the Y (or X) axis at the cusp point. But Γ has contact of order $\frac{3}{2}$ there, which is smaller than 2. Therefore the branches of the cusp Γ straddle the Y (or X) axis.

The following proposition summarizes what we have determined so far about the degeneracy locus Δ of the potential function $V_{\overline{\sigma}_3,\overline{\sigma}_4}$.

Proposition 7.8.4.13. *The normal form* (142) *shows that the part of the degeneracy locus* Δ*, which comes from the local behavior of the potential* $V_{\overline{\sigma}_3,\overline{\sigma}_4}$ *at* $\sigma_1 = 0$ *with the parameters* $(\overline{\sigma}_3, \overline{\sigma}_4)$ *near* $(\overline{\sigma}_3^*, \overline{\sigma}_4^*)$*, is a nondegenerate analytic cusp* Γ*, which points away from the origin of parameter space toward the cusp point* $(\overline{\sigma}_3^*, \overline{\sigma}_4^*)$ *and has branches, which straddle the relevant coordinate axis. Thus the degeneracy locus* Δ *has four cusps.*

The function f_x with f the family given by (142) has one critical point, which is a nondegenerate minimum if $27x_1^2 + 8x_2^3 > 0$, that is, if x is in the complement of the region bounded by Γ_0 (146) and containing the negative x_2 semiaxis. If $27x_1^2 + 8x_2^3 < 0$, that is, x lies in the region bounded by Γ_0 containing the negative x_2 semiaxis, then f_x has three critical points: two nondegenerate local minima with a local maximum in between. If $x \in \Gamma_0$ but $x \neq (0,0)$, then one of the local minima of f_x has merged with the local maximum into a third order degenerate critical point; whereas the other local minimum survives as a nondegenerate critical point. Near the third order critical point the family f (142) is equivalent to the family

$$\mathbb{R}^2 \times \mathbb{R} \rightarrow \mathbb{R} : \big((x_1, x_2), \alpha\big) \mapsto \alpha^3 + x_1\, \alpha + x_2\, \alpha^2,$$

which is the A_2 *singularity*. This follows by checking that the hypotheses of proposition 7.8.3.10 hold for $p = 1$.

Using the local analytic diffeomorphism $x \mapsto (\overline{\sigma}_3(x), \overline{\sigma}_4(x))$ of proposition 7.8.3.12, we see that near $\sigma_1 = 0$, the potential $V_{\overline{\sigma}_3,\overline{\sigma}_4}$ has one critical point, which is a nondegenerate local minimum if $(\overline{\sigma}_3, \overline{\sigma}_4)$ lies near one of the four cusp points of the degeneracy locus Δ but not on the cusp Γ nor in the region bounded by the branches of the cusp which contains a coordinate axis. If $(\overline{\sigma}_3, \overline{\sigma}_4)$ lies inside the region bounded by the cusp, but is near the cusp point, then $V_{\overline{\sigma}_3,\overline{\sigma}_4}$ has three critical points σ_1, two of which are nondegenerate minima with a nondegenerate maximum in between. If $(\overline{\sigma}_3, \overline{\sigma}_4)$ is near a cusp point and lies on the cusp curve Γ less its cusp point, then one of the local minima has merged with the local maximum into a third order degenerate critical point. During this merger the other local minimum survives.

Thus we have proved

Proposition 7.8.4.14. *On the degeneracy locus Δ near, but not equal to a cusp point, the family $(\overline{\sigma}_3, \overline{\sigma}_4) \mapsto V_{\overline{\sigma}_3,\overline{\sigma}_4}$ has an A_2-*singularity.

7.9 The global geometry of the degeneracy locus

In this section we give a *global description of the degeneracy locus* Δ. Recall that Δ is the set of degenerate critical points of $V_{\overline{\sigma}_3,\overline{\sigma}_4}$.

7.9.1 *The circle of degenerate critical points*

Let $\widetilde{\Delta}$ be the set of points $(\sigma_1, \overline{\sigma}_3, \overline{\sigma}_4) \in (-1,1) \times \mathbb{R}^2$ such that

$$\text{the potential } V_{\overline{\sigma}_3, \overline{\sigma}_4} \text{ has a degenerate critical point at } \sigma_1. \tag{147}$$

Then the degeneracy locus Δ is the image of $\widetilde{\Delta}$ under the projection map $\pi_2 : (-1,1) \times \mathbb{R}^2 \to \mathbb{R}^2 : (\sigma_1, \overline{\sigma}_3, \overline{\sigma}_4) \mapsto (\overline{\sigma}_3, \overline{\sigma}_4)$, see (80). The goal of this subsection is to prove

Proposition 7.9.1.15. *The locus $\widetilde{\Delta}$ is an analytically embedded circle, for which the set $\widetilde{\Delta} \cap \{\sigma_1 = 0\}$ is four points and the connected components I of $\widetilde{\Delta} \cap \{\sigma_1 \neq 0\}$ are mapped consecutively onto each other by the sequence of reflections*

$$(\sigma_1, \overline{\sigma}_3, \overline{\sigma}_4) \mapsto (-\sigma_1, \overline{\sigma}_3, -\overline{\sigma}_4) \mapsto (\sigma_1, -\overline{\sigma}_3, -\overline{\sigma}_4) \mapsto (-\sigma_1, -\overline{\sigma}_3, \overline{\sigma}_4).$$

We make some preliminary observations. Recall the partial differential operator $\mathfrak{C} = \frac{\partial}{\partial z} + y\,\frac{\partial}{\partial x} + c\frac{1}{1-z^2}\,\frac{\partial}{\partial y}$ (96). The equations defining the locus of degenerate critical points of the rescaled potential $W_{X,Y}$ (95) are given by $\mathfrak{C}\widetilde{W} = \mathfrak{C}^2\widetilde{W} = 0$. Here $\widetilde{W}$ is the function of (z, x, y) given by equation (94). A straightforward calculation shows that

$$\mathfrak{C}\widetilde{W} = d\,\frac{z}{1-z^2}\,x^2 + (c+d)\,\frac{1}{1-z^2}\,xy - \frac{z}{\sqrt{1-z^2}} \tag{148}$$

and

$$\begin{aligned}\mathfrak{C}^2\widetilde{W} = &\left[4d\,\frac{z^2}{(1-z^2)^3} + (d+cd+c^2)\frac{1}{(1-z^2)^2}\right]x^2 \\ &+ 2(c+2d)\,\frac{z}{(1-z^2)^2}\,xy + (c+d)\,\frac{1}{1-z^2}\,y^2 \\ &- \frac{1}{(1-z^2)^{3/2}}.\end{aligned} \tag{149}$$

First we look at a special solution of $\mathfrak{C}\widetilde{W} = \mathfrak{C}^2\widetilde{W} = 0$.

Lemma 7.9.1.16. *When $x = z = 0$, then there are functions $z = \phi_{\pm}(x)$, which are analytic near 0 with $\phi_{\pm}(0) = 0$, $\phi'_{\pm}(0) = \pm(c+d)^{1/2}$ such that*

$$\mathfrak{C}\widetilde{W}(\phi_{\pm}(x), x, \pm(c+d)^{-1/2}) = \mathfrak{C}^2\widetilde{W}(\phi_{\pm}(x), x, \pm(c+d)^{-1/2}) = 0.$$

Proof. When $x = z = 0$ holds, the equation $\mathfrak{C}^2\widetilde{W} = 0$ (149) gives $y = \pm(c+d)^{-1/2}$. At the point $(z, x, y) = (0, 0, \pm\frac{1}{\sqrt{c+d}})$ we see that

$$\operatorname{grad}\mathfrak{C}\widetilde{W} = (-1, \pm(c+d)^{1/2}, 0) \quad \text{and} \quad \operatorname{grad}\mathfrak{C}^2\widetilde{W} = (0, 0, \pm(c+d)^{-1/2})$$

are linearly independent. Therefore from the implicit function theorem, it follows that near each of the points $p_\pm = (0,0,\pm(c+d)^{-1/2})$, the set of solutions of $\mathfrak{C}\widetilde{W} = \mathfrak{C}^2\widetilde{W} = 0$ is a smooth analytic curve, whose tangent space at $p_\pm$ is $\mathrm{span}\{(\pm(c+d)^{1/2},1,0)\}$. Under the projection map $(z,x,y) \mapsto (z,x)$ the image of these curves is the graph of two analytic functions $z = \phi_\pm(x)$ such that $\phi_\pm(0) = 0$ and $\phi'_\pm(0) = \pm(c+d)^{1/2}$. □

In general, solving $\mathfrak{C}\widetilde{W} = 0$ (148) for y as a function of z and x and then substituting the result into (149) gives

$$\begin{aligned}\mathfrak{C}^2\widetilde{W} &= \left[4d\,\frac{z^2}{(1-z^2)^3} + (d+cd+c^2)\frac{1}{(1-z^2)^2} - \frac{d(2c+3d)}{c+d}\,\frac{z^2}{(1-z^2)^3}\right]x^2 \\ &\quad + \frac{2z^2-1}{(1-z^2)^{3/2}} + \frac{1}{c+d}z^2\frac{1}{x^2} \end{aligned} \tag{150}$$

Multiplying (150) by $(c+d)(1-z^2)^{3/2}x^2$ and then replacing x by the expression $(1-z^2)^{3/4}\,\widetilde{x}$, the equation $\mathfrak{C}^2\widetilde{W} = 0$ becomes

$$c_2(z)(\widetilde{x}^2)^2 + c_1(z)\widetilde{x}^2 + z^2 = 0, \tag{151}$$

where

$$c_1(z) = (c+d)(2z^2-1) \quad \text{and} \quad c_2(z) = c\big(d-(c+d)^2\big)z^2 + (c+d)(d+cd+c^2).$$

We now analyze (151). From the equation above we see that c_2 is an inhomogeneous linear function λ of z^2 with $\lambda(0) = (c+d)(d+cd+c^2) > 0$ and $\lambda(1) = d(2c+d) > 0$. Consequently, $\lambda(z^2) > 0$ when $z^2 \in [0,1]$. Therefore $c_2(z) > 0$ when $z \in [-1,1]$. So (151) holds for $\widetilde{x} \ne 0$ and $z \in (-1,1)$ only if $c_1(z) < 0$. From the defining equation for $c_1(z)$ this implies that $z^2 < \frac{1}{2}$. As equation (151) is quadratic in $\widetilde{x}^2$, it has complex solutions given by

$$y = y_\pm(z) = \frac{-c_1(z) \pm \sqrt{\mathrm{discr}\,(z)}}{2c_2(z)}, \tag{152}$$

where

$$\begin{aligned}\mathrm{discr}\,(z) &= c_1^2(z) - 4\,z^2c_2(z) \\ &= 4(2c^2+3cd+2d^2)z^4 - 4(c+d)(2c+3d+cd+c^2)z^2 + (c+d)^2.\end{aligned}$$

Let δ be the function defined by $\mathrm{discr}(z^2) = \delta(z^2)$. Then $\delta(0) = (c+d)^2 > 0$ and $\delta(\frac{1}{2}) = -2c_2(\frac{1}{\sqrt{2}}) = -2\lambda(\frac{1}{2}) < 0$. Thus there is a unique $\zeta \in (0, \frac{1}{2})$ such that $\delta(\zeta) = 0$. Let $\widetilde{z} = \sqrt{\zeta}$. Then $\mathrm{discr}\,(z) > 0$ when $z \in (-\widetilde{z}, \widetilde{z})$, $\mathrm{discr}\,(\pm\widetilde{z}) = 0$, and $\mathrm{discr}\,(z) < 0$ when $\widetilde{z}^2 < z^2 < \frac{1}{2}$. Suppose that $z \in (-\widetilde{z}, \widetilde{z})$, then $c_1(z) < 0$, $c_2(z) > 0$, and $0 < \mathrm{discr}\,(z) \le c_1(z)^2$.

These conditions imply that the functions $y = y_{\pm}(z)$ defined by (152) are nonnegative and depend analytically on $z \in (-\widetilde{z}, \widetilde{z})$. Because $c_1(z)$, $c_2(z)$, and $\mathrm{discr}\,(z)$ are polynomials in z^2, it follows that $y_{\pm}(z) = y_{\pm}(-z)$. Also $0 < y_-(z) < y_+(z)$ when $0 < |z| < \widetilde{z}$. For $z \in (-\widetilde{z}, \widetilde{z})$ the function $y_-(z)$ equals 0 only when $z = 0$. Thus we have proved

Lemma 7.9.1.17. *When $z^2 \in (0, \widetilde{z})$, equation (151) has four real solutions $\widetilde{x} = \pm\sqrt{y_+(z)}$ and $\pm\sqrt{y_-(z)}$, which depend analytically on z. The solutions $\widetilde{x} = \pm\sqrt{y_+(z)}$ extend to nonzero analytic functions on $(-\widetilde{z}, \widetilde{z})$; while the solutions $\widetilde{x} = \pm\sqrt{y_-(z)}$ converge to 0 as $z \to 0$. When either $z \uparrow \widetilde{z}$ or $z \downarrow -\widetilde{z}$, we have* $\mathrm{discr}\,(z) \downarrow 0$. *There the solution $\widetilde{x} = \pm\sqrt{y_-(z)}$ merges with the solution $\widetilde{x} = \pm\sqrt{y_+(z)}$.*

An immediate consequence of the above lemma is

Corollary 7.9.1.18. *The set of solutions $(z, \widetilde{x})$ of (151) in the strip $(-1, 1) \times \mathbb{R}$ is connected and compact.*

Proof. This follows because this solution set is a closed subset of $(-1,1)\times\mathbb{R}$ which stays away from $\{\pm 1\} \times \mathbb{R}$. □

We now give a

Proof of proposition 7.9.1.15. First we rewrite equation (151) as

$$A(\widetilde{x})\, z^2 = \widetilde{x}^2 B(\widetilde{x}), \tag{153}$$

where

$$A(\widetilde{x}) = c(d - (c+d)^2)\widetilde{x}^4 + 2(c+d)\widetilde{x}^2 + 1 \tag{154}$$

and

$$B(\widetilde{x}) = (c+d)(1 - (d + cd + c^2)\widetilde{x}^2). \tag{155}$$

This leads to two solutions

$$z = \pm w(\widetilde{x}) = \pm \widetilde{x}\sqrt{\frac{B(\widetilde{x})}{A(\widetilde{x})}},$$

which depend analytically on $\widetilde{x}$ as long as $A(\widetilde{x}) > 0$ and $\frac{B(\widetilde{x})}{A(\widetilde{x})} > 0$. Because $A(0) = 1 > 0$ and $B(0) = c + d > 0$, there is a maximal open interval J containing 0 such that $A(\widetilde{x}) > 0$ and $B(\widetilde{x}) > 0$. Since $z = 0$ when $\widetilde{x} = 0$, it follows that $\phi_{\pm}(x) = \pm w(\widetilde{x})$, where $x = (1 - z^2)^{3/4}\widetilde{x}$. Here $\phi_{\pm}$ are the functions given by lemma 7.9.1.16.

Now $B(\widetilde{x}) > 0$ when $|\widetilde{x}| < (d + cd + c^2)^{-1/2}$. Consequently, $A(\widetilde{x}) > 0$ when $|\widetilde{x}| < (d + cd + c^2)^{-1/2}$ because for solutions $(z, \widetilde{x}) \in (-1, 1) \times \mathbb{R}$ of equation (151), the value of z stays away from ± 1. Therefore $|z| < 1$ in equation (153). So

$$|A(\widetilde{x})| > |A(\widetilde{x})z^2| = \widetilde{x}^2 B(\widetilde{x}).$$

This implies that A is positive when B is. On the other hand, the value of A at $(d + cd + c^2)^{-1/2}$ equals

$$\left[cd + 2d(c + d) + c(c + d)^2 + (d + cd + c^2)^2\right] / (d + cd + c^2)^2,$$

which is positive. Therefore the set of solutions $(z, \widetilde{x}) \in (-1, 1) \times \mathbb{R}$ of (153) is connected. Let J be the interval $|\widetilde{x}| < (d + cd + c^2)^{-1/2}$. Then $(\pm w(\widetilde{x}), \widetilde{x}) \in (-1, 1) \times \mathbb{R}$ satisfies equation (151) if and only if $\widetilde{x} \in J$ and $z = \pm w(\widetilde{x})$.

Consequently, the solution set of (151) in the $(z, \widetilde{x})$-strip $(-1, 1) \times \mathbb{R}$ is an analytically immersed closed curve, which consists of the following successively overlaping analytically embedded curves: 1) $(w(\widetilde{x}), \widetilde{x})$ for $\widetilde{x} \in J$; 2) $(z, \sqrt{y_+(z)})$ for $z \in (-\widetilde{z}, \widetilde{z})$; 3) $(-w(\widetilde{x}), \widetilde{x})$ for $\widetilde{x} \in J$; 4) $(z, -\sqrt{y_+(z)})$ for $z \in (-\widetilde{z}, \widetilde{z})$ and finally returning to 1). This analytically immersed curve has possibly one transverse self intersection at $\widetilde{x} = 0$. But in lemma 7.9.1.16 we have shown that the solution set of $\mathfrak{C}\widetilde{W} = \mathfrak{C}^2\widetilde{W} = 0$ is the union of two disjoint analytic curves, which project to the curves $(\phi_\pm(\widetilde{x}), \widetilde{x})$ near $\widetilde{x} = 0$. Therefore the analytic curve 1)–4) is embedded.

The last statement of proposition 7.9.1.15 follows by observing that the reflections $(z, x, y) \mapsto (-z, x, -y) \mapsto (z, -x, -y) \mapsto (-z, -x, y)$ correspond to $w|J_+ \to -w|J_- \to w|J_- \to -w|J_+$, where w is the curve $\widetilde{x} \mapsto \big(w(\widetilde{x}), \widetilde{x}\big)$ and $J_\pm = \{\widetilde{x} \in J \big| \pm \widetilde{x} > 0\}$. This proves proposition 7.9.1.15. □

7.9.2 *A global description of the degeneracy locus*

In this subsection we give a global description of the degeneracy locus Δ.

We begin by proving

Proposition 7.9.2.19. *If at $\sigma_1^0 \in (-1, 1)$, the first, second and third derivatives of $V_{\overline{\sigma}_3, \overline{\sigma}_4}$ vanish, then $\sigma_1^0 = 0$ and $(\overline{\sigma}_3, \overline{\sigma}_4)$ is one of the four cusp points of Δ found in proposition* 7.8.4.13.

Proof. We need only show that the equations

$$\mathfrak{C}\widetilde{W} = \mathfrak{C}^2\widetilde{W} = \mathfrak{C}^3\widetilde{W} = 0 \tag{156}$$

imply that $z = 0$. Straightforward calculations show that (156) is equivalent to

$$0 = dz\,\widetilde{x}^2 + (c+d)\widetilde{x}\widetilde{y} - z \tag{157}$$

$$\begin{aligned} 0 &= \left[(3d - c(c+d))z^2 + d + cd + c^2\right]\widetilde{x}^2 \\ &\quad + 2(c+2d)z\,\widetilde{x}\widetilde{y} + (c+d)\widetilde{y}^2 - 1 \end{aligned} \tag{158}$$

$$\begin{aligned} 0 &= 2\left[(6d - 4cd - 3c^2)z^3 + (6d + 4cd + 3c^2)z\right]\widetilde{x}^2 \\ &\quad + 2\left[(3c + 9d - 2cd - 2c^2)z^2 + (c + 3d + 2cd + 2c^2)\right]\widetilde{x}\widetilde{y} \\ &\qquad + 2(2c+3d)z\,\widetilde{y}^2 - 3z, \end{aligned} \tag{159}$$

where we have replaced y by $(1-z^2)^{-1/4}\widetilde{y}$. Equations (157)–(159) are inhomogeneous linear equations in $\widetilde{u} = \widetilde{x}^2$, $\widetilde{v} = \widetilde{x}\widetilde{y}$, and $\widetilde{w} = \widetilde{y}^2$. Using (157) we can eliminate $\widetilde{v}$ from equations (158) and (159). Using the equation which results from (158), we can eliminate $\widetilde{w}$ from (159). The resulting equation for $\widetilde{u}$ has a common factor of z, which we divide out under the hypothesis that $z \neq 0$. This leads to the equation $2c\,D(z)\widetilde{u} = U(z)$, where

$$D(z) = (c+d)(3d + c^2 - d^2)(1 - z^2) + 2d^2 + 4cd\,z^2$$

and

$$U(z) = -(c+d)\big(3c + 9d + 4cd + 4c^2\big)(1 - z^2) - (3d^2 + 8cd + c^2)z^2.$$

Clearly, $U(z) < 0$, when $z^2 \le 1$. Moreover, $d^2 = (c+d)d - cd \le 2d - cd$, since $c + d \le 2$. Therefore $3d + c^2 - d^2 \ge d + c^2 + cd > 0$. So $D(z) > 0$ when $z^2 \le 1$. Hence $\widetilde{u} < 0$ when $z^2 \le 1$ and $z \neq 0$. But this contradicts the definition of $\widetilde{u}$, namely, $\widetilde{u} = \widetilde{x}^2 \ge 0$. Therefore equation (156) implies that $z = 0$. □

In §9.1 we have shown that the degeneracy locus Δ in the $(\overline{\sigma}_3, \overline{\sigma}_4)$-plane is the image of an analytically embedded circle $\hat{\Delta}$ in $(-1,1) \times \mathbb{R}^2$ under the projection map $\pi : (-1,1) \times \mathbb{R}^2 \to \mathbb{R}^2 : (\sigma_1, \overline{\sigma}_3, \overline{\sigma}_4) \mapsto (\overline{\sigma}_3, \overline{\sigma}_4)$. The circle $\widetilde{\Delta}$ intersects the plane $\{\sigma_1 = 0\}$ in exactly four points, which map to the four cusp points $(0, \pm\overline{\sigma}_4^*)$ and $(\pm\overline{\sigma}_3^*, 0)$ described in §8.2. Let $\widetilde{I}$ be a connected component of $\widetilde{\Delta} \cap \{\sigma_1 = 0\}$. Then the restriction of the projection map π to $\widetilde{I}$ is an embedding of $\widetilde{I}$ onto an analytic curve I in an open quadrant Q of the $(\overline{\sigma}_3, \overline{\sigma}_4)$-plane.

Proposition 7.9.2.20. *The curve I is transverse to the radial direction and is traced out from a cusp point of Δ on half axis adjacent to Q to the cusp point on the other half axis adjacent to Q. The other connected components of $\Delta \cap \{\sigma_1 \neq 0\}$ are obtained by applying the reflections $(\overline{\sigma}_3, \overline{\sigma}_4) \mapsto (\overline{\sigma}_3, -\overline{\sigma}_4) \mapsto (-\overline{\sigma}_3, -\overline{\sigma}_4) \mapsto (-\overline{\sigma}_3, \overline{\sigma}_4)$ to I.*

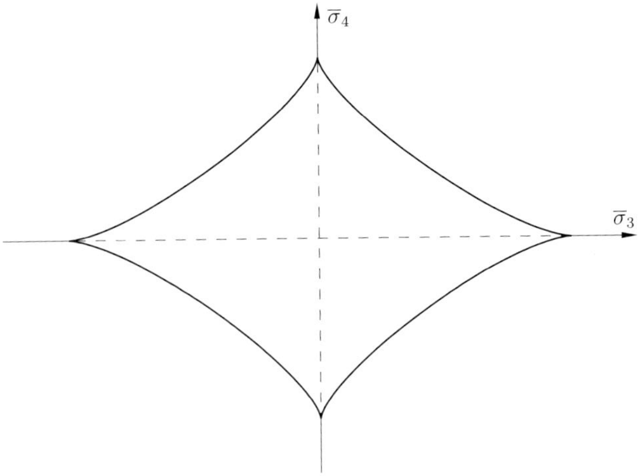

Fig. 7.2 The degeneracy locus Δ for the uniform disk. Here we have $I_1 = I_2 = \frac{1}{4}\mathrm{mr}^2$, $I_3 = \frac{1}{2}\,\mathrm{mr}^2$, $\frac{\mathrm{g}}{\mathrm{r}} = \frac{5}{4}$, $c = \frac{4}{3}$, and $d = \frac{2}{3}$. Using (132) and (133), we get the cusp points $(\pm 1, 0)$ and $(0, \pm\sqrt{\frac{5}{12}})$.

Thus the degeneracy locus Δ is formed by four cusp points together with four connected components of $\Delta \cap \{\sigma_1 \neq 0\}$. Therefore Δ is a continuously embedded circle in the $(\overline{\sigma}_3, \overline{\sigma}_4)$-plane, which is an analytic set with four cusp points as its set of singular points.

Proof. Let

$$G : (-1,1) \times \mathbb{R}^2 \to \mathbb{R}^2 : (z,x,y) \mapsto \big(\mathfrak{C}\widetilde{W}(z,x,y), \mathfrak{C}^2\widetilde{W}(z,x,y)\big). \tag{160}$$

Then the degeneracy locus Δ is defined by $G(z,x,y) = (0,0)$. In addition to the partial differential operator $\mathfrak{C}$ coming from Chaplygin's equations (96), we consider the Euler operator $\mathfrak{E} = x\,\frac{\partial}{\partial x} + y\,\frac{\partial}{\partial y}$. Because Chaplygin's equations (76) are linear, the operators $\mathfrak{C}$ and $\mathfrak{E}$ commute when $(x,y) \neq (0,0)$. Note that $\mathfrak{C}^2\widetilde{W} = 0$ (149) implies that $(x,y) \neq (0,0)$. When $G = (0,0)$ we have

$$(\mathfrak{E}G,\ \mathfrak{C}G) = \begin{pmatrix} \mathfrak{E}\mathfrak{C}\widetilde{W} & \mathfrak{C}^2\widetilde{W} \\ \mathfrak{E}\mathfrak{C}^2\widetilde{W} & \mathfrak{C}^3\widetilde{W} \end{pmatrix} = \begin{pmatrix} \mathfrak{E}\mathfrak{C}\widetilde{W} & 0 \\ \mathfrak{E}\mathfrak{C}^2\widetilde{W} & \mathfrak{C}^3\widetilde{W} \end{pmatrix},$$

which has determinant equal to $(\mathfrak{E}\mathfrak{C}\widetilde{W})(\mathfrak{C}^3\widetilde{W})$. From (148) we see that $\mathfrak{C}\widetilde{W}$ is equal to the sum of $(1-z^2)^{-1/2}$ and a homogeneous polynomial q of degree 2 in (x, y). Therefore

$$\mathfrak{E}\mathfrak{C}\widetilde{W} = 2q = 2(\mathfrak{C}\widetilde{W} + 2(1-z^2)^{-1/2}) = 2z(1-z^2)^{-1/2}, \tag{161}$$

when $G = (0,0)$. Therefore $\mathfrak{E}\mathfrak{C}\widetilde{W} \neq 0$ when $z \neq 0$. From proposition 7.9.2.19 it follows that $\mathfrak{C}^3\widetilde{W} \neq 0$ when $G = (0,0)$ and $z \neq 0$. Therefore when $G = (0,0)$ and $z \neq 0$, the derivative of G, restricted to the plane Π spanned by the vectors $\mathfrak{E}$ and $\mathfrak{C}$, is invertible. Consequently, $\{G = (0,0)\} \cap \{z \neq 0\}$ is a smooth one dimensional manifold, whose tangent space $\ker DG$ is complementary to Π. Now a fiber of the map

$$\widetilde{\pi} : \mathbb{R}^3 \to \mathbb{R}^2 : (z, x(z; X, Y), y(z; X, Y)) \mapsto (X, Y),$$

where $z \mapsto (x(z; X, Y), y(z; X, Y))$ is an integral curve of the rescaled Chaplygin vector field on $\mathbb{R}^2$ with initial condition (X, Y), are integral curves of the vector field defined by the operator $\mathfrak{C}$. Moreover, the map $\widetilde{\pi}$ intertwines $\mathfrak{E}$ with the Euler vector field in the (X, Y)-plane. Therefore $\widetilde{\pi}$ maps the tangent space to $\{G = (0,0)\} \cap \{z \neq 0\}$ to a one dimensional linear subspace, which is complementary to $\mathfrak{E}$. So $\widetilde{\pi}|(\{G = (0,0)\} \cap \{z \neq 0\})$ is an immersion from $\{G = (0,0)\} \cap \{z \neq 0\}$ to the (X, Y)-plane, whose image is transverse to the radial direction.

Let J be a connected component of $\{G = (0,0)\} \cap \{z \neq 0\}$. Then $\widetilde{\pi}|J$ traces out the rescaled degeneracy locus from a cusp point on one of the half axes to another cusp point on the adjacent half axis in such a way that it stays in the complement of the origin of the (X, Y)-plane and always has a nonzero angular velocity with respect to the origin. If we continue on to an adjacent component J' of $\{G = (0,0)\} \cap \{z \neq 0\}$, then $\widetilde{\pi}|J'$ still encircles the origin in the same direction, because the branches of the cusps straddle the half axes. Note that J' is the image of J under one of the reflections $(z, x, y) \mapsto (-z, -x, y)$ or $(z, x, y) \mapsto (-z, x, -y)$. Moreover the map $\widetilde{\pi}$ intertwines each of these reflections with the reflections $(X, Y) \mapsto (-X, Y)$ or $(X, Y) \mapsto (X, -Y)$, respectively.

If $\widetilde{\pi}|J$ intersects a nonnegative integer number k of half axes, then so does $\widetilde{\pi}|J'$, because this latter map is obtained from the former by a reflection. Therefore $\widetilde{\pi}|(\{G = (0,0)\} \cap \{z \neq 0\})$ intersects $4 + 4k$ half axes in the same direction. Hence the winding number of $\widetilde{\pi}|\{G = (0,0)\}$ about the origin has absolute value $k+1$. This winding number is equal to the linking number of $\{G = (0,0)\}$ with the z-axis in $(-1,1) \times \mathbb{R}^2$ with coordinates

(z, x, y). In turn this linking number is equal to the winding number of $\widetilde{p}|\{G = (0,0)\}$, where

$$\widetilde{p} : (-1,1) \times \mathbb{R}^2 \to \mathbb{R}^2 : (z, x, y) \mapsto (x, y).$$

If $x = 0$ then $z = 0$, because $\mathfrak{C}\widetilde{W} = 0$, see (148). If $y = 0$, $G = (0,0)$, and $z \neq 0$, then from equations (148) and (149) we find that both $d\,x^2$ and $[4d\,z^2 + (d + cd + c^2)(1 - z^2)]x^2$ are equal to $(1 - z^2)^{3/2}$. Therefore $x \neq 0$ and

$$d = 4d\,z^2 + (d + cd + c^2)(1 - z^2) > d\,z^2 + d(1 - z^2) = d,$$

which is a contradiction. Therefore $\widetilde{p}|\{G = (0,0)\}$ intersects a half axis only when $z = 0$. Because $\{G = (0,0)\} \cap \{z = 0\}$ is four points, it follows that $\widetilde{p}|\{G = (0,0)\}$ intersects four consecutive half axes. Therefore the absolute value of the winding number of $\widetilde{p}|\{G = (0,0)\}$ is equal to 1, that is, $k = 0$. This completes the proof of proposition 7.9.2.21. □

We now prove the following supplement to proposition 7.9.1.15.

Proposition 7.9.2.21. *Let $\widetilde{I}$ be a connected component of $\widetilde{\Delta}$ (147) less the four points where $\sigma_1 = 0$. For each $\widetilde{i} = (\sigma_1, i) \in \widetilde{\Delta}$, where $i \in I$, there is an open neighborhood $U \times A$ of $(0,0) \in \mathbb{R}^2 \times \mathbb{R}$, an analytic substitution of variables $\mathbb{R} \to \mathbb{R} : \alpha \mapsto \sigma_1(x, \alpha)$, depending analytically on $x = (x_1, x_2) \in U$, a local analytic diffeomorphism $\Phi : x \mapsto (\overline{\sigma}_3(x), \overline{\sigma}_4(x))$ of $\mathbb{R}^2$ near $(0,0)$, and an analytic function $h : U \subseteq \mathbb{R}^2 \to \mathbb{R}$ such that $\sigma_1(0,0) = \sigma_1$, $(\overline{\sigma}_3(0), \overline{\sigma}_4(0)) = i$, and*

$$V_{\overline{\sigma}_3(x), \overline{\sigma}_4(x)}(\sigma_1(x, \alpha)) = \alpha^3 + x_1\,\alpha + h(x)$$

for every $(x, \alpha) \in U \times A$.

Moreover, the preimage i' of the outward radial vector i under $D\Phi(0)$ of the local diffeomorphism Φ has nonzero x_1-component and $Dh(0)i' > 0$.

Proof. This follows immediately from proposition 7.9.2.19, proposition 7.8.4.14, and corollary 7.8.3.11. □

7.10 Falling flat

In this section we discuss the behavior of the disk when it falls flat.

7.10.1 *When the disk does not fall flat*

In this subsection we give the exact conditions when the disk does *not fall flat*.

Because the kinetic energy T (82) is nonnegative and the total energy $\overline{E}$ (81) is a constant of the motion with value E, it follows that during a motion on $(-1,1) \times \mathbb{R}$ we have $V_{\overline{\sigma}_3,\overline{\sigma}_4}(\sigma_1) \le E$. Therefore if $\lim_{\sigma_1 \to \pm 1} \sup V_{\overline{\sigma}_3,\overline{\sigma}_4}(\sigma_1) \le E$, then σ_1 remains bounded away from ± 1. In other words, the angle between the plane of the rim of the rolling disk and the horizontal plane remains bounded away from 0. Thus the disk does not fall flat.

We now prove a much stronger statement about the asymptotic behavior of the potential $V_{\overline{\sigma}_3,\overline{\sigma}_4}$ (83) as $\sigma_1 \to \pm 1$, provided that $(\overline{\sigma}_3, \overline{\sigma}_4) \in \mathbb{R}^2$ does not lie on one of the one dimensional linear subspaces $\ell_\pm$ defined by

$$I_1 r'(0)\, \overline{\sigma}_3 \mp I_3 r(0)\, \overline{\sigma}_4 = 0. \tag{162}$$

Here $r'(0)$ and $r(0)$ are positive constants given by (121) and (124). In what follows the quantity

$$\eta_\pm = I_1 r'(0)\, \overline{\sigma}_3 \mp I_3 r(0)\, \overline{\sigma}_4 \tag{163}$$

will be used as a coordinate relative to the line $\ell_\pm$. If $\eta_\pm \neq 0$, then $(\overline{\sigma}_3, \overline{\sigma}_4)$ lies on one side of $\ell_\pm$; while if $\eta_\pm = 0$, then $(\overline{\sigma}_3, \overline{\sigma}_4)$ lies on $\ell_\pm$. The distance of $(\overline{\sigma}_3, \overline{\sigma}_4)$ from $\ell_\pm$ is of the same order as $|\eta_\pm|$.

Lemma 7.10.1.22. *If $(\overline{\sigma}_3, \overline{\sigma}_4)$ does not lie on $\ell_\pm$, then*

$$V_{\overline{\sigma}_3,\overline{\sigma}_4}(\sigma_1) \sim \frac{1}{4I_1}\, \eta_\pm^2\, \frac{1}{1 \mp \sigma_1} \quad \text{as } \sigma_1 \to \pm 1 \tag{164}$$

and

$$V'_{\overline{\sigma}_3,\overline{\sigma}_4}(\sigma_1) \sim \frac{1}{4I_1}\, \eta_\pm^2\, \frac{1}{(1 \mp \sigma)_1^2} \quad \text{as } \sigma_1 \to \pm 1. \tag{165}$$

These asymptotic relations are locally uniform for $(\overline{\sigma}_3, \overline{\sigma}_4) \in \mathbb{R}^2 \setminus \ell_\pm$.

Proof. Write $\frac{1}{r(0)} X = a$ and $r'(0)\, X \pm r(0)\, Y = \epsilon$, where $a = \frac{I_1}{I_3 \widetilde{u} r(0)}$ and $\epsilon = -\eta_\pm \frac{1}{I_3 \widetilde{u}}$ with $\widetilde{u} = \sqrt{\frac{\mathrm{mgr}}{I_3 + \mathrm{mr}^2}}$, see (98). Then the hypothesis that $(\overline{\sigma}_3, \overline{\sigma}_4) \notin \ell_\pm$ implies that $\epsilon \neq 0$. From lemma 7.7.2.9 it follows that

$$x(z) = a\, r(\mp z) + \epsilon \left[\tfrac{1}{2}\, c\, r(\mp z) \ln(1 \mp z) + w(\mp z) \right],$$

where $r(z) \sim 1 \mp z$ and $w(z) \sim 1$ as $z \rightarrow \pm 1$. Therefore $x(z) \sim \epsilon$ as $z \rightarrow \pm 1$. Similarly,

$$x'(z) = \mp\Big(a\, r'(\mp z) + \epsilon\left[\tfrac{1}{2}\, c\left(r'(\mp z)\ln(1 \pm z) + \frac{r(\pm z)}{1 \mp z}\right) + w'(\mp z)\right]\Big),$$

where $r'(z) \sim 1$, $\frac{r(z)}{1+z} \sim 1$, and $w'(z) = \mathrm{O}(1)$ as $z \downarrow -1$. Therefore $x'(z) \sim \mp\frac{1}{2}\, c\epsilon\, \ln(1 \pm z)$ as $z \rightarrow \pm 1$. Because

$$1 - z^2 = (1-z)(1+z) \sim 2(1 \pm z) \quad \text{as } z \rightarrow \pm 1, \tag{166}$$

it follows that the rescaled potential $W_{X,Y}$ (95) satisfies

$$\begin{aligned} W_{X,Y}(z) &\sim \tfrac{1}{4} d\epsilon^2 \frac{1}{1 \mp z} + \tfrac{1}{8}\epsilon^2 c^2 \ln(1 \mp z) + \mathrm{O}(1) \\ &\sim \tfrac{1}{4} d\epsilon^2 \frac{1}{1 \mp z} \quad \text{as } z \rightarrow \pm 1. \end{aligned} \tag{167}$$

This implies (164), because $d = \frac{I_3^2}{I_1(I_3+\mathrm{mr}^2)}$, $\epsilon = -\eta \pm \frac{1}{I_3}\sqrt{\frac{I_3+\mathrm{mr}^2}{\mathrm{mgr}}}$, and $V_{\overline{\sigma}_3,\overline{\sigma}_4} = \mathrm{mgr}\, W_{X,Y}$.

Now consider the function $\mathfrak{C}\widetilde{W} = W'_{X,Y}(z)$. Then

$$\begin{aligned} \mathfrak{C}\widetilde{W} &= d\,\frac{z}{1-z^2}\,x + (c+d)\,\frac{1}{1-z^2}\,xy - \frac{z}{(1-z^2)^{1/2}} \\ &\sim \pm\tfrac{1}{4} d\epsilon^2 \frac{1}{(1 \mp z)^2} + \mathrm{O}\Big(\frac{\ln^2(1 \mp z)}{1 \mp z}\Big) + \mathrm{O}((1 \mp z)^{1/2}) \\ &\sim \pm\tfrac{1}{4} d\epsilon^2 \frac{1}{(1 \mp z)^2} \quad \text{as } z \rightarrow \pm 1. \end{aligned} \tag{168}$$

This implies (165). □

Let $(\overline{\sigma}_3, \overline{\sigma}_4) \in \mathbb{R}^2 \setminus (\ell_- \cup \ell_+)$. From (164) it follows that a level set of the energy $\overline{E}$ (81) is a compact subset of $(-1,1) \times \mathbb{R}$. Since $\overline{E}$ is a constant of motion, this implies that the vector field $X_{\overline{\sigma}_3,\overline{\sigma}_4}$ (86) is complete. Thus its integral curves are either equilibrium points, or asymptotic to unstable equilibrium points, or nonconstant periodic functions of t with $\sigma_1(t)$ oscillating between a minimum and a maximum value in $(-1,1)$. Note that the hypothesis that $(\overline{\sigma}_3, \overline{\sigma}_4) \notin \ell_\pm$ and equation (164) imply that there are no relative equilibria with the plane of the rim of the disk almost horizontal.

7.10.2 When the disk falls flat

We now discuss when the *disk falls flat*.

The next lemma shows that we have a strong dichotomy.

Lemma 7.10.2.23. *Suppose that* $(\overline{\sigma}_3, \overline{\sigma}_4) \in \ell_\pm$. *Then the potential function*

$$V : (-\pi/2, \pi/2) \to \mathbb{R} : \varphi \mapsto V(\varphi) = V_{\overline{\sigma}_3, \overline{\sigma}_4}(\sin \varphi)$$

has an extension to an analytic function in some neighborhood of $\pm\pi/2$ *such that*

$$V(\pm\pi/2) = \frac{I_1^2(I_3 + \mathrm{mr}^2)}{2I_3^2 r(0)^2}\, \overline{\sigma}_3^2 \tag{169}$$

and

$$V'(\pm\pi/2) = \mp \mathrm{mgr}. \tag{170}$$

In particular, $V(\varphi)$ *has a strict local minimum at* $\pm\pi/2$.

Proof. We use the notation of the proof of lemma 7.10.1.22. The hypothesis $(\overline{\sigma}_3, \overline{\sigma}_4) \in \ell_\pm$ implies that $r'(0)\, X \mp r(0)\, Y = 0$. From (104) and (105) it follows that $x(z) = \frac{X}{r(0)}\, r(\mp z)$ and $x'(z) = \mp \frac{X}{r(0)}\, r'(\mp z)$. Here $r(z)$ is analytic and $r(z) = (1 + z) + \mathrm{O}((1 + z)^2)$ for z in an open neighborhood of -1. If we substitute $z = \sin \varphi$, then for φ near $\pi/2$ we have

$$1 \mp z = 1 \mp \sin \varphi = 1 - (1 - \cos^2 \varphi)^{1/2} \sim \tfrac{1}{2} \cos^2 \varphi. \tag{171}$$

This implies that $r(\mp z) \sim \frac{1}{2} \cos^2 \varphi$. Therefore $\varphi \mapsto \frac{r(\mp z)^2}{\cos^2 \varphi}$ extends to an analytic function with a double zero at $\pm\pi/2$. In view of (171) we find that

$$r'(\mp z)^2 = 1 + \mathrm{O}(1 \mp z) = 1 + \mathrm{O}(\cos^2 \varphi),$$

which shows that $\varphi \mapsto W_{X,Y}(\sin \varphi)$ has an analytic extension in a neighborhood of $\mp\pi/2$. Moreover, $W_{X,Y}(\mp\pi/2) = \frac{X^2}{2r(0)^2}$. The value of $W'_{X,Y}$ at $\pm\pi/2$ is equal to the value of the derivative of $\cos \varphi$ at $\pm\pi/2$, namely, ∓ 1. Using (98) the lemma follows. □

Suppose that $(\overline{\sigma}_3, \overline{\sigma}_4) \in \ell_- \setminus \{(0,0)\}$ and that $E > V(-\pi/2)$. From the fact that $V'(-\pi/2) > 0$, see (170), it follows that the set $\{\varphi \in (-\pi/2, \pi/2) \,|\, V(\varphi) < E\}$ has a connected component of the form $(-\pi/2, \varphi_+)$ for some $\varphi_+ \in (-\pi/2, \pi/2)$. Note that the hypothesis $(\overline{\sigma}_3, \overline{\sigma}_4) \notin \ell_+$ implies that $\varphi_+ < \pi/2$. Because $\lim_{\varphi \downarrow -\pi/2} V(\varphi) = V(-\pi/2) < E$, from the description of the solutions of the conservative Newtonian system in §5.3, every solution of energy E, which starts at $\varphi_0 \in (-\pi/2, \varphi_+)$ and for which $\dot{\varphi}_0 < 0$, will *fall flat in finite time.* More precisely, suppose that $\varphi_+ < \pi/2$ (which implies $E = V(\varphi_+)$) and $V'(\varphi_+) \neq 0$, which means that $\sigma_1^+ = \sin\varphi_+$ is not a critical point of $V_{\overline{\sigma}_3,\overline{\sigma}_4}$, then for every solution of the conservative Newtonian system (85) of energy E with $\varphi_0 \in (-\pi/2, \varphi_+)$ there exist finite times T_1 and T_2 with $T_1 < T_2$ such that $\lim_{t\downarrow T_1} \varphi(t) = -\pi/2$, $\lim_{t\uparrow T_2} \varphi(t) = \pi/2$, and $\varphi(t)$ increases monotonically from $-\pi/2$ to φ_+ as t goes from T_1 to $\frac{1}{2}(T_1 + T_2)$ and decreases monotonically from φ_+ to $-\pi/2$ as t goes from $\frac{1}{2}(T_1 + T_2)$ to T_2. In other words, the plane of the rim of the disk rises monotonically from the horizontal to a maximum angle with the horizontal plane and then falls back to the horizontal — all in finite time.

Now suppose that $\varphi_+ < \pi/2$ and $\sigma_1^+ = \sin\varphi_+$ is a critical point of $V_{\overline{\sigma}_3,\overline{\sigma}_4}$. If $\varphi_0 \in (-\pi/2, \varphi_+)$, then there is a finite time T such that the solution of the conservative Newtonian system (85) is defined for $t \in (-\infty, T)$. Moreover, we have $\lim_{t\downarrow -\infty} \varphi(t) = \varphi_+$, $\lim_{t\uparrow T} \varphi(t) = -\pi/2$, and $\varphi(t)$ decreases monotonically from φ_+ to $-\pi/2$.

Similar conclusions can be drawn if $(\overline{\sigma}_3, \overline{\sigma}_4) \in \ell_+ \setminus \{(0,0)\}$ by replacing φ with $-\varphi$.

If $(\overline{\sigma}_3, \overline{\sigma}_4) \in \ell_+ \cap \ell_- = (0,0)$, then Chaplygin's equations (76) have a trivial solution, that is, $\sigma_3(\sigma_1; 0, 0) = 0 = \sigma_4(\sigma_1; 0, 0)$. Therefore the potential $V_{\overline{\sigma}_3,\overline{\sigma}_4}$ (83) becomes $V(\varphi) = \mathrm{mgr}\cos\varphi$. In this situation the conservative Newtonian system (85) is the half of the mathematical pendulum where $\cos\varphi \geq 0$. If $E > \mathrm{mgr}$, then φ traverses $[-\pi/2, \pi/2]$ monotonically in finite time, increasing or decreasing depending on the sign of $\dot{\varphi}_0$, which is constant during the motion. If $E = \mathrm{mgr}$, then there is a finite time T and a solution, which is defined for $t \in (-\infty, T)$ such that $\varphi(t) \uparrow \pi/2$ as $t \uparrow T$ and $\varphi(t) \downarrow 0$ as $t \downarrow -\infty$. The other solutions are obtained by reversing the sign of t or the sign of φ. If $E < \mathrm{mgr}$, then $\varphi(t)$ increases in finite time from $-\pi/2$ to $-\cos^{-1}\frac{E}{\mathrm{mgr}}$ and then decreases symmetrically back to $-\pi/2$, or decreases in finite time from $\pi/2$ to $\cos^{-1}\frac{E}{\mathrm{mgr}}$ and then increases symmetrically back to $\pi/2$.

If we consider $(\overline{\sigma}_3, \overline{\sigma}_4)$ as a mapping from the fully reduced space $(-1,1) \times \mathbb{R}^3$ (with coordinates $(\sigma_1, \sigma_2, \sigma_3, \sigma_4)$) to $\mathbb{R}^2$, then the preimage $L_\pm$ of the lines $\ell_\pm$ under this map is a nonempty open *codimension one analytic subset* of $(-1,1) \times \mathbb{R}^3$. The set of initial conditions in the fully reduced space which lead to motions of the disk that fall flat in finite time contains $L_\pm$ and therefore has codimension one. It is quite surprising that this set of initial conditions has such a high dimension.

7.10.3 *Limiting behavior when falling flat*

In this subsection we discuss the *limiting behavior* of the disk when it falls flat. Up until now we have only considered the angle φ between the oriented plane of the rim of the disk and the vertical direction. Here we treat other aspects of the motion.

Proposition 7.10.3.24. *Suppose that the disk falls flat or rises up from the flat position when $t \uparrow T_*$ or $t \downarrow T_*$. The disk falls onto or rises from the flat position with a bang in the sense that $\lim_{t \to T_*} \frac{\mathrm{d}\varphi}{\mathrm{d}t}$ is finite and nonzero. Moreover, the solution on the constraint manifold $\mathcal{C}$ (3) and the point of contact converge as $t \to T_*$. Also the vector $\frac{\mathrm{d}u}{\mathrm{d}t}$ converges to a nonzero vector in the horizontal plane as $t \uparrow T_*$ or $t \downarrow T_*$.*

Proof. Since the disk falls flat as $t \to T_*$, we know that $u(t) \to \pm e_3$. Suppose that $u(t) \to -e_3$. This corresponds to $\varphi \to -\pi/2$, or equivalently, $\sigma_1 \downarrow -1$ as $t \to T_*$. In other words, we have assumed that $(\overline{\sigma}_3, \overline{\sigma}_4) \in \ell_-$. The condition $\varphi \to -\pi/2$ means that the energy E of the solution is greater than $V(-\pi/2)$. Because $E = \frac{1}{2} M\,(\dot{\varphi})^2 + V(\varphi)$ is constant on a solution, using lemma 7.10.2.23 we find that

$$\dot{\varphi} \to \dot{\varphi}_* = \pm\Big(\frac{2}{M}(E - V(-\pi/2))\Big)^{1/2} \neq 0, \tag{172}$$

as $t \to T_*$. The minus sign holds in the above limit, when the disk falls flat and the plus sign when it rises from being flat. Because $\sigma_1(t) = \sin\varphi(t)$, we get

$$\sigma_2(t) = \dot{\sigma}_1(t) = (\cos\varphi(t))\dot{\varphi} = (1 - \sigma_1^2(t))^{1/2}\dot{\varphi},$$

which implies

$$(1 - \sigma_1^2(t))^{-1/2}\sigma_2(t) \to \dot{\varphi}_* \quad \text{as } t \to T_*. \tag{173}$$

The condition that $(\overline{\sigma}_3, \overline{\sigma}_4) \in \ell_-$ implies $r'(0)X - r(0)Y = 0$, using (162) and (98), and is equivalent to

$$\sigma_3(\sigma_1; \overline{\sigma}_3, \overline{\sigma}_4) = \frac{\overline{\sigma}_3}{r(0)}\, r(\sigma_1).$$

Therefore

$$\sigma_4(\sigma_1; \overline{\sigma}_3, \overline{\sigma}_4) = -\frac{I_1}{I_3}\frac{\mathrm{d}\sigma_3}{\mathrm{d}\sigma_1} = -\frac{I_1\overline{\sigma}_3}{I_3 r(0)}\, r'(\sigma_1).$$

Because $r(\sigma_1)$ is analytic in a neighborhood of -1 with $r(-1) = 0$ and $r'(-1) = 1$ we find that

$$\frac{\sigma_3(t)}{1 - \sigma_1^2(t)} \to \frac{1}{2r(0)}\,\overline{\sigma}_3 \quad \text{as } t \to T_* \tag{174}$$

and

$$\sigma_4(t) \to -\frac{I_1}{I_3 r(0)}\,\overline{\sigma}_3 \quad \text{as } t \to T_*. \tag{175}$$

If we combine the equation $\dot{\psi} = \frac{\sigma_3}{1-\sigma_1^2}\sigma_1 - \sigma_4$ for $\dot{\psi}$ with the above two equations we find that

$$\dot{\psi} \to (-\tfrac{1}{2} + \frac{I_1}{I_3})\frac{\overline{\sigma}_3}{r(0)} \quad \text{as } t \to T_*.^{1} \tag{176}$$

Because $\psi(t)$ is obtained from $\dot{\psi}$ by integrating a differential equation, whose right hand side converges as $t \to T_*$, it follows that $\psi(t)$ converges to angle ψ_* as $t \to T_*$. In view of $u_{\mathrm{hor}} = \cos\varphi(\cos\psi, \sin\psi, 0)$, and $u(t) \to -e_3$, we get

$$\dot{u} \to \dot{\varphi}_*(\cos\psi_*, \sin\psi_*, 0) \neq 0 \quad \text{as } t \to T_*.$$

Combining the reconstruction equation (26) for ω with $u(t) \to -e_3$ as $t \to T_*$, and using (173) and (174), we find that

$$\omega_1 \to -\dot{\varphi}_* \sin\psi_*, \;\; \omega_2 \to \dot{\varphi}_* \cos\psi_*, \;\; \omega_3 \to -\frac{I_1}{I_3 r(0)}\,\overline{\sigma}_3, \tag{177}$$

as $t \to T_*$. From $u(t) \to -e_3$ and (177), using the differential equation $\dot{u} = u \times \omega$ (6) from the E(2)-reduced equations of motion, we recover the fact that $\dot{u}$ has a limit as $t \to T_*$. By definition of the angular velocity vector ω, the rotational motion of the disk obtained by integrating $\dot{A} = A(\omega(t))^{\times}$ (29). Because $\omega(t)$ converges to (177) as $t \to T_*$, it follows that $A(t)$ converges to $A_* \in \mathrm{SO}(3)$. Similarly the position $a(t)$ of the center of mass of the disk, which is obtained by integrating the differential equation (33) whose right hand side converges. Thus $a(t)$ converges to some element a_* of $\mathbb{R}^3$. Therefore, the motion of the disk in the constraint manifold $\mathcal{C} \subseteq TE(3)$ converges. Finally, equation (38) implies that the point of contact $p(t)$ converges to $-\mathrm{r}A_*(\cos\psi_*, \sin\psi_*, 0) + a_*$ as $t \to T_*$.

The case when $\sigma_1(t) \to 1$, that is, $(\overline{\sigma}_3, \sigma_4) \in \ell_+$, follows by applying the symmetry $(\sigma_1, \overline{\sigma}_3, \overline{\sigma}_4) \mapsto (-\sigma_1, -\overline{\sigma}_3, \overline{\sigma}_4)$ or the symmetry $(\sigma_1, \overline{\sigma}_3, \overline{\sigma}_4) \mapsto (-\sigma_1, \overline{\sigma}_3, -\overline{\sigma}_4)$. □

The angular velocity vector ω in body coordinates converges when the disk fall flat. Also the length of its horizontal component converges and is nonzero. Moreover, its limiting direction is perpendicular to the limit of u. This is not surprising in view of the falling rotation of the disk. The

[1]Note that this limit is 0 in the case of a hoop.

vertical component of the limiting angular velocity is $-\frac{I_1}{I_3 r(0)}\overline{\sigma}_3$, which is nonzero as soon as $\overline{\sigma}_3 \neq 0$. Because $\overline{\sigma}_4$ is proportional to $\overline{\sigma}_3$ on $\ell_\pm$, the condition $\overline{\sigma}_3 \neq 0$ is equivalent to $(\overline{\sigma}_3, \overline{\sigma}_4) \neq (0,0)$. In other words, we are *not* in the special case of falling flat described in §5.4. Thus, at the same time that the disk is rotating about a horizontal axis when it falls flat, it is also spinning about its vertical symmetry axis with an angular speed equal to $-\frac{I_1}{I_3 r(0)}\overline{\sigma}_3$.

7.11 Near falling flat

In this section we investigate the *asymptotic behavior* of the solutions where $p = (\overline{\sigma}_3, \overline{\sigma}_4) \notin \ell_\pm$ approaches $q \in \ell_\pm$, when the disk falls flat. For σ_1 bounded away from ± 1, the potential function $V_p(\sigma_1)$ is close to $V_q(\sigma_1)$. Accordingly, the solutions of the Newtonian system (85) in $(-1,1)\times\mathbb{R}$ and the reconstructed full motions are close to each other. However, when σ_1 is close to ± 1, the motions differ greatly, because $V_p(\sigma_1) \to \infty$ as $\sigma_1 \to \pm 1$, see lemma 7.10.1.22; whereas $\varphi \mapsto V_q(\sin\varphi)$ has an analytic extension to a neighborhood of $\pm\pi/2$, when the disk falls flat. See lemma 7.10.2.23.

7.11.1 *Elastic reflection*

The next lemma implies that as p approaches the point q, the motion $\varphi(t)$ converges to a motion in a potential well with potential as given in lemma 7.10.2.23, but with an *elastic reflection* at $\pm\pi/2$.

Lemma 7.11.1.25. *Let B be a bounded subset of $\mathbb{R}$ with $\overline{\sigma}_3 \in B$. Let $q = (\overline{\sigma}_3, \pm\frac{I_1 r'(0)}{I_3 r(0)}\overline{\sigma}_3)$ be a point on $\ell_\pm$, which we have parametrized by $\overline{\sigma}_3$. Suppose that $p = (\overline{\sigma}_3, \overline{\sigma}_4) \notin \ell_\pm$, is near q, and $E > V_q(\pm 1)$. Finally suppose that $\sigma_1 = \sin\varphi$ is close to ± 1. If $V_q(\pm 1) < V_p(\sigma_1) \leq E$ and $V_p(\sigma_1)$ is* not *close to $V_q(\pm 1)$, then $\cos\varphi$ is of order $|\eta_\pm|$. In addition, during the time interval that $V_p(\sigma_1)$ is not close to $V_q(\sigma_1)$, the velocity $\frac{\mathrm{d}\varphi}{\mathrm{d}t}$ changes monotonically from $\pm\left[\frac{2}{M}(E - V_q(\pm 1))\right]^{-1/2}$ to its negative in time of order $|\eta_\pm|$.*

Proof. All the asymptotic statements below are for $\eta_\pm \to 0$ and are locally uniform in $\overline{\sigma}_3$ and E.

We use the notation of the proof of lemma 7.10.1.22. This time we have two small variables: ϵ and $1 \mp z$. Because $r(z)(1+z) \sim (1+z)\ln(1+z)$ and $w(z) \sim 1$ as $z \downarrow -1$, we have $x(z) \sim a(1 \mp z) + \epsilon$ as $z \to \pm 1$. Since

$r'(z) \sim 1$, $\frac{r(z)}{(1+z)} \sim 1$, and $w'(z) = \mathrm{O}(1)$ as $z \downarrow 1$, we have $y(z) = x'(z) \sim \mp\big(a + \frac{1}{2}\,\epsilon c\,\ln(1 \mp z)\big)$ as $z \to \pm 1$. From (94) and the above it follows that

$$\begin{aligned}\widetilde{W} &\sim \frac{d}{4(1 \pm z)}\big(a(1 \mp z) + \epsilon\big)^2 + \tfrac{1}{2}\left(a + \tfrac{1}{2}\,\epsilon c \ln(1 \mp z)\right)^2 + \sqrt{2(1 \mp z)} \\ &\sim \frac{d\epsilon^2}{4}\,\frac{1}{(1 \mp z)} + \tfrac{1}{2}\,a^2 + \tfrac{1}{2}\,ac\epsilon\,\ln(1 \mp z). \end{aligned} \tag{178}$$

In (178) we have deleted all terms which are small — in particular the term $\frac{1}{8}c^2\epsilon^2\ln^2(1 \mp z)$, since it is small compared to $\frac{1}{4}d\epsilon^2\,\frac{1}{(1\mp z)}$. If $\frac{\epsilon^2}{(1 \mp z)}$ is small, then $1 \mp z >> \epsilon^2$. Therefore

$$|\epsilon\,\ln(1 \mp z)| << |\epsilon \ln(\epsilon^2)| << 1.$$

Hence $|\widetilde{W} - \frac{1}{2}\,a^2|$ is *not* small. It is bounded, only if $1 \pm z$ is of order ϵ^2. In this situation we have

$$x(z) \sim \epsilon \quad \text{and} \quad y(z) = x'(z) \sim \mp a, \tag{179}$$

because $\epsilon\ln(1 \mp z)$ is small. Therefore $\widetilde{W} \sim \frac{1}{2}\,a^2 + \frac{1}{4}d\epsilon^2\,\frac{1}{1\mp z}$. Since $z = \sigma_1 = \sin\varphi$ is close to ± 1, using (171) we see that $\cos\varphi$ is the same order as $|\epsilon|$, and

$$\widetilde{W} \sim \tfrac{1}{2}\,a^2 + \tfrac{1}{2}\,d\epsilon^2\,\frac{1}{\cos^2\varphi}. \tag{180}$$

To estimate the derivative of the potential $V_p(\varphi) = V_{\overline{\sigma}_3,\overline{\sigma}_4}(\sin\varphi)$, we substitute $x(z) \sim a(1 \mp z) + \epsilon$ and $y(z) \sim \mp\big(a + \frac{1}{2}\,c\epsilon\,\ln(1 \mp z)\big)$ into the equation (97) for $\mathfrak{C}\widetilde{W}$ and get

$$\begin{aligned}\mathfrak{C}\widetilde{W} \sim \tfrac{1}{2}\left(a + \frac{\epsilon}{1 \mp z}\right)\Big[\pm\tfrac{1}{2}\,d(a + \frac{\epsilon}{1 \mp z}) \\ \mp(c+d)(a + \tfrac{1}{2}\,c\epsilon\,\ln(1 \mp z)\Big] \pm \frac{1}{\sqrt{2(1 \mp z)}}.\end{aligned} \tag{181}$$

If we use the fact that $1 \mp z$ is of the same order as ϵ^2, then the above equation becomes

$$(\cos\varphi)\mathfrak{C}\widetilde{W} \sim \pm\frac{d\epsilon^2}{\cos^3\varphi}. \tag{182}$$

Because (166) implies that $\cos\varphi$ is of the same order as ϵ, we conclude that $\pm V'(\varphi)$ is positive and of order $|\epsilon|^{-1}$. Using the Newtonian equations of motion (85), from the preceding statement we conclude that $\mp\frac{\mathrm{d}^2\varphi}{\mathrm{d}t^2}$ is positive and of order $|\epsilon|^{-1}$ during the time interval that $V_p(\sigma_1)$ is not close to $V_q(-1)$. From the expression for the energy $\overline{E}$ (84) we see that this is the time interval when $\dot{\varphi}$ is not close to $C_{\pm} = \pm[\frac{2}{M}(E - V_q(-1))]^{-1/2}$. Let τ be the time needed to go from $C_{\pm}$ to $C_{\mp}$. Then

$$2C_{\pm} = \dot{\varphi}(\tau) - \dot{\varphi}(0) = \int_0^{\tau} \frac{\mathrm{d}^2\varphi}{\mathrm{d}t^2}\,\mathrm{d}t.$$

So $2C_{\pm} \sim \frac{\tau}{|\epsilon|}$. Therefore τ is of order $|\epsilon|$, which is of order $|\eta_{\pm}|$. This proves the lemma. □

If $p \notin \ell_{\pm}$ is close to $q \in \ell_{\pm}$ and the energy $E > V_q(\pm 1)$ is close to $V_q(\pm 1)$, then $\varphi(t)$ stays close to $\pm\pi/2$. In addition, if we take $|\eta_{\pm}| > 0$ and sufficiently small compared to $E - V_q(\pm 1)$, then for most of the time

$$\frac{\mathrm{d}^2\varphi}{\mathrm{d}t^2} = -\frac{\partial V_p(\sin\varphi)}{\partial\varphi} \sim -\frac{\partial V_q(\sin\varphi)}{\partial\varphi} \sim \pm\mathrm{mgr},$$

see (170). If we start at the point where $\varphi(0)$ is farthest away from $\pm\pi/2$, which occurs when $V_p(\varphi(0)) = E$ and $\dot{\varphi} = 0$, then $\dot{\varphi}(t) \sim \pm\mathrm{mgr}$, until $\varphi(t) \sim \pi/2$, where $\dot{\varphi}(t) \sim \pm\big(\frac{2}{M}(E - V_q(\pm 1)\big)^{1/2}$, see (87). The period of the periodic solution of the Newtonian system (85), that is, the time needed for $\varphi(t)$ to go to the nearest $\pm\pi/2$ and then return back to its initial position, is asymptotically equal to

$$2^{3/2}\frac{1}{\mathrm{mgr}\sqrt{I_1 + \mathrm{mr}^2}}(E - V_q(\pm 1))^{1/2}.$$

Here we first let $\eta_{\pm} \to 0$ and then let $E \downarrow V_q(\pm 1)$. The fact that this period goes to zero, that is, the frequency of the oscillations in $\varphi(t)$ goes to infinity, is quite *uncommon for oscillations* in a potential well.

If $\eta_{\pm} \to 0$, then the motion of $\varphi(t)$ converges to that of the conservative Newtonian system (85) with $(\overline{\sigma}_3, \overline{\sigma}_4) = q \subset \ell_{\pm}$ together with the condition that we have an elastic reflection $\dot{\varphi} \mapsto -\dot{\varphi}$ when $\varphi(t) = \pm\pi/2$, see §10.2. For $E > V_q(\pm 1)$ and E close to $V_q(\pm 1)$, the motion of $\varphi(t)$ resembles that of a ball which is dropped on the floor and bounces back up elastically. The frequency of the bouncing tends to infinity as the height from which the ball falls goes to zero.

7.11.2 *The increase of the angles ψ and χ*

In order to reconstruct the motion of the disk during the short time of order $|\eta_{\pm}|$ that $V_p(\sigma_1)$ differs markedly from $V_q(\sigma_1)$, we look at what the angles χ and ψ in (35) and (27) do. The following lemma shows that in this short time they increase very rapidly from an initial limiting value to a finite final limiting value by an amount which in absolute value is equal to a constant whose sign depends on which side the point p approaches the line $\ell_\pm$ from.

Lemma 7.11.2.26. *We retain the assumptions of lemma 7.11.1.25. If $\cos\varphi << |\eta_\pm|^{1/2}$, then the time derivatives of $\chi(t)$ and $\psi(t)$ are asymptotically equal to $\frac{\eta_\pm}{I_1}\frac{1}{\cos^2\varphi}$ and $\pm\frac{\eta_\pm}{I_1}\frac{1}{\cos^2\varphi}$, respectively. Moreover, they are of order $\pm\frac{1}{\eta_\pm}$ and $\frac{1}{\eta_\pm}$, respectively. However, the change is $\chi(t)$ and $\psi(t)$ is small over any time interval for which $1 >> \cos\varphi(t) >> |\eta_\pm|$.*

Proof. According to equation (36) we have

$$\frac{\mathrm{d}\chi}{\mathrm{d}t} = \frac{\sigma_3}{1-\sigma_1^2} = \frac{\sigma_3}{\cos^2\varphi}.$$

It follows from the beginning of the proof of lemma 7.11.1.25 and (171) that $x \sim a(1\mp z)+\epsilon \sim \frac{1}{2}a\cos^2\varphi+\epsilon$. This implies

$$\frac{\sigma_3}{\cos^2\varphi} = -\frac{I_3}{I_1}\widetilde{u}x\,\frac{1}{\cos^2\varphi} \sim -\frac{I_3}{I_1}\widetilde{u}\epsilon\,\frac{1}{\cos^2\varphi} = \frac{\eta_\pm}{I_1}\,\frac{1}{\cos^2\varphi}, \tag{183}$$

as long as $\cos\varphi << \epsilon^{1/2}$, see (91) and the beginning of the proof of lemma 7.10.2.23. This shows that $\dot\chi \sim \frac{\eta_\pm}{I_1}\frac{1}{\cos^2\varphi}$, provided $\cos\varphi << |\eta_\pm|^{1/2}$.

If $\cos\varphi >> |\epsilon|$, then (181) implies that $V_p(\sin\varphi) \sim V_q(\pm 1)$. Therefore

$$\dot\varphi = \pm\big(\frac{2}{M}(E-V(\varphi))\big)^{1/2} \sim \pm\big(\frac{2}{M}(E-V_q(\pm 1))\big)^{1/2} \tag{184}$$

remains bounded away from 0. As long as $\epsilon^{1/2} >> \cos\varphi >> |\epsilon|$, then

$$\chi(t_1)-\chi(t_0) = \int_{t_0}^{t_1}\dot\chi\,\mathrm{d}t = \int_{\varphi(t_0)}^{\varphi(t_1)}\frac{\sigma_3(\varphi)}{\cos^2\varphi}\big(\frac{2}{M}(E-V(\varphi))\big)^{-1/2}\,\mathrm{d}\varphi \tag{185}$$

is of order $\frac{|\epsilon|}{\cos\varphi} << 1$, because, using $\gamma = \cos\varphi$ as the integration variable, the integral becomes $\int_{\gamma_0}^{\gamma_1}\gamma^{-2}\,\mathrm{d}\gamma = \gamma_0^{-1}-\gamma_1^{-1}$. If $1 >> \cos\varphi$ and $|\epsilon| = \mathrm{O}(\cos^2\varphi)$, then $\sigma_3 = -\frac{I_3}{I_1}\widetilde{u}\,x = \mathrm{O}(\cos^2\varphi)$. Therefore $\frac{\mathrm{d}\chi}{\mathrm{d}t} = \frac{\sigma_3}{\cos^2\varphi} = \mathrm{O}(1)$ and the change in $\chi(t)$ is small during the small time interval $[\varphi(t_0),\varphi(t_1)]$ when $\cos\varphi(t)$ remains small.

The statements about $\psi(t)$ follow because (28) implies

$$\frac{\mathrm{d}\psi}{\mathrm{d}t} = \frac{\sigma_3}{1-\sigma_1^2}\,\sigma_1 - \sigma_4 \sim \pm\frac{\sigma_3}{\cos^2\varphi}.$$

Here we have used the facts that $\sigma_1 = \sin\varphi \sim \pm 1$ and σ_4 remains bounded in view of (19). This proves the lemma. □

Proposition 7.11.2.27. *Let* $\Delta\chi = \sqrt{\frac{I_1+\mathrm{mr}^2}{I_1}}\pi$ *and let* $0 < \mu < \nu$ *be positive constants. Consider the solutions of the Newtonian system* (85) *with constants of motion* $p = (\overline{\sigma}_3, \overline{\sigma}_4) \notin \ell_\pm$ *near* $q \in \ell_\pm$ *and energy* $E \in [V_q(\pm 1) + \mu, V_q(\pm 1) + \nu]$. *Then for every* δ *and* μ *positive there are numbers* $0 < \widetilde{\mu} \le \mu$ *and* $\widetilde{\eta}$ *such that if* 1) $0 < |\eta_\pm| \le \widetilde{\eta}$ *and* 2) $[t_0, t_1]$ *is a time interval such that the value* $V_p(\sigma_1(t))$ *grows from* $V_q(\pm 1) + \widetilde{\mu}$ *to* E *and then falls back to* $V_q(\pm 1) + \widetilde{\mu}$ *as* t *traces out* $[t_0, t_1]$, *then*

$$|\chi(t_1) - \chi(t_0) - (\mathrm{sgn}\,\eta_\pm)\,\Delta\chi| \le \delta \tag{186}$$

and

$$|\psi(t_1) - \psi(t_0) - (\mathrm{sgn}\,\eta_\pm)\,\Delta\chi| \le \delta. \tag{187}$$

Here $\mathrm{sgn}\,\eta_\pm$ *is the sign of* $\eta_\pm$.

Proof. We begin by estimating the integral on the right hand side of equation (185), where $t_0 < t_1$ and $\varphi(t_1)$ is the angle which is closest to ± 1. In addition, $V(\varphi(t_1)) = E$, $\cos\varphi(t_1) = \mathrm{O}(|\epsilon|)$, and $\cos\varphi(t_0) >> |\epsilon|$. Let $V_0 = V_q(\pm 1)$. Then

$$\big((E - V(\varphi))\big)^{-1/2} = (E - V_0)^{-1/2}\big(1 - \Big(\frac{V(\varphi) - V_0}{E - V_0}\Big)\big)^{-1/2}. \tag{188}$$

It follows from $V = \mathrm{mgr}\,\widetilde{W}$, (95), and (182) that

$$V'(\varphi) \sim \pm \mathrm{mgr}\, d\epsilon^2 \,\frac{1}{\cos^3\varphi}. \tag{189}$$

This implies that $V'(\varphi)$ has constant sign. Therefore there is an analytic substitution of variables $\varphi = \Phi(w)$ with $w > 0$ such that

$$\frac{V(\varphi) - V_0}{E - V_0} = w^2, \tag{190}$$

where $\varphi(t_1) = \Phi(1)$. From $V = \mathrm{mgr}\,\widetilde{W}$, (95), and (180) it follows that

$$V(\varphi) - V_0 \sim \tfrac{1}{2}\,\mathrm{mgr}\, d\epsilon^2\, \frac{1}{\cos^2\varphi}. \tag{191}$$

Therefore (190) and (191) imply that $w \sim C\frac{|\epsilon|}{\cos\varphi}$, where C is a positive constant. Consequently, $\varphi(t_0) = \Phi(w_0)$, where $0 < w_0 << 1$. If we differentiate (190) with respect to w, we obtain

$$V'(\varphi)\,\Phi'(w) = 2(E - V_0)w = 2\sqrt{(E - V_0)(V(\varphi) - V_0)}, \tag{192}$$

using (190) to obtain the second equality. Therefore using (188) and (192) we get

$$(E - V(\varphi))^{-1/2}\,\Phi'(w) = 2\frac{\sqrt{V(\varphi) - V_0}}{V'(\varphi)\sqrt{1-w^2}}.$$

In view of (191) and (189) we obtain

$$\pm\Big(\frac{2}{M}(E - V(\varphi))\Big)^{-1/2}\Phi'(w) \sim \sqrt{\frac{M}{\text{mgrd}}}\,\frac{\cos^2\varphi}{|\epsilon|}\,\frac{1}{\sqrt{1-w^2}}.$$

Combining the above relation with $\sigma_3(\varphi) = -\frac{I_3}{I_1}\,\widetilde{u}x \sim -\frac{I_3}{I_1}\,\widetilde{u}\epsilon$ and using the fact that $\operatorname{sgn}\epsilon = -\operatorname{sgn}\eta_\pm$, we see that the right hand side of (185) is asymptotically equal to

$$(\operatorname{sgn}\eta_\pm)D\int_{w_0}^{1}\frac{\mathrm{d}w}{\sqrt{1-w^2}} \sim (\operatorname{sgn}\eta_\pm)D\,\frac{\pi}{2},$$

where $D = \sqrt{\frac{M}{\text{mgrd}}}\,\frac{I_3}{I_1}\,\widetilde{u}$. Using $M = I_1 + \text{mr}^2$, $d = \frac{I_3^2}{I_1(I_3+\text{mr}^2)}$, and $\widetilde{u} = \sqrt{\frac{\text{mgr}}{I_3+\text{mr}^2}}$, we obtain $D = \sqrt{\frac{I_1+\text{mr}^2}{I_1}}$.

During the time interval that $V_p(\sigma_1(t))$ decreases from E to being close to $V_q(\pm 1)$, we have the same increase in the angle $\chi(t)$. This follows because due to the elastic reflection at $\pm\pi/2$, the angle $\varphi(t)$ performs its time reversed motion during the time interval that $V_p(\sigma_1(t))$ increases from close to $V_q(\pm 1)$ to E. Note that $\frac{\mathrm{d}\chi(t)}{\mathrm{d}t}$ does not change sign during this reversed motion, see the estimate for $\frac{\mathrm{d}\chi}{\mathrm{d}t}$ in (183). The proof of the statement about the increase of $\chi(t)$ is complete.

Again the statements about $\psi(t)$ follow because (28) implies

$$\frac{\mathrm{d}\psi}{\mathrm{d}t} = \frac{\sigma_3}{1-\sigma_1^2}\,\sigma_1 - \sigma_4 \sim \pm\frac{\sigma_3}{\cos^2\varphi}.$$

Here we have used the facts that $\sigma_1 = \sin\varphi \sim \pm 1$ and σ_4 remains bounded in view of (19). □

During the elastic reflection described by proposition 7.11.2.27, the limiting angles χ and ψ increase by $(\operatorname{sgn}\eta_\pm)\,\Delta\chi$ and $(\operatorname{sgn}\eta_\pm)\,\Delta\psi$, respectively. Note that $\Delta\chi$ is greater than π and depends only on $\frac{\text{mr}^2}{I_1}$ (and *not* on the initial conditions of the solution). For the uniform disk and the hoop we have $\Delta\chi = \sqrt{5}\pi$ and $\Delta\chi = \sqrt{3}\pi$, respectively.

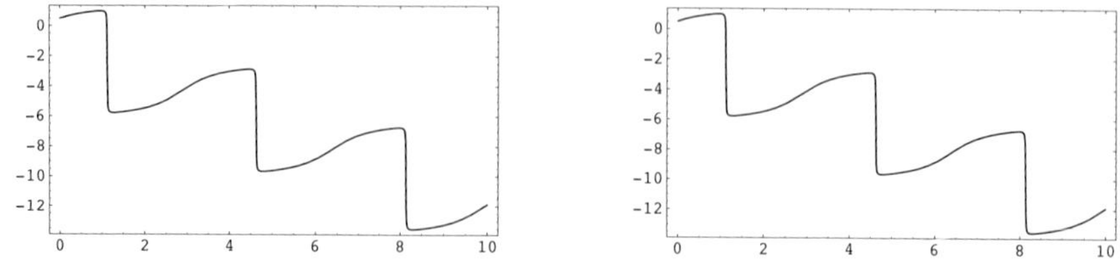

(a) (b)

Fig. 7.3 The angle $\chi(t)$ for several periods of the motion of the uniform disk. In the figure (a) $0 < \eta_+ << 1$; while in figure (b) $-1 << \eta_+ < 0$.

7.11.3 *Motions near falling flat*

In this subsection we describe the *motions* of the disk which are *near* to those where it *falls flat*.

Because the E(2)-reduced energy (7)

$$\overline{E}(u,\omega) = \tfrac{1}{2}\langle I\omega,\omega\rangle + \tfrac{1}{2}\,\mathrm{m}\,\langle \omega\times s, \omega\times s\rangle + \mathrm{mg}\,\langle -s, u\rangle$$

is the sum of positive terms and is conserved, being the total energy E of the disk, it follows that $\langle I\omega,\omega\rangle \le E$. Therefore the angular velocity vector ω and the rotational angular velocity $\nu = A\omega$ are bounded during the short time interval that the rim of the disk is close to the horizontal position. From $\dot{A} = A(\omega(t))^{\times}$ it follows that the rotational motion $A(t)$ of the near flat falling disk, during the time interval from the moment that the disk rises from close to the horizontal position until it falls close to the flat position, converges to the rotational motion A_* of the disk falling flat, as described in §10.3.

For $q \in \ell_\pm$ the solution of the conservative Newtonian system is discontinuous at the instant when the disk has fallen flat. However, nearby solutions for $p \notin \ell_\pm$ but near q, exist for all time – rising up from being near flat after having fallen to being nearly flat, and repeating this periodically for all time. To understand the rotational motion for all time, we look at the rotational angular velocity $\nu(t)$ as $\eta_\pm \to 0$.

Lemma 7.11.3.28. *We make the same hypotheses as in lemma* 7.11.1.25. *In addition, suppose that for* $\eta_\pm = 0$, *the disk falls flat as* $t \uparrow T_*$ *and rises from the flat position as* t *increases from* T_*. *Let* χ *and* ν *be the angle and the vector defined in* (35) *and* (21), *respectively. When* $\eta_\pm \neq 0$ *but* $\eta_\pm \to 0$, *then* $\chi(t) \to \begin{cases} \chi_\downarrow & \text{as } t\uparrow T_* \\ \chi_\uparrow & \text{as } t\downarrow T_* \end{cases}$ *and* $\nu(t) \to \begin{cases} \nu_\downarrow & \text{as } t\uparrow T_* \\ \nu_\uparrow & \text{as } t\downarrow T_* \end{cases}$. *Moreover, these limiting* *values satisfy*

$$\chi_\uparrow = \chi_\downarrow + (\operatorname{sgn}\eta_\pm)\,\Delta\chi \tag{193}$$

$$\nu_\downarrow = (\pm A\sin\chi_\downarrow, \mp A\cos\chi_\downarrow, B) \tag{194}$$

$$\nu_\uparrow = (\mp A\sin\chi_\uparrow, \pm A\cos\chi_\uparrow, B), \tag{195}$$

where

$$A = \sqrt{\frac{2}{I_1 + \mathrm{mr}^2}}\,\big(E - V_q(\pm 1)\big)^{1/2} \text{ and } B = \frac{I_1}{I_3 r(0)}\,\overline{\sigma}_3. \tag{196}$$

Proof. Equation (193) comes from (186).

Consider the situation when the disk is close to being flat, but $|\eta_\pm| << \cos^2\varphi << 1$. Then the disk is *not* in the time interval where the angle χ in proposition 7.11.2.27 increases by order 1, that is, when $|\epsilon| << 1 \mp z << 1$, see (171). Therefore $x(z) \sim a(1 \mp z) = \mathrm{O}(1 - z^2)$ and $y(z) \sim \mp a$. In view of (91) this implies that $\sigma_3(t) = \mathrm{O}(1 - \sigma_1^2)$ and

$$\sigma_4(t) \sim \pm\frac{I_1}{I_3 r(0)}\,\overline{\sigma}_3 = \pm B = (\sigma_4)_*. \tag{197}$$

In other words, $\sigma_4(t)$ is close to its limiting value (175) when the disk is falling flat or rising up from being flat. From $\sigma_1(t) = \sin\varphi(t)$ and $\sigma_2(t) = \frac{\mathrm{d}\sigma_1}{\mathrm{d}t}$ it follows that

$$(1 - \sigma_1^2)^{-1}\sigma_2 = (\cos^{-1}\varphi)\dot{\varphi},$$

which together with $v_1 = \cos\varphi\,\cos\chi$ and $v_2 = \cos\varphi\sin\chi$, see (35), gives

$$(1 - \sigma_1^2)^{-1}\sigma_2\, v_1 = \dot{\varphi}\,\cos\chi \quad \text{and} \quad (1 - \sigma_1^2)^{-1}\sigma_2\, v_2 = \dot{\varphi}\,\sin\chi.$$

Let χ_* be the limit of the angle χ just before or just after the small time interval where the angle increases by order 1. Then from

$$\nu_1 = \big(\sigma_4 - \frac{\sigma_3}{1 - \sigma_1^2}\,\sigma_1\big)\,v_1 + \frac{\sigma_2}{1 - \sigma_1^2}\,v_2$$

and

$$\nu_2 = -\big(\frac{\sigma_2}{1 - \sigma_1^2}\big)\nu_1 + \big(\sigma_4 - \frac{\sigma_3}{1 - \sigma_1^2}\,\sigma_1\big)\,\nu_2$$

see (34), it follows that

$$\nu \sim (\dot{\varphi}\,\sin\chi_*, -\dot{\varphi}\,\cos\chi_*, B). \tag{198}$$

To obtain the third coordinate in (198) we used $v_3 = \sigma_3 + \sigma_1\sigma_4 \sim \pm(\sigma_4)_*$, see (197). Just before the disk falls flat we have $\dot{\varphi} \sim \pm A$, see (172), which changes sign to $\dot{\varphi} \sim \mp A$ just when the disk starts rising again. This proves the lemma. □

As $\eta_\pm \downarrow 0$ the rotational motion $A(t)$ of the disk converges to a rotation A_* as if it were falling flat, that is, when $\eta_\pm = 0$. This motion is continued in time as follows.

Lemma 7.11.3.29. *Every time the disk falls flat, there are two cases. When $\eta_\pm \downarrow 0$, it rises up with a jump in the rotational velocity vector ν from $\nu_\downarrow$ to $\nu_\uparrow$. Here $\nu_\uparrow$ is obtained from $\nu_\downarrow$ by applying a rotation around the vertical axis through an angle $\Delta\chi - \pi$. When $\eta_\pm \uparrow 0$, then $\nu_\uparrow$ is obtained from $\nu_\downarrow$ by applying a rotation around the vertical axis through an angle $-(\Delta\chi - \pi)$.*

Because ω and $(1-\sigma_1^2)^{-1/2} u_{\mathrm{hor}} = (\cos\psi, \sin\psi, 0)$ are bounded, from

$$\dot{a} = \mathrm{r}(1-u_3^2)^{-1/2}\, A(\omega \times u_{\mathrm{hor}})$$

it follows that $\dot{a}$ remains bounded. Therefore the motion of the center of mass $a(t)$ converges to the motion of the center of mass of the disk falling flat as described in §10.3, during the time interval that the disk rises up from being close to the flat position until it falls back close to the flat position. In order to understand how the limiting motion of the center of mass, which exists for all time, is a continuation of the motion of the center of mass of the flat falling disk, we investigate the limiting behavior of $\frac{\mathrm{d}a}{\mathrm{d}t}$.

Lemma 7.11.3.30. *We use the same hypotheses as in lemma 7.11.1.25. For $\eta_\pm \neq 0$ and $\eta_\pm \to 0$ we have $\dot{a} \to \begin{cases} \dot{a}_\downarrow & \text{as } t \uparrow T_* \\ \dot{a}_\uparrow & \text{as } t \downarrow T_*, \end{cases}$ where*

$$\dot{a}_\downarrow = \mathrm{r}(\pm B \,\sin\chi_\downarrow, \mp B\, \cos\chi_\downarrow, A)$$

and

$$\dot{a}_\uparrow = \mathrm{r}(\pm B \,\sin\chi_\uparrow, \mp B\, \cos\chi_\uparrow, A),$$

and A and B are given by (196).

Proof. To estimate the right hand side of the equation

$$\dot{a} = \mathrm{r}\big(\sigma_4 \,\sin\chi - \dot{\varphi}\,\cos\varphi\cos\chi, -\sigma_4\,\cos\chi - \dot{\varphi}\,\cos\varphi\sin\chi, -\dot{\varphi}\,\sin\varphi\big),$$

see (37), we observe that

$$\sigma_4\,\sin\chi - \dot{\varphi}\,\cos\varphi\cos\chi \sim \pm B\,\sin\chi$$

and

$$-\sigma_4\,\cos\chi - \dot{\varphi}\,\cos\varphi\sin\chi \sim \mp B\,\cos\chi.$$

The asymptotic behavior of the component $-\dot{\varphi}\,\sin\varphi$ follows from lemma 7.11.1.25 and (196). □

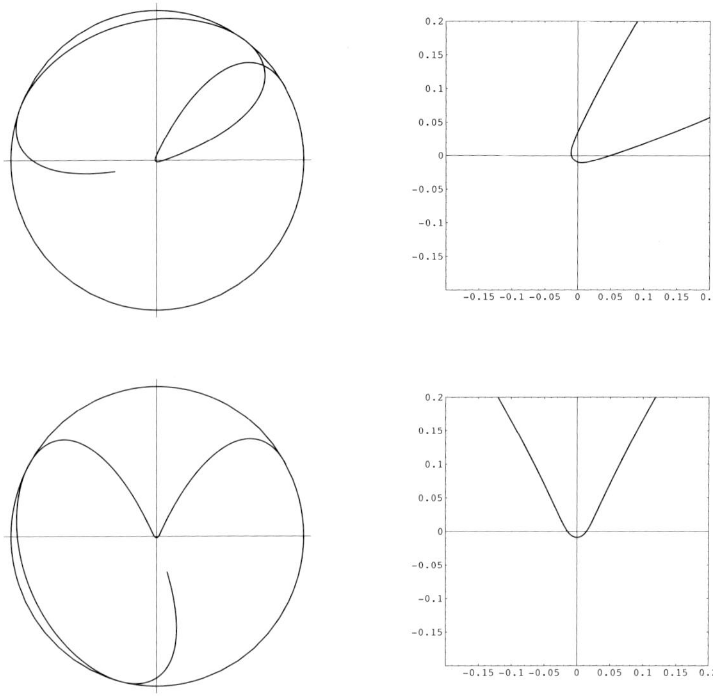

Fig. 7.4 The horizontal projection of $v(t) = A(t)\,e_3$ for the uniform disk, during 1.4 period. In the upper left figure $0 < \eta_+ << 1$; while in the lower left figure $-1 << \eta_+ < 0$. Enlargement of the movement near the origin is given in the figures on the right.

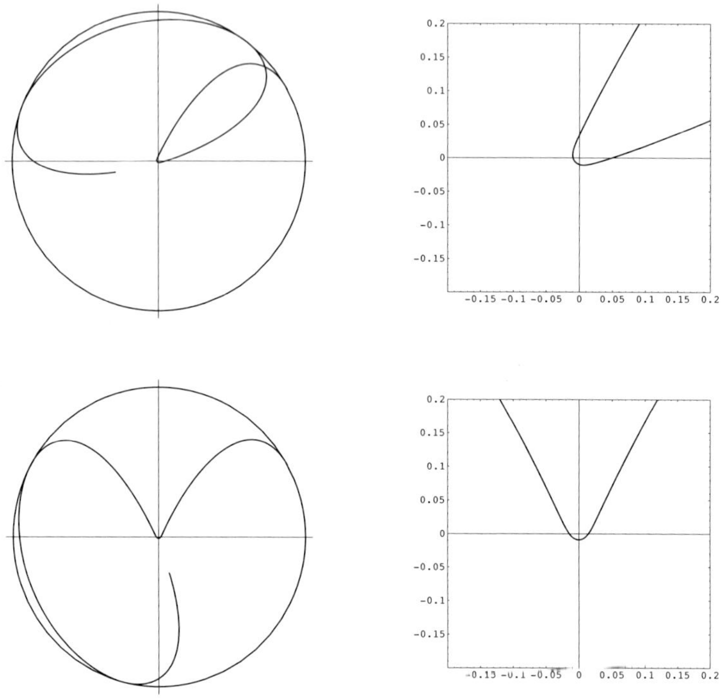

Fig. 7.5 The horizontal projection of $v(t) = A(t)\, e_3$ for the hoop, during 1.4 period. In the upper left figure $0 < \eta_+ << 1$; while in the lower left figure $-1 << \eta_+ < 0$. Enlargement of the movement near the origin are given in the figures on the right.

Lemma 7.11.3.31. *Every time the disk falls flat, it rises up again with the velocity $\dot{a}$ there are two cases for continuing its motion a_* when it is flat, see §10.3. First when $\eta_\pm \downarrow 0$, then the velocity $\dot{a}$ of its center of mass jumps from $\dot{a}_\downarrow$ to $\dot{a}_\uparrow$. Here $\dot{a}_\uparrow$ is obtained from $\dot{a}_\downarrow$ by applying a rotation about the vertical axis through an angle $\Delta\chi$ to $(\dot{a}_\downarrow)_{\mathrm{hor}}$ and changing the sign of $(\dot{a}_\downarrow)_3$. As $\eta_\pm \uparrow 0$ then $(\dot{a}_\uparrow)_{\mathrm{hor}}$ is obtained from $(\dot{a}_\downarrow)_{\mathrm{hor}}$ by applying a rotation about the vertical axis through an angle $\Delta\chi$.*

If a_* denotes the limiting position of the center of mass when the disk is flat, it follows that

$$p(t) = \mathrm{r}\,\frac{\sigma_1}{\sqrt{1-\sigma_1^2}}\,v_{\mathrm{hor}} + a_{\mathrm{hor}},$$

see (38), gives the position of the point of contact of the rolling disk with the horizontal plane. During the time interval when the disk is near the flat position, we have

$$p(t) \sim \pm\mathrm{r}(\cos\chi(t), \sin\chi(t), 0) + a_*,$$

since v is given by (35).

From lemma 7.11.2.26 and proposition 7.11.2.27 it follows that during this time interval the *point of contact races around the rim* of the disk with velocity of order $\frac{1}{\eta_\pm}$. The angle $p(t) - a_*$ increases from the limiting angle $\chi_\downarrow \pm \pi$, after falling down, to the limiting angle $\chi_\uparrow \pm \pi = \chi_\downarrow \pm \pi + (\operatorname{sgn}\eta_\pm)\,\Delta\chi$ before rising up. The point of contact has the limiting position

$$p_\downarrow = \pm\mathrm{r}(\cos\chi_\downarrow, \sin\chi_\downarrow, 0) + a_*,$$

after falling down, and

$$p_\uparrow = \pm\mathrm{r}(\cos\chi_\uparrow, \sin\chi_\uparrow, 0) + a_*,$$

when it is about to rise up. From (194) and (195) it follows that $(\nu_\downarrow)_{\mathrm{hor}}$ is orthogonal to $p_\downarrow - a_*$ and $(\nu_\uparrow)_{\mathrm{hor}}$ is orthogonal to $p_\uparrow - a_*$ with orientation so that $p_\downarrow$ and $p_\uparrow$ are the respective hinge points about which the disk is falling down or rising up, respectively. The vertical part ν_3 of ν, which does not jump, is the spinning part of the limiting motion. The component ν_3 vanishes if and only if $\overline{\sigma}_3 = 0$, see (196). In view of the fact that $\eta_\pm = 0$, the condition $\overline{\sigma}_3 = 0$ implies that $\overline{\sigma}_4 = 0$. Thus we are looking at limiting solutions near the special case of falling flat treated in §5.4. Even in this situation, nearby solutions exhibit the same jumps in the angles $\chi(t)$ and $\psi(t)$ as do all the other near falling flat motions.

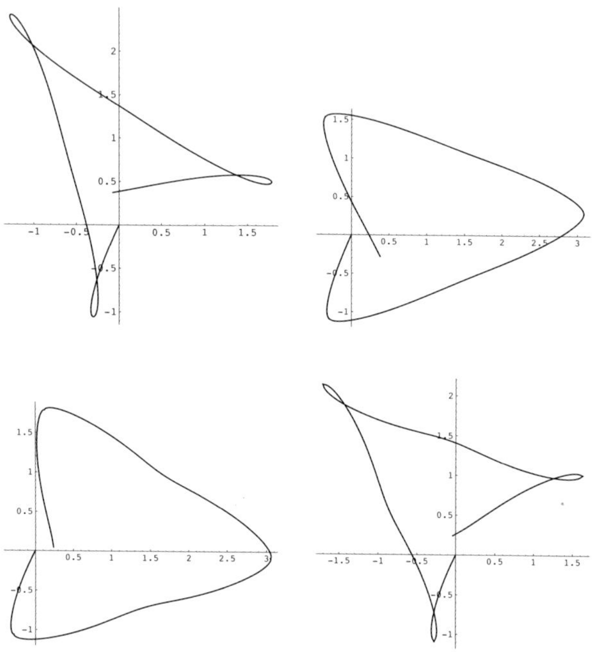

Fig. 7.6 The upper left and right figures give the trajectory of the center of mass of the uniform disk when it nearly falls flat, as in figure 7.4. In the upper left figure $\eta_\pm \downarrow 0$; while in the upper right figure $\eta_\pm \uparrow 0$. The lower left and right figures give the trajectory of the center of mass of the hoop when it nearly falls flat, as in figure 7.5. In the lower left figure $\eta_\pm \downarrow 0$; while in the lower right figure $\eta_\pm \uparrow 0$.

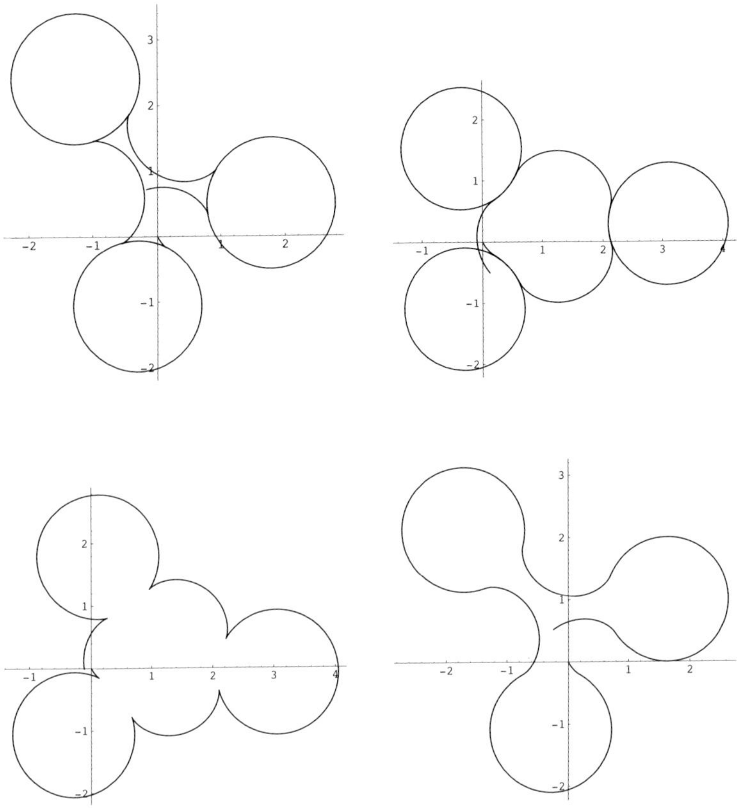

Fig. 7.7 The upper left and upper right figures give the trajectory of the point of contact of the uniform disk as in the lower right and lower left figures of figure 7.4, respectively. The lower left and right figures give the trajectory of the point of contact of the hoop as in the upper right and upper left figures of 7.5, respectively.

7.12 The bifurcation diagram

In this section we study in detail the behavior of the *critical points of the family* $(\overline{\sigma}_3, \overline{\sigma}_4) \mapsto V_{\overline{\sigma}_3,\overline{\sigma}_4}$ *of potential functions* on $(-1,1) \times \mathbb{R}$, see (83).

7.12.1 *The bifurcation set* $\mathcal{B}$

Even though for points $(\overline{\sigma}_3, \overline{\sigma}_4)$ off the degeneracy locus Δ, the potential $V_{\overline{\sigma}_3,\overline{\sigma}_4}$ is a *Morse function*, that is, all of its critical points are nondegenerate, it can happen that as $(\overline{\sigma}_3, \overline{\sigma}_4)$ varies over a connected component of $\mathbb{R}^2 \setminus \Delta$ the *Morse data*[2] of $V_{\overline{\sigma}_3,\overline{\sigma}_4}$ changes.

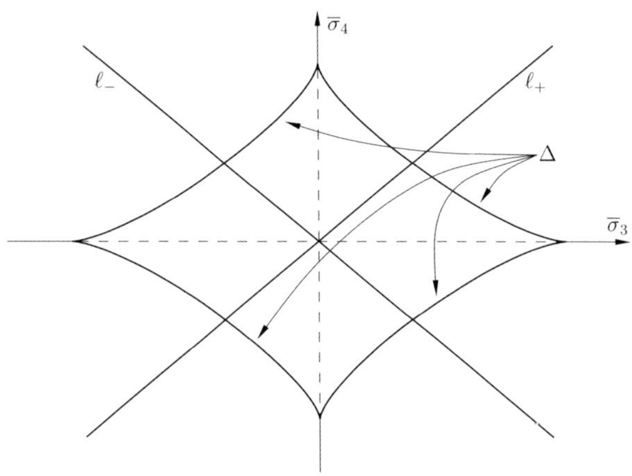

Fig. 7.8 The bifurcation set $\mathcal{B}$ of the family $(\overline{\sigma}_3, \overline{\sigma}_4) \mapsto V_{\overline{\sigma}_3,\overline{\sigma}_4}$ of potentials for the uniform disk.

Proposition 7.12.1.32. *If* $(\overline{\sigma}_3, \overline{\sigma}_4)$ *varies in a connected component of* $\mathbb{R}^2 \setminus (\Delta \cup \ell_+ \cup \ell_-)$*, then the Morse data of* $V_{\overline{\sigma}_3,\overline{\sigma}_4}$ *does not change.*

Proof. From (165) it follows from that as $(\overline{\sigma}_3, \overline{\sigma}_4)$ varies in $\mathbb{R}^2 \setminus (\ell_+ \cup \ell_-)$, no critical point σ_1^0 of $V_{\overline{\sigma}_3,\overline{\sigma}_4}$ can escape to or come in from the boundary of $(-1,1)$. Therefore as $(\overline{\sigma}_3, \overline{\sigma}_4)$ varies in $\mathbb{R}^2 \setminus (\ell_+ \cup \ell_-)$, all critical points

[2]Namely, the number of critical points or their Morse index, that is, the number of negative eigenvalues of the second derivative at a critical point.

of $V_{\overline{\sigma}_3,\overline{\sigma}_4}$ remain in a compact subset of $(-1,1)$. As every critical point is then nondegenerate, it is isolated. Because its set of critical points is compact, it follows that $V_{\overline{\sigma}_3,\overline{\sigma}_4}$ has only finitely many critical points for each $(\overline{\sigma}_3,\overline{\sigma}_4) \in \mathbb{R}^2 \setminus (\Delta \cup \ell_+ \cup \ell_-)$. From the implicit function theorem we see that these critical points vary in an analytic way as the parameters $(\overline{\sigma}_3,\overline{\sigma}_4)$ vary over $\mathbb{R}^2 \setminus (\Delta \cup \ell_+ \cup \ell_-)$. Therefore their number is constant on each connected component of $\mathbb{R}^2 \setminus (\Delta \cup \ell_+ \cup \ell_-)$. Since each of the subspaces ℓ_+ and ℓ_- intersect the degeneracy locus Δ transversely at two diametrically opposite points, the set $\mathbb{R}^2 \setminus (\Delta \cup \ell_+ \cup \ell_-)$ has eight connected components: four in the exterior of Δ and four in the interior.

The *Morse type* of each nondegenerate critical point, indicating whether it is a local minimum or maximum, that is, of *Morse index* 0 or 1, respectively, does not change under smooth perturbations of $V_{\overline{\sigma}_3,\overline{\sigma}_4}$. Therefore on a connected component of $\mathbb{R}^2 \setminus (\Delta \cup \ell_+ \cup \ell_-)$ the Morse data of $V_{\overline{\sigma}_3,\overline{\sigma}_4}$ is constant. □

The conclusion of proposition 7.12.1.32 is the reason we call $\mathcal{B} = \Delta \cup \ell_+ \cup \ell_-$ the *bifurcation set* of the rolling disk in $(\overline{\sigma}_3,\, \overline{\sigma}_4)$–space.

If $p = (\overline{\sigma}_3,\overline{\sigma}_4)$ is a smooth point of the degeneracy locus Δ, where it is permitted that $p \in \ell_+ \cup \ell_-$, then at p the function V_p has a third order degenerate critical point σ_1^0. This critical point disappears if p moves to the exterior of Δ; whereas two nondegenerate critical points appear near σ_1^0 if p moves into the interior of Δ: one of which is a local minimum and the other is a local maximum. Therefore, if we know the number of critical points of V_p and their Morse types for some p in each connected component of $\mathbb{R}^2 \setminus \mathcal{B}$ in the exterior of Δ, then we know these data for every $p \in \mathbb{R}^2 \setminus \mathcal{B}$.

7.12.2 *Off the bifurcation set $\mathcal{B}$*

First we consider the case when $(\overline{\sigma}_3,\overline{\sigma}_4)$ does not lie on $\mathcal{B}$.

Lemma 7.12.2.33. *Let $(\overline{\sigma}_3^*,\overline{\sigma}_4^*)$ be one of the four cusp points of the degeneracy locus Δ, see proposition 7.9.2.19. If either $|\overline{\sigma}_4| \ge |\overline{\sigma}_4^*|$ or $|\overline{\sigma}_3| \ge |\overline{\sigma}_3^*|$, then $V_{0,\overline{\sigma}_4}$ or $V_{\overline{\sigma}_3,0}$, respectively, has $\sigma_1 = 0$ as its only critical point, where each of the functions attains its minimal value.*

Proof. Using the notation of §6 and §7.3, if $X = 0$ and $z \ne 0$, then the solution of the rescaled Chaplygin equations (92) is given by $(x(z), y(z)) =$

$(Y\, o(z), Y\, o'(z))$. Therefore

$$\widetilde{\mathfrak{C}W} = \frac{z}{\sqrt{1-z^2}}\left[dY^2 \frac{1}{(1-z^2)^{3/2}}\, o(z)^2 \right.$$
$$\left. +(c+d)Y^2 \frac{1}{\sqrt{1-z^2}} \frac{o(z)}{z}\, o'(z) - 1 \right], \tag{199}$$

see (148). From §7.3 we see that the normalized odd solution of the rescaled Chaplygin equation satisfies $o(z)/z > 1$ and $o'(z) > 1$. Because $\frac{1}{\sqrt{1-z^2}} > 1$ and the first term in the square brackets in (199) is positive; while the second term is greater than or equal to 1 if $(c+d)\, Y^2 \geq 1$. Therefore the expression between the square brackets in (199) is strictly positive if

$$(c+d)\, Y^2 \geq 1. \tag{200}$$

Since $\overline{\sigma}_4 = \widetilde{u}\, Y$ (98) and $\widetilde{u} = \sqrt{\frac{\mathrm{mgr}}{I_3+\mathrm{mr}^2}}$, equation (200) is equivalent to

$$\overline{\sigma}_4^2 \geq \frac{1}{c+d}\, \frac{\mathrm{mgr}}{I_3+\mathrm{mr}^2} = (\overline{\sigma}_4^*)^2,$$

using (132). Therefore $V_{0,\overline{\sigma}_4}$ has no critical points other than $\sigma_1 = 0$, if $|\overline{\sigma}_4| \geq |\overline{\sigma}_4^*|$. From (164) it follows that this critical point is a global minimum of $V_{0,\overline{\sigma}_4}$.

If $Y = 0$ and $z \neq 0$, then $(x(z), y(z)) = (X\, e(z), X\, e'(z))$ is a solution of the rescaled Chaplygin equations. Therefore

$$\widetilde{\mathfrak{C}W} = \frac{z}{\sqrt{1-z^2}}\left[dX^2 \frac{1}{(1-z^2)^{3/2}}\, e(z)^2 \right.$$
$$\left. +c(c+d)X^2 \frac{1}{\sqrt{1-z^2}}\, e(z)\, \frac{e'(z)}{z} - 1 \right]. \tag{201}$$

But the normalized even solution of the rescaled Chaplygin equations satisfies $e(z) > 1$ and $e'(z)/z > c$, see §7.3. Because $\frac{1}{\sqrt{1-z^2}} > 1$ and $\frac{1}{(1-z^2)^{3/2}} > 1$, the expression in (201) between square brackets is strictly positive if

$$(d+cd+c^2)\, X^2 \geq 1. \tag{202}$$

Since $\overline{\sigma}_3 = -\frac{I_3}{I_1}\widetilde{u}\, X$ (98) and $\widetilde{u} = \sqrt{\frac{\mathrm{mgr}}{I_3+\mathrm{mr}^2}}$, we see that

$$\overline{\sigma}_3^2 = \frac{I_3^2}{I_1^2}\, \frac{\mathrm{mgr}}{I_3+\mathrm{mr}^2}\, X^2 \geq \frac{1}{d+cd+c^2}\, \frac{I_3^2}{I_1^2}\, \frac{\mathrm{mgr}}{I_3+\mathrm{mr}^2} = (\overline{\sigma}_3^*)^2,$$

using (133). Therefore $V_{\overline{\sigma}_3,0}$ has no critical points other than $\sigma_1 = 0$, if $|\overline{\sigma}_3| \geq \sigma_3^*$. Again from (164) it follows that this critical point is the global minimum of $V_{\overline{\sigma}_3,0}$. □

Because each connected component of $\mathbb{R}^2 \setminus \mathcal{B}$ in the exterior of Δ contains points where lemma 7.12.2.33 holds, we conclude that if $(\overline{\sigma}_3, \overline{\sigma}_4)$ lies in the exterior of Δ and does not lie on either of the lines ℓ_+ or ℓ_-, then $V_{\overline{\sigma}_3,\overline{\sigma}_4}$ has precisely one critical point, which is a global minimum. If $(\overline{\sigma}_3, \overline{\sigma}_4) \in \mathbb{R}^2 \setminus (\ell_+ \cup \ell_-)$ is in the interior of Δ, then $V_{\overline{\sigma}_3,\overline{\sigma}_4}$ has three critical points, one of which is a local maximum and the other two are local minima, which are separated by the local maximum. This follows because on $\Delta \setminus (\ell_+ \cup \ell_-)$ less its four cusp points, the family $(\overline{\sigma}_3, \overline{\sigma}_4) \to V_{\overline{\sigma}_3,\overline{\sigma}_4}$ has an A_2-singularity. In other words, $V_{\overline{\sigma}_3,\overline{\sigma}_4}$ has two critical points, one of which is a third order degenerate critical point and the other a nondegenerate global minimum, see proposition 7.8.4.14. If $(\overline{\sigma}_3, \overline{\sigma}_4)$ is a cusp point of Δ, then from proposition 7.8.3.12 and lemma 7.12.2.33 it follows that $V_{\overline{\sigma}_3,\overline{\sigma}_4}$ has only one critical point $\sigma_1 = 0$, which is degenerate of degree four, and is a global minimum of $V_{\overline{\sigma}_3,\overline{\sigma}_4}$.

7.12.3 *On a coordinate axis or in an open quadrant*

In this subsection we consider the case when $(\overline{\sigma}_3, \overline{\sigma}_4)$ lies on either the $\overline{\sigma}_3$-axis or the $\overline{\sigma}_4$-axis or in an open quadrant of the $(\overline{\sigma}_3, \overline{\sigma}_4)$-plane.

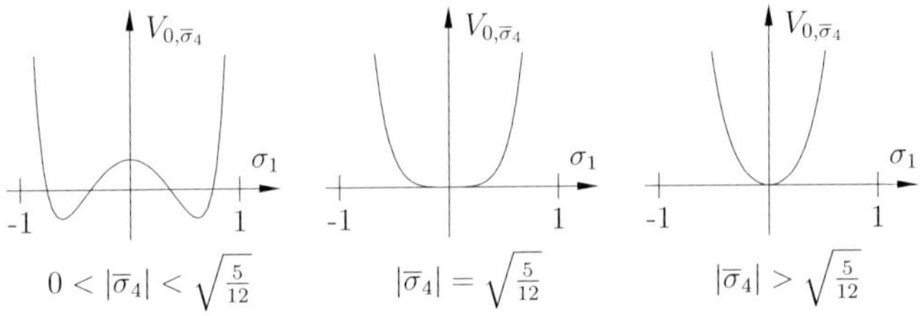

Fig. 7.9 The qualitative behavior of the potential $V_{0,\overline{\sigma}_4}$ along the $\overline{\sigma}_4$-axes for the uniform disk, when it crosses the bifurcation set $\mathcal{B}$ at one of the cusps $(0, \pm\sqrt{\frac{5}{12}})$.

Suppose that either $\overline{\sigma}_3 = 0$ or $\overline{\sigma}_4 = 0$. Then under either of the symmetries $(\sigma_1, \overline{\sigma}_3, \overline{\sigma}_4) \mapsto (-\sigma_1, -\overline{\sigma}_3, \overline{\sigma}_4)$ or $(\sigma_1, \overline{\sigma}_3, \overline{\sigma}_4) \mapsto (-\sigma_1, \overline{\sigma}_3, -\overline{\sigma}_4)$ the potential $V_{\overline{\sigma}_3,\overline{\sigma}_4}$ is invariant, that is, $V_{\overline{\sigma}_3,\overline{\sigma}_4}(\sigma_1) = V_{\overline{\sigma}_3,\overline{\sigma}_4}(-\sigma_1)$. Therefore, if σ_1^0 is a critical point of $V_{\overline{\sigma}_3,\overline{\sigma}_4}$, then so is $-\sigma_1^0$. Moreover, if σ_1^0 is nondegenerate, then the Morse index of $-\sigma_1^0$ is the same as the Morse index of σ_1^0. If $(\overline{\sigma}_3, \overline{\sigma}_4)$ is outside Δ, then $V_{\overline{\sigma}_3,\overline{\sigma}_4}$ has only one critical point, which must be $\sigma_1 = 0$, see lemma 7.12.2.33. If $(\overline{\sigma}_3, \overline{\sigma}_4)$ is inside Δ, then $V_{\overline{\sigma}_3,\overline{\sigma}_4}$ has a unique local maximum, which is equal to $\sigma_1 = 0$, and two local minima σ_1^-, σ_1^+ with $\sigma_1^- < \sigma_1^+$, for which $\sigma_1^- = -\sigma_1^+$. If $(\overline{\sigma}_3, \overline{\sigma}_4) = (0,0)$ then $V_{0,0}(\sigma_1) = \mathrm{mgr}\, \sqrt{1-\sigma_1^2}$, whose graph is the upper semicircle. The situation when $(\overline{\sigma}_3, \overline{\sigma}_4)$ is on one of the coordinate axes is illustrated in figure 7.9.

Conversely, if $\sigma_1^0 = 0$ is a critical point of $V_{\overline{\sigma}_3, \overline{\sigma}_4}$, then from (45) it follows that $\sigma_3\, \sigma_4 = 0$. Because $\sigma_1^0 = 0$ implies that $\sigma_3 = \overline{\sigma}_3$ and $\sigma_4 = \overline{\sigma}_4$ (see (78) and (79)), we obtain $\overline{\sigma}_3\, \overline{\sigma}_4 = 0$. In other words, if $(\overline{\sigma}_3, \overline{\sigma}_4)$ is not on one of the coordinate axes, then every critical point σ_1^0 of $V_{\overline{\sigma}_3, \overline{\sigma}_4}$ is nonzero.

Lemma 7.12.3.34. *If $\overline{\sigma}_3\, \overline{\sigma}_4 > 0$ and $\sigma_1 > 0$, then*

$$V_{\overline{\sigma}_3, \overline{\sigma}_4}(-\sigma_1) > V_{\overline{\sigma}_3, \overline{\sigma}_4}(\sigma_1) \text{ and } V'_{\overline{\sigma}_3, \overline{\sigma}_4}(-\sigma_1) < -V'_{\overline{\sigma}_3, \overline{\sigma}_4}(\sigma_1). \tag{203}$$

The other cases can be obtained by applying either of the symmetries $(\sigma_1, \overline{\sigma}_3, \overline{\sigma}_4) \mapsto (-\sigma_1, -\overline{\sigma}_3, \overline{\sigma}_4)$ or $(\sigma_1, \overline{\sigma}_3, \overline{\sigma}_4) \mapsto (-\sigma_1, \overline{\sigma}_3, -\overline{\sigma}_4)$.

Proof. We insert

$$x(z) = X\, e(z) + Y\, o(z) \text{ and } y(z) = x'(z) = X\, e'(z) + Y\, o'(z),$$

see (110), into equation (95), which defines $W_{X,Y}$. Then we use the fact that

$$\begin{aligned} x(-z) &= X\, e(-z) + Y\, o(-z) = X\, e(z) - Y\, o(z) \\ y(-z) &= X\, e'(-z) + Y\, o'(-z) = -X\, e'(z) + Y\, o'(z), \end{aligned}$$

which implies

$$\begin{aligned} x(-z)^2 - x(z)^2 &= -4XY\, e(z)\, o(z) \\ y(-z)^2 - y(z)^2 &= -4XY\, e'(z)\, o'(z). \end{aligned} \tag{204}$$

In this way we obtain

$$\begin{aligned} &W_{X,Y}(-z) - W_{X,Y}(z) \\ &\qquad = -2XY \left[d\, \frac{1}{1-z^2}\, e(z)\, o(z) + e'(z)\, o'(z) \right]. \end{aligned} \tag{205}$$

Because the functions $e(z)$, $o(z)$, $e'(z)$, and $o'(z)$ are positive when $z > 0$. From the second inequality in (203) it follows that $W_{X,Y}(-z) > W_{X,Y}(z)$, when $z > 0$ and $XY < 0$. In view of the fact that $\overline{\sigma}_3 = -\frac{I_3}{I_1} \widetilde{u}\, X$ and $\overline{\sigma}_4 = \widetilde{u}\, Y$ (98), where $z = \sigma_1$, the first inequality in (203) follows.

If we insert

$$-z\, x(-z)^2 + z\, x(z)^2 = 4XY\, e(z)\, o(z)$$

and

$$x(-z)\,y(-z) + x(z)\,y(z) = 2XY\,(e(z)\,o'(z) + o(z)\,e'(z)),$$

into the expression for $\mathfrak{C}\widetilde{W}(-z) + \mathfrak{C}\widetilde{W}(z)$ we get

$$(\mathfrak{C}\widetilde{W})(-z) + (\mathfrak{C}\widetilde{W})(z) = 2XY\left[2d\,\frac{z}{(1-z^2)^2}\,e(z)o(z) \right.$$
$$\left. + e(z)o'(z) + o(z)e'(z)\right]. \qquad (206)$$

This yields the second inequality in (203). □

Suppose that $(\overline{\sigma}_3,\,\overline{\sigma}_4) \notin \ell_+ \cup \ell_-$ and that $\overline{\sigma}_3\,\overline{\sigma}_4 > 0$. If $V_{\overline{\sigma}_3,\overline{\sigma}_4}$ attains its global minimum at σ_1^0, then $\sigma_1^0 \neq 0$ However, the first inequality in (203) shows that $\sigma_1^0 > 0$.

From proposition 7.8.3.12 it follows that if $(\overline{\sigma}_3,\,\overline{\sigma}_4)$ is near one of the four cusp points of Δ and inside Δ, then $V_{\overline{\sigma}_3,\overline{\sigma}_4}$ has two local minima σ_1^- and σ_1^+ with $\sigma_1^- < 0 < \sigma^+$. Because these critical points do not pass through 0, when $(\overline{\sigma}_3,\,\overline{\sigma}_4)$ varies in an open quadrant of the $(\overline{\sigma}_3,\,\overline{\sigma}_4)$-plane and lies in that part of $\mathbb{R}^2 \setminus \mathcal{B}$ which is inside Δ, it follows that $V_{\overline{\sigma}_3,\overline{\sigma}_4}$ has two local minima σ_1^- and σ_1^+ with $\sigma_1^- < 0 < \sigma^+$. Since there are no other local minima, σ_1^- is a global minimum of $V_{\overline{\sigma}_3,\overline{\sigma}_4}|(-1,0]$ and σ_1^- is a global minimum of $V_{\overline{\sigma}_3,\overline{\sigma}_4}|[0,1)$. Now suppose that $\overline{\sigma}_3\overline{\sigma}_4 > 0$. Because $V_{\overline{\sigma}_3,\overline{\sigma}_4}$ attains its global minimum at $\sigma_1^0 > 0$, it follows that $\sigma_1^0 = \sigma_1^+$. So $V_{\overline{\sigma}_3,\overline{\sigma}_4}(\sigma_1^+) < V_{\overline{\sigma}_3,\overline{\sigma}_4}(\sigma_1^-)$. From (203) we see that

$$0 > V'_{\overline{\sigma}_3,\overline{\sigma}_4}(-\sigma_1^-) + V'_{\overline{\sigma}_3,\overline{\sigma}_4}(\sigma_1^-) = V'_{\overline{\sigma}_3,\overline{\sigma}_4}(-\sigma_1^-),$$

since σ_1^- is a critical point of $V_{\overline{\sigma}_3,\overline{\sigma}_4}$. This in combination with (164) implies that $V_{\overline{\sigma}_3,\overline{\sigma}_4}$ has a local minimum in $(-\sigma_1^-, 1)$. But σ_1^+ is the unique minimum of $V_{\overline{\sigma}_3,\overline{\sigma}_4}|[0,1)$. Therefore $\sigma_1^+ > -\sigma_1^-$, assuming that $\overline{\sigma}_3\overline{\sigma}_4 > 0$.

If $(\overline{\sigma}_3, \overline{\sigma}_4) \notin \ell_+ \cup \ell_-$ is inside Δ, $V_{\overline{\sigma}_3,\overline{\sigma}_4}$ has one more critical point σ_1^*, which is a local maximum. If $(\overline{\sigma}_3, \overline{\sigma}_4)$ approaches a point of Δ in the positive quadrant, then σ_1^* merges with a local minimum to form a third degree degenerate critical point, which is not the global minimum of $V_{\overline{\sigma}_3,\overline{\sigma}_4}$. Therefore σ_1^* merges with σ_1^-, which implies that $\sigma_1^* < 0$. Because the critical points of $V_{\overline{\sigma}_3,\overline{\sigma}_4}$ do not cross 0 as long as $(\overline{\sigma}_3, \overline{\sigma}_4)$ stays in the same open quadrant, we obtain that $\sigma_1^* < 0$ for the local maximum σ_1^* of $V_{\overline{\sigma}_3,\overline{\sigma}_4}$ as long as $\overline{\sigma}_3\,\overline{\sigma}_4 > 0$. □

7.12.4 *Near* $\ell_\pm$

In this subsection we look at the behavior of $V_{\overline{\sigma}_3,\overline{\sigma}_4}$ where $(\overline{\sigma}_3, \overline{\sigma}_4)$ is near one of the lines $\ell_\pm$.

Proposition 7.12.4.35. *If* $p = (\overline{\sigma}_3, \overline{\sigma}_4) \notin \ell_\pm$ *approaches a point* $q = (\overline{\sigma}_3, \overline{\sigma}_4)$ *of* $\ell_\pm$, *then there is a critical point of* $V_{\overline{\sigma}_3,\overline{\sigma}_4}$ *which converges to* ± 1.

Proof. Let $\mu > 0$ be given. From lemma 7.10.2.23 it follows that there exist numbers a and b with $1-\mu \le a < b < 1$ such that $V_q(a) - V_q(b) > 0$. Because the potential $V_{\overline{\sigma}_3,\overline{\sigma}_4}$ depends continuously on the parameters $(\overline{\sigma}_3, \overline{\sigma}_4)$, there is a $\delta > 0$ such that if $|p-q| < \delta$, then $V_p(a) - V_p(b) > 0$. If $p \notin \ell_+$, then from (164) it follows that there is a number $\alpha \in (b, 1)$ such that $V_p(\alpha) > V_p(b)$. Because V_p is continuous on $[a, \alpha]$, it attains its minimum at some point $\sigma \in [a, \alpha]$. So $V_p(\sigma) \le V_p(b)$. Therefore $\sigma \neq a$ and $\sigma \neq \alpha$, which implies that $V_p'(\sigma) = 0$. Thus for every $p \notin \ell_+$ such that $|p - q| < \delta$, the potential V_p has a critical point $\sigma \in (-1, 1)$ such that $|\sigma - 1| < \mu$.

The case that $q \in \ell_-$ is proved similarly. □

If $\overline{\sigma}_3\overline{\sigma}_4 > 0$, then $p = (\overline{\sigma}_3, \overline{\sigma}_4)$ can only approach a point $q \in \ell_+$. In this situation V_p has only one critical point σ_1^0 in $[0, 1)$, at which V_p attains its global minimum. We conclude that $\sigma_1^0 \to 1$ if $p \to q$. Using lemma 7.10.2.23, we know that $V_q(\sigma_1)$ has a continuous extension to $(-1, 0]$ and it attains it global minimum at $\sigma_1 = 1$ with minimal value $V_q(1) = \frac{I_1^2(I_3+\mathrm{mr}^2)}{2I_3^2 r(0)}\overline{\sigma}_3$, see (169).

If σ_1^0 stays away from ± 1, then the only bifurcations which can happen are the creation or annihilation of critical points at degenerate critical points, that is, when $q \in \Delta \cap \ell_+$. In this situation there is a unique point $\sigma_1^0(q)$ at which V_q has a third degree degenerate critical point. For $p \in \Delta \backslash \ell_+$ close to q, the degenerate critical point $\sigma_1^0(p)$ satisfies $\sigma_1^0(p) < 0$. Because $\sigma_1^0(p)$ depends analytically on $p \in \Delta$, when $(\sigma_1^0(p), p) \in \widetilde{\Delta}$, see proposition 7.9.1.15, it follows that $\sigma_1^0(p) < 0$ as well. Therefore, if $q \in \ell_+$ but is outside of Δ, then V_q has no critical points at all and is montonically decreasing from $+\infty$ at $\sigma_1 = -1$ to its global minimum $V_q(1)$ at $\sigma_1 = 1$. If $q \in \ell_+$ is inside of Δ, then V_q has a local minimum at σ_1^- and a local maximum at σ_1^* such that $\sigma_1^- < \sigma_1^* < 0$ and $V_q(\sigma_1^*) > V_q(\sigma_1^-) > V_q(1)$.

7.12.5 Global qualitative description of $V_{\overline{\sigma}_3,\overline{\sigma}_4}$

In this subsection we summarize the results we have obtained about the *family* $(\overline{\sigma}_3, \overline{\sigma}_4) \rightarrow V_{\overline{\sigma}_3,\overline{\sigma}_4}$ of *potential functions*. Figure 7.10 illustrates this family of potentials for the uniform disk.

Proposition 7.12.5.36. *Write* $p = (\overline{\sigma}_3, \overline{\sigma}_4)$.

a) *Suppose that* $p = (0,0)$. *Then* $V_p(\sigma_1) = \mathrm{mgr}\,\sqrt{1-\sigma_1^2}$.

b) *Suppose that* $p \neq (0,0)$ *but* $\overline{\sigma}_3\overline{\sigma}_4 = 0$.

   i) *If* p *is outside* Δ, *then* V_p *has one critical point at* $\sigma_1 = 0$, *which is a nondegenerate global minimum of* V_p.

   ii) *If* p *is inside* Δ, *then* V_p *has three nondegenerate critical points* σ_1^-, 0, *and* σ_1^+ *such that* $\sigma_1^- < 0 < \sigma_1^+$. *Here* σ_1^- *and* σ_1^+ *are local minima and* 0 *is a local maximum. Moreover* $\sigma_1^- = -\sigma_1^+$, $V_p(\sigma_1^-) = V_p(\sigma_1^+)$, *and* V_p *attains its global minimum at both* σ_1^- *and* σ_1^+. *As* $\sigma_1 \rightarrow \pm 1$, *we have* $V_p(\sigma_1) \uparrow \infty$.

   iii) *If* $p \in \Delta$, *then* V_p *has a unique critical point, which is degenerate of fourth degree. Moreover,* $V(\sigma_1) \uparrow \infty$ *as* $\sigma_1 \rightarrow \pm 1$.

c) *Suppose that* $p \notin \ell_+$ *and* $\overline{\sigma}_3\overline{\sigma}_4 > 0$.

   i) *If* p *is outside* Δ, *then* V_p *has only one critical point* σ_1^0, *which is nondegenerate and satisfies* $\sigma_1^0 > 0$. *At* σ_1^0 *the potential* V_p *attains its global minimum value.*

   ii) *If* p *is inside* Δ, *then* V_p *has three nondegenerate critical points,* σ_1^-, σ_1^*, *and* σ_1^+ *such that* $\sigma_1^- < \sigma_1^* < 0 < \sigma_1^+$ *with* $\sigma_1^+ > -\sigma_1^-$ *and* $V_p(\sigma_1^+) < V_p(\sigma_1^-) < V_p(\sigma_1^*)$. *The critical points* σ_1^- *and* σ_1^+ *are local minima of* V_p, *while* σ_1^* *is a local maximum.* V_p *attains its global minimum value at* σ_1^+ *and as* $\sigma_1 \rightarrow \pm 1$, *we have* $V_p(\sigma_1) \rightarrow \infty$.

   iii) *If* $p \in \Delta$, *then* V_p *has two critical points* σ_1^- *and* σ_1^* *with* $\sigma_1^- < \sigma_1^* < 0$ *and* $V_p(\sigma_1^*) > V_p(\sigma_1^-) > V_p(1)$. *The critical point* σ_1^- *is a nondegenerate local minimum and* σ_1^* *is a degenerate critical point of third degree.* $V_p|[\sigma_1^*, 1)$ *monotonically decreases to* $V_p(1)$.

d) *Suppose that* $p \in \ell_+$ *and* $p \neq (0,0)$. *Then* $V_p(\sigma_1) \rightarrow \infty$ *as* $\sigma_1 \rightarrow -1$ *and* $V_p(\sigma_1) \rightarrow V_p(1) = \frac{I_1^2(I_3+\mathrm{mr}^2)}{2I_3^2 r(0)^2}\,\overline{\sigma}_3$ *as* $\sigma_1 \uparrow 1$. *For every* $\sigma_1 \in (-1,1)$ *we have* $V_p(1) < V_p(\sigma_1)$.

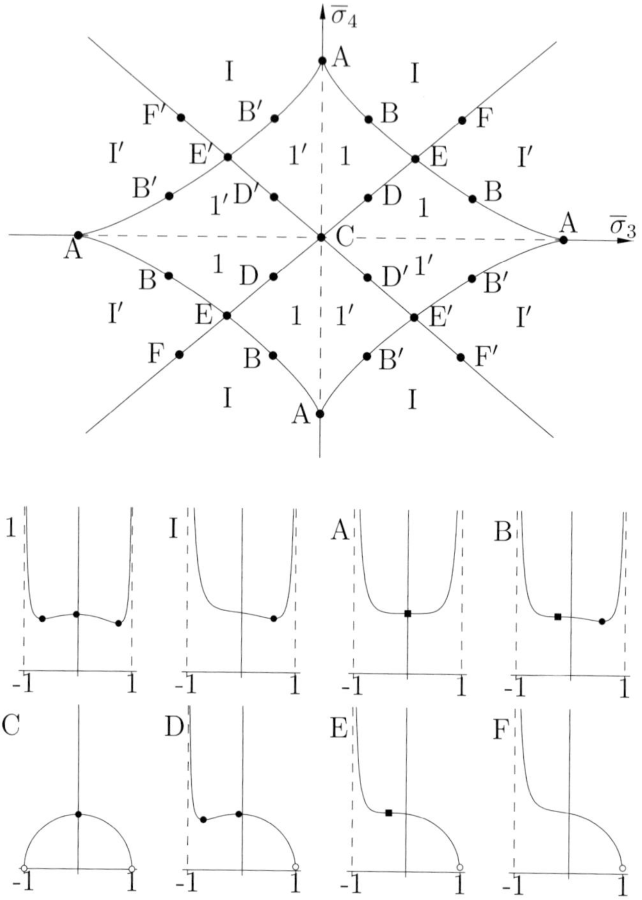

Fig. 7.10 The horizontal axis is σ_1. The vertical axis gives the values of the potential $V_{\overline{\sigma}_3,\overline{\sigma}_4}$, where $(\overline{\sigma}_3, \overline{\sigma}_4)$ is in the numbered region or lettered position. $\circ$ is a limiting value of the potential; $\bullet$ is a nondegenerate critical point; a black box is a degenerate critical point; vertical dotted line is a vertical asymptote. The bifurcation diagram for the family $(\overline{\sigma}_3, \overline{\sigma}_4) \to V_{\overline{\sigma}_3,\overline{\sigma}_4}$ of potential functions. The insets labeled with prime are reflections in the vertical axis of the corresponding unprimed inset.

i) If p is outside Δ, then V_p has no critical points in $(-1, 1)$.

ii) If p is inside Δ, then V_p has two nondegenerate critical points σ^- and σ_1^* such that $\sigma_1^- < \sigma_1^* < 0$ and $V_p(\sigma_1^-) < V_p(\sigma_1^*)$. Moreover, σ_1^- is a local minimum and σ_1^* is a local maximum of V_p.

iii) If $p \in \Delta$, then σ_1^- and σ_1^* merge to form a third degree degenerate critical point $\sigma^0 < 1$. Also V_p decreases monotonically from ∞ to $V_p(1)$ as σ_1 increases from -1 to 1.

7.12.6 *Global description of the orbits of $X_{\overline{\sigma}_3,\overline{\sigma}_4}$*

In this subsection we give a *global qualitative description of the solutions of the conservative Newtonian system* (85) in the $(\varphi, \dot{\varphi})$-plane, which arises from the potential $V(\varphi) = V_{\overline{\sigma}_3,\overline{\sigma}_4}(\sin(\varphi))$ (83). This is the same as giving a description of the orbits in (σ_1, σ_2)-space, where $\frac{\mathrm{d}\sigma_1}{\mathrm{d}t} = \sigma_2$. We use the results we have obtained in §5.3. The results of this discussion for the uniform disk are summarized in figure 7.11 below.

When $p = (\overline{\sigma}_3, \overline{\sigma}_4)$ is in the first quadrant with $\overline{\sigma}_3 > 0$ and $\overline{\sigma}_4 > 0$, is inside Δ, and does not lie on ℓ_+, then the potential V_p has three nondegenerate critical points σ_1^-, σ_1^*, and σ_1^+ such that $\sigma_1^- < \sigma_1^* < 0 < \sigma_1^+$ and $V_p(\sigma_1^*) < V_p(\sigma_1^-) < V_p(\sigma_1^+)$. When the total energy E (81) is greater than $V_p(\sigma_1^*)$, then the level set of the energy is a closed curve on which the integral curve of $X_{\overline{\sigma}_3,\overline{\sigma}_4}$ is periodic, that is, σ_1 oscillates between two values, where the value of the potential V_p is E. If $E = V(\sigma_1^*)$, then because σ_1^* is a nondegenerate critical point of V_p of Morse index 1, it follows that the level set of the energy is a figure eight, consisting of a *hyperbolic equilibrium point* $(\sigma_1^*, 0)$ and its stable and unstable manifolds (which are equal). If $E \in (V_p(\sigma_1^-), V_p(\sigma_1^*))$, then the energy level set consists of two closed curves, on which the motion is *periodic*. More precisely, the set $\{\sigma_1 \in (-1, 1) \big| V_p(\sigma_1) \le E\}$ consists of two intervals I^+ and I^-, where $\sigma_1^{\pm} \in I^{\pm}$. Because $V_p(\sigma_1^+) < V_p(\sigma_1^-)$ we call I^+ the *lowest potential well* and I^- the *highest*. The periodic orbits of $X_{\overline{\sigma}_3,\overline{\sigma}_4}$ correspond to the angle $\varphi(t)$ oscillating in one of the potential wells. When $E \downarrow V(\sigma_1^-)$, the highest potential well shrinks to the point σ_1^- and the corresponding periodic orbit shrinks to the *elliptic equilibrium point* $(\sigma_1^-, 0)$. The periodic orbit in the lowest potential well still exists. When $E \in (V_p(\sigma_1^+), V_p(\sigma_1^-))$ the level set of the energy is one closed curve, where the angle oscillates in the lowest

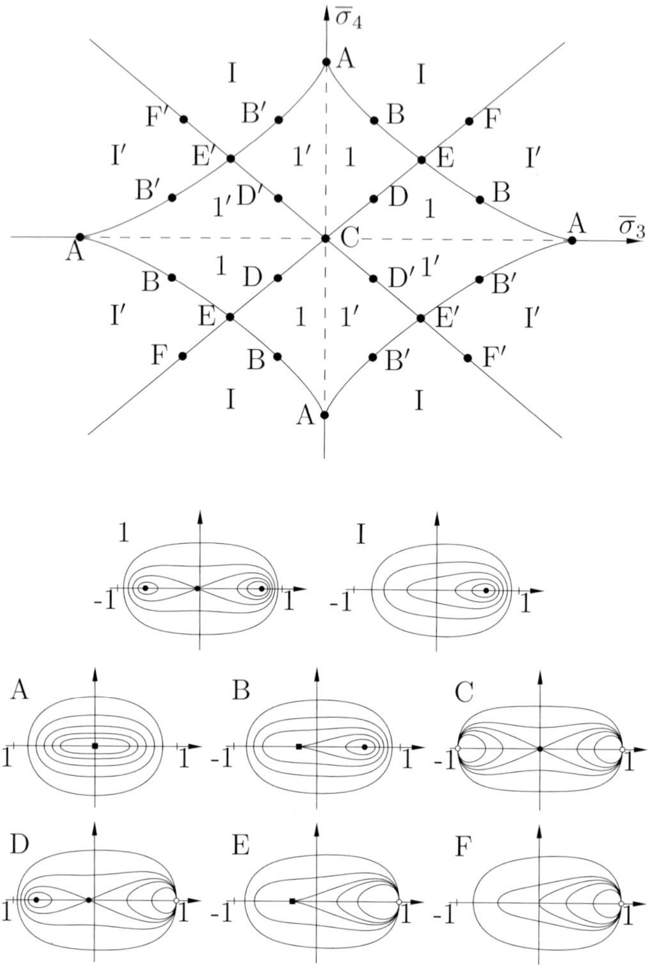

Fig. 7.11 The horizontal axis is σ_1. The vertical axis is σ_2. The curves are the level lines of the Hamiltonian $H_{\overline{\sigma}_3,\overline{\sigma}_4}$, where $(\overline{\sigma}_3, \overline{\sigma}_4)$ is in the numbered region or lettered position. • is a nondegenerate critical point; a black box is a degenerate critical point; ∘ is a limit point of a level set. The level sets of the total energy $H_{\overline{\sigma}_3,\overline{\sigma}_4}$ on the (σ_1, σ_2)-plane. The insets labeled with prime are reflections in the vertical axis of the corresponding unprimed set.

potential well. As $E \downarrow V_p(\sigma_1^+)$, the lowest potential well shrinks to the elliptic equilibrium point $(\sigma_1^+, 0)$. Because V_p attains its global minimum at $\sigma_1 = \sigma_1^+$, the energy level set is empty when $E < V_p(\sigma_1^+)$. When $p = (\overline{\sigma}_3, \overline{\sigma}_4)$ lies on ℓ_+ but is outside Δ, then all solutions rise from the flat position $\sigma_1 = 1$ and fall back to the flat position in finite time. When $p \in \ell_+$ but is inside Δ, then V_p has three nondegenerate critical points σ_1^-, σ_1^*, and σ_1^+, as described above, but σ_1^+ has moved to 1. If $E > V_p(\sigma_1^*)$, then the energy level set is an open interval with end points at $(1, 0)$ in the (σ_1, σ_2)–plane. The solution leaves $(1, 0)$ and returns there in finite time. If $E \in (V_p(\sigma_1^-), V_p(\sigma_1^+))$, then the highest potential well survives; whereas the solution which was in the lowest well now comes from and goes to $(1, 0)$ in finite time.

When $E \in (V_p(1), V_p(\sigma_1^-))$ only the solution rising from the flat position survives. It shrinks to $(1, 0)$ as $E \downarrow V_p(1)$. Because $V_p(1)$ is a global minimum of V_p, the energy level set is empty when $E < V_p(1)$. When $p \in \Delta \cap \ell_+$, then σ_1^* is a degree three degenerate critical point of V_p; whereas 1 is an absolute minimum. If $E > V_p(\sigma_1^*)$ or $E \in \big(V_p(\sigma_1^*), V_p(1)\big)$, then the energy level set is an open interval whose ends are at the point $(1, 0)$. The disk rises from being flat and reverses its motion as it falls flat. If $E = V_p(\sigma_1^*)$, then the energy level set is a closed curve with a cusp at $\sigma_1 = \sigma_1^*$. As $E \downarrow V_p(1)$ the energy level set shrinks to the point $\sigma_1 = 1$; while for $E < V_p(1)$ it is empty, because $V_p(1)$ is the absolute minimum value of V_p.

The other cases follow using the symmetries

$$(\sigma_1, \overline{\sigma}_3, \overline{\sigma}_4) \mapsto (-\sigma_1, -\overline{\sigma}_3, \overline{\sigma}_4) \text{ and } (\sigma_1, \overline{\sigma}_3, \overline{\sigma}_4) \mapsto (-\sigma_1, \overline{\sigma}_3, -\overline{\sigma}_4),$$

reflect the graph of the potential $V_{\overline{\sigma}_3, -\overline{\sigma}_4}$ in the vertical coordinate axis and reflect the orbits in the first quadrant to those in the second or the fourth quadrant, respectively.

7.13 The integral map

In this section we consider $\overline{\sigma}_3$ and $\overline{\sigma}_4$ as functions on the fully reduced phase space $(-1, 1) \times \mathbb{R}^3$. In other words, $(\overline{\sigma}_3, \overline{\sigma}_4)$ are components of the map $\pi_1 \circ \psi^{-1} \circ P$ introduced in §5.1. By construction, the functions $\overline{\sigma}_3$ and $\overline{\sigma}_4$ are integrals, or constants of motion, of the fully reduced vector field $\mathcal{V}$ (13). These integrals are not defined algebraically, but instead are found by solving Chaplygin's equations (76).

If we combine these integrals with the energy integral $\overline{E}$ (19), we obtain the *integral map*

$$\mathcal{I} : (-1,1) \times \mathbb{R}^3 \to \mathbb{R}^3 : (\sigma_1, \sigma_2, \sigma_3, \sigma_4) \mapsto \big(\overline{\sigma}_3,\, \overline{\sigma}_4,\, E_{\overline{\sigma}_3,\overline{\sigma}_4}(\sigma_1,\sigma_2)\big), \quad (207)$$

where

$$E_{\overline{\sigma}_3,\overline{\sigma}_4}(\sigma_1,\sigma_2) = \overline{E}\big(\sigma_1, \sigma_2, \sigma_3(\sigma_1; \overline{\sigma}_3, \overline{\sigma}_4), \sigma_4(\sigma_1; \overline{\sigma}_3, \overline{\sigma}_4)\big).$$

Because the derivative of the map $(\sigma_1,\sigma_2,\sigma_3,\sigma_4) \mapsto (\overline{\sigma}_3,\overline{\sigma}_4)$ has rank two everywhere, (in fact this map is an analytic fibration), it follows that the derivative of the integral map $\mathcal{I}$ has rank less than three if and only if (σ_1,σ_2) is a critical point of $E_{\overline{\sigma}_3,\overline{\sigma}_4}$. This in turn is equivalent to saying that $\sigma_2 = 0$ and σ_1 is a critical point of the potential $V_{\overline{\sigma}_3,\,\overline{\sigma}_4}$ (83).

7.13.1 *Regular values of* $\mathcal{I}$

If $(\overline{\sigma}_3, \overline{\sigma}_4, E)$ is a *regular value* of $\mathcal{I}$, which lies in its image, then the fiber $\mathcal{I}^{-1}(\overline{\sigma}_3,\overline{\sigma}_4,E)$ is a smooth one dimensional submanifold of (σ_1,σ_2)-space $(-1,1)\times\mathbb{R}$. This submanifold is a union of integral curves of the vector field $X_{\overline{\sigma}_3,\overline{\sigma}_4}$ (86). If $(\overline{\sigma}_3,\overline{\sigma}_4) \notin \ell_+ \cup \ell_-$, then the fiber $\mathcal{I}^{-1}(\overline{\sigma}_3,\overline{\sigma}_4,E)$ consists of one or two periodic orbits of $X_{\overline{\sigma}_3,\overline{\sigma}_4}$; whereas if $(\overline{\sigma}_3,\overline{\sigma}_4) \in \ell_+ \cup \ell_-$, then the fiber also has orbits which are open intervals corresponding to motions where the disk falls flat in finite time. Let $\mathcal{R}$ be the set of *regular values of the integral map* $\mathcal{I}$, which lie *in its image*. Let C be a connected component of the complement of the bifurcation set $\mathcal{B} = \Delta \cup \ell_+ \cup \ell_-$. Let π be the projection map $\pi : \mathbb{R}^3 \to \mathbb{R}^2 : (\overline{\sigma}_3,\overline{\sigma}_4,E) \mapsto (\overline{\sigma}_3,\overline{\sigma}_4)$.

The set $\mathcal{R}$ has three connected components. One is

$$\mathcal{R}_2 = \{(\overline{\sigma}_3,\overline{\sigma}_4,E) \big| \, p = (\overline{\sigma}_3,\overline{\sigma}_4) \in C,\ p \text{ inside } \Delta,\ \mathrm{v}^-(p) < E < \mathrm{v}^*(p)\}.$$

We now define what v^- and v^* are. If σ_1^0 is a nondegenerate critical point of V_{p^0} with $p^0 = (\overline{\sigma}_3^0, \overline{\sigma}_4^0)$, then by the implicit function theorem there is a neighborhood U of p^0 in $\mathbb{R}^2$ and a neighborhood W of σ_1^0 in $(-1,1)$ such that the potential V_p has exactly one critical point $\sigma_1(p)$. Let $\mathrm{v}(p) = V_p(\sigma_1(p))$. From proposition 7.12.5.36 b) case ii) for every $p = (\overline{\sigma}_3,\overline{\sigma}_4) \in \mathcal{R}_2$, the potential V_p has three nondegenerate critical points $\sigma_1^-(p) < \sigma_1^*(p) < \sigma_1^+(p)$ with corresponding critical values $\mathrm{v}^-(p) < \mathrm{v}^* < \mathrm{v}^+(p)$. Therefore $\mathcal{R}_2$ is a connected component of the set $\mathcal{R}$ of regular values of the integral map $\mathcal{I}$, which lie in $\pi^{-1}(C)$. The fiber of $\mathcal{I}$ over a point in $\mathcal{R}_2$ consists of two periodic orbits of $X_{\overline{\sigma}_3,\overline{\sigma}_4}$. The other two connected components are

$$\mathcal{R}_1^i = \{(\overline{\sigma}_3,\overline{\sigma}_4,E) \big| \, p = (\overline{\sigma}_3,\overline{\sigma}_4) \in C,\ p \text{ inside } \Delta,\ \mathrm{v}^-(p) < E < \mathrm{v}^+(p)\},$$

and

$$\mathcal{R}_1^o = \{(\overline{\sigma}_3, \overline{\sigma}_4, E) \big| \, p = (\overline{\sigma}_3, \overline{\sigma}_4) \in C, \; p \text{ outside } \Delta, \; \mathrm{v}^0(p) < E\},$$

where $\mathrm{v}^0(p) = V_p(\sigma_1^0(p))$ and $\sigma_1^0(p)$ is the unique global minimum of V_p. The fiber of $\mathcal{I}$ over a point in either $\mathcal{R}_1^i$ or $\mathcal{R}_1^o$ consists of one periodic orbit of $X_{\overline{\sigma}_3, \overline{\sigma}_4}$. This completes the description of the set of regular values of the integral map $\mathcal{I}$.

7.13.2 *The global geometry of the critical value surface*

In this subsection we define the *surface of critical values of the integral map* $\mathcal{I}$, find its singularities, and describe its global geometry.

7.13.2.1 *The critical value surface*

Let Σ be the set of singular values of the integral map $\mathcal{I}$ (207), that is, the image under $\mathcal{I}$ of the set of points in $(-1, 1) \times \mathbb{R}^3$ where the derivative of $\mathcal{I}$ has rank less than three. Then

$$\Sigma = \big\{(\overline{\sigma}_3, \overline{\sigma}_4, \mathrm{v}) \in \mathbb{R}^3 \big| \, \mathrm{v} \text{ is a critical value of } V_{\overline{\sigma}_3, \overline{\sigma}_4}\big\}.$$

Let S be the set of critical points of the family $V_{\overline{\sigma}_3, \overline{\sigma}_4}$, that is,

$$S = \big\{(\sigma_1, \overline{\sigma}_3, \overline{\sigma}_4) \in (-1, 1) \times \mathbb{R}^2 \big| \, V'_{\overline{\sigma}_3, \overline{\sigma}_4}(\sigma_1) = 0\big\}.$$

Then Σ is the image of S under the map

$$\Phi : (-1, 1) \times \mathbb{R}^2 \to \mathbb{R}^3 : (\sigma_1, \overline{\sigma}_3, \overline{\sigma}_4) \mapsto \big(\overline{\sigma}_3, \overline{\sigma}_4, V_{\overline{\sigma}_3, \overline{\sigma}_4}(\sigma_1)\big). \qquad (208)$$

Above we showed that the set S_{nd} of points in $(\sigma_1, \overline{\sigma}_3, \overline{\sigma}_4) \in (-1, 1) \times \mathbb{R}^2$, where σ_1 is a nondegenerate critical point of $V_{\overline{\sigma}_3, \overline{\sigma}_4}$, is a smooth analytic surface in $(-1, 1) \times \mathbb{R}^3$. Moreover, the restriction of the map Φ (208) to S_{nd} is an analytic immersion into $\mathbb{R}^3$ with $\Phi(S_{\mathrm{nd}}) \subseteq \Sigma$. The set of critical points S is the disjoint union of S_{nd} and the set S_{deg} of points $(\sigma_1, \overline{\sigma}_3, \overline{\sigma}_4) \in (-1, 1) \times \mathbb{R}^2$ where σ_1 is a degenerate critical point of $V_{\overline{\sigma}_3, \overline{\sigma}_4}$. Therefore $\Sigma = \Phi(S_{\mathrm{nd}}) \cup \Phi(S_{\mathrm{deg}})$. From proposition 7.9.1.15 it follows that there is a continuous function $\Delta \to (-1, 1) : p \mapsto \sigma_1(p)$ on the degeneracy locus Δ, which is analytic on the smooth part of Δ such that $\{(\sigma_1(p), p) \in (-1, 1) \times \mathbb{R}^2 \big| \, p \in \Delta\} = \Phi(S_{\mathrm{deg}})$. Below in §13.2.3 we show that $\Phi(S_{\mathrm{deg}}) \cap \Phi(S_{\mathrm{nd}}) = \emptyset$. Therefore we call $\Phi(S_{\mathrm{nd}})$ the *nondegenerate part* Σ^{nd} of Σ and $\Phi(S_{\mathrm{deg}})$ the *degenerate part* Σ^{deg} of Σ.

For $p = (\overline{\sigma}_3, \overline{\sigma}_4)$ on the smooth part of the degeneracy locus Δ, every degenerate critical point of V_p is of third degree and can be approximated by two nondegenerate critical points of $V_{p'}$, where p' lies inside of Δ and close to p. The corresponding critical values of $V_{p'}$ and V_p depend continuously on p' and p. Therefore $\Sigma^{\mathrm{deg}} \subseteq \overline{\Sigma^{\mathrm{nd}}}$. Because Σ is the union of the analytically immersed surface Σ^{nd} and the analytic curve Σ^{deg}, we call Σ the *critical value surface* of the family $(\overline{\sigma}_3, \overline{\sigma}_4) \mapsto V_{\overline{\sigma}_3, \overline{\sigma}_4}$ of potential functions.

7.13.2.2 *The singularities of* Σ

In order to describe the critical value surface Σ near the singular points Σ^{nd}, we use the normal form given in §8.3.

We start with the observation that the local diffeomorphism

$$\mathbb{R}^2 \times \mathbb{R} \to \mathbb{R}^3 : (x, \mathrm{v}) \mapsto (\overline{\sigma}_3(x), \overline{\sigma}_4(x), \mathrm{v})$$

maps a local piece of the set of critical values of the normal form family onto the corresponding set of critical values of the family of potential functions $(\overline{\sigma}_3, \overline{\sigma}_4) \mapsto V_{\overline{\sigma}_3, \overline{\sigma}_4}$. The set of critical values of the normal forms is the image under the map $\mathbb{R}^2 \times \mathbb{R} \to \mathbb{R}^2 \times \mathbb{R} : (x, \mathrm{v}) \mapsto (x, \mathrm{v} + h(x))$ of the set of critical values of the normal form (136) for the A_2 and A_3-singularities, namely, $p = 1$ and $p = 2$, respectively.

The set of critical values of the A_2 normal form family $((x_1, x_2), \alpha) \mapsto \alpha^3 + x_1\,\alpha$ is

$$\{(-3\,\alpha^3, x_2, -2\,\alpha^2) \in \mathbb{R}^3 \big| \alpha \in \mathbb{R},\ x_2 \in \mathbb{R}\}.$$

The $x_2 =$ const. slice of the set of critical values is a standard cusp. Therefore the critical value surface of the A_2-family is a surface with a cusped edge along the x_2-axis. Transfering this to the critical value surface Σ, we find that along the smooth part of Σ^{deg}, the critical value surface has a cusped edge with a tangent plane Π which contains the tangent plane of the smooth part of the curve Σ^{deg}. The fact that Dh on the cusp edge is increasing along the axis tangent to the branches of the cusp at the cusp point, see corollary 7.8.3.11, implies that Π tilts upward if we move outward in a radial direction in the $(\overline{\sigma}_3, \overline{\sigma}_4)$-plane. The upper and lower sheets of the smooth part of Σ near Σ^{deg} correspond to nondegenerate local maxima and minima of our family of potential functions. These sheets meet at the cusped edge, which corresponds to the set of third degree degenerate critical points.

Because Σ is singular along Σ^{deg}, we call Σ^{deg} the *singular part* of Σ. Even though Σ^{nd} is an analytically immersed surface in $\mathbb{R}^3$ with self intersections, we will call Σ^{nd} the *regular part* of Σ.

The set of critical values of the A_3-normal form family $((x_1, x_2), \alpha) \mapsto \alpha^4 + x_1\,\alpha + x_2\,\alpha^2$ is

$$\big\{(-4\alpha^3 - 2x_2\,\alpha, x_2, -3\alpha^4 - x_2\,\alpha^2) \in \mathbb{R}^3 \big|\, \alpha \in \mathbb{R},\ x_2 \in \mathbb{R}\big\},$$

which is the *standard swallow tail* in (x_1, x_2, x_3)-space with its axis along the x_2-axis and its tail upward, see figure 7.12. For negative x_2 the x_1-x_3 planar cross sections of the swallow tail has two cusps lying over the degeneracy locus Γ_0 (146) of the normal form, which is a cusp. As x_2 moves toward the origin, the cusped edges merge at the origin, which is called the organizing center of the swallow tail. The upper sheet of the tail, which connects the cusped edges, corresponds to the nondegerate local maxima of the normal form; whereas the remaining smooth part corresponds to the nondegenerate minima. Note that the transverse self-intersection of the two sheets, corresponding to the local minima, lies below the sheet corresponding to the local maxima. With the geometry of the standard

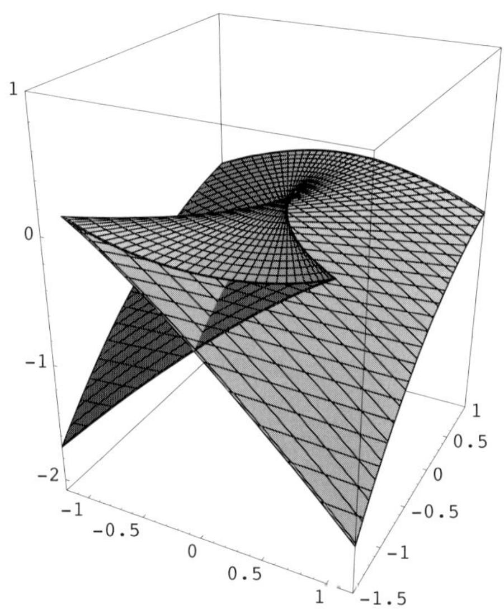

Fig. 7.12 The standard swallow tail.

swallow tail in hand, we look at the corresponding singularities of the critical value surface Σ of our family of potentials. Let $p^0 = (\overline{\sigma}_3^0, \overline{\sigma}_4^0)$ be a cusp point of the degeneracy locus Δ and let $q^0 = (p^0, V_{p^0}(0))$ be the corresponding point of Σ. Using proposition 7.9.2.19 we see that the part Σ^0 of

Σ lying over a neighborhood of p_0 is analytically diffeomorphic to the standard swallow tail surface with q^0 as its organizing center. The axis of Σ^0 lies in the vertical plane over the coordinate axis straddled by the branches of the cusp in Δ with cusp point p^0. The fact that $Dh(0)$ is positive along the positive direction of this axis implies that the swallow tail surface is tilted upward if we move radially outward in the $(\overline{\sigma}_3, \overline{\sigma}_4)$-plane, see proposition 7.9.2.21. The two cusped edges of Σ, which meet at p^0, lie over the branches of the cusp in Δ, which meet at p^0. The cusped edges of Σ, which form part of Σ^{deg} near q^0, meet in a cusped fashion at q^0. Over the interior of Δ near p^0, the surface Σ^0 has three sheets: the upper one, corresponding to the nondegenerate local maxima of the potential functions, and two lower ones, which correspond to the nondegenerate minima. At the edge $\Sigma^{\mathrm{deg}} \setminus \{q^0\}$ near q^0, the upper sheet merges with the highest of the lower sheets. In addition, the lower sheets meet transversely over the coordinate axis straddled by the cusp of Δ at p_0, because the potential functions with parameter values along this axis are even functions.

7.13.2.3 *The global structure of* Σ

Using proposition 7.12.5.36 we now describe the *global structure of the surface of critical values*.

The smooth part $\Sigma^{\mathrm{nd}} = \Phi(S_{\mathrm{nd}})$ of Σ is the disjoint union of the *stable part* Σ^s, consisting of critical values of local nondegenerate minima of members of our family of potential functions, and an *unstable part* Σ^u, consisting of critical values corresponding to nondegenerate maxima. Σ^s is the graph of an analytic function v^*, defined on the region inside of Δ. The closure $\overline{\Sigma^u}$ of Σ^u is equal to $\Sigma^u \cup \Phi(S_{\mathrm{deg}}) = \Sigma^u \cup \Sigma^{\mathrm{deg}}$, which is a compact subset of the interior of the image of the integral map $\mathcal{I}$.[3] On each closed quadrant $\overline{Q}$ in the $(\overline{\sigma}_3, \overline{\sigma}_4)$-plane bounded by coordinate semiaxes, there is an analytic function v^+ defined on the complement of $\ell_\pm$, whose graph is a subset of Σ^s. This subset of Σ corresponds to the global minimum of a member of our family of potentials. The function v^+ can be extended continuously to $p \in \ell_\pm$ by defining $\mathrm{v}^+ = V_p(\pm 1)$. Therefore v^+ is continuous on $\overline{Q}$. The image of the integral map $\mathcal{I}$ is the subset of $\mathbb{R}^3$ which lies on or above the graph of v^+, excluding the points $(p, V_p(\pm 1))$ with $p \in \ell_\pm$. Thus

[3]This leads to the following global form of gyroscopic stabilization. If the energy is larger than that of the restriction of the integral map $\mathcal{I} = (\overline{\sigma}_3, \overline{\sigma}_4, \mathrm{v})$ to $\Sigma^u \cup \Sigma^{\mathrm{deg}}$, then for every $(\sigma_1, \overline{\sigma}_3, \overline{\sigma}_4)$ such that $V'_{\overline{\sigma}_3, \overline{\sigma}_4}(\sigma_1) = 0$ and $V_{\overline{\sigma}_3, \overline{\sigma}_4}(\sigma_1) > E$, the point σ_1 is a nondegenerate local minimum of $V_{\overline{\sigma}_3, \overline{\sigma}_4}(\sigma_1)$.

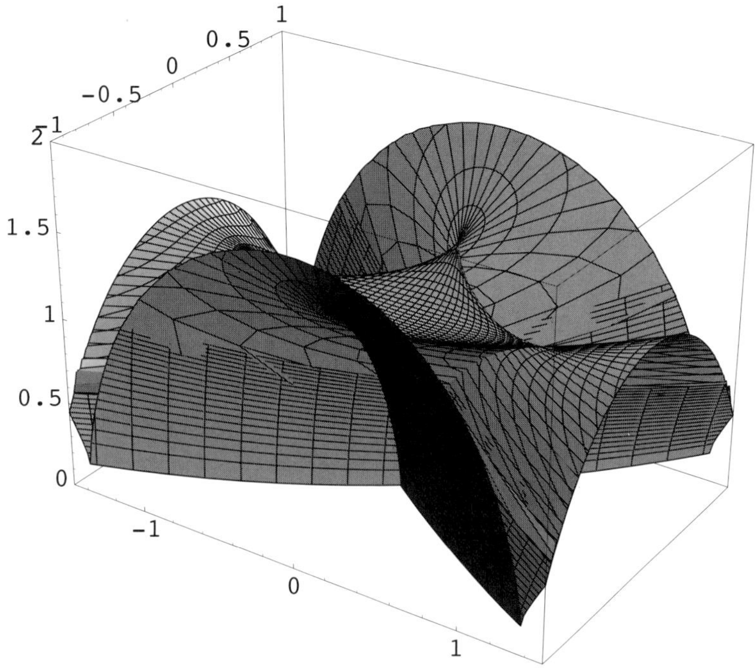

Fig. 7.13 The critical value surface.

it is not a closed set. Over the exterior of Δ, the function v^+ extends to an analytic function whose graph is equal to Σ. The remaining subset of Σ^s is the graph of an analytic function v^-, defined on the inside of Δ less the coordinate axes. For every p in the domain of v^+ we have $\mathrm{v}^+(p) < \mathrm{v}^-(p) < \mathrm{v}^*(p)$, because $\mathrm{v}^+(p) = V_p(\sigma^+(p))$, $\mathrm{v}^-(p) = V_p(\sigma^-(p))$, and $\mathrm{v}^*(p) = V_p(\sigma^*(p))$ when $p = (\overline{\sigma}_3, \overline{\sigma}_4)$ and $\overline{\sigma}_3\overline{\sigma}_4 > 0$, see proposition 7.12.5.36 c) case ii). This description shows that $\Sigma^{\mathrm{deg}} \cap \Sigma^{\mathrm{nd}} = \Phi(S_{\mathrm{deg}}) \cap \Phi(S_{\mathrm{nd}}) = \emptyset$. The function v^- extends to an analytic function on each closed quadrant in the inside of Δ so that on its bounding half-axis we have $\mathrm{v}^- = \mathrm{v}^+$. This corresponds to the self intersections of Σ, which are a continuation of the self intersections of the swallow tail surface near the swallow tail point (= its organizing center). If p is in the open quadrant inside of Δ, then $\mathrm{v}^+(p)$ and $\mathrm{v}^-(p)$ are the values of the two nondegenerate local minima of the potential V_p, which lie symmetrically in different connected components of $\mathbb{R} \setminus \{0\}$. When p moves to an adjacent quadrant by crossing a bounding semiaxis, the nondegenerate local minima persist and do not cross 0, but the respective values of V_p get interchanged. In other words, the function v^+ in quadrant Q_1 analytically continues to v^- in an adjacent quadrant Q_2. Similarly, v^- in Q_1 becomes v^+ in Q_2. The graphs of v^+ and v^* merge at the cusped edge of $\Phi(S_{\mathrm{nd}}) = \Sigma^{\mathrm{nd}}$. At the swallow tail points all the sheets of Σ merge together, see figure 7.13.

7.14 Constant energy slices

7.14.1 *Numerical pictures of the constant energy slices*

All the figures for the *constant energy slice of the critical value surface* have been computed numerically for the uniform disk with $m\,r^2 = m\,g\,r = 1$, and therefore $I_1 = 1/4 = I_2$, $I_3 = 1/2$.

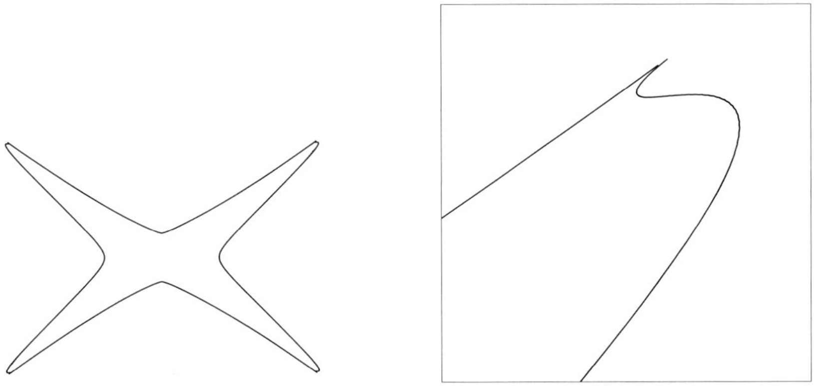

(a) (b)

Fig. 7.14 (a) is the $h = 1.251$ slice of the critical value surface in the $(\overline{\sigma}_3, \overline{\sigma}_4)$-plane. (b) is an enlargement of the region $3.6 < \overline{\sigma}_3 < 3.9$ and $3.6 < \overline{\sigma}_4 < 3.9$.

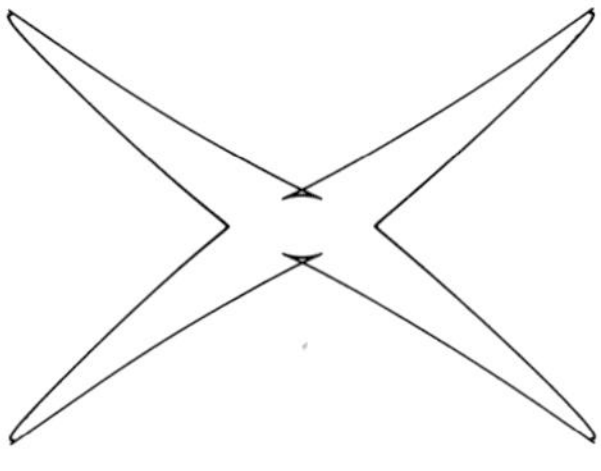

Fig. 7.15 The $h = 1.11$ slice of the critical value surface.

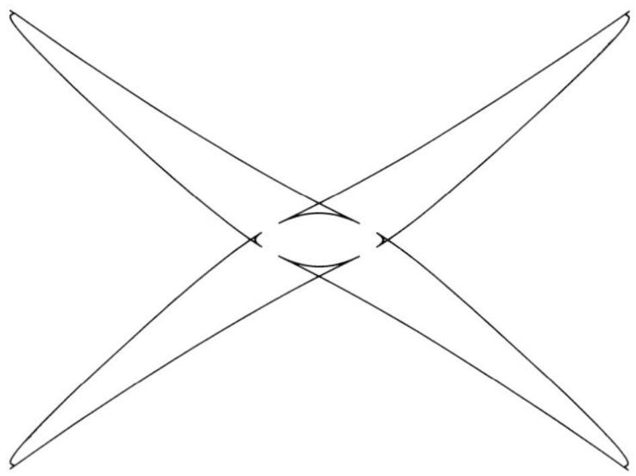

Fig. 7.16 The $h = 1.065$ slice of the critical value surface.

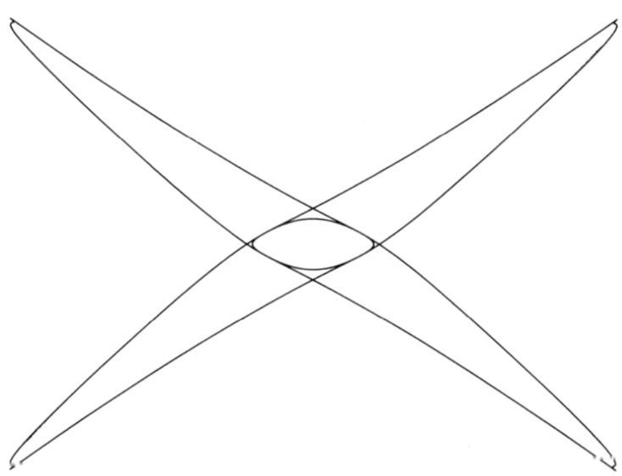

Fig. 7.17 The $h = 1.061$ slice of the critical value surface.

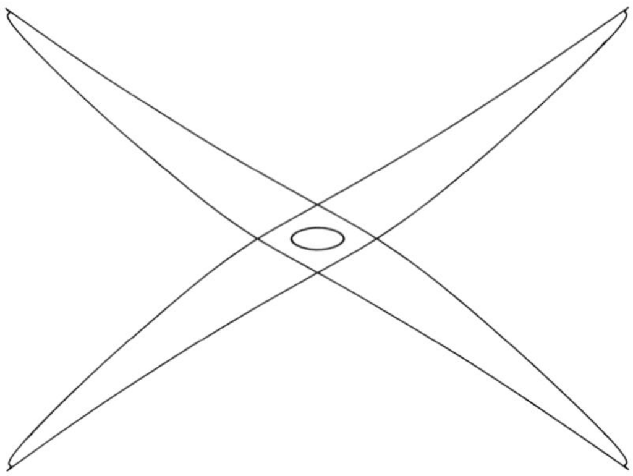

Fig. 7.18 The $h = 1.01$ slice of the critical value surface.

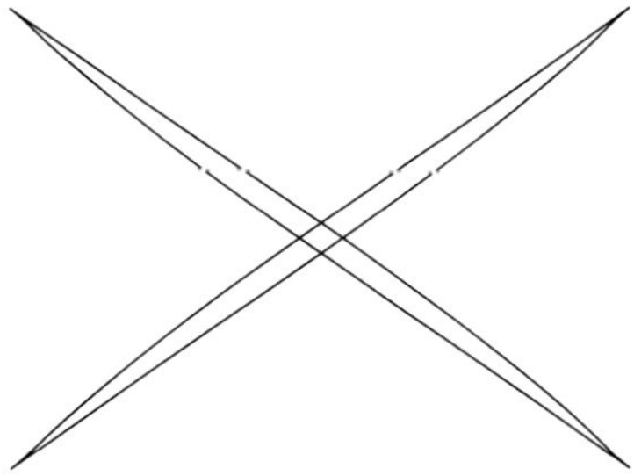

Fig. 7.19 The $h = .05$ slice of the critical value surface.

In figures 7.14–7.19 we have calculated the horizontal slice of the critical value surface at the respective heights 1.251, 1.1, 1.065, 1.061, 1.01, and 0.5.

The scenario is as follows. When the height, corresponding to the energy level, is larger than 1.25 (= the energy at critical straight rolling), then the horizontal slice is a closed curve, which is smooth except at the four points over $\ell_{\pm}$, where it has cusps pointing outward. These cusps are practically invisible in figure 7.14, but the enlargement indicates that the cusp sits like a nail on an otherwise round fingertip, where the round fingertip recedes gradually as the energy decreases. The interior of the closed curve is the domain of $(\overline{\sigma}_3, \overline{\sigma}_4)$-values for which we have one periodic orbit, whereas the exterior does not belong to the image of the integral map. In proposition 7.14.3.37 we show that at each energy level the $(\overline{\sigma}_3, \overline{\sigma}_4)$-values form a bounded set.

If the energy is between 1.1 (= the energy at critical rectilinear rolling) and 1.25, then two opposite swallow tails have formed in the horizontal slice, see figure 7.15. The interiors of the swallow tails is the domain of $(\overline{\sigma}_3, \overline{\sigma}_4)$-values for which we have two periodic orbits.

If the energy is in between approximately 1.061... and 1.1, see figures 7.15 and 7.16, then the horizontal slice has four swallow tails. The interior of each of these is the domain of $(\overline{\sigma}_3, \overline{\sigma}_4)$-values for which we have two periodic orbits. When the energy is approximately equal to 1.061..., see figure 7.17 then the cusps of the swallow tails meet in the beak to beak singularity, which is discussed in §14.5 and the paragraph preceding it.

If the energy is between .5 and 1.061..., see figures 7.17–7.19, then the horizontal slice consists of a crossing of two railway tracks, which when going out meet in cusps over $l_{\pm}$; whereas in the interior of the railway track crossing there is another component which is a smooth oval. The region within the crossing and outside the oval is the domain of $(\overline{\sigma}_3, \overline{\sigma}_4)$-values for which we have two periodic orbits, whereas the other bounded components of the complement of the horizontal slice form the domain of $(\overline{\sigma}_3, \overline{\sigma}_4)$-values for which we have one periodic orbit. Note that in the interior of the oval the energy is larger than the local maximum of the potential function (= the energy at the hyperbolic relative equilibrium). The oval decreases with decreasing energy, see figures 7.17–7.18. It shrinks to the origin when the energy reaches the value 1.

For energy values between 0 and 1, see figure 7.19 we only have the aforementioned railway crossing. The railway tracks become extremely narrow when the energy shrinks to and the figure looks like a cross of very thin needles. Indeed, it follows from lemma 7.14.7.44 below that for small energy the width of each needle is of the order of magnitude of the third power of the length of the needle.

7.14.2 *Geometric features of the constant energy slices*

Here we note the prominent *geometric features of the constant energy slices* of the critical value surface Σ.

If one intersects Σ with the horizontal plane $E_{\overline{\sigma}_3, \overline{\sigma}_4} = h$ and then projects the result to the $(\overline{\sigma}_3, \overline{\sigma}_4)$-plane, one obtains the constant energy slice Σ_h.

We note the following geometric features shown for the unform disk in figures 7.14–7.19 of §14.1.

1. For each h the slice Σ_h is a union of compact analytic curves which expand as h increases.

2. There is a "finger nail" at the ends of the extremes of Σ_h when $1.0.1 < h < 1.251$.

3. There are sections of four swallow tails when $1.01 < h < 1.061$, whose cusps merge to form an oval when $h = 1.061\ldots$.

4. The oval shrinks to a point when $h = 1$ and disappears when $h < 1$.

5. A slice of Σ, which is transverse to one of the lines $\ell_{\pm}$, is curve with a cusp point at the point where $\ell_{\perp}$ intersects the slice. This point is not in the image of the integral map.

In the next few subsections, we prove some of the geometric features noted above.

7.14.3 *Outward radial growth*

In this section we study the critical values of the potential $V_{\overline{\sigma}_3, \overline{\sigma}_4}$ as the parameter $(\overline{\sigma}_3, \overline{\sigma}_4)$ increases in a radial direction.

We prove

Proposition 7.14.3.37. *We have*

$$\inf_{\sigma_1 \in (-1,1)} V_{\overline{\sigma}_3,\overline{\sigma}_4} \to \infty \quad as \ \|(\overline{\sigma}_3,\overline{\sigma}_4)\| \to \infty.$$

Proof. Suppose that the rescaled potential $W_{X,Y}$ (95) stays bounded as $\|(\overline{\sigma}_3,\overline{\sigma}_4)\| \to \infty$. Then the solutions $(x(z), y(z))$ of the rescaled Chaplygin equations (92) stay bounded. Because the Wronskian of the normalized even and odd solutions of the rescaled Chaplygin equations, see §7.3, is

$$e(z)o'(z) - o(z)e'(z) = 1 \cdot 1 - 0 \cdot 0 = 1,$$

we may solve $\begin{pmatrix} x(z) \\ y(z) \end{pmatrix} = \begin{pmatrix} e(z) & e'(z) \\ o(z) & o'(z) \end{pmatrix} \begin{pmatrix} X \\ Y \end{pmatrix}$ to obtain

$$\begin{aligned} X &= x(z)\, o'(z) - y(z)\, o(z) \\ Y &= -x(z)\, e'(z) + y(z)\, e(z). \end{aligned}$$

Since $e(z)$ and $o(z)$ are bounded as a consequence of our hypothesis, $\|(X,Y)\|$ can only become large if either $|x(z)o'(z)|$ or $|x(z)e'(z)|$ becomes large as $z \to \pm 1$. However, we know that $o'(z) \sim -\frac{1}{2}\, cr(0)\, \ln(1 \mp z)$ and $e'(z) \sim \pm\frac{1}{2}\, cr'(0)\, \ln(1 \mp z)$ as $z \to \pm 1$. Consequently, $\frac{o'(z)}{e'z} \sim \frac{\pm r(0)}{r'(0)}$ as $z \to \pm 1$. Therefore both $|x(z)o'(z)|$ and $|x(z)e'(z)|$ become large as $z \to \pm 1$. In particular, X becomes large as $z \to \pm 1$. But then

$$\begin{aligned} \frac{2}{d}\, W_{X,Y}(z) \geq \frac{x(z)^2}{1-z^2} &= \frac{1}{1-z^2}\left(\frac{X + y(z)o(z)}{o'(z)}\right)^2 \sim X^2\, \frac{1}{(1-z^2)(o'(z))^2} \\ &\sim X^2 \left(\frac{2}{cr(0)}\right)^2 \frac{1}{(1-z^2)\ln^2(1-z^2)}, \end{aligned} \tag{209}$$

which tends to ∞ as $z \to \pm 1$. This contradicts the hypothesis that $W_{X,Y}$ is bounded as $\|(X,Y)\| \to \infty$. □

An immediate consequence of the above proposition is that all constant energy slices Σ_h of the critical value surface are compact. This proves half of observation 1.

Lemma 7.14.3.38. *If $\sigma_1(p)$ is a nondegenerate critical point of V_p with $p = (\overline{\sigma}_3,\overline{\sigma}_4) \neq (0,0)$, then the derivative of the function $p \mapsto \sigma_1(p)$ is strictly positive.*

Proof. Because $\sigma_1(p)$ is a nondegenerate critical point of V_p, the function $p \mapsto \sigma_1(p)$ is analytic. Therefore the function $p \mapsto V_p(\sigma_1(p))$ is analytic. Using the Euler operator $\mathfrak{E} = x\,\frac{\partial}{\partial x} + y\,\frac{\partial}{\partial y}$ and

$$\widetilde{W} = \tfrac{1}{2}\, d\, \frac{1}{1-z^2}\, x^2 + \tfrac{1}{2}\, y^2 + \sqrt{1-z^2},$$

see (94), we obtain $\mathfrak{E}\widetilde{W} = d\,\frac{1}{1-z^2}\,x^2 + y^2 > 0$, when $(X, Y) \neq (0, 0)$ (which implies $(x, y) \neq (0, 0)$). Because the rescaled Chaplygin equations (92) are linear, the diffeomorphism ψ (79), giving its flow leaves $\mathfrak{E}$ invariant. This proves the lemma. □

Because every ray from the origin intersects Δ, lemma 7.14.3.38 implies

Corollary 7.14.3.39. *The supremum of the critical values $V_p(\sigma_1(p))$, in which $\sigma_1(p)$ is a local maximum of V_p is equal $V_q(\sigma_1(q))$ for some $q \in \Delta$. Here $\sigma_1(q)$ denotes the degenerate critical point of V_q.*

It follows that the maximum of the critical values on $\Sigma^{\mathrm{unst}} \cup \Sigma^{\mathrm{deg}}$ is attained at the edge Σ^{deg}. The minimum of the critical values on $\Sigma^{\mathrm{unst}} \cup \Sigma^{\mathrm{deg}}$ is attained over the origin, where it is equal to mgr. This leads to the following sharpening of the *gyroscopic stabilization* principle.

Proposition 7.14.3.40. *The energy E at any unstable or degenerate relative relative equilibrium satisfies* $\mathrm{mgr} \leq E \leq \max_{p\in\Delta} V_p(\sigma_1(p))$. *Here $\sigma_1(p)$ is the degenerate critical point of V_p.*

Proof of observation 1. Lemma 7.14.3.38 implies that in the complement of the origin the h-slice Σ_h^{nd} of the analytic surface Σ^{nd} of critical values of nondegenerate critical points of $V_{\overline{\sigma}_3, \overline{\sigma}_4}$ is a collection of immersed analytic curves, which are transverse to every radial direction. Moreover, as h increases, these curves move outward in the $\overline{\sigma}_3 - \overline{\sigma}_4$-plane away from the origin. If h is larger than all the values of $V_p(\sigma_1)$, where $(\sigma_1, p) \in S_{\mathrm{nd}}$, then Σ_h is a simple closed analytic curve which is everywhere transverse to the radial directions and has the degeneracy locus Δ in its interior. This proves observation 1. □

Remark 7.14.3.41. It follows from the equation $\mathfrak{E}\widetilde{W} - d\,(1-z^2)^{-1}\,x^2 + y^2 = 2\widetilde{W} - 2(1-z^2)^{1/2}$ that $2\widetilde{W} - 2 \leq \mathfrak{E}\widetilde{W} < 2\widetilde{W}$.

The second inequality implies that, if $(X,\,Y) \neq (0,\,0)$ and $r \geq 1$, then

$$\inf W_{r\,X,\,r\,Y} \leq W_{r\,X,\,r\,Y}(z) \leq r^2\,W_{X,\,Y}(z) \quad \text{for all} \quad z \in (-1,\,1),$$

and therefore

$$\inf W_{r\,X,\,r\,Y} \leq r^2\,\inf W_{X,\,Y} \quad \text{if} \quad r \geq 1.$$

The first inequality implies that $\mathfrak{E}(\widetilde{W} - 1) \geq (\widetilde{W} - 1)$, and therefore $W_{r\,X,\,r\,Y}(z) - 1 \geq r^2\,(W_{X,\,Y}(z) - 1)$ when $W_{X,\,Y}(z) > 1$ and $r \geq 1$. This implies that

$$\inf W_{r\,X,\,r\,Y} \geq 1 + r^2\,(\inf W_{X,\,Y} - 1) \quad \text{if} \inf W_{X,\,Y} > 1,\; r \geq 1.$$

In combination with proposition 7.14.3.37 this shows that the global minimum of V_p grows of order $|p|^2$ as $|p| \to \infty$.

7.14.4 *The swallow tail sections*

In this subsection we explain why sections of a swallow tail appear in the constant energy slices Σ_h.

Let q_0 be a swallow tail point of the critical value surface Σ. Then q_0 lies over a cusp point p_0 of the degeneracy locus Δ, see §13.2.2. The positivity of the outward radial directional derivative of the function $p \mapsto V_p$, see proposition 7.14.3.37, implies that the swallow tail tilts upward in the radial direction. Therefore the local piece of Σ_h, when h is slightly smaller than $V_{p_0}(0)$, exhibits cross sections of a swallow tail surface with its tail pointing inward and whose self intersections lie along an axis pointing away from the origin. When h is larger than $V_{p_0}(0)$, the swallow tail surface disappears and the local piece of Σ_h becomes smooth. This proves observation 3 except for the merger of the cusps of the swallow tail sections at $h = 1.061\ldots$ for the uniform disk, for which we have no proof.

7.14.5 *Behavior of cusp points*

As h decreases further, the cusp points p in Σ_h move on Δ, away from the cusp point p_0 of Δ. Whereas Σ_h near p lies inside Δ. This situation continues as long as the derivative of the degenerate critical value function $v(p)$ is negative, as p moves on Δ away from p_0. If for $p \in \Delta$, the function $v(p)$ has a nondegenerate local minimum at $p_{\mathrm{m}} \in \Delta$ and h is slightly larger than $v(p_{\mathrm{m}})$, then there are two points $p_1, p_2 \in \Delta$ close to p_{m} such that $v(p_1) = v(p_2) = h$, with p_{m} in between p_1 and p_2 on Δ. Here p_1 and p_2 are cusp points of Σ_h, where the cusps are approximately parallel to Δ, lie in the interior of Δ, and point approximately in the direction from p_1 and p_2 to p_{m}, respectively. When $h = v(p_{\mathrm{m}})$ then the cusps meet beak to beak. When h is slightly smaller than $v(p_{\mathrm{m}})$ then two of the branches of the two cusps in Σ_h closest to Δ have merged to form a smooth curve. The other branches of the two cusps have merged to form another smooth curve. Both curves near p_{m} lie in the interior of Δ, approximately parallel to Δ.

Computer pictures for the uniform disk, see figures 7.14–7.19 in §14.1, suggest that each smooth interval I of Δ curves outward, but we have no proof of this.

Lemma 7.14.5.42. *If I curves outward, then the degenerate critical value function dcv* $: \Delta \to \mathbb{R} : p \mapsto v(p)$ *has only nondegenerate local minima as critical points. This implies that the function dcv has only one critical point p_{m} which is a nondegenerate global minimum.*

Proof. Write $V_p(z)$ for $V(z, p)$, $(\partial_z V)(z, p)$ for $\frac{\partial V_p}{\partial z}(z)$ and $(\partial_p V)(z, p)$ for $\frac{\partial V_p(z)}{\partial p}$. Let $J \subseteq \mathbb{R} \to I \subseteq \Delta : s \mapsto p(s)$ be a parametrization of a smooth interval I on Δ. Write $z(s)$ for the degenerate critical point of $V_{p(s)}$, which depends smoothly on s. In other words, $(\partial_z V)(z(s), p(s)) = 0$ and $(\partial_z^2 V)(z(s), p(s)) = 0$ for every $s \in J$. It follows that

$$\begin{aligned} 0 &= \frac{\mathrm{d}}{\mathrm{d}s}(\partial_z V)(z(s), p(s)) \\ &= (\partial_z^2 V)(z(s), p(s))\, z'(s) + (\partial_p\, \partial_z V)(z(s), p(s))\, p'(s) \\ &= (\partial_p\, \partial_z V)(z(s), p(s))\, p'(s). \end{aligned} \tag{210}$$

Moreover,

$$\begin{aligned} \frac{\mathrm{d}}{\mathrm{d}s} V(z(s), p(s)) &= (\partial_z V)(z(s), p(s))\, z'(s) + (\partial_p V)(z(s), p(s))\, p'(s) \\ &= (\partial_p V)(z(s), p(s))\, p'(s), \end{aligned}$$

and therefore

$$\begin{aligned} \frac{\mathrm{d}^2}{\mathrm{d}^2 s} V(z(s), p(s)) &= (\partial_z\, \partial_p V)(z', p') + (\partial_p^2 V)\,(p', p') + \partial_p V\, p'' \\ &= (\partial_p^2 V)\,(p', p') + \partial_p V\, p'' \end{aligned}$$

in view of (210). Now $V_{\overline{\sigma}_3, \overline{\sigma}_4}(\sigma_1)$ is equal to mgr $(1-{\sigma_1}^2)^{1/2}$ plus a positive definite quadratic form in the parameters $(\overline{\sigma}_3, \overline{\sigma}_4)$, see (83) and therefore $\partial_p^2 V\,(p', p') > 0$. At a critical point of $s \mapsto V(z(s), p(s))$ we have $\partial_p V\, p' = 0$, and because the critical value function has a postive outward radial derivative, we conclude that $\partial_p V p'' \geq 0$ if Δ curves outward. □

If the degenerate critical value function $dcv : \Delta \to \mathbb{R} : p \mapsto v(p)$ has only one critical point p_{m}, which is a nondegenerate global minimum, then we have the beak to beak configuration for Σ_h where $h = v(p_{\mathrm{m}})$, see figure 7.17. This also implies that dcv attains its maximum at the cusp points of Δ.

In view of corollary 7.14.3.39 and remark 7.14.3.41 this leads to the conclusion that if E is the energy at an unstable or degenerate relative equilibrium then

$$1 \leq \frac{E}{\mathrm{mgr}} \leq \max\left\{1 + \frac{I_1}{2I_3},\ 1 + \frac{I_1}{2(I_1 + \mathrm{mr}^2)}\right\}. \tag{211}$$

This can be viewed as a further specification of the global gyroscopic stabilization principle.

7.14.6 *Over the coordinate axes in the $(\overline{\sigma}_3, \overline{\sigma}_4)$-plane*

Regarding the self-intersections of Σ over the coordinate axes, we have

Lemma 7.14.6.43. *If $\sigma_1 \neq 0$ and $\overline{\sigma}_4 \neq 0$, then $\left.\frac{\partial V_{\overline{\sigma}_3,\,\overline{\sigma}_4}(\sigma_1)}{\partial \overline{\sigma}_3}\right|_{\overline{\sigma}_3=0} \neq 0$. If $\sigma_1 \neq 0$ and $\overline{\sigma}_3 \neq 0$, then $\left.\frac{\partial V_{\overline{\sigma}_3,\,\overline{\sigma}_4}(\sigma_1)}{\partial \overline{\sigma}_4}\right|_{\overline{\sigma}_4=0} \neq 0$.*

Proof. Using (95) and (125) we have

$$\left.\frac{\partial W_{X,\,Y}(z)}{\partial X}\right|_{X=0} = Y\,[d\,(1 - z^2)\,o(z)\,e(z) + o'(z)\,e'(z)].$$

The expression between square brackets is an odd function of z, which for $z > 0$ is strictly positive because $e(z)$ and $o(z)$ is an even and odd power series with positive coefficients, respectively, and therefore the same holds for $o'(z)$ and $e'(z)$, respectively.

Similarly we have

$$\left.\frac{\partial W_{X,\,Y}(z)}{\partial X}\right|_{Y=0} = X\,[d\,(1 - z^2)\,e(z)\,o(z) + e'(z)\,o'(z)],$$

which is nonzero when $X \neq 0$ and $z \neq 0$. The statement of the lemma follows in view of the scaling (98). □

From §14 recall the analytic functions v and w on the closed quadrants in the interior of Δ defined by $v(p) = V_p(\sigma_1^+(p))$ and $w(p) = V_p(\sigma_1^-(p))$, where $p = (\overline{\sigma}_3, \overline{\sigma}_4)$ is and $\overline{\sigma}_3\,\overline{\sigma}_4 > 0$, see case c) in proposition 7.12.5.36. Note that $\sigma_1^-(p) < 0 < \sigma_1^+(p)$, even if p is on one of the half-axes bounding the open quadrant, whereas $\sigma_1^*(p) = 0$ when that happens.

Let us consider v and w in the positive quadrant. In view of the fact that both v and w are analytic, it follows from lemma 7.14.6.43 that if p is on the $\overline{\sigma}_4$-axis, then the partial derivative v' of v with respect to $\overline{\sigma}_3$ at $\overline{\sigma}_3 = 0$ is nonzero. Because w is equal to the image of the analytic extension of v to the adjacent quadrant under the reflection about the $\overline{\sigma}_4$-axis, the partial derivative w' of w with respect to $\overline{\sigma}_3$ at $\overline{\sigma}_3 = 0$ satisfies $w' = -v$. Because $v < w$ in the interior of the positive quadrant, it follows that $v' < 0$

and $w' = -v > 0$. This implies that the self-intersection of Σ over the part of the $\overline{\sigma}_4$-axis inside Δ is transversal.

In view of the positivity of the radial derivative, the derivative with respect to $\overline{\sigma}_4$ of $v = w$ on the positive $\overline{\sigma}_4$-axis is positive. It follows that near the $\overline{\sigma}_4$-axis, the level set Σ_h consists of two analytic curves which intersects each other and the positive $\overline{\sigma}_4$-axis transversally. Moreover, the two curves are each other mirror images under the reflection about the $\overline{\sigma}_4$-axis. At the $\overline{\sigma}_3$-axis we have a similar behavior.

If $\overline{\sigma}_3\,\overline{\sigma}_4 = 0$, then $\sigma_1 = 0$ is a critical point of $V_{\overline{\sigma}_3,\,\overline{\sigma}_4}$, a nondegenerate local maximum when $(\overline{\sigma}_3,\,\overline{\sigma}_4)$ is inside Δ and the nondegenerate global minimum when $(\overline{\sigma}_3,\,\overline{\sigma}_4)$ is outside Δ. Because $\sigma_3 = \overline{\sigma}_3$ and $\sigma_4 = \overline{\sigma}_4$ when $\sigma_1 = 0$, we have $V_{\overline{\sigma}_3,\,0}(0) = \frac{1}{2}\,I_1\,\overline{\sigma}_3{}^2 + \mathrm{mgr}$ and $V_{0,\,\overline{\sigma}_4}(0) = \frac{1}{2}\,(I_3 + \mathrm{mr}^2)\,\overline{\sigma}_4{}^2 + \mathrm{mgr}$.

For p inside Δ we saw that V_p had the nondegenerate local maximum at $\sigma_1^*(p)$ and the corresponding critical value $v^*(p) = V_p(\sigma_1^*(p))$. Note that v^* attains its minimum at the origin, because of the positivity of the radial derivatives. Because $\sigma_1^*(p) = 0$ when p is on a coordinate axis, the above identities imply that $\partial_1^2 v^*(0,\,0) = I_1 > 0$ and $\partial_2^2 v^*(0,\,0) = I_3 + \mathrm{mr}^2 > 0$, where ∂_j denotes the partial derivative with respect to the j-th variable. On the other hand it follows from the symmetry $v^*(\overline{\sigma}_3,\,\overline{\sigma}_4) = v^*(-\overline{\sigma}_3,\,\overline{\sigma}_4)$ that $\partial_1 v^*(0,\,\overline{\sigma}_4) = -\partial_1 v^*(0,\,\overline{\sigma}_4)$, hence $\partial_1 v^*(0,\,\overline{\sigma}_4) = 0$. Therefore $(\partial_2\,\partial_1 v^*)(0,\,0) = 0$. We conclude that the Hessian of v^* at the origin is a diagonal matrix with positive entries. Consequently, the minimum of v^* at the origin is nondegenerate.

Because $v^*(0,\,0) = \mathrm{mgr}$, we find for h slightly larger than mgr a connected component of Σ_h which looks like a small ellipse around the origin, which shrinks to the origin as $h \downarrow \mathrm{mgr}$, and disappears when $h < \mathrm{mgr}$. The points p inside and outside the ellipse correspond to cases where the energy E is larger and smaller than the value $V_p(\sigma_1^*(p))$ of the potential energy V_p at the local maximum $\sigma_1(p)$, respectively. The description of the solutions of the conservative Newtonian system in §5.3 tells what this means for the dynamics of the reduced system. This proves observation 4 in §14.2. □

7.14.7 Σ *over* $\ell_\pm$

We discuss the behavior of the critical value surface Σ near the lines $\ell_\pm$, corresponding to the solutions which fall flat.

Lemma 7.14.7.44. *For $p \notin \ell_\pm$ but near $q \in \ell_\pm$ let $\sigma_1(p) = \sin\varphi(p)$ be the critical point of V_p near ± 1 and let $v(p) = V_p(\sigma_1(p))$ be its corresponding critical value. Then*

$$\cos\varphi(p) \sim (I_1\,\mathrm{mgr})^{-1/3}\,|\eta_\pm|^{2/3} \tag{212}$$

and

$$v(p) \sim v(q) + \frac{3}{2}\,(\mathrm{mgr})^{2/3}\,I_1{}^{-1/3}\,|\eta_\pm|^{2/3} \tag{213}$$

as $p = (\overline{\sigma}_3, \overline{\sigma}_4) \to q$. Here the asymptotics is locally uniform in q. Moreover, $v(q)$ is equal to the right hand side of (169) and $\eta_\pm$ is defined in (163).

Proof. If ϵ and $1 \mp z$ are small, then we have $(1-z^2)^{-1}\,x(z) \sim (1/2)(a + \epsilon/(1 \mp z))$, $y(z) \sim \mp\{a + \epsilon\,(c/2)\,\ln(1 \mp z)\}$ and (167) as $z \to \pm 1$, see the beginning of the proof of lemma 7.11.1.25.

If we substitute $\cos\varphi = C\,|\epsilon|^{2/3}$, where C is a positive constant, into (171), it follows that $1 \mp z = \mathcal{O}(|\epsilon|^{4/3}) = o(|\epsilon|)$ and therefore $x(z) \sim \epsilon$ and $y(z) \sim \mp a$. Then (97) leads to

$$\begin{aligned}\mathfrak{C}\widetilde{W} &\sim \pm\, d\, C^{-4}\,|\epsilon|^{-8/3}\,\epsilon^2 \mp (d+c)\,C^{-2}\,|\epsilon|^{-4/3}\,\epsilon\, a \mp C^{-1}\,|\epsilon|^{-2/3} \\ &\sim \pm\,|\epsilon|^{-2/3}\,C^{-4}\,(d - C^3).\end{aligned}$$

Because $d - C^3 > 0$ when $C < d^{1/3}$ and < 0 when $C > d^{1/3}$, it follows that for small $|\epsilon|$ we have $\pm\,\mathfrak{C}\widetilde{W} > 0$ and $\pm\mathfrak{C}\widetilde{W} < 0$ when $\cos\varphi = C\,|\epsilon|^{2/3}$ with $C < d^{1/3}$ and $C > d^{1/3}$, respectively. Using the continuity of $\mathfrak{C}\widetilde{W}$, we obtain the existence of a family of zeros $z = z(p) = \sin\varphi(p)$ of $\mathfrak{C}\widetilde{W}$, which are critical points of V_p, such that

$$\cos\varphi(p) \sim d^{1/3}\,|\epsilon|^{2/3} \quad \text{as} \quad \epsilon \to 0. \tag{214}$$

In view of the relations $d = \frac{I_3^2}{I_1\,(I_3+\mathrm{mr}^2)}$ and $\epsilon = -\eta_\pm \frac{1}{I_3}\sqrt{\frac{I_3+\mathrm{mr}^2}{\mathrm{mgr}}}$, we obtain that $d\,\epsilon^2 = |\eta_\pm|^2/(I_1\,\mathrm{mgr})$. Therefore (214) implies (212).

Inserting (214) and $x \sim \epsilon$, $y \sim \mp a$ into (94), we obtain

$$\widetilde{W} = (d/2)\,d^{-2/3}\,|\epsilon|^{-4/3}\,\epsilon^2 + (1/2)a^2 + d^{1/3}\,|\epsilon|^{2/3} \sim (1/2)\,a^2 + (3/2)\,d^{1/3}\,|\epsilon|^{2/3},$$

which implies (213) in view of the rescalings in (94) and $d\,\epsilon^2 = |\eta_\pm|^2/I_1\,\mathrm{mgr}$. Moreover, we need the observation that when $\eta_\pm = 0$ we have $\epsilon = 0$. Thus $\frac{1}{2}\,\mathrm{mgr}\,a^2 = v(q)$. □

From lemma 7.14.7.44 it follows that, for p near $q \in \ell_\pm$, the global minimum $v(p)$ of V_p behaves asymptotically like

$$v(p) \sim v(q) + \text{const.}\;|p - q|^{2/3},$$

which means that *the lowest part of the critical value surface has a downward pointing cusped edge over* $\ell_\pm$, where the edge itself does not belong to the image of the integral map.

Because $v(q)$, $q \in \ell_\pm$ has a positive outward radial derivative, it follows that the corresponding horizontal slices Σ_h have a cusp at $l_\pm \setminus \{(0,0)\}$, straddling $\ell_\pm$ and pointing away from the origin.

7.15 The spatial rotational shift

7.15.1 *The shift*

The action of the symmetry group $\mathrm{E}(2) \times S^1$ on the constraint manifold $\mathcal{C}$ (3) of the rolling disk is free and proper. Therefore its orbit space $\mathcal{M} = \mathcal{C}/(\mathrm{E}(2) \times S^1)$ is a smooth manifold, called the *fully reduced phase space* and the projection $\pi : \mathcal{C} \to \mathcal{M} : c \mapsto m = \pi(c)$ is an $\mathrm{E}(2) \times S^1$-principal bundle over $\mathcal{M}$. Because most solutions of the *fully reduced vector field* $\overline{\mathcal{V}}$ (13) on $\mathcal{M}$ are periodic, we can apply the theory of §2.2 of chapter 4.

Let $c \in \mathcal{C}$ and let $\gamma_c : \mathbb{R} \to \mathcal{C}$ be an integral curve starting at c of the vector field V (4), which governs the motion of the rolling disk. The curve γ_c is a *relative periodic solution* of V if its projection

$$\overline{\gamma}_m = \pi \circ \gamma_c : \mathbb{R} \to \mathcal{M} : t \mapsto \pi(\gamma_c(t)) = \overline{\gamma}_m(t)$$

is a nonconstant periodic integral curve of the reduced vector field $\overline{\mathcal{V}}$ on $\mathcal{M}$. Let $\tau = \tau(c)$ be the smallest positive period of $\overline{\gamma}_m$ with $m = \pi(c)$.

Let $\mathcal{C}^{\mathrm{per}}$ be the collection of all $c \in \mathcal{C}$ such that $\overline{\gamma}_m$ with $m = \pi(c)$ is a nonconstant periodic integral curve of the reduced vector field $\overline{\mathcal{V}}$. It follows from the description in §5.3 of the solutions of a conservative one degree of freedom Newtonian system that for the rolling disk $\mathcal{C}^{\mathrm{per}}$ is a dense open subset of $\mathcal{C}$ and that the function $\tau : \mathcal{C} \to \mathbb{R} : c \mapsto \tau(c)$ is real analytic.

Because

$$\pi(\gamma_c(\tau)) = \pi(\gamma_c(0)) = \pi(c) = m,$$

the fibers of the map π are the $\mathrm{E}(2) \times S^1$ orbits on $\mathcal{C}$. Since the $\mathrm{E}(2) \times S^1$ action on $\mathcal{C}$ is free, there is a unique element $g = g(c) \in \mathrm{E}(2) \times S^1$ such that

$$\gamma_c(\tau(c)) = g(c) \cdot \gamma_c(0) = g(c) \cdot c,$$

where the $E(2) \times S^1$ action on $\mathcal{C}$ is denoted by $\cdot$. The element $g(c)$ in $E(2) \times S^1$ is called the *shift element* of $E(2) \times S^1$ at c.

The $E(2) \times S^1$ invariance of the vector field V on $\mathcal{C}$ implies that for any $h \in E(2) \times S^1$ the curve $h \cdot \gamma_c : \mathbb{R} \rightarrow \mathcal{C} : t \mapsto h \cdot \gamma_c(t)$ is an integral curve of V. Moreover, $\pi(h \cdot \gamma_c) = \pi(\gamma_c) = \overline{\gamma}_m$. From $h \cdot \gamma_c(0) = h \cdot c$, it follows that

$$h \cdot \gamma_c(t) = \gamma_{h \cdot c}(t), \tag{215}$$

that is, the flow of V commutes with the action of h on $\mathcal{C}$. Applying (215) when $t = \tau = \tau(c)$ we get

$$(h \cdot g(c)) \cdot c = h \cdot (g(c) \cdot c) = g(h \cdot c) \cdot (h \cdot c) = (g(h \cdot c) \cdot h) \cdot c.$$

Therefore $h \cdot g(c) = g(h \cdot c) \cdot h$, because the $E(2) \times S^1$ action on $\mathcal{C}$ is free. In other words, $g(h \cdot c) = h \cdot g(c) \cdot h^{-1}$ is conjugate to $g(c)$ by the element $h \in E(2) \times S^1$.

7.15.2 *Quasiperiodic motion*

Fix $c \in \mathcal{C}^{\mathrm{per}}$ and let T be a 2-dimensional torus subgroup of $E(2) \times S^1$ such that $g = g(c) \in T$. Let $\mathfrak{t}$ be the lie algebra of T and let $\exp : \mathfrak{t} \rightarrow T$ be the exponential mapping. Because exp is a surjective homomorphism of abelian groups, it induces an isomorphism $\mathfrak{t}/ \ker \exp \rightarrow T$. Since T is compact, the additive discrete group $\ker \exp$ has a $\mathbb{Z}$-basis, which is also an $\mathbb{R}$-basis of $\mathfrak{t}$. Our assumption that $g(c) \in T$ implies that there is a $\xi = \xi(c) \in \mathfrak{t}$ such that $g(c) = \exp \xi(c)$. Of course $\xi(c)$ is determined modulo the lattice $\ker \exp$.

Now consider the analytic map

$$\Phi : \mathfrak{t} \times \mathbb{R} \rightarrow \mathcal{C} : (\eta, t) \mapsto \mathrm{e}^{\eta} \cdot \gamma_c(t). \tag{216}$$

We have $\Phi(\eta, t) = \Phi(\zeta, s)$ if and only if

$$\gamma_c(t) = \mathrm{e}^{\zeta - \eta} \cdot \gamma_c(s) = \gamma_{\mathrm{e}^{\zeta-\eta} \cdot c}(s). \tag{217}$$

In view of the fact that the flow $\varphi_t : \mathcal{C} \rightarrow \mathcal{C}$ of $\mathcal{V}$ given by $\varphi_t(c) = \gamma_c(t)$ gives rise to a homomorphism from $\mathbb{R}$ into the group of diffeomorphisms of $\mathcal{C}$, it follows from (217) that $\gamma_c(t - s) = \mathrm{e}^{\zeta - \eta} \cdot c$. Thus

$$\overline{\gamma}_{\overline{c}}(t - s) = \pi(\gamma_c(t - s)) = \pi(\mathrm{e}^{\zeta - \eta} \cdot c) = \pi(c) = m.$$

So there is a unique $k \in \mathbb{Z}$ such that $t - s = k\tau(c)$. Because $g(g(c) \cdot c) = g(c) \cdot g(c) \cdot g(c)^{-1} = g(c)$, by induction we find that $\gamma_c(k\tau) = (g(c))^k \cdot c$ for every $k \in \mathbb{Z}$. Using the fact that the action of $E(2) \times S^1$ on $\mathcal{C}$ is free, we obtain $\mathrm{e}^{\zeta - \eta} = g(c)^k = \mathrm{e}^{k\xi}$, that is, $\zeta - \eta = k\xi \bmod (\ker \exp)$. Therefore

$\Phi(\eta, t) = \Phi(\zeta, s)$ if and only if there is a $k \in \mathbb{Z}$ and a $\sigma \in \ker \exp$ such that $t - s = k\tau(c)$ and $\zeta - \eta = k\xi + \sigma$. In other words, $\Phi(\eta, t) = \Phi(\zeta, s)$ if and only if $(\zeta - \eta, t - s) \in \Lambda = \Psi(\ker \exp \times \mathbb{Z})$, where

$$\Psi : \mathfrak{t} \times \mathbb{R} \to \mathfrak{e} \times \mathbb{R} : (\sigma, k) \mapsto (k\xi + \sigma, k\tau). \qquad (218)$$

Now $\Psi|(\ker \exp \times \mathbb{Z}) : \ker \exp \times \mathbb{Z} \to \Lambda$ is a surjective homomorphism of additive groups. Because Ψ is a bijective affine transformation of $\mathfrak{t} \times \mathbb{R}$ into itself, it maps any $\mathbb{Z}$-basis of $\ker \exp \times \mathbb{Z}$ onto a $\mathbb{Z}$ basis of Λ, which is an $\mathbb{R}$ basis of $\mathfrak{t} \times \mathbb{R}$. Therefore Λ is a discrete subgroup of $\mathfrak{t} \times \mathbb{R}$ of full rank. This implies that $T' = (\mathfrak{t} \times \mathbb{R})/\Lambda$ is a 3-dimensional torus, where $\dim T = 2$. Choosing a $\mathbb{Z}$-basis $\mathbf{b}$ of $\mathfrak{t} \times \mathbb{R}$ gives rise to a $\mathbb{Z}$-basis $\Psi(\mathbf{b})$ of Λ. This identifies T' with the standard 3-dimensional torus $\mathbb{R}^3/\mathbb{Z}^3$.

Therefore the analytic map $\Phi : \mathfrak{t} \times \mathbb{R} \to \mathcal{C}$ (216) induces an analytic map $\widehat{\Phi} : T' = (\mathfrak{t} \times \mathbb{R})/\Lambda \to \mathcal{C}$, which is an injective immersion. Because T' is compact, the map $\widehat{\Phi}$ is an embedding. Therefore $\Phi(\mathfrak{t} \times \mathbb{R}) = \widehat{\Phi}(T')$ is a compact analytic submanifold of $\mathcal{C}$, which is analytically diffeomorphic to the 3-dimensional torus $\mathbb{R}^3/\mathbb{Z}^3$. From (216) it follows that

$$\Phi(\xi, t + s) = \gamma_{e^{\xi}\cdot c}(t + s) = \gamma_{\gamma_{e^{\xi}\cdot c}(t)}(s) = \gamma_{\Phi(\xi,t)}(s).$$

Consequently, the manifold $\widehat{\Phi}(T') \subseteq \mathcal{C}$ is invariant under the flow of V on $\mathcal{C}$. Moreover, Φ intertwines the straight line flow on T', defined by the *constant* vector field $(0, 1) \in \mathfrak{t} \times \mathbb{R}$, with the flow of V on $\widehat{\Phi}(T')$. In other words, the restriction of the flow of V to $\widehat{\Phi}(T')$ is *quasiperiodic* and Φ is the change of variables which exhibits an integral curve of $V|\widehat{\Phi}(T')$ as a straight line on the 3-dimensional torus T'.[4]

7.15.3 *The spatial rotational shift*

Recall that we have denoted an element of $\mathrm{E}(2) \times S^1$ by $((B, b), C)$, where $B, C \in S^1 = \mathrm{SO}(2, \mathbb{R})$, viewed as those elements of $\mathrm{SO}(3, \mathbb{R})$ which leave the vertical vector e_3 in $\mathbb{R}^3$ fixed, $b \in \mathbb{R}^2 = \mathbb{R}^2 \times \{0\}$, and (B, b) is the affine transformation of $\mathbb{R}^2$ into itself given by $x \mapsto Bx + b$. We call B the *spatial part* of the element $((B, b), C) \in \mathrm{E}(2) \times S^1$. Note that the mapping

$$\mu : \mathrm{E}(2) \times S^1 \to \mathrm{SO}(2, \mathbb{R}) : ((B, b), C) \mapsto B$$

[4]In general a smooth function $f : \mathbb{R} \to \mathbb{R}$ is *quasiperiodic with k frequencies*, if there is a smooth function $F : \mathbb{R}^k/\mathbb{Z}^k \to \mathbb{R}$ and constants $\nu_1, \ldots, \nu_k$, called the frequencies, such that $f(t) = F(t\nu_1, \ldots, t\nu_k)$ for all $t \in \mathbb{R}$.

is a group homomorphism. If $((B, b), C)$ is the shift element at $c \in \mathcal{C}$ of a relative periodic solution, then B is the spatial part of the shift element, which is called the *spatial rotational shift* of the relative periodic orbit.

Because μ is a group homomorphism and shift elements on an $\mathrm{E}(2) \times S^1$-orbit on $\mathcal{C}$ are conjugate, the rotational shift is constant on the $\mathrm{E}(2) \times S^1$-orbit, since $\mathrm{SO}(2, \mathbb{R})$ is commutative. Therefore the spatial rotational shift defines a real analytic $\mathrm{SO}(2, \mathbb{R})$-valued map $\mathcal{B} : \mathcal{C}^{\,\mathrm{per}} \rightarrow \mathrm{SO}(2, \mathbb{R})$ on the space $\mathcal{C}^{\,\mathrm{per}}$ of relative periodic orbits of the vector field V (4) on the constraint manifold $\mathcal{C}$.

Suppose that we are in the following situation. Let

$$\gamma_c : \mathbb{R} \rightarrow \mathcal{C} : t \mapsto \big((A(t), a(t)), (\omega(t), \dot{a}(t))\big) = \gamma_c(t),$$

be an integral curve starting at $c \in \mathcal{C}$ of the vector field V, which governs the motion of the rolling disk. Suppose that the projection of γ_c under the reduction map π gives an integral curve

$$\overline{\gamma}_m : \mathbb{R} \rightarrow \mathcal{M} : t \mapsto \overline{\gamma}_m$$

of the fully reduced vector field $\mathcal{V}$ on the fully reduced phase space $\mathcal{M}$, which is periodic of period $\tau = \tau(c)$. We now prove

Lemma 7.15.3.45. *Suppose that the spatial rotation part of $\gamma_c(\tau)$ is equal to* id, *that is, $A(\tau) = A(0)C^{-1}$, $a(\tau) = a(0) + b$, $\omega(\tau) = C\omega(0)$, and $\dot{a}(\tau) = \dot{a}(0)$ for some $b \in \mathbb{R}^2$ and $C \in \mathrm{SO}(2, \mathbb{R})$. Then*

1. *The map*

$$\mathbb{R} \rightarrow \mathbb{R}^3 : t \mapsto a(t) - t\tau^{-1}b$$

is τ-periodic. In other words, if $b = 0$, then the motion of the center of mass of the disk is periodic with period τ. If $\dot{a}(\tau) \neq 0$, then $a(t)$ moves rectilinearly with constant velocity plus a τ-periodic perturbation.
2. *The rotational motion $t \mapsto A(t)$ of the disk is quasiperiodic on a 2-dimesional torus, while the vectors $v(t) = A(t)e_3$ and $\nu(t)$, see (25), perform a periodic motion.*

Proof. Recall that the map

$$\begin{aligned}(\mathrm{E}(2) \times S^1) \times \mathcal{C} \rightarrow \mathcal{C} : (g, c) &= \big(((B, b), C), ((A, a), (\omega, \dot{a}))\big) \\ &\mapsto \big((BAC^{-1}, Ba + b), (C\omega, B\dot{a})\big)\end{aligned}$$

defines the action of $\mathrm{E}(2) \times S^1$ on the constraint manifold $\mathcal{C}$. If $g = g(c)$ is the shift element at c, then after time τ the vector $u = A^{-1}e_3$ has become $(BAC^{-1})^{-1}e_3 = -CA^{-1}B^{-1}e_3$ and the vector $A(\omega \times e_3)$ has become

$$\begin{aligned}(BAC^{-1})(C\omega \times (CA^{-1}B^{-1}e_3)_{\mathrm{hor}}) &= BAC^{-1}(C\omega \times C(A^{-1}B^{-1}e_3)_{\mathrm{hor}}) \\ &= BA(\omega \times (A^{-1}B^{-1}e_3)_{\mathrm{hor}}) . \end{aligned} \tag{219}$$

If $B = \mathrm{id}$, then the right hand side of (219) is $A(\omega \times (A^{-1}e_3))_{\mathrm{hor}} = A(\omega \times u_{\mathrm{hor}})$. Because σ_1 is periodic with period τ on the fully reduced space, it follows that the right hand side of equation (33) for $\dot{a}$ is periodic of period τ. Since the time integral of a periodic function is a linear function of t plus a τ-periodic function, we have proved point 1 of the lemma with b equal to the integral of the right hand side of (33) over $t \in [0, \tau]$.

Point 2 follows if we observe that disregarding the translational part of the motion of the disk yields a symmetry group $\mathrm{SO}(2,\mathbb{R}) \times S^1$, where the first component of the shift element is id. □

From results of §15.4 below we see that the motion of the disk is quasiperiodic on a 2-dimensional torus. If we disregard the internal S^1 symmetry of the disk, then the reduced phase space $\mathcal{C}/S^1$ is parametrized by v and ν. We obtain τ-periodic motion because the disk's motion has returned to the same point in the fiber.

Let $p \in \mathbb{R}^2$. If $B \in \mathrm{SO}(2,\mathbb{R})$, then let $B_p : x \mapsto B(x-p)+p$ be the *rotation about the point* p. The mapping $B \mapsto B_p$ is an injective homomorphism, whose image is the set of all rotations about p, denoted $\mathrm{SO}(2,\mathbb{R})^p$. It is straightforward to see that $\mathrm{SO}(2,\mathbb{R})^p$ is a circle subgroup of $\mathrm{E}(2)$, which depends analytically on p.

Lemma 7.15.3.46. *Let* $g = ((B,b),C) \in \mathrm{E}(2) \times S^1$. *Suppose that the spatial rotational part of* g *is nontrivial, that is,* $B \neq \mathrm{id}$. *(Since* $B \in \mathrm{SO}(2,\mathbb{R})$, *the condition* $B \neq \mathrm{id}$ *implies* $B - \mathrm{id}$ *is invertible.) Then the centralizer* $C_{\mathrm{E}(2)\times S^1}(g)$ *of* g *in* $\mathrm{E}(2) \times S^1$ *is* $\mathrm{SO}(2,\mathbb{R})^{(\mathrm{id}-B)^{-1}b} \times S^1$.

Proof. Because every element in the second S^1 factor of $\mathrm{E}(2) \times S^1$ commutes with every element of $\mathrm{E}(2) \times S^1$, it suffices to show that $C_{\mathrm{E}(2)}(g) = \mathrm{SO}(2,\mathbb{R})^{(\mathrm{id}-B)^{-1}b}$. Now the equation $(A\, a)\,(B,\, b) = (AB, Ab + a)$ is the same as the equation $(B, b)\,(A, a) = (BA,\, Ba + b)$ if and only if $AB = BA$ and $Ab + a = Ba + b$. The first equality holds because $\mathrm{SO}(2,\mathbb{R})$ is commutative. Whereas the second equality is equivalent to $(1-A)b = (1-B)a$, or $a = (1-B)^{-1}(1-A)b = (1-A)(1-B)^{-1}b$, since $AB = BA$. Therefore

$$(A,a)(x) = Ax + (1-A)(1-B)^{-1}\,a = A(x-(1-B)^{-1}b) + (1-B)^{-1}b,$$

that is, $(A,\, a)$ is equal to the rotation about the point $(1-B)^{-1}\,b$. □

For $c \in \mathcal{C}^{\,\mathrm{per}}$ let $\mathcal{B}(c) \in \mathrm{SO}(2,\mathbb{R})$ be the spatial rotational shift at c, that is, the spatial rotational part B of the shift element $g(c) = ((B,b),C) \in \mathrm{E}(2) \times S^1$ at c. Note that $\mathcal{B} : \mathcal{C}^{\,\mathrm{per}} \rightarrow \mathrm{SO}(2,\mathbb{R}) : c \mapsto \mathcal{B}(c)$ is a real analytic mapping. In addition, let $\mathcal{C}^{\,\mathrm{per}}_{\mathrm{nt}} = \{c \in \mathcal{C}^{\,\mathrm{per}} \big| \, \mathcal{B}(c) \neq \mathrm{id}\}$ be the set of all $c \in \mathcal{C}^{\,\mathrm{per}}$ with nontrivial spatial rotational shift. Let U be a connected component of the open subset $\mathcal{C}^{\,\mathrm{per}}$ of $\mathcal{C}$. Then $U \setminus \mathcal{C}^{\,\mathrm{per}}_{\mathrm{nt}}$ is the set of all $c \in U$ such that the rotational shift $g(c)$ is the identity. Because the mapping $\mathcal{C}^{\,\mathrm{per}} \rightarrow \mathrm{E}(2) \times S^1 : c \mapsto g(c)$ is real analytic, the image of $U \setminus \mathcal{C}^{\,\mathrm{per}}_{\mathrm{nt}}$ under the group homomorphism

$$\mu : \mathrm{E}(2) \times S^1 \rightarrow \mathrm{SO}(2,\mathbb{R}) : ((B,b),C) \mapsto B$$

is either equal to U or is a proper real analytic subvariety of U. In §15.4 below we show that for every connected component U of $\mathcal{C}^{\,\mathrm{per}}$ there is an element c of U such that $g(c)$ has a nontrivial spatial rotational part. □

This proves

Proposition 7.15.3.47. *The set $\mathcal{C}^{\,\mathrm{per}}_{\mathrm{nt}}$ is the complement of a proper closed analytic subvariety of $\mathcal{C}^{\,\mathrm{per}}$. Thus it is a dense open subset of $\mathcal{C}^{\,\mathrm{per}}$. Consequently, the set $\mathcal{C}^{\,\mathrm{per}}_{\mathrm{nt}}$ admits a real analytic fibration with fibers which are 3-tori that are invariant under the flow of the vector field V and on which the motion of the disk is quasiperiodic.*

We can say more. If we disregard the internal rotation of the disk, that is, if we pass to the reduced phase space $\mathcal{C}/S^1$ of §?.?, which has coordinates $((v,a),(\nu,\dot{a}))$, then $\big(\mathcal{C}/S^1\big)^{\mathrm{per}}_{\mathrm{nt}} = \mathcal{C}^{\,\mathrm{per}}_{\mathrm{nt}}/S^1$ admits an analytic fibration with 2-tori as fibers that are invariant under the S^1 reduced flow of V on $\mathcal{C}^{\,\mathrm{per}}_{\mathrm{nt}}/S^1$ and on which the motion is quasiperiodic. Therefore $\mathcal{C}^{\mathrm{per}}_{\mathrm{nt}}$ admits an analytic fibration with 3-tori as fibers which are invariant under the flow of V and on which the motion is quasiperiodic. Also the point of contact of the disk, which is a smooth function of σ_1, v, and a, see (38), is a quasiperiodic function of t with two frequencies that depends analytically on the initial conditions. Looking at the reduced space $\mathcal{C}/\mathrm{E}(2) = (S^2 \setminus \{\pm e_3\}) \times \mathbb{R}$, which is parametrized by (u,ω), see (6), we find that $\big(\mathcal{C}/\mathrm{E}(2)\big)^{\mathrm{per}}_{\mathrm{nt}} = \mathcal{C}^{\,\mathrm{per}}_{\mathrm{nt}}/\mathrm{E}(2)$ admits a real analytic fibration with 2-tori as fibers that are invariant under the flow of V and on which the motion is quasiperiodic.

Proof. To prove the proposition and the above observations we argue as follows. The statement about the motion on the S^1-reduced phase space follows from the observation that $\mathcal{C}/S^1$ has an E(2) symmetry group and that the E(2)-reduced space $\big(\mathcal{C}/S^1\big)/\mathrm{E}(2) = \mathcal{C}/(\mathrm{E}(2)\times S^1)$. From proposition 7.15.4.48 below it follows that if $\mathcal{B} \neq \mathrm{id}$ then the centralizer of (B,b) in E(2) is the circle group $(S^1)^* = \mathrm{SO}(2,\mathbb{R})^{(\mathrm{id}-B)^{-1}b}$. The theory of §15.2 then yields the existence of a quasiperiodic motion on 2-tori on $\big(\mathcal{C}/S^1\big)^{\mathrm{per}}_{\mathrm{nt}}/(S^1)^*$. The construction in §15.2 leads to 2-tori in $(\mathcal{C}/S^1)^{\mathrm{per}}_{\mathrm{nt}}$, which are invariant under the $(S^1)^*$-action. The 2-tori in $\big(\mathcal{C}/S^1\big)^{\mathrm{per}}_{\mathrm{nt}} = \mathcal{C}^{\mathrm{per}}_{\mathrm{nt}}/S^1$ are quotients of 3-tori in $\mathcal{C}^{\mathrm{per}}_{\mathrm{nt}}$. The situation with the E(2) reduced phase space is simpler. The space $\mathcal{C}/\mathrm{E}(2)$ is an S^1 principal bundle over the fully reduced phase space $\mathcal{C}/(\mathrm{E}(2)\times S^1) = \big(\mathcal{C}/\mathrm{E}(2)\big)/S^1$. Here S^1 is a 1-dimensional torus and we do not need any conditions on the shift element in S^1. □

7.15.4 *Near elliptic relative equilibria*

In this subsection we discuss the behavior of the rolling disk near an *elliptic relative periodic orbit*.

Let $c \in \mathcal{C}^{\,\mathrm{per}}$ be the starting point of a relative periodic solution

$$t \mapsto c(t) = ((A(t), a(t)), (\omega(t), \dot{a}(t))) = \gamma_c(t)$$

of the equations of motion of the disk, that is, $t \mapsto \gamma_c(t)$ is an integral curve of V starting at c. With $m = \pi(c)$ let $\tau = \tau(c)$ be the smallest positive period of the integral curve $t \mapsto \overline{\gamma}_m(t) = \pi(\gamma_c(t))$ of the fully reduced vector field $\mathcal{V}$. Then, using the notation of §4.2, there are η, $\zeta \in \mathbb{R}$ such that

$$A(\tau) = \mathrm{e}^{\tau\eta\, E_3}\, A(0)\, \mathrm{e}^{-\tau\zeta\, E_3} e_3 = \mathrm{e}^{\tau\eta\, E_3}\, A(0) e_3 = \mathrm{e}^{\beta\, E_3} v(0),$$

where $\mathrm{e}^{\beta\, E_3}$ is the spatial rotational shift and $\beta = \tau\eta$. In view of (21) it follows that

$$v(\tau) = A(\tau)e_3 = \mathrm{e}^{\beta\, E_3}\, A(0)\mathrm{e}^{-\tau\zeta\, E_3} e_3 = \mathrm{e}^{\beta\, E_3}\, A(0)e_3 = \mathrm{e}^{\beta\, E_3} v(0).$$

Using (35) we get $\chi(\tau) = \chi(0) + \beta = \chi(0) + \tau\eta$. Integrating (36) gives

$$\beta = \chi(\tau) - \chi(0) = \int_0^\tau \frac{\mathrm{d}\chi}{\mathrm{d}t}\,\mathrm{d}t = \int_0^\tau (1 - \sigma_1^2(t))^{-1}\sigma_3(t)\,\mathrm{d}t$$

and therefore

$$\mathcal{B}(c) = \mathrm{e}^{\beta\, E_3}, \quad \text{where } \beta = \textstyle\int_0^\tau (1 - \sigma_1^2(t))^{-1}\sigma_3(t)\,\mathrm{d}t. \tag{220}$$

The set E^{ell} of *elliptic relative periodic orbits* of V on $\mathcal{C}$ is a proper real analytic subvariety of $\mathcal{C}$. Now $\mathcal{C} \cup E^{\mathrm{ell}}$ is an open neighborhood of E^{ell} in $\mathcal{C}$.

Lemma 7.15.4.49 below implies that the period function $\tau : \mathcal{C}^{\,\mathrm{per}} \to \mathbb{R}_{>0} : c \mapsto \tau(c)$ extends to an analytic function on $\mathcal{C}^{\,\mathrm{per}} \cup E^{\mathrm{ell}}$, which we also denote by τ. It follows that the spatial rotation map $\mathcal{B} : \mathcal{C}^{\,\mathrm{per}} \to \mathrm{SO}(2,\mathbb{R})$ extends to an analytic map $\mathcal{B} : \mathcal{C}^{\,\mathrm{per}} \cup E^{\mathrm{ell}} \to \mathrm{SO}(2,\mathbb{R})$. If c_0 is the starting point of a relative periodic orbit of V in $\mathcal{C}$, then $m_0 = \pi(c_0)$ is an equilibrium point of the reduced vector field $\mathcal{V}$ on $\mathcal{M}$. Its coordinates σ_1^0, σ_3^0 in $\mathcal{M}$ are constant functions of time. Therefore from (220) it follows that

$$\mathcal{B}(c_0) = \mathrm{e}^{\tau\eta\, E_3}, \quad \text{where } \eta = (1-(\sigma_1^0)^2)^{-1}\sigma_3^0. \tag{221}$$

Let U be a connected component of $\mathcal{C}^{\,\mathrm{per}}$ in $\mathcal{C}$. Recall that $\mathcal{C}^{\,\mathrm{per}} \cup E^{\mathrm{ell}}$ is an open neighborhood of E^{ell} in $\mathcal{C}^{\,\mathrm{per}}$. Suppose that U' is another connected component of $\mathcal{C}^{\,\mathrm{per}}$. Then $\overline{U} \cap \overline{U'}$ in $\mathcal{C}$ consists of hyperbolic or degenerate relative equilibria. Hence $U \cup (E^{\mathrm{ell}} \cap \overline{U})$ is an open neighborhood of $E^{\mathrm{ell}} \cap \overline{U}$ in $\mathcal{C}$.

We now show that for each connected component U of $\mathcal{C}^{\,\mathrm{per}}$ the right hand side of (221) is a nonconstant real analytic function on U. This implies

Proposition 7.15.4.48. *The set of points $c \in U$ at which the spatial rotational shift is equal to the identity is a proper analytic subvariety of U.*

Before proving this proposition we consider the conservative one degree of freedom Newtonian system

$$M\ddot{\varphi} + V(\varphi) = 0 \tag{222}$$

with mass M and potential V, which is real analytic on an open interval $I \subseteq \mathbb{R}$. Suppose that V has a nondegenerate minimum at φ_0, that is, $V'(\varphi_0) = 0$ and $V''(\varphi_0) > 0$. If $E \in J = \big(V(\varphi_0) - \eta, V(\varphi_0)\big)$ and $\eta > 0$ is sufficiently small, then the solutions of (222) with energy E near $\varphi = \varphi_0$ and $\frac{\mathrm{d}\varphi}{\mathrm{d}t}(\varphi_0) = 0$ oscillate in a small interval around φ_0 and hence are periodic of period $\tau = \tau(E)$.

Lemma 7.15.4.49. *The function $\tau : J \to \mathbb{R} : E \mapsto \tau(E)$, which gives the period of the periodic solution of (222) near the nondegenerate critical point φ_0 of the real analytic potential V, can be extended to a* complex *analytic function in a complex neighborhood $\widetilde{J}$ of J. Moreover,*

$$\lim_{E \downarrow V(\varphi_0)} \tau(E) = 2\pi M^{1/2}\, V''(\varphi_0)^{-1/2}. \tag{223}$$

If the potential V depends real analytically on parameters p near p_0, then the period function τ extends to an analytic function of (p, E) in a neighborhood of $(p_0, V(\varphi_0, p))$.

Proof of lemma 7.15.4.49. Even though the limit formula (223) can be proved in an elementary fashion, we will give a sketch of its proof.

From equation (88) recall that

$$\tau(E) = 2\int_{\varphi_-}^{\varphi_+} \Big(\frac{2}{M}(E - V(\varphi))\Big)^{-1/2} \,\mathrm{d}\varphi, \tag{224}$$

where $\varphi_- = \varphi_-(E)$ and $\varphi_+ = \varphi_+(E)$ are close to φ_0 with $\varphi_- < \varphi_0 < \varphi_+$. Moreover, $V(\varphi_-) = E = V(\varphi_+)$ with $V'(\varphi_-) < 0$, $V'(\varphi_+) > 0$, and $V(\varphi) < E$ for $\varphi \in I = [\varphi_-, \varphi_+]$. Thus the right hand side of (224) is convergent.

Because V is real analytic on $I \subseteq \mathbb{R}$, it has an extension to a complex analytic function on an open complex neighborhood U of I. Let $\varepsilon = \frac{1}{2}(\varphi_+ - \varphi_-)$. Consider the curve γ_ε in $\mathbb{C}$, which consists of the concatenation of following four curves: 1) φ traverses the closed interval $[\varphi_- + \varepsilon, \varphi_+ - \varepsilon]$, followed by 2) the semicircular arc $t \mapsto \varphi_+ + \varepsilon\,\mathrm{e}^{it}$ with $t \in [-\pi, \pi]$, followed by 3) φ traversing the interval $(\varphi_- + \varepsilon, \varphi_+ - \varepsilon)$ backward, and finished by 4) the semicircular arc $t \mapsto \varphi_- + \varepsilon\,\mathrm{e}^{it}$ with $t \in [0, \pi]$. Because $V(\varphi_\pm) = E$ and $V'(\varphi_\pm) = 0$, from the Taylor expansion of V at $\varphi_\pm$

$$\begin{aligned} V(\varphi) &= V(\varphi_\pm) + V'(\varphi_\pm)(\varphi - \varphi_\pm) + \mathrm{O}((\varphi - \varphi_\pm)^2) \\ &= E + V'(\varphi_\pm)(\varphi - \varphi_\pm) + \mathrm{O}((\varphi - \varphi_\pm)^2) \end{aligned}$$

we get

$$(E - V(\varphi))^{1/2} = \sqrt{\pm V'(\varphi_\pm)}\,\sqrt{\mp(\varphi - \varphi_\pm)} + \mathrm{O}(|\varphi - \varphi_\pm|). \tag{225}$$

Therefore the complex analytic function $F = \big(\frac{2}{M}(E - V(\varphi))\big)^{-1/2}$ is double valued near $\varphi_\pm$. If $\varepsilon > 0$ is sufficiently small, then the two values of the integrand F (224) are interchanged along segments 2) and 4) of the curve γ_ε. While traversing all of γ_ε, the function F returns to its original value.

Taking I and U sufficiently small we can arrange that $\varphi_\pm$ are the only zeroes of the function $\varphi \mapsto E - V(\varphi)$ in U. Consider the Riemann surface $X(E)$ given by

$$\{(\varphi, \psi) \in (U \setminus \{\varphi_\pm\}) \times \mathbb{C} \,|\, \frac{2}{M}(E - V(\varphi))\,\psi^2 = 1\}$$

with projection map $\rho : X(E) \to U \setminus \{\varphi_\pm\} : (\varphi, \psi) \mapsto \varphi$. Note that ρ is a twofold covering. Because the integrand F on the right hand side of (224) returns to its original value after traversing the path γ_ε, each of the two lifts $\widetilde{\gamma}_\varepsilon$ of γ_ε to $X(E)$ is a closed curve. Also

$$\int_{\widetilde{\gamma}_\varepsilon} \big(\frac{2}{M}(E - V(\varphi))\big)^{-1/2} \, \mathrm{d}\varphi = \int_{\widetilde{\gamma}_\varepsilon} \rho^*(F \, \mathrm{d}\varphi) = \int_{\widetilde{\gamma}_\varepsilon} \psi \, \mathrm{d}\varphi,$$

where $\psi \, \mathrm{d}\varphi$ is a holomorphic 1-form on $X(E)$. Because $X(E)$ has complex dimension 1, the 1-form $\psi \, \mathrm{d}\varphi$ is closed. Since $\int_{\widetilde{\gamma}_\varepsilon} \psi \, \mathrm{d}\varphi$ is homotopy invariant, it does not depend on ε and is therefore equal to $\int_\gamma \psi \, \mathrm{d}\varphi$, where γ is a closed path in $U \setminus \{\varphi_\pm\}$. Because γ avoids $\varphi_\pm$, it follows that $\tau(E) = \int_\gamma \psi \, \mathrm{d}\varphi$ is a complex analytic function of E in a complex neighborhood $\widetilde{J}$ of $V(\varphi_0)$. From (225) it follows that $\psi = \big(\frac{2}{M}(E - V(\varphi))\big)^{-1/2}$ is a meromorphic function on $X(E)$ with a simple pole at φ_0. The value of the integral $\int_\gamma \psi \, \mathrm{d}\varphi$ is $2\pi i$ times the residue Res_{φ_0}, which we compute as follows.

$$\begin{aligned}
\mathrm{Res}_{\varphi_0} &= \lim_{\varphi \to \varphi_0} (\varphi - \varphi_0) \frac{1}{i\big(\frac{2}{M}(V(\varphi) - V(\varphi_0))\big)^{1/2}} \\
&= \frac{1}{i}\sqrt{\frac{M}{2}} \lim_{\varphi \to \varphi_0} (\varphi - \varphi_0) \frac{1}{\sqrt{\frac{1}{2} V''(\varphi_0)(\varphi - \varphi_0)^2 + \mathrm{O}((\varphi - \varphi_0)^3)}},
\end{aligned}$$

which comes from taking the square root of the Taylor series of the function $\varphi \mapsto V(\varphi) - V(\varphi_0)$ to second order about φ_0 and using the fact that φ_0 is a nondegenerate critical point of V

$$= \frac{1}{i}\sqrt{\frac{M}{V''(\varphi_0)}}. \tag{226}$$

This gives (223). □

Proof of proposition 7.15.4.48. To prove that the spatial rotational shift map $\mathcal{B}$ is *not* constant on every connected component of $\mathcal{C}^{\,\mathrm{per}}$, we begin by describing these components. Let $\mathcal{C}^{\,\mathrm{flat}}$ be the set of starting points of integral curves of V, where the disk rises from and/or falls to a flat position in finite time. The value of the integral map $\mathcal{I}$ (207) on $\mathcal{C}^{\,\mathrm{flat}}$ is contained in $\widetilde{\pi}^{-1}(\ell_+ \cup \ell_-)$, where $\widetilde{\pi} = (\overline{\sigma}_3, \overline{\sigma}_4) : \mathcal{C} \to \mathbb{R}^2$. The set $\widetilde{\pi}^{-1}(\ell_+ \cup \ell_-)$ contains all the periodic integral curves corresponding to the part of $\ell_+ \cup \ell_-$, which lies in the interior of the region bounded by the discriminant locus Δ. Here $\varphi(t)$ oscillates in the highest occurring potential well described in proposition 7.12.5.36. The boundary $\partial\mathcal{C}^{\,\mathrm{flat}}$ of $\mathcal{C}^{\,\mathrm{flat}}$ consists of unstable (= hyperbolic) and degenerate relative equilibria for which $(\overline{\sigma}_3, \overline{\sigma}_4) \in \ell_+ \cup \ell_-$.

The set $\mathcal{C} \setminus (\mathcal{C}^{\,\mathrm{flat}} \cup \partial\mathcal{C}^{\,\mathrm{flat}})$ is the largest open subset of $\mathcal{C}$ where the flow of V is complete.

The set $\mathcal{C} \setminus (\mathcal{C}^{\,\mathrm{flat}} \cup \partial\mathcal{C}^{\,\mathrm{flat}})$ is connected. For if $c, c' \in \mathcal{C}$ and $\widetilde{\pi}(c)$, $\widetilde{\pi}(c')$ lie in different connected components U and U' of $\mathbb{R}^2 \setminus (\ell_+ \cup \ell_-)$, then we first move in $\widetilde{\pi}^{-1}(U)$ from c to c_0 such that $\widetilde{\pi}(c_0) \in U$ and $\widetilde{\pi}(c_0)$ lies in the interior of the region bounded by Δ. Moreover, $\widetilde{\pi}(c_0)$ is a periodic orbit for which $\varphi(t)$ oscillates in the highest potential well. Because the set of all such c_0 is connected, we can move c_0 to c_0' such that $\widetilde{\pi}(c_0') \in U'$. Then we move c_0' to c' in $\widetilde{\pi}^{-1}(U')$. However, the open subset $\mathcal{C}^{\,\mathrm{per}}$ of $\mathcal{C} \setminus (\mathcal{C}^{\,\mathrm{flat}} \cup \partial\mathcal{C}^{\,\mathrm{flat}})$ is *not* connected. In fact, it has four connected components. Each of the four connected components U of $\mathbb{R}^2 \setminus (\ell_+ \cup \ell_-)$, where $\varphi(t)$ oscillates in the lowest potential well each form a connected component of $\mathcal{C}^{\,\mathrm{per}}$; while those which oscillate in the highest well form the fifth connected component.

Because an elliptic relative periodic orbit corresponds to a nondegenerate local minimum of the potential function V, we have $\mathcal{I}(E^{\,\mathrm{ell}}) = \Sigma^{\mathrm{nd}}$. Moreover, for each connected component U of $\mathcal{C}^{\,\mathrm{per}}$, we have $\mathcal{I}(E^{\,\mathrm{ell}} \cap \overline{U})$ is a nonempty open subset of $\Sigma^{\,\mathrm{stabl}}$, which contains one of the connected components of $\Sigma^{\,\mathrm{deg}} \setminus (\ell_+ \cup \ell_-)$ in its closure.

Let a point in $\Sigma^{\,\mathrm{stabl}}$ approach a fixed point in $\Sigma^{\,\mathrm{deg}}$, which is not one of the two swallow tail points on the $\overline{\sigma}_4$-axis. In other words, we do not approach a rectilinear motion of the disk rolling at constant angular velocity. Because $\sigma_3 = 0$ for a relative equilibrium, this implies $\sigma_1 = 0$, see (45). Hence $\overline{\sigma}_3 = \sigma_3 = 0$, which we have just excluded. Therefore $\sigma_3 \neq 0$ at our degenerate critical point. Because $V''(\varphi) \to 0$ as we approach a degenerate critical point, in view of (223) it follows that $\tau \to \infty$. Thus the spatial rotational shift $\mathrm{e}^{\beta\, E_3}$ in (220) winds faster and faster around the circle $\mathrm{SO}(2,\mathbb{R})$. This implies that the spatial rotational shift map $\mathcal{B}$ is *not* constant on $\mathcal{I}(E^{\,\mathrm{ell}} \cap \overline{U})$. Thus $\mathcal{B}$ is *not* constant on U. □

Corollary 7.15.4.50. *The set $\mathcal{C}_{\mathrm{t}}^{\mathrm{per}}$ of all $c \in \mathcal{C}^{\,\mathrm{per}}$ such that $\mathcal{B}(c) = \mathrm{id}$ is a proper real analytic subvariety of $\mathcal{C}^{\,\mathrm{per}}$. On its complement $\mathcal{C}_{\mathrm{nt}}^{\mathrm{per}}$ the motion is quasiperiodic and forms a dense subset of $\mathcal{C}^{\,\mathrm{per}}$.*[5]

Let $c \in \mathcal{C}^{\,\mathrm{per}}$ and let $g(c) = \big((B(c), b(c)), C(c)\big) \in \mathrm{E}(2) \times S^1$ be the shift element. It is an interesting problem to determine the asymptotic behavior

[5]The observation that the spatial rotational shift runs infintely fast on $\mathrm{SO}(2,\mathbb{R})$ when approaching a degenerate relative equilibrium implies that we have an accumulation of infinitely many sheets of $\mathcal{C}_{\mathrm{t}}^{\mathrm{per}}$.

of $(B(c), b(c))$ for c near $E^{\,\mathrm{ell}}$. Recall that $\mathcal{C}^{\mathrm{per}}_{\mathrm{t}}$ is the set of all $c \in \mathcal{C}^{\,\mathrm{per}}$ such that $\mathcal{B}(c) = \mathrm{id}$. Asymptotic analysis should give information about the structure of $\mathcal{C}^{\mathrm{per}}_{\mathrm{t}}$ near $E^{\,\mathrm{ell}}$. If $c \in \mathcal{C}^{\mathrm{per}}_{\mathrm{t}}$ and $b(c) \neq 0$, then the motion of the disk is a superposition of rectilinear motion with nonzero constant velocity $\tau(c)^{-1}b(c)$ and a periodic motion, see lemma 7.15.3.45.

We now give an alternative proof of proposition 7.15.4.48.

Proof. If $\overline{\sigma}_4 = 0$ and $|\overline{\sigma}_3| > \frac{\mathrm{mgr}}{(I_1+\mathrm{mr}^2)^{1/2}}$, which is the critical angular speed of spinning of the vertical disk, see case 2 in §8.2, then $\sigma_1 = 0$ is the only critical point of $V_{\overline{\sigma}_3,\overline{\sigma}_4}$. It is a nondegenerate minimum and corresponding to the disk spinning vertically in a stable way. Because $\sigma_1 = 0$, we have $\sigma_3 = \overline{\sigma}_3$ and $\sigma_4 = \overline{\sigma}_4 = 0$. From (129) we obtain $V''_{\overline{\sigma}_3,\overline{\sigma}_4}(0) = (I_1 + \mathrm{mr}^2)\overline{\sigma}_3^2 - \mathrm{mgr}$. Because $M = I_1 + \mathrm{mr}^2$, from (223) it follows that the limiting period of nearby periodic solutions of the reduced system is

$$\tau = 2\pi\big((\overline{\sigma}_3^2 - \mathrm{mgr})/(I_1 + \mathrm{mr}^2)\big)^{-1/2}.$$

Thus from (220) we get

$$\beta = \tau\eta = \tau\overline{\sigma}_3 = 2\pi\big(1 - \big(\frac{\mathrm{mgr}}{I_1 + \mathrm{mr}^2}\big)\frac{1}{\overline{\sigma}_3^2}\big)^{-1/2}. \qquad (227)$$

As $|\overline{\sigma}_3| \downarrow \big(\frac{\mathrm{mgr}}{I_1+\mathrm{mr}^2}\big)^{1/2}$, the right hand side of (227) increases to ∞ on $\Sigma^{\,\mathrm{ell}}$. Because the set of elliptic relative equilibria is connected and meets the boundary of each connected component of $\mathcal{C}^{\,\mathrm{per}}$, this computation shows that the spatial rotational shift is a nonconstant real analytic function on each connected component of $\mathcal{C}^{\,\mathrm{per}}$. □

7.15.5 *Nearly flat solutions*

We use the notation of §10.1. Let $p = (\overline{\sigma}_3, \overline{\sigma}_4) \notin \ell_\pm$ be near $q \in \ell_\pm$. Suppose that $\sigma_1(p)$ is a critical point of V_p, which is near ± 1. Then $\sigma_1(p)$ is a nondegenerate local minimum of V_p.

Lemma 7.15.5.51. *Let τ be the limit of the period of periodic solutions of the reduced equations of motion near an equilibrium point in the reduced phase space with $\sigma_1 = \sigma_1(p)$. Then*

$$\tau \sim \frac{2\pi}{\sqrt{3}}\,(I_1 + \mathrm{mr}^2)^{1/2} I_1^{-1/6}(\mathrm{mgr})^{-2/3}|\eta_\pm|^{1/3} \quad \textit{as } \eta_\pm \to 0. \qquad (228)$$

The limit of the spatial rotational shift of periodic solutions near the equilibrium point in the reduced phase space is $e^{\beta E_3}$, *where*

$$\beta = \tau\eta = (\operatorname{sgn}\eta_\pm)\frac{2\pi}{\sqrt{3}}(1+\mathrm{mr}^2/I_1)^{1/2}$$
$$= (\operatorname{sgn}\eta_\pm)\frac{2\pi}{\sqrt{3}}\Delta\chi \quad \text{as } \eta_\pm \to 0. \tag{229}$$

Here $\Delta\chi$ *is defined in proposition* 7.11.2.27.

Proof. With $V(\varphi) = V_{(\overline{\sigma}_3,\overline{\sigma}_4)}(\sin\varphi) = V_p(\sin\varphi)$ and $M = I_1 + \mathrm{mr}^2$ we obtain

$$V'(\varphi) = V_p'(\sigma_1)\,\cos\varphi \;\text{ and }\; V''(\varphi) = V_p''(\sigma_1)\cos^2\varphi - V_p'(\sigma_1)\sin\varphi.$$

Using the notation of (95), we see that φ is a critical point of V if and only if $V_p'(\sigma_1) = 0$ if and only if $\mathfrak{C}\widetilde{W} = 0$. In our situation we find that

$$V''(\varphi) = \mathrm{mgr}\,\mathfrak{C}^2\widetilde{W}\cos^2\varphi.$$

From the proof of lemma 7.14.7.44 we find that $x \sim \epsilon$, $y \sim \mp a$, and $\cos\varphi \sim d^{1/3}|\epsilon|^{2/3}$, where $d\epsilon^2 = |\eta_\pm|^2/(I_1+\mathrm{mgr})$. If we substitute these asymptotic relations into $\mathfrak{C}^2\widetilde{W}$ (149) we obtain

$$\mathfrak{C}^2\widetilde{W}\cos^2\varphi \sim 4d\,d^{-4/3}\,|\epsilon|^{-8/3}\,\epsilon^2 - d^{-1/3}\,|\epsilon|^{-4/3}\,\epsilon(4d+2c)a$$
$$+ (d+c)^2a^2 - d^{-1/3}|\epsilon|^{-2/3}$$
$$\sim 3d^{-1/3}|\epsilon|^{-2/3}.$$

In view of the formula for τ given in lemma 7.15.4.49 with $M = I_1 + \mathrm{mr}^2$ we obtain (228).

According to (221) β is obtained by multiplying (228) by

$$(1-\sigma_1^2)^{-1}\sigma_3 = -\frac{I_1\,ux}{I_3\cos^2\varphi} \sim (\operatorname{sgn}\eta_\pm)\frac{I_1 u}{I_3}d^{-2/3}|\epsilon|^{-1/3}, \tag{230}$$

where we have used (91),

$$\cos\varphi \sim d^{1/3}|\epsilon|^{2/3},\;\; x \sim \epsilon, \;\text{ and }\; \operatorname{sgn}\epsilon = -\operatorname{sgn}\eta_\pm.$$

For the last equality above see the proof of lemma 7.10.1.22. In the product of (228) and (230) the powers of ϵ cancel. Using $u = (\mathrm{mgr})^{1/2}/(I_3+\mathrm{mr}^2)^{1/2}$ and $d = I_3^2/(I_1(I_3+\mathrm{mr}^2))$ we obtain (229). □

From (228) it follows that if the disk is close to an almost flat relative equilibrium, then its angle with the vertical oscillates very rapidly with a frequency of order $|\eta_\pm|^{-1/3}$. This oscillation is in combination with the rapid motion around the rim by the point of contact which we observed in §4.5.

It is quite remarkable that the spatial rotational shift approaches a constant, whose absolute value equal to $(2/\sqrt{3})\,\Delta\chi$, where $\Delta\chi$ is the absolute value of the increase of the angle χ when the disk rises from and falls to the flat position, see (186). Note that β changes sign together with $\eta_\pm$.

For the uniform disk and the hoop the absolute value of the right hand side of (229) is $2\pi\,\sqrt{5/3}$ and 2π, respectively. When the mass distribution becomes concentrated near the center of mass, that is, when $I_1/\mathrm{mr}^2 \downarrow 0$, then the absolute value of the right hand side of (229) increases to ∞.

When the right hand side of (229) is not an integral multiple of 2π, that is, when $\mathrm{mr}^2/I_1 \neq 3k^2 - 1$ for every $k \in \mathbb{Z}$, then all the motions of the disk which are sufficiently close to the nearly flat relative equilibrium are quasiperiodic as described in proposition 7.15.3.47. This holds for the uniform disk, whereas for the hoop further asymptotic analysis is needed in order to decide whether there are nearly flat motions, which have trivial spatial rotational shift and nonzero translational shift, and hence run off to infinity as described in lemma 7.15.3.45.

The fact that the absolute value of the right hand side of (229) is not equal to $\Delta\chi$ implies that the spatial rotational shift does not have a unique limit if we let both $\eta_\pm$ and $E - V_q(\pm 1)$ tend to zero, keeping the sign of $\eta_\pm$ constant. To see this note that if $\eta_\pm = 0$, that is, $p = q \in \ell_\pm$ and $E - V_q(\pm 1)$ is small, then the time interval I between rising from and falling to the flat position is small, see the last paragraph in §11.1. As observed in (174), the quantity $(1 - {\sigma_1}^2)^{-1}\,\sigma_3 = \frac{\mathrm{d}\chi}{\mathrm{d}t}$ remains bounded. Therefore in the time interval I the increase of χ is small. From (186) if follows that if $E - V_q(\pm 1)$ is small, $\eta_\pm \neq 0$, and $|\eta_\pm|$ is small enough compared to $E - V_q(\pm 1)$, then the increase β of $\chi(t)$ during a period of the periodic solution of the reduced system is close to $(\operatorname{sgn}\eta_\pm)\,\Delta\chi$.

Let β be a real number between $(\operatorname{sgn}\eta_\pm)\,\Delta\chi$ and $(\operatorname{sgn}\eta_\pm)\,(2/\sqrt{3})\,\Delta\chi$ and fix the sign of $\eta_\pm$. By the intermediate value theorem, it follows that in any continuous one parameter family of solutions, which goes from a falling flat solution, that remains near the flat position to an elliptic relative equilibrium, that remains near the flat position, there is at least one for which the increase $\chi(t)$ during a period of the periodic solution of the reduced system is equal to β. Because the spatial rotational shift is equal to $e^{\beta E_3}$, the spatial rotational shift does not have a unique limit if we let both $\eta_\pm$ and $E - V_q(\pm 1)$ tend to zero, keeping the sign of $\eta_\pm$ fixed.

The argument in the preceding paragraph shows that if there is an integer k such that $\Delta\chi < 2\pi k < (2/\sqrt{3})\,\Delta\chi$, that is, $3k^2 - 1 < \mathrm{m}\,\mathrm{r}^2/I_1 < 4k^2 - 1$, then for each choice of sign of $\eta_\pm$ in each continuous family as described above there is a solution for which the spatial rotational shift is the identity. Since $3(k+1)^2 - 1 < 4k^2 - 1$ if $k \ge 7$, this last conclusion holds when

$$\mathrm{mr}^2/I_1 \in (2,\,3)\cup(11,\,15)\cup(26,\,35)\cup(47,\,63)\cup(74,\,99)\cup(107,\,143)\cup(146,\,\infty).$$

In particular this conclusion holds if there is more mass inside the rim than for the hoop, that is, if mr^2/I_1 is slightly larger than 2.

7.16 Notes

The first papers on the rolling disk considered only the uniform disk. In 1872 Ferrers [43] obtained the fully reduced equations of motion (13) for the first time using Euler angles. The first serious analysis of these equations was given by Vierkandt [114] in 1892, who observed that almost all of its solutions were periodic. In 1897 Chaplygin [23] derived what we call Chaplygin's equations (76) for the rolling disk. In §244 Routh [96] noted that these equations define Legendre functions and in 1899 Korteweg [62] observed that hypergeometric functions appear in the solutions of the fully reduced equations of motion of the disk. Later in 1900 Appell [3] made the same observation. In the past one hundred years there has been little interest in the rolling disk. In this period, many authors were content only to repeat the derivation of the equations of motion and the proof of integrability by quadratures and leave it at that. Only recently has interest arisen in the geometric behavior of the solutions of the rolling disk in phase space. Kemppainen [57] was the first to construct the two parameter family of one degree of freedom Hamiltonian systems (81) and (82) from the fully reduced system. Figures 7.10 and 7.11 give all the possible potential functions and level sets for the corresponding Hamiltonian. In [84] O'Reilly has an interesting discussion of the motion of the rolling disk. He finds several bifurcations numerically. See Srinivasan [110] for a discussion of some rocking and rolling motions of cans.

The two frequency quasiperiodic behavior of the point of contact of the disk, shown in §15.3, is nicely illustrated in figure 8 of Borisov et al. [18].

This figure also exhibits some special cases where the spatial rotational shift is the identity and the motion of the disk is a combination of constant velocity rectilinear and periodic motion.

The motion of the rolling disk is incomplete because it runs off the constraint manifold in finite time by falling flat. A complete analogue of the disk is the body of revolution rolling without slipping on a horizontal plane under the influence of a uniform vertical gravitational field. The reduced dynamics of this problem was first analyzed by Chaplygin [23]. In [81] Moshchuk obtained the full motion $t \mapsto (A(t), a(t))$ of the body of revolution by integrating equation (29) and then (33). Because the rotational motion is quasiperiodic on 2-tori, we are confronted with the following difficulty: integrals of quasiperiodic functions need not be linear functions of time plus quasiperiodic functions. The reason for this is the appearance of *small divisors* in the coefficients of the Fourier series of a quasiperiodic function leads to the divergence of the series. If we use reconstruction from the periodic solutions in the fully reduced phase space $\mathcal{M}$ as in §15.4, then we do *not* meet the problem of small divisors. Moreover, we are able to obtain the stronger conclusion that the motion of the disk is quasiperiodic on 3-tori, if the spatial rotational shift is not the identity. If the spatial rotational shift is the identity, then the motion of the disk is a periodic perturbation of a rectilinear motion with constant velocitiy, see §15.3.

Figure 4.4 in [31] gives pictures of the cusps in a constant energy slice of the set of critical values of the integral map $\mathcal{I}$ (207) appear to be rounded off compared with our figures in §14.1.

One can find a picture of the critical surface in Borisov et al. [18]. Its downward cusps over the lines $\ell_{\pm}$ are clearly visible in their figures 3, 4, and 5. Their explanation of these cusps in terms of the singularities of the fully reduced differential equation at $\sigma_1 = \pm 1$ is a bit too soft for us. By the way the authors of this paper do not prove what one sees in their computer generated figures.

Bibliography

[1] Abraham, R. and Robbin, J. (1967). *Transversal mappings and flows*, (W.A. Benjamin).

[2] Abramowitz, M. and Stegun, I. (1972). *The Handbook of Mathematical Functions*, (Dover).

[3] Appell, P. (1900). Sur l'intégration des équations du mouvement d'un corps pesant de révolution roulant par une arête circulaire sur un plan horizontal; cas particular du cerceau, *Rend. Palermo* **14** pp. 1–6.

[4] Appell, P. (1896). *Traité de Méchanique rationelle*, Tome I, II, Paris. (1904). 2nd edn.

[5] Arms, J.M., Cushman, R., and Gotay, M.J. (1991). A universal reduction procedure for Hamiltonian group actions, in *The geometry of Hamiltonian systems*, T.S. Ratiu, ed., (Birkhäuser) pp. 31–51.

[6] Arms, J.M., Marsden, J.E., and Moncrief, V. (1975). Symmetry and bifurcations of the momentum mappings, *Commun. Math. Phys.* **78**, pp. 455–478.

[7] Arnol'd, V.I., Kozlov, V.V., and Neishtadt, A.I. (1988). Mathematical aspects of classical and celestial mechanics, in *Dynamical Systems, III*, ed. V.I. Arnol'd, (Springer Verlag).

[8] Arnol'd, V.I., Gusein-Zade. S.M., and Varchenko, A.N. (1985). *Singularities of Differentiable Maps*, Vol. I, (Birkhäuser).

[9] Aronszjan, N. (1967). Subcartesian and subriemannian spaces, *Notices Amer. Math. Soc.* **14** p. 111.

[10] Bates, L. (1998). Examples of singular nonholonomic reduction, *Rep. Math. Phys.* **42**, pp. 231–247.

[11] Bates, L., Graumann, H., and MacDonnell, C. (1996). Examples of gauge conservation laws in nonholonomic systems, *Rep. Math. Phys.* **37**, pp. 295–308.

[12] Bates, L. and Lerman, E. (1997). Proper group actions and symplectic stratified spaces, *Pacific J. Math.* **191**, pp. 201–229.

[13] Bates, L. and Śniatycki, J. (1993). Nonholonomic reduction, *Rep. Math. Phys.* **32**, pp. 99–115.

[14] Bierstone, E. (1975). Lifting isotopies from orbit spaces, *Topology* **14**, pp.

245–252.

[15] Bierstone, E. (1980). *The structure of orbit spaces and the singularities of equivariant mappings*, Monografias de Matematica, **35**, (Instituto de Matematica Pura e Aplicada, Rio de Janeiro).

[16] Bloch, A.M., Khrishnaprasad, P.S., Marsden, J.E., and Murray, R.M. (1996). Nonholonomic systems with symmetry, *Arch. Ration. Mech. Anal.* **138**, pp. 21–99.

[17] Bocharov, A.V. and Vinogradov, A.M. (1997). The Hamiltonian form of mechanics with friction, nonholonomic mechanics, invariant mechanics, the theory of refraction and impact, appendix II in The structure of Hamiltonian mechanics, B.A. Kuperschmidt and A.M. Vinogradov (authors), *Russ. Math. Surv.* **42** pp. 177–243.

[18] Borisov, A.V., Manaev, I.S., and Kilin, A.A. (2003). Dynamics of rolling disk, *Regular and Chaotic Dynamics* **8** pp. 201–212.

[19] Cantrijn, F., de León, M., and Martin de Diego, D. (1999). On almost Poisson structures in nonholonomic mechanics, *Nonlinearity* **12**, pp. 721–737.

[20] Cartan, H. (1967). Généralités sur les espaces fibrés, I, Séminaire Henri Cartan, E.N.S. 1949/1950, No. 6. in *Séminair Cartan*, Vol. I, (W.A. Benjamin).

[21] Carathéodory, C. (1933). Der Schlitten, *Z. für Angew. Math. und Mech.*, **13**, pp. 71–76.

[22] Cendra, H., Marsden, J.E., and Ratiu, T. (2001). Geometric Mechanics, Lagrangian Reduction, and Nonholonomic Systems, in *Mathematics Unlimited - 2001 and beyond*, B. Engquist and W. Schmid, eds., (Springer Verlag) pp. 221–273.

[23] Chaplygin, S.A. (1897). On the motion of a heavy body of revolution on a horizontal plane, Physics Section of the Imperial Friends of Physics, Anthropology and Ethnographics, (Moscow). (1954). *Selected Works on Mechanics*, (Moscow) pp. 413–425 (in Russian). English translation (2002) *Regular and chaotic dynamics* **7**, pp. 119–130.

[24] Chaplygin, S.A. (1911). On the theory of nonholonomic systems. The reducing multiplier theorem, *Mathematical Collection* **XXVII** (1954). *Selected Works on Mechanics*, (Moscow) pp. 413–425 (in Russian).

[25] Chaplygin, S.A. (1903). On a sphere rolling on a horizontal plane, *Mathematical Collection*, **XXIV**. (1897) Physics Section of the Imperial Friends of Physics, Anthropology and Ethnographics, (Moscow). (1954). *Selected Works on Mechanics*, (Moscow) pp. 413–425 (in Russian). English translation (2002). *Regular and chaotic dynamics* **7**, pp. 131–149.

[26] Chow, W.L. (1940/41). Über Systeme von linearen partiellen Differentialgleichungen erster Ordnung, *Math. Ann.* **117**, pp. 98–115.

[27] Coddington, E.A. and Levinson, N. (1955). *Theory of Ordinary Differential Equations*, (McGraw-Hill).

[28] Copson, E.T. (1935). *An Introduction to the Theory of Functions of a Complex Variable*, (Oxford University Press).

[29] Courant, T.J. (1990). Dirac manifolds, *Trans. Am. Math. Soc.* **319**, pp.

631–661.

[30] Cushman, R. and Bates, L. (1997). *Global aspects of classical integrable systems*, (Birkhäuser).

[31] Cushman, R., Hermans, J., and Kemppainen, D. (1996). The rolling disc, in *Nonlinear dynamical systems and chaos* (Groningen, 1995), Progress in Nonlinear Differential Equations and Their Applications, volume **19**, pp. 21–60, (Birkhäuser).

[32] Cushman, R., Kemppainen, D., Śniatycki, J., and Bates, L. (1995). Geometry of nonholonomic constraints, *Rep. Math. Phys.* **36**, pp. 275–268.

[33] Cushman, R. and Śniatycki, J. (2001). Differential structure of orbit spaces, *Canad. J. Math.* **53**, pp. 715–755.

[34] Cushman, R. and Śniatycki, J. (2002). A nonholonomic oscillator, *Rep. Math. Phys.* **50**, pp. 85–98.

[35] Cushman, R. and Śniatycki, J. (2002). Nonholonomic reduction for free and proper actions, *Regular and Chaotic Dynamics* **7**, pp. 61–72.

[36] Dalsmo, M. and van der Schaft, A. (1999). On representations and integrability of mathematical structures on energy-conserving physical systems, *SIAM J. Optimal Control* **37**, pp. 54–91.

[37] Dieudonné, J. (1944), Une généralisation des espaces compacts, *J. Math. Pures Appl.* **23**, pp. 65–76.

[38] Dieudonné, J. (1970). *Éléments d'analyse, III*, (Gauthiers-Villars).

[39] Dirac, P.A.M. (1950). Generalized Hamiltonian systems, *Canad. J. Math.* **2**, pp. 129–148.

[40] Duistermaat, J.J. (1974). Oscillatory integrals, Lagrange immersion and unfoldings of singularities, *Comm. Pure Appl. Math.* **27** pp. 207–281.

[41] Duistermaat, J.J. (2004). Lectures on dynamical systems with symmetry given at Utrecht spring school on Lie groups, unpublished.

[42] Duistermaat, J.J and Kolk, J.A.C. (1999). *Lie groups*, (Springer Verlag).

[43] Ferrers, N.M. (1872). Extension of Lagrange's equations, *Quart. J. Math.* **12** pp. 1–5.

[44] Field, M. (1980). Equivariant dynamical systems, *Trans. Amer. Math. Soc.* **259** pp. 185–205.

[45] Foote, R.L. (1998) Geometry of Prytz planimeter, *Rep. Math. Phys.* **42** pp. 248–271.

[46] Friedman, A. (1963). *Generalized functions and partial differential equations*, (Prentice Hall).

[47] Gibbs, J. Willard. (1879). On the fundamental formulæ of dynamics, *Amer. J. Math.* **II** pp. 49–64. (1928). in *Collected Works*, vol. II, (New York). (1961). reprint *The scientific papers of J. Willard Gibbs*, (Dover).

[48] Goresky, M. and MacPherson, R. (1988). *Stratified Morse Theory*, (Springer Verlag).

[49] Guillemin, V. and Sternberg, S. (1984). *Symplectic Techniques in Physics*, (Cambridge University Press).

[50] Gunning, R.C. (1966). *Lectures on Riemann surfaces*, (Princeton University Press).

[51] Hamel, G. (1949). *Theoretische Mechanik*, (Springer Verlag).

[52] Helgason, S. (1962). *Differential Geometry and Symmetric Spaces*, (Academic Press).

[53] Hermans, J. (1995). *Rolling rigid bodies with and without symmetries*, Ph.D. thesis, University of Utrecht, Utrecht, the Netherlands.

[54] Hermans, J. (1995). A symmetric sphere rolling on a surface, *Nonlinearity* **8**, pp. 295–317.

[55] Hirsch, M. (1976). *Differential Topology*, (Springer Verlag).

[56] Karapetian, V.A. (1980). On the problem of steady motions stability of nonholonomic systems, *J. Appl. Math. Mech.* **44**, pp. 295–300.

[57] Kemppainen, D. (1996). *Geometry of Rolling Disk*, MSc. Thesis, University of Calgary, Alberta, Canada.

[58] Kobayashi, S. and Nomizu, K. (1963). *Foundations of Differential Geometry*, volume 1, (Interscience Publishers).

[59] Koiller, J. (1992). Reduction of some classical nonholonomic systems with symmetry, *Arch. Rat. Mech. Anal.* **118**, pp. 113–148.

[60] Kolár, I., Michor, P.W., and Slovák, J. (1993). *Natural Operations in Differential Geometry*, (Springer Verlag).

[61] Koon, W.S. and Marsden, J.E. (1988). Poisson reduction for nonholonomic mechanical systems with symmetry, *Rep. Math. Phys.* **42** 101–134.

[62] Korteweg, D.J. (1899). Über eine ziemlich verbreitete unrichtige Behandlungsweise eines Problems der rollenden Bewegung, *Nieuw Archief voor Wiskunde* **4**, pp. 130–155.

[63] Kronecker, L. (1884). Näherungsweise ganzzahlige Auflösung linearer Gleichungen *Monatsber. Kön. Preuss. Akad. Wiss. zu Berlin* pp. 1179–1193, 1271–1299. (1899). *Werke*, Band 3:1, (Teubner) pp. 47–109.

[64] Krupa, M. (1990). Bifurcation of relative equilibria, *SIAM J. Math. Anal.* **21**, pp. 1453–1486.

[65] Lagrange, J.L. (1788). *Méchanique Analytique*, (Paris). (1965). reprint with notes from the third and fourth edition, (A. Blanchard).

[66] Lee, J.M. (2003). *Introduction to Smooth Manifolds*, Graduate Texts in Mathematics, vol. 218, (Springer-Verlag).

[67] Lewis, A.D. and Murray, R.M. (1995). Variational principles for constrained systems: theory and experiment, *Internat. J. Non-Linear Mech.* **30**, pp. 793–815.

[68] Libermann, P. and Marle, C.M. (1987). *Symplectic Geometry and Analytical Mechanics*, (D. Reidel).

[69] Lie, S. (1888). *Theorie der Transformationsgruppen*, Unter Mitwirkung van Prof. Dr. F. Engel, Erster Abschnitt, (Teubner).

[70] Lindelöf, E. (1895). Sur le mouvement d'un corps de revolution roulant sur un plan horizontal, *Acta Societatis Scientiarum Fennicae* **XX**, number 10.

[71] Lusala, T. and Śniatycki, J., Stratified subcartesian spaces, `http://arXiv.org/abs/0805.4807v1`.

[72] Lusala, T. and Śniatycki, J. Consequences of the Palais slice theorem for differential spaces, in preparation.

[73] Marle, C.M. (1995). Reduction of constrained mechanical systems and stability of relative equilibria, *Commun. Math. Phys.* **174**, pp. 295–318.

[74] Marle, C.M. (1998). Various approaches to conservative and nonconservative nonholonomic systems, *Rep. Math. Phys.* **42** pp. 211–229.

[75] Marsden, J.E. (1992). *Lectures on Mechanics*, (Cambridge University Press).

[76] Marsden, J.E., Misiołek, G., Ortega, J.-P., Perlmutter, M. and Ratiu, T. (2007). Hamiltonian Reduction by Stages, *Lecture Notes in Mathematics* **1913**, (Springer Verlag).

[77] Marsden, J.E. and Weinstein, A. (1974). Reduction of symplectic manifolds with symmetry, *Rep. Math. Phys.* **5**, pp. 121–130.

[78] Mather, J. (1967). Differential invariants, *Topology* **16**, pp. 145–155.

[79] Meyer, K. (1973). Symmetries and integrals in mechanics, in *Dynamical Systems*, M. Piexoto (ed.), pp. 259–272, (Academic Press).

[80] Michel, L. (1974). Points critiques des fonction invariantes sur une *G*-variété, *Comptes Rendus Acad. Sci. Paris* **272** pp. 433–436.

[81] Moshchuk, N.K., (1988). A qualitative analysis of the motion of a heavy solid of revolution on an absolutely rough plane, *J. Appl. Math. Mech.* **52** pp. 203–210.

[82] Neimark, J.I. and Fufaev, N.A. (1972). *Dynamics of nonholonomic systems*, Trans. of Moscow Math. Soc. **33**, (Amer. Math. Soc.).

[83] Newton, I. (1686). Philosophiæ Naturalis Principia Mathematica, (London). (1999). *The principia: mathematical principles of natural philosophy*, translated by I.B. Cohen and A. Whitman, (University of California Press).

[84] O'Reilly, O.M. (1996). The dynamics of rolling disks and sliding disks, *Nonlinear Dynamics* **10** pp. 287–305.

[85] Ortega, J.P. (1998). *Symmetry, Reduction, and Stability in Hamiltonian Systems*, PhD. Thesis, University of California, Santa Cruz, CA.

[86] Ortega, J.P., private communication.

[87] Ortega, J.P. and Ratiu, T.S. (2004). *Reduction of Hamiltonian systems with symmetry*, (Birkhäuser).

[88] Palais, R. (1961). On existence of slices for actions of noncompact Lie groups, *Ann. Math.* **73** pp. 295–323.

[89] Pars, L.A. (1965). *A Treatise on Analytical Dynamics*, (Wiley). (1979). reprint (Oxbow Press).

[90] Patrick, G. (1995). Relative equilibria of Hamiltonian systems with symmetry: linearization, smoothness and drifts, *J. Nonlinear Sci.* **5**, pp. 2089–2105.

[91] Pflaum, M.J. (2001). *Analytic and geometric stratified spaces*, Lecture Notes in Mathematics, **1768**, (Springer Verlag).

[92] Poincaré, H. (1901). Sur une forme nouvelle des équations de la méchanique, *Comptes Rendus Acad. Sci. Paris* **132** pp. 369–371. *Oevres* t. VII pp. 218–219.

[93] Procesi, C. and Schwarz, G. (1985). Inequalities defining orbit spaces, *Invent. math.* **81** pp. 539–554.

[94] Roberts, M., Wulff, C. and Lamb, J.S.W. (2002). Hamiltonian systems near relative equilibria, *J. Diff. Eq.* **179**, pp. 562–604.

[95] Rosenberg, R. (1977). *Analytical Dynamics*, (Plenum Press).

[96] Routh, E. (1860). *A Treatise on the Dynamics of a System of Rigid Bodies*, (MacMillan and Co.) (1960). reprint *Advanced Rigid Body Dynamics*, parts I and II, (Dover).

[97] Sard, A. (1942). The measure of critical values of differentiable maps, *Bull. Amer. Math. Soc.* **48**, pp. 883–890.

[98] Schwarz, G.W. (1975). Smooth functions invariant under the action of a compact Lie group, *Topology* **14**, pp. 63–68.

[99] Schwarz, G.W. (1980). Lifting smooth homotopies of orbit spaces, *Inst. Hautes. Études Sci. Publ. Math.* **51**, pp. 37–135.

[100] Sikorski, R. (1967). Abstract covariant derivative. *Colloq. Math.* **18**, pp. 251–272.

[101] Sikorski, R. (1972). *Wstep do geometrii różniczkowej*, (PWN).

[102] Sjamaar, R. and Lerman, E. (1991). Stratified symplectic spaces and reduction, *Ann. Math.* **134**, pp. 375–422.

[103] Śniatycki, J. (1998). Non-holonomic Noether theorem and reduction of symmetries, *Rep. Math. Phys.* **42**, pp. 5–23.

[104] Śniatycki, J. (2001). Almost Poisson spaces and non-holonomic singular reduction, *Rep. Math. Phys.* **48**, pp. 235–248.

[105] Śniatycki, J. (2002). The momentum equation and the second order differential equation condition, *Rep. Math. Phys.* **49**, pp. 371–394.

[106] Śniatycki, J. (2002). Integral curves of derivations on locally semi-algebraic differential spaces, in *Proceedings of the Fourth International Conference on Dynamical Systems and Differential Equations, May 24-27, 2002* (Wilmington, NC, USA), eds. S. Hu and X. Lu, pp. 825–831.

[107] Śniatycki, J. (2003). Orbits of families of vector fields on subcartesian spaces, *Ann. Inst. Fourier* (Grenoble) **53**, pp. 2257–2296.

[108] Śniatycki, J. and Cushman, R. (2007). Non-holonomic reduction, symmetries, constraints, and integrablility, *Reg. and Chaotic Dyn.* **12**, pp. 615–621.

[109] Souriau, J.M. (1997). *Structure of Dynamical Systems*, (Birkhäuser). translation of (1970). *Structure des Systémes Dynamique*, (Dunod).

[110] Srinivasan, M. (2008). Rocking and rolling: A can that appears to rock might actually roll, *Phys. Rev. E.* **78**, 066609.

[111] Stefan, P. (1974). Accessible sets, orbits and foliations with singularities, *Proc. London Math. Soc.* **29**, pp. 699–713.

[112] Sussmann, H. (1973). Orbits of families of vector fields and integrability of distributions, *Trans. Amer. Math. Soc.* **180**, pp. 171–188.

[113] van der Schaft, A.J. and Maschke, B.M. (1994). On the Hamiltonian formulation of nonholonomic mechanical systems, *Rep. Math. Phys.* **34**, pp. 225–233.

[114] Vierkandt, A. (1892). Über gleitende und rollende Bewegung, *Monatsh. f. Math. u. Physik*, **3**, pp. 31–54, 97–134.

[115] Walker, G.T. (1895). On a curious dynamical property of celts, *Proc. Cambridge Phil. Soc.* **8**, pp. 305–306.

[116] Walker, G.T. (1896). On a dynamical top, *Quart. J. Pure Appl. Math.* **28**, pp. 175–184.

[117] Whitney, H. (1965). Local properties of analytic varieties, in *Differential and combinatorial topology*, ed. S.S. Cairns, (Princeton University Press) pp. 205–244.
[118] Whittaker, E.T. (1964). *A Treatise on the Analytical Dynamics of Particles and Rigid Bodies*, 4th edn. (Cambridge Univ. Press).
[119] Yoshimura, H. and Marsden, J.E. (2007). Reduction of Dirac structures and the Hamilton-Pontryagin principle, *Rep. Math. Phys* **60**, pp. 381–426.

Index